MICROELECTRONIC RELIABILITY

VOLUME II

Integrity Assessment and Assurance

The Artech House Materials Science Library

GaAs FET Principles and Technology by J.V. DiLorenzo and D. Khandelwal

Microelectronic Reliability, Vol. I: Reliability, Test and Diagnostics, Edward B. Hakim, ed.

Materials Handbook for Hybrid Microelectronics, J.A. King, ed.

Microelectronic Reliability, Vol. II: Integrity Assessment and Assurance, Emiliano Pollino, ed.

Electrical and Magnetic Properties of Materials, by Philippe Robert

GaAs MESFET Circuit Design, Robert A. Soares, ed.

Gallium Arsenide Processing Techniques by Ralph Williams

MICROELECTRONIC RELIABILITY

VOLUME II

Integrity Assessment and Assurance

Emiliano Pollino
CSELT

Artech House

Library of Congress Cataloging-in-Publication Data

Affidabilità dei componenti elettronici a semiconduttore. English.
Microelectronic reliability/Emiliano Pollino, editor.

p. cm.
Translation of: L'affidabilità dei componenti elettronici a semiconduttore.
Bibliography: p.
Includes index.
Contents: — v. 2. Integrity assessment and assurance.
ISBN 0-89006-350-8: $78.00
1. Semiconductors — Reliability. 2. Electronic apparatus and appliances — Reliability. I. Pollino, Emiliano. II. Title.

621.381—dc19 88-36803
CIP

ARTECH HOUSE, INC.
685 Canton Street
Norwood, MA 02062

International Standard Book Number: 0-89006-350-8
Library of Congress Catalog Card Number: 88-36803

10 9 8 7 6 5 4 3 2 1

Originally published in the Italian, *L'Affidabilità dei Componenti Elettronici a Semiconduttore*, by Scuola Superiore G. Reiss Romoli, 1987.

CONTENTS

Preface

The importance of reliability during the second half of the twentieth century is rising to unprecedented levels, irrespective of the object and the application. In the electronic era, due to the continuing improvment in technological processes, miniaturization and yield—which mean high performance, high packing density, and affordable cost—semiconductor devices play a key role in all applications, from consumer electronics to highly sophisitcated and strategic fields.

In the 1960s, when the work on the reliability of components and systems was started in CSELT (Centro Studi E Laboratori Telecomunicazioni, a corporate research and development laboratory operating in every branch of telecommunication, located in Torino, Italy), there was already a strong indication that this activity was to become significant.

Today, after years devoted to the investigation of semiconductor device failure mechanisms, searching for solutions, developing tools and techniques, discussing and interacting with a number of parties involved at different levels in the problem, both directly and in attending the technical seminar held every year at SSGRR (Scuola Superiore Guglielmo Reiss Romoli, a school for postgraduate training in telecommunication, electronics, economics, and management established by the STET Group, which is the main holding company for electronics and telecommunication in Italy), we concluded that an attempt should be made to concentrate, in a single book, the major issues and to provide solutions or at least suggestions on how to cope with the problems.

Thus, this book was intended to provide guidelines to anyone who faces everyday problems with the reliability of solid-state devices as well as to those who just need a reference or tutorial to find a somewhat detailed explanation of what might happen or has just happened in their electronic system. The task proved to be anything but easy to accomplish, but the ongoing support by Professor Luigi Bonavoglia, President of SSGRR, whose experience and enthusiasm contributed to overcoming difficulties of every kind, and the committment of all the people involved, turned an idea into reality.

The authors contributing to this effort were selected according to their professional experience in the fields covered; furthermore, the different areas where they operate (i.e., industry, academic and corporate research), enabled us to show different approaches and points of view tied to specific needs and applications. Other than CSELT, device (SGS-Thomson) and equipment (ITALTEL-SIT and TELETTRA) manufacturers as well as research centers (LAMEL and CSATA) and universities (Modena, Pisa, and Bologna) are represented.

Because *Microelectronic Reliability,* Vol. 1, edited by E.B. Hakim, is mainly concerned with diagnostics and testing, Vol. 2 (which is the revised and updated English version of the book earlier published in Italy), focuses on *Reliability Assessment and Assurance*. This basically means that great emphasis is placed on the physics of failures, both mechanisms and for investigation tools, as the key for the development of an effective approach to the improvement of the reliability of solid-state devices, which is then based on the knowledge of all technological aspects and processes, an understanding of the basic degradation phenomena, *ad-hoc* designed tests aimed to stress the weakest points, suitable analytical tools, and ultimately specifications and standards. All of these activities are closely linked in a flow chart established over the years as a fixed route.

The book is organized as follows: Chapters 1 and 2 are introductory, providing the reader with an historical overview of the evolution of the approaches to the reliability of semiconductor devices and with the basics of semiconductor technology. The major degradation mechanisms of silicon devices are analyzed in Chapters 3 to 7: dielectric and metallization related problems and latch-up are particularly taken into account due to their effects on VLSI and ULSI. Some chapters are devoted to the analysis of the reliability of some types of devices (i.e., solid-state emitters and receivers for fiber optics communications, microwave compound semiconductor FETs and ICs, and power devices), due to their growing importance and strategic issues. Since the statistics of reliability are provided in Vol. 1, here we supply the practical tools, tests and analytical techniques, used to squeeze quickly but effectively all the information needed from the device to assess its reliability. These topics are covered in Chapters 11, 12, and 13, while Chapter 14 gives an overview of existing standards and standardization activities in progress.

Acknowledgements

I wish to thank all of the authors, whose cooperation and efforts made this book possible. Particular thanks to Ing. Basilio Catania and Ing. Mario Muccini, General Manager and Vice General Manager of CSELT, to Prof. Saverio Rotella, General Manager of Scuola Superiore Guglielmo Reiss Romoli, and to Ing. Rocco Casale, in charge of the Transmission Engineering of SIP (Societa' Italiana per l'esercizio delle Telecomunicazioni) for continuous support and constructive criticism during the development of the work.

Thanks to Dr. Danilo Riva to whom I am particularly indebted for his friendly help and useful quarreling about the organization and editing of the book. Thanks also to Mr. Gianni Turinetti for most of the photographic documentation and for his patience and cooperation in ensuring the availability of all the documentation needed, to Mr. Piero Capella for helpful discussion and up-to-date information, to Mr. Francesco Rosso for the preparation of the artwork, and to Miss Loredana Giralda who undertook the considerable task of typing the manuscript.

Chapter 1
THE TECHNOLOGICAL REVOLUTION AND RELIABILITY

E. Pollino

1.1 RELIABILITY ENGINEERING

There was a time when the prevailing opinion among designers of equipment was that the product of their imagination, once realized, was intrinsically reliable. It was enough to take "good" components and to put them togther in the "correct" way to make the equipment so conceived function long and well.

The economic effect of failures, as the following will show, removed this illusion very quickly. Today, faced with more complex systems, with demands for more sophisticated performance and very long mission times, and in particular, with a process of frenetic technological innovation, it is no longer possible to expect reliability in the product itself using only good common sense. It is necessary before anything else that the concepts of *good, correct,* and therefore *reliable* be defined unequivocally. A systematic treatment of the concepts of *quality, reliability, safety,* and others can be found in [26].

Many activities have been proposed to ensure the good functioning over time of a product, and these activities have been generically called *reliability engineering*. The official definition of such an aggregate of activities is the following: reliability engineering is the branch of human knowledge that has the aim of describing, predicting, and dominating the future behavior of objects, which can be, for example, electronic components, telecommunication systems, *et cetera.*

The term *behavior* can mean for example, the functional performance under certain environmental conditions for a certain period of time and within certain limits of the cost, weight, encumbrance, and so on. The definition given refers, as explained, to the group of theories, techniques, and procedures with an essentially economic aim (when safety is not involved, of course); the same term, *reliability,* is used, as the following will show, to indicate both a property of the objects and a probabilistic quantity tied to the expected life.

This definition of *reliability engineering* is, at the same time, all-comprehensive and nebulous, formidable and ambiguous, and certainly it is not possible through it to immediately identify what, in reality, it represents.

Figure 1.1 attempts to classify the branches of activity that constitute reliability. This example can be useful in showing the vast interests involved and the complexity of the problems faced in "describing, predicting, and dominating" the behavior over time of certain objects. In this same figure, the fields this book intends to cover, at least in part, are defined; or, in other words, its place with respect to the taxonomy reported [1].

1.2 HISTORIC EVOLUTION OF RELIABILITY ENGINEERING

The resources and the emphasis dedicated to the various aspects of reliability vary notably over the course of the years in response to the development of technology in equipment manufacturing. Before the Second World War, most equipment and systems were mechanical. What distinguished a reliable product was essentially the safety margin in either stress-strength, wear, or fatigue conditions for the basic design. The care with which such margins came to be inserted into the design was considered more a good engineering practice than a real and true precaution of reliability; and therefore the two were mixed together in a general "product ruggedness."

In fact, reliability, in the period under consideration, was inherent in the product and was automatically expected by the final user. Rather, it can be said, the good name of a single manufacturer or all the manufacturers in a certain geographical area was tied to such an idea of durability. In particular, for example, the myth of German mechanics comes to mind. The only quantitative measure tied to reliability was, in substance, the useful life of the equipment before wearout.

At the beginning of the 1940s, the U.S. military created a committee to manage studies on vacuum tubes (Vacuum Tube Development Committee) and began to issue *joint army and navy* (JAN) standards on such devices. Following this, the initiatives multiply even those of industry (generally tied to supply or collaboration contracts). In 1949 Aeronautical Radio Inc. and Cornell University performed extensive analysis on tubes that failed in field, later followed by Vitro Laboratories and Bell Laboratories. In 1952 the US Department of Defense formed an ad hoc committee on reliability, the Advisory Group on the Reliability of Electronic Equipment (AGREE); and in 1957 they issued their monumental report on "Reliability of Military Electronic Equipment." These years saw a real flowering of initiative and publications the most important of which are listed in the references [5–12].

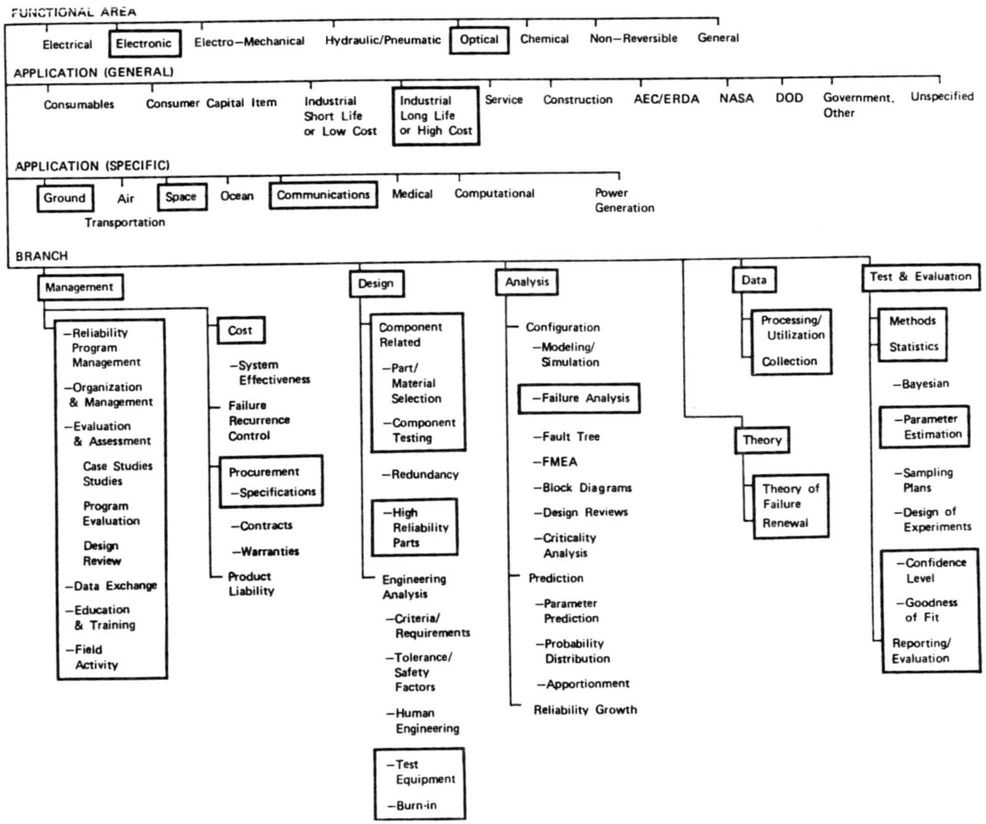

Figure 1.1 Reliability taxonomy.

Toward the middle of the 1950s, transistors and diodes became available to industry. Thanks to their low price and functional characteristics, they quickly replaced thermionic tubes. Their simplicity was regarded as a promise of reliability, and in fact, transistor-based equipment were found to be 1 or 2 degrees more reliable than tube-based equipment [2].

The semiconductors could then be produced in greater quantities and at lower costs than the tubes, which further accelerated the popularity of the equipment. In the second half of the 1950s, the space race began. In 1957 the U.S.S.R. successfully launched Sputnik, and in the same year, the United States failed to put Vanguard TV3 in orbit. In the following years, too, several American satellites were destroyed, and the federal government decided to substantially support initiatives to improve the reliability of space systems [19]. The result of this new emphasis on reliability is shown in Figure 1.2 [13], where it is noted how in just a few years the probability of a successful launch had come close to 100 percent.

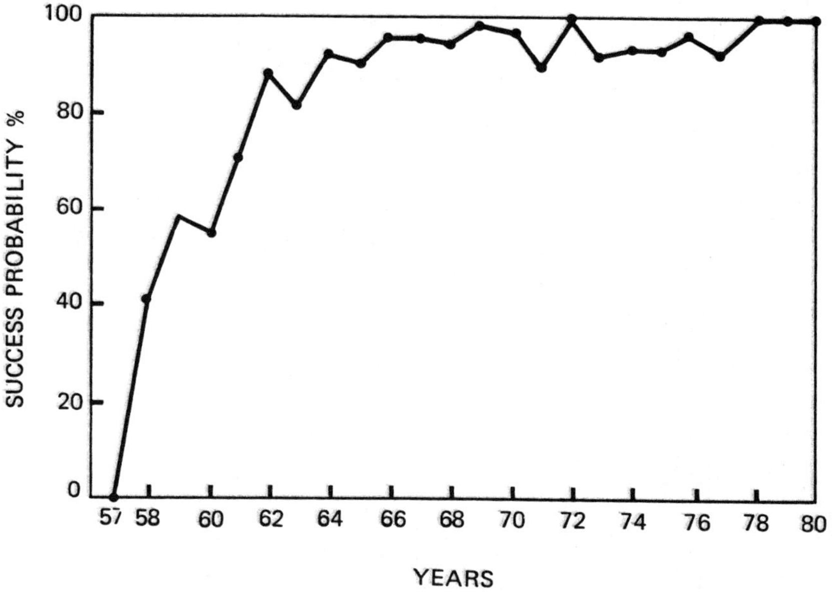

Figure 1.2 Success of U.S. space launches.

Most papers and books on reliability written during this period developed mathematical concepts and statistics, with special emphasis on calculations on the reliability of systems or verification tests and demonstrations of reliability of the equipment. The component was regarded as a black box, and it was a common tendency to view reliability as an electrical parameter. An outline of the historical evolution of reliability mathematics, which is outside the scope of this book, can be found in [20].

At the beginning of the 1960s, the first integrated circuits were born, the manufacturing technology of semiconductor devices became more complex and sophisticated, thus it became obvious that to improve reliability it was necessary to get inside the component. Solid-state chemistry and physics joined statistics and metrology as fundamental constituents of the new engineering. The greatest difficulties during these years were obtaining models of failures, for both components and systems, and obtaining realistic data on failure in terms of components quickly enough to make use of them.

In 1962 the first manual reporting the failure rate of components was issued, the famous MIL-HDBK-217 [15]. Also, the first Physics of Failure in Electronics Symposium was held, later this would become the annual Reliability Physics Symposium [16]. In 1963, NASA published NPC 250-1, *Reliability Program Provision for Space System Contractors* [17], which laid down guidelines for reliability in contracts.

These three events were fundamental to the evolution of reliability engineering. The first one, the issue of official failure rates imposed a little order on the "game of numbers," providing a common baseline for the computation of prediction of reliability in systems.

The congress on the physics of failure (the second event) represented the formation of a body of knowledge focused on improving the reliability of the product. The damaging effects of failure no longer could be predicted or circumscribed solely by the "game of numbers," the causes had to be recognized and eliminated at the microscopic level.

With the third event, the issue of reliability guidelines in contracts, reliabililty officially entered into procurement contracts. To get a reliable product, the users had to thoroughly understand the requirements and know how to ask for them in the right way.

The interaction between deeper knowledge of the weak points in components and the necessity of having standardized testing methods, in order to be able to take them into consideration in supply contracts, led to the MIL-STD-883 test methods and procedures for microelectronics [18] in 1968. The event can be regarded as the end of the "historic" phase of reliability engineering.

The MIL-STD-883 compiled by the Rome Aircraft Development Center (RADC), as was the MIL-HDBK-217, was updated more frequently to underline the ever greater difficulty with which the standards succeed in keeping pace with the technological evolution.

1.3 CHARACTERISTICS OF SEMICONDUCTOR TECHNOLOGY AND ITS INFLUENCE ON RELIABILITY

The phenomenal growth in semiconductor technology (which will be examined in detail in Chapter 2) has given rise to technological evolution so frenetic as to be compared to a real revolution, given the speed and complexity of changes involved. The main factors characterizing the technological revolution follow.

1.3.1 Growth in Device Complexity

Every year the complexity of integrated circuits, intended as the number of elementary components (transistors or gates) per chip, doubles. This trend is known as Moore's Law (Moore, the president of Intel, which for a long time has been in the vanguard of the development of very complex integrated circuits). Moore's Law is the result of the miniaturization made possible by the evolution of lithographic techniques, and of the increase in the dimensions of the chip, resulting from the greater automation and cleanliness of the process.

1.3.2 Improvement in Function and Efficiency

The electric performance of the devices is constantly improving and so are the factors of merit (speed-power product) of the main technological families. The growth in sophistication and performance of the devices allow increasingly complex functions to be integrated on silicon.

1.3.3 Reduction of Costs

The cost per unit function of integrated circuits has reduced constantly, as Figure 1.3 shows, as the chip memory has grown. The actual price of DRAM during 1988 was higher than that indicated in Figure 1.3. Shifting of production from 256 kbit to 1 Mbit and the low assembly yield of the latter, associated with a strong demand for memory devices, caused an abnormal rise in prices. This results both from technical factors (reduction of geometry, increase in yield, *et cetera*) and economic factors (enormous productive capability, market pressure, *et cetera*).

1.3.4 Change in Authority

The availability of an increasingly advanced hardware technology and its massive interaction with software generates new concepts of system. On the other hand, new concepts of system require the development of new technologies. That is,

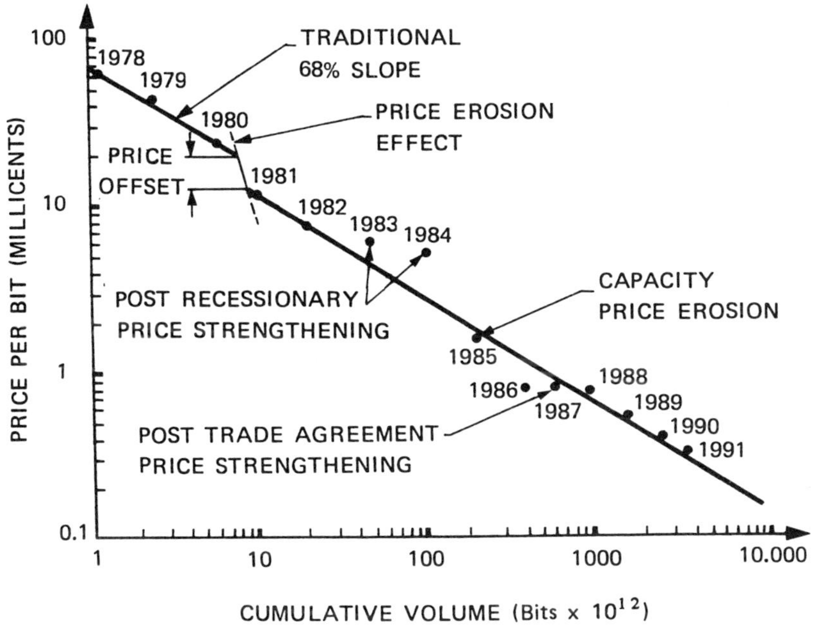

Figure 1.3 Price curve for MOS dynamic RAM.

system and component designers run after each other continuously, with the need to have software technology available at all levels.

Until recently, the designer of equipment or systems was able to find needed components easily in the marketplace, because these components were widely available and known, and therefore easily controllable in terms of quality. Today, similar production meets great difficulties. Modern ingredients, in fact, are so sophisticated and complex (and often more difficult to find) as to constitute a roadblock in innovation, or even in the final product. For example, it is considered that in the 1990s, 90 percent of the cost of an electronic telephone exchange will go for a quite restricted number of integrated VLSI circuits, of which at least one-quarter is custom-made. These facts have great consequences from the point of view of reliability.

1.3.5 Increase in Risk

The market's aggressiveness and the necessity of getting there first result in the components being commercialized when their technology of realization is far from settled. Figure 1.4 examines the failure rate of some new devices at their first appearance in the market [23].

To face this problem the manufacturers of components offer a wide range of devices at different levels of quality and, naturally, at different prices, but the evaluation of the relationship between benefits and costs is problematical and left to the user. The various programs of improvement in quality become instruments of marketing, in the face of which the user is disconcerted.

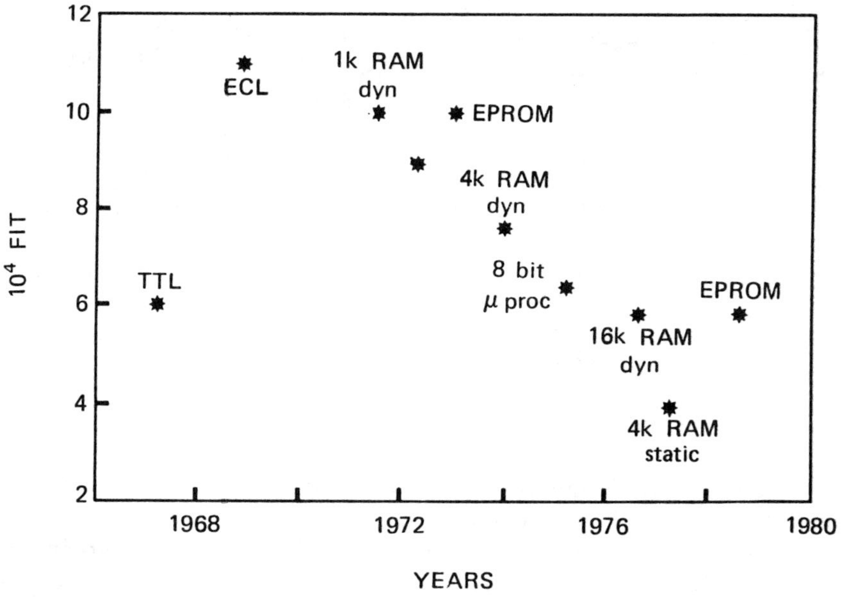

Figure 1.4 Failure rate of some devices as they appeared on the market.

1.3.6 Importance of the Learning Factor

In the last 20 years, the failure rate of devices has constantly diminished as illustrated in Figure 1.5 [24]. Learning at all levels, production and use of components, plays a fundamental role. As has been already said, devices as they first appear on the market are unreliable; their reliability grows little by little during manufacture and use. This fact is by now taken for granted, but the problem is to determine how much time is needed to reach a level of reliability. In some cases, for example the CMOS integrated circuits, it took years before reliability reached acceptable values (25).

The designer, having to choose between a settled component that is old and a brand new one that therefore is unknown in reliability but more powerful in

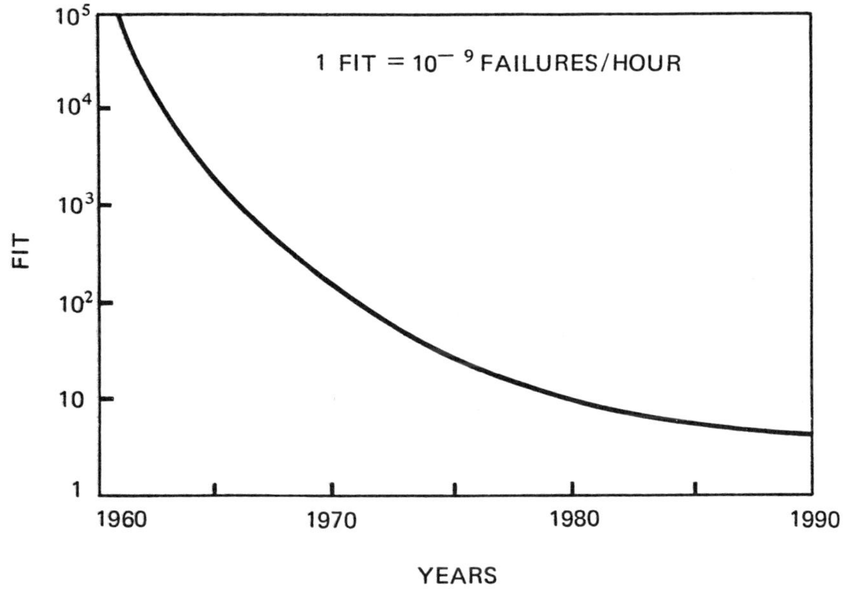

Figure 1.5 Trend of bipolar integrated circuit failure rate.

performance will prefer the second item nine times out of ten. Good performance, in fact, is safe, objective, and allows the designer to do a better job. It is therefore necessary to reduce, as much as possible, the learning period; and this has become one of the main tasks of reliability engineering.

1.3.7 Change in Relationships Inside and Outside the Firm

As the equipment has come to contain more know-how within the components, which become more complex, by now integrated systems, become less understandable and controllable. This revolutionizes the division of competence inside the firm. The designer must know how to deal with semiconductor technology, software, and so on. The quality control department cannot confine itself anymore to the end of the production process with drastic interventions, but must intervene to check the quality of the product during manufacturing, and so on.

Furthermore, the power relationship between manufacturer and user has also changed. Once, users of components held the power, playing on competition and setting the standards in quality. For certain devices, for example standard integrated circuits, it was possible to choose from among ten or more suppliers.

Today, the new components, custom or strategic (for example VLSI integrated circuits or laser devices), may come only from one supplier. It is not possible anymore to play the competition card, there must be a collaboration, but the dialogue on which this is founded can only come about if both parties possess the weapon of technological knowledge.

1.3.8 Crisis in Standards

Even deciding the optimum level of quality for equipment to be produced and, in particular, knowing how to turn such needs into written requirements has become a problem more and more difficult to solve. The standards for electronic components, which are the basis of common understanding and discussion, have difficulty in keeping abreast of the technological evolution.

There are many motivations for setting standards. In technology, a standard presupposes a minimum standardization of the equipment, and, as was said, this is valid, at least in part, only for the methods. In economics big investments would be necessary to produce in the necessary standards. In politics, the standards constitute an enormous amount of free know-how, accessible to everybody. At the present, faced with a widespread reprivatization of services and increasingly fierce competition, the know-how itself is defended more and more vigilantly.

1.4 EVOLUTION OF INVESTIGATION IN RELIABILITY

If the complexity of the components renders their control arduous, the growing complexity of the systems renders the choices relative to the component manufacturer irreversible. It is enough to think of the first US electronic switching system: the project choices were made in 1960, the first installations in 1966, and the first results were available in 1967. At this point, a good 50 million semiconductor components were installed and any corrective action on the choices made would have been impossible. It is obvious that field data could not wait for confirmation of the choices made, but even other methods of investigation used for solving reliability problems would lose their effectiveness if applied to technology in rapid evolution.

Analyzing the main methods of assessing reliability, which favor the settlement of the technology of semiconductors, several factors can be examined.

1.4.1 Field Data

Gathering field data was the first reliability methodology. It consists of analyzing what happened in the past to predict the future. The past is accurately examined with the help of mathematical and statistical projections for the future. In a situation of rapid change, like the present, this method can seem too slow: in some cases, as when the first field data comes in, the components have already been surpassed technologically. If in most cases, such data cannot be used to modify the project because these data are not available in time, they are still, however, fundamental as a final answer (and one with no appeal) for the decisions taken earlier.

1.4.2 Accelerated Tests

The principle is to apply stress to a limited quantity of devices, under bias or environmental conditions above the normal ratings, to speed up the failure mechanisms that would occur in actual operating conditions and then extrapolate the behavior observed on one lot to an entire population. The emphasis is then on reducing the time necessary to obtain data, which is in line with the imperative of getting it soon.

However, two problems exist that sometimes make this approach difficult, too. The first is economical: due to the complexity of the devices, such tests are very expensive in terms of people and tools involved. The second is scientific: for new technologies it is not sure at all that the effect of the applied stress really accelerates the phenomena already latent in the conditions of normal operation and does not introduce new ones.

1.4.3 Screening

The principle of screening is to submit the entire population to low intensity stress with the aim of inducing failures only in the devices already weak and without damaging those, hoped to be in the majority, that are "strong." As will be clearly illustrated in the following chapters, the basic hypothesis is that the same lot contains both populations, one strong and the other weak, characterized by different failure mechanisms. The elimination of the weak population through screening increases the global reliability of the treated lot, especially in the initial part of the life cycle of the equipment.

This empirical method of improving reliability continues to retain its efficacy on the devices of new technology, too, essentially because the failure mechanisms influenced by screening are relative to gross errors of processing (bad bonds, nonhermetic encapsulation, heavy contamination, *et cetera*) that, at least in the first approach, do not vary much with the technological evolution. Besides, the availability of better performing devices tends to speed up the obsolescence of the equipment, thus failure in the first years of life of the equipment will play a fundamental role in its reliability. The techniques of screening, therefore, are well adapted to a situation of rapid technological evolution.

1.4.4 Failure Analysis

Failure analysis entails the careful study of each individual anomaly with the aim of determining the dynamic of physiochemical mechanism that brought it on and locating the cause, outside or inside the component, that set it off. Once the mechanism and the cause of failure are identified, it is possible to propose the action to eliminate it at the most convenient level.

As was mentioned in the historic introduction, the failure analysis carried out intensively, implemented with its relative methodology of observation and analysis, and the development of a new discipline called the physics of failure have exercised a decisive role in the evolution of the technology of semiconductors.

Its use has borne the gradual passage from one concept of reliability engineering, understood essentially as a measure and knowledge of reliable parameters for prediction, to another way of understanding it, more consonant with the times, which tends to favor and improve the reliability of devices apart from their value.

1.4.5 Technological Characterization

In the methods cited some detailed knowledge of the technology of devices on the user's side is proving to be essential.

The techniques of investigation applied to failure analysis therefore have been applied to the study of new devices, with the aim both of knowing the process and trying to identify eventual weak points present in the new components on the basis of the accumulated experience in the physics of failure. Once more, we are going back to the beginning and extracting information that will help improve the object of our study.

In synthesis, from what was said earlier, we can conclude that the technological evolution and reliability engineering influence each other. Reliability in the wake of technology, favors settlement and therefore accelerates the process. In turn, it is pushed toward techniques that favor improvement rather than knowledge for prediction.

In order to take into account the tendency toward this incremental approach to reliability (in counterposition to the statistical-predictional one), the definition given at the beginning of this chapter on reliability engineering could be brought up to date and completed in the following way: reliability engineering is a branch of human knowledge that has the aim of describing, predicting, and improving the behavior over time of components and systems.

REFERENCES

1. *International Conference on Reliability and Maintainability,* Paris, June 1978.
2. K.L. Wong, "A New Direction for Electronic Reliability Engineering in the 80's," *Proceeding Eurocon 82,* Copenhagen, pp. 3, 10, 1982.

3. W. Weibull, "A Statistical Distribution Function of Wide Applicability," *Journal of Applied Mechanics,* September 1951.
4. D.L. Davis, "An Analysis of Some Failure Data," *Journal Of American Statistical Association,* June 1952.
5. P. Klass, "Search is on for Electronic Reliability," *Aviation Week,* November 1952.
6. R.R. Carhart, "A Survey of Current Status of the Electronic Reliability Problem," *Rand Research Memorandum, 1131.*
7. B. Epstein and M. Sobel, "Life Testing," *Journal of American Statistical Association,* September 1953.
8. U.S. Navy Electronics Labs, *NEL Reliability Design Handbook,* San Diego, March 1954.
9. *National Symposium on Quality Control and Reliability in Electronics,* 1954.
10. G.R. Herd, *Estimation of Equipment Reliability,* Los Angeles, (ARINC), 1955.
11. National Bureau of Standards, *Handbook of Preferred Circuits for Navy Aeronautical Electronic Equipment,* U.S. Navy Weapons Department, September 1955.
12. Agree, "Reliability of Military Electronic Equipment," *Report of the Advisory Group on Reliability of Electronic Equipment,* Office of the Assistant Secretary of Defense (Research and Engineering), Washington D.C., June 1957.
13. TRW Defence and Space Systems Group, *TRW Space Log,* Vol. 18, 1981.
14. H.S. Blanks, "A Review of New Methods and Attitudes in Reliability Engineering," *Microelectronics and Reliability,* Vol. 12, 1973.
15. "Reliability Prediction of Electronic Equipment," *MIL-HDBK-217,* 1962 and successive editions compiled by the RADC.
16. *Reliability Physics Symposium,* Electron Devices and Reliability Society, Proceedings published by the IEEE since 1962.
17. NPC 250-1, *Reliability Program Provision for Space System Contractors.*
18. "Test Methods and Procedures for Microelectronics," *MIL-STD-883,* 1968 and successive editions compiled by the RADC.
19. H. Cohen "Space Reliability Technology, A Historical Persepctive," *IEEE Transaction on Reliability,* Vol. R-33, No. 1, April 1984.
20. R.E. Barlow, "Mathematical Theory of Reliability: A Historical Perspective," *IEEE Transaction on Reliability,* Vol. R-33, No. 1, April 1984.
21. L. Basile and G. Mattana, "Reliability and the Evolution of Components," *Telettra Review,* September 1982.
22. B. Catania, workshop on "Impact of VLSI in Telecommunication," ISCAS/IEEE, Rome, May, 1982.
23. *Electronic Product Magazine,* Microtest System, Inc., February, 1980.
24. C.G. Peatty, J.D. Adams, S.L. Carrel, T.D. George, and M.H. Valek, "Elements of Semiconductor Device Reliability," *Proceedings of IEEE,* February 1974.
25. P. Brambilla, F. Fantini, P. Malberti, and G. Mattana, "CMOS Reliability, a Useful Case History to Revise Extrapolation Effectiveness Length and Slope of the Learning Curve," *Microelectronics and Reliability,* Vol. 21, February 1981.
26. G. Mattana, *Qualità, Affidabilità, Certificazione,* Franco Angeli Editore, 1986.

Chapter 2
BASICS OF SEMICONDUCTOR DEVICE TECHNOLOGY

V. Ghergia and D.E. Riva

2.1 INTRODUCTION

Over the last decade, the evolution of semiconductor technology has made great strides in the fields of telecommunication, *electronic data processing* (EDP), and automation, taking on greater and greater importance in all sectors affected by electronics. The main reason for this growth is the revolution brought about by the development and diffusion of semiconductor integrated circuits.

In the field of telecommunication, in particular, the sophisticated performances afforded by monolithic ICs has led to a dramatic modification of signal switching techniques [1–3], which are the very heart of telecommunication systems. This was accomplished by centralizing addressing, control, and signal generation and by transferring to the field of telecommunication procedures typical of EDP. Nowadays these circuits can be used to handle any kind of data in digital form at extraordinarily high speeds, paving the way to new applications in space communication, optical fiber transmission, the distribution of wideband videomatic services, and the exploitation of artificial intelligence.

To cope with the highly complex problems encountered in these extraordinary application domains, the performance of integrated circuits must be continuously improved; and this can be done only by pushing the integration process further and further through the use of innovative materials, the development of even smaller geometries, and the adoption of new techniques.

All this entails tremendous efforts in the field of technological research, to be carried out by semiconductor manufacturers and research labs worldwide, mainly focused on the following issues:

- Material preparation.
- Material analysis and characterization.
- Processes.
- Interconnections and packaging.
- Design for testability.
- Testing.

These issues not only affect device performance but also another key parameter: reliability.

The so-called integrated circuit, in which the functions of many discrete electronic devices are integrated on a single substrate, can be either monolithic or hybrid. The latter involves the use of thick and thin film depositon techniques as well as semiconductor device assembly techniques. Most integrated circuits in use today are monolithic semiconductor devices, and nearly 99 percent of them are manufactured starting from a silicon substrate.

Recently, however, there has been increasing interest in the development of monolithic integrated circuits made of compound semiconductors, such as gallium arsenide and related compounds. Thanks to higher electron mobility, GaAs devices may reach theoretical working speeds about 1 order of magnitude higher than their silicon counterparts [4] with lower power dissipation, which is of major importance both for system design and packing density (see Figure 2.1).

Until a short time ago it had not been possible to exploit these inherent advantages of GaAs, as the basic material preparation techniques, such as crystal growth and further processing techniques, had not yet reached a satisfactory development level and the results obtained, both in terms of reproducibility and performance, have failed to confirm the potential superiority of GaAs over silicon.

Nevertheless, the latest studies performed on this compound have resulted in new, very fast logic circuits whose complexity is reaching higher and higher levels of integration. This follows the development of nearly defect-free single crystal growth techniques, exhibiting satisfactory uniformity and reproducibility characteristics, as well as the progress made in ion implantation and, consequently, annealing techniques. The latter two developments were necessary both to eliminate the defects caused by particle bombardment and to activate the implanted species.

At present GaAs-integrated circuits can be manufactured with over 10,000 elementary devices on one chip, and market projections forecast the turnover volumes of these logic ICs to exceed $2.5 billion by 1992, which corresponds roughly to 5–10 percent of the estimated total world market for integrated circuits.

Such a rapid growth in the use of GaAs-integrated circuits, however, is not expected to challenge the overwhelming predominance of silicon circuits in the near future. Substantial progress has been made in this area, too; for example, 1 Mbit dynamic RAMs are now commercially available, and 4 Mbit DRAMs have

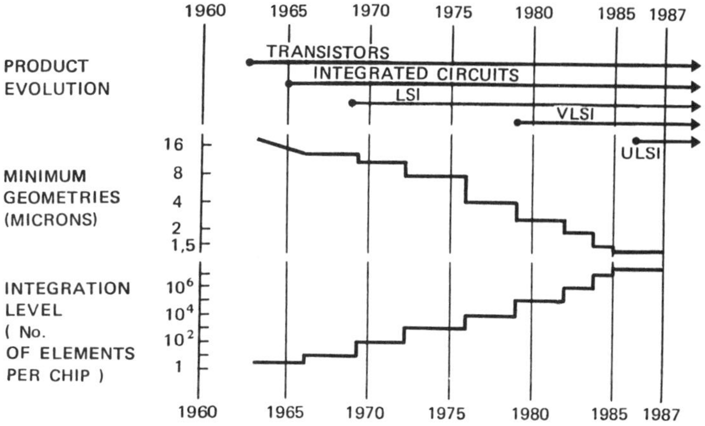

Evolution of silicon ICs: device types, minimum geometries, and integration level.

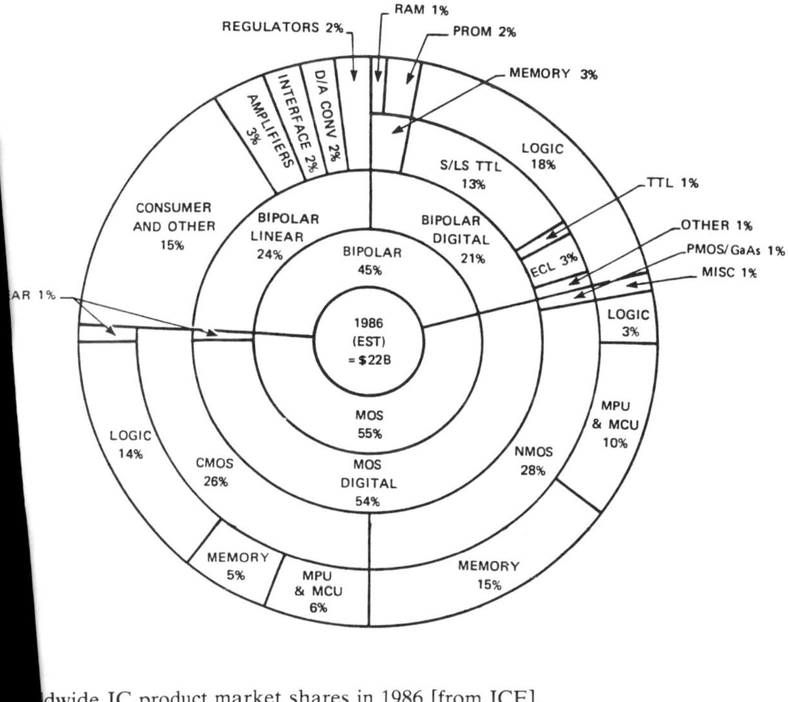

dwide IC product market shares in 1986 [from ICE].

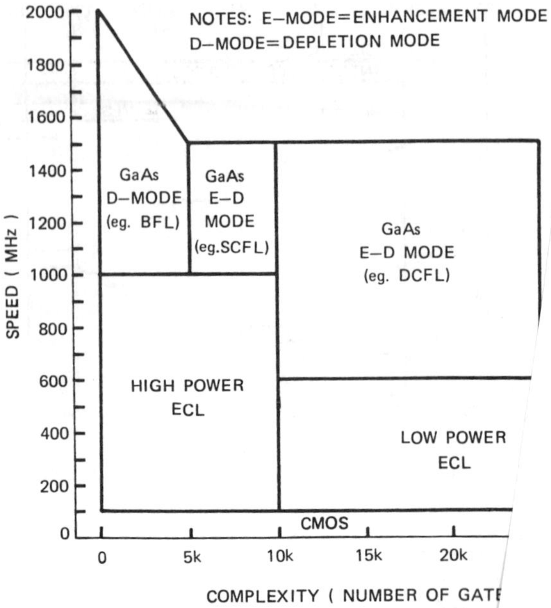

Figure 2.1 In this approximate comparison, the performance levels fo digital technologies overlap. The maximum power dissipa point that allows the greatest complexity times the gate- ECL devices, the maximum power dissipation is 15 W;

already been announced. These results have been obta in the preparation of the basic materials, in the lit growth and deposition of dielectric layers, in the me tilevel interconnections, and so on.

Figure 2.2 illustrates the evolution of silicon i both product type and minimum feature size. The l in integration level.

Silicon circuits have been traditionally subdiv elementary device into bipolar and *metal oxide se* circuits. In 1986 the market share for MOS ICs v and they seem still to be the most popular.

Both MOS and bipolar integrated circuits digital and linear circuits. Digital circuits are in r but although MOS devices with linear circuits a market, bipolar digital circuits add up to a 24 p

Linear circuits find numerous, widely dive circuits (amplifiers, voltage regulators, compa oscillators, filters, *et cetera*) to the wide market of ad-hoc designed circuits (application specific and multiuser specific integrated circuits) used in both consumer and professional applications.

Digital circuits are classified more easily as they can be grouped into memory and data or signal processing circuits (see Figure 2.4) other than logic circuits.

TYPES OF MEMORY :

- DYNAMIC (DRAM)
- STATIC (SRAM)
- READ– ONLY (ROM)
- PROGRAMMABLE (PROM)
- ERASABLE PROGRAMMABLE (EPROM)
- ELECTRICALLY ERASABLE PROGRAMMABLE (E^2 PROM)

VLSI CIRCUITS FOR INFORMATION PROCESSING:

- DATA PROCESSING
- SIGNAL PROCESSING

Figure 2.4 Classification of standard and special application VLSI digital ICs.

Memory is performed by the large scale integrated circuits, which can be standardized more easily and grouped into specific families by technology and data storage characteristics; that is, dynamic read-write memories (DRAMs), static read-write memories (SRAMs), *read only memories* (ROMs), off-line *programmable read only memories* (PROMs), off-line *programmable and erasable read only memories* (EPROMs), and finally in-line *programmable and erasable read only memories* (EEPROMs).

Dynamic memories are now being manufactured with both N-MOS depletion load and CMOS technology (almost all one Mbit DRAMs are CMOS); these memories feature the highest packing density, achieved thanks to the miniaturization of the memory cell. Data is stored by charging a stray capacitor, thus periodical refreshing is needed to avoid information loss.

Static memories, mostly CMOS (but some combination bipolar and CMOS do exist), instead do not require any refresh cycles. However, these memories cannot store the data for an indefinite period of time either, as they are "volatile"; that is, when the bias is turned off all information is lost.

Successful solutions to the problem of data storage in a nonvolatile memory include the use of ROMs, either mask or electrically programmable; a promising approach is to combine very fast *random access memory* (RAM) areas for data handling at high speed with programmable-erasable read only memory blocks for nonvolatile information storage.

The evolution of memory devices is illustrated in Figure 2.5. As far as technology is concerned, ROMs are mainly manufactured using n-channel or complementary MOS transistors, and the configuration is determined by the metal mask, whereas programmable memories can be either bipolar or MOS. In bipolar memories, programming is achieved by blowing a fuse (fusible link) or shorting a transistor junction with a proper electrical pulse, so that these devices can be programmed only once. MOS programmable memories instead are programmed charging a floating gate or trapping charge carriers at an oxide-nitride interface. Upon exposure to UV radiation or by applying a proper bias, the information stored can be erased and the device is ready for reprogramming. Tough technological (and reliability) problems, mainly related to the size of the memory cell and the thickness and physical characteristics of dielectric layers had to be solved to ensure proper operation of these devices.

Data processing ICs (microprocessors) are implemented with combination logic circuits (such as arithmetic logic units), sequential logic control circuits (counters, registers, *et cetera*), and data transfer circuits. If these circuits are together on the same chip, the resulting circuit is called a *microcontoller*.

Signal processing circuits are actually CPUs, where control circuitry is reduced to a minimum to enhance operating speed and performance of arithmetic units, thus limiting flexibility. They are used for specific applications such as modems, filtering, signal compression, voice synthesis, and so on. (See Figure 2.6).

This family of circuits, which of course cannot be easily standardized, includes mostly dedicated (custom) integrated circuits whose design, fabrication, and reliability characteristics are peculiar: they are usually large scale integration circuits manufactured with several tens of thousand elementary devices (10^4 to 10^6), and from the technological point of view, such complexity means high packing density and miniaturization. Thus, considerable work has to be done to accommodate these two fundamental prerequisites: new solutions must constantly be sought where numerous technological aspects are concerned, from lithographic processes to doping, from the thickness and quality of dielectrics to the metallurgy of interconnections and the type of bonding; and this constant drive for new, more advanced solutions obviously means reliability problems.

Designing these circuits may prove time consuming and costly. It may take from 12 to 24 months to design a circuit with a complexity on the order of 10^6 transistors with an investment as high as $10 million or more.

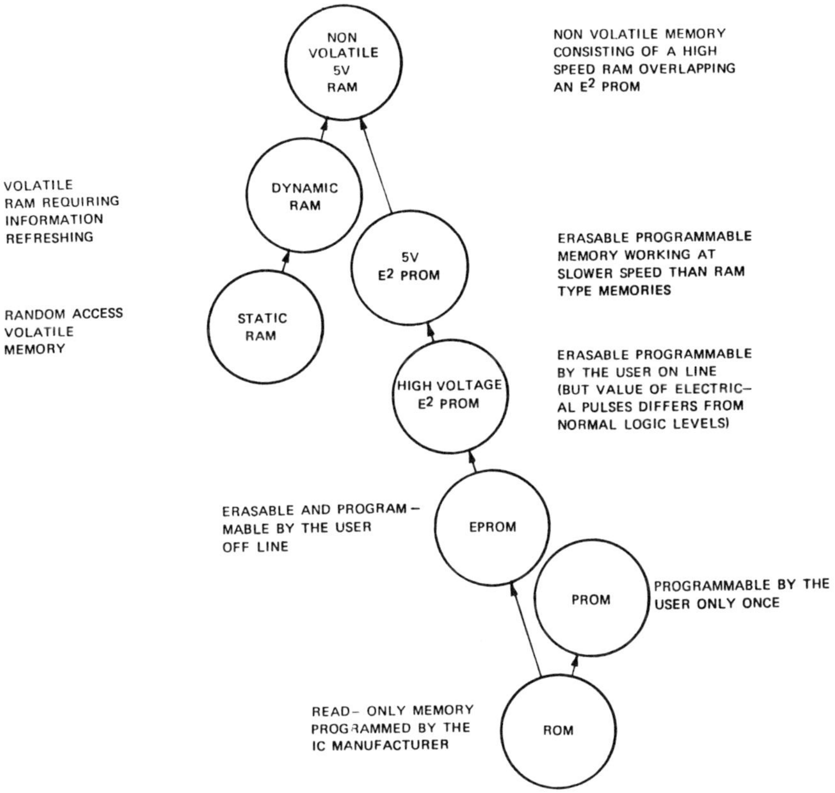

Figure 2.5 Evolution of semiconductor memory ICs.

In order to maintain the design stage within acceptable limits on time and costs, which cannot be easily justified by the limited production typical of dedicated circuits, two main approaches are now being followed, based on the utilization of gate arrays and standard cells. In the first case, a prediffused slice is customized by a proper design of the interconnection (metalization) mask. In the second, which exhibits a higher degree of flexibility and allows a better utilization of the semiconductor area, the user can design the whole circuit, taking advantage of a library of precharacterized functional blocks or cells whose physical location on the chip (placement) and connection patterns (routing) are optimized by CAD tools according to the required performance. This method is being used more and more widely by IC manufacturers to shorten the design times of complex standard circuits.

Let us now classify the different families of bipolar and MOS logic circuits currently used:

- *Transistor-transistor logic* (TTL) seems still to be the most popular bipolar family: the latest versions, advanced Schottky and advanced low power Schottky, allow high switching speed with reasonable power consumption. When very high switching speeds are required a nonsaturating bipolar logic family, *emitter coupled logic* (ECL), seems to be the best choice, even though the high speed brings on high power dissipation and related problems.
- Complementary MOS logic families (HC, HCT, AC, and ACT) are now challenging TTLs, thanks mostly to its reduced power consumption and high integration levels, possible through sophisticated fabrication processes.
- Linear circuits can be either bipolar or MOS or both (mainly bipolar and CMOS together).

DATA PROCESSING CIRCUITS

APPLICATIONS:

- MICROPROCESSORS
- MICROCONTROLLERS
- I/O – PERIPHERALS
- ETC.

SIGNAL PROCESSING CIRCUITS

APPLICATIONS:

- COMMUNICATIONS (COMPRESSION AND FILTERING OF DATA– VIDEO– AUDIO, WIDE BAND PROCESSING).
- RECORDING (DATA– VIDEO– AUDIO).
- VOICE SYNTHESIZING.
- NATURAL LANGUAGE RECOGNITION
- HIGH RESOLUTION IMAGE PROCESSING
- RADAR (COMPRESSION OF DIGITAL PULSES– RADAR SIGNAL PROCESSING)
- ETC.

Figure 2.6 Applications of VLSI data and signal processing circuits [from SGS].

Figure 2.7 shows the market shares of the different types of semiconductor integrated circuits forecast for 1991.

2.2 TECHNOLOGY OF SILICON INTEGRATED CIRCUITS

The structure of an integrated circuit is very complex and its three-dimensional realization in the upper layers of a wafer (planar process) must be performed with

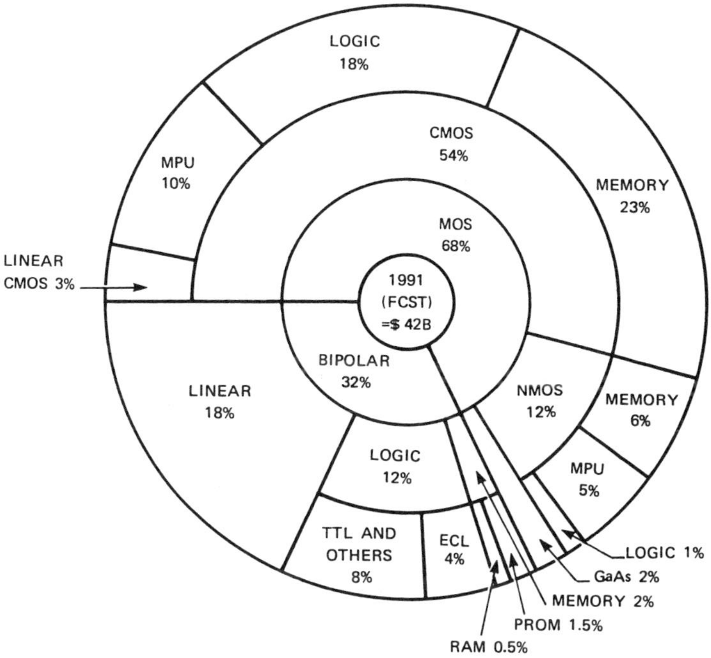

Figure 2.7 Worldwide IC product market shares forecast for 1991 [from ICE].

the utmost accuracy. Each layer in the structure has a well-defined position: some layers are located inside the silicon substrate and others are fabricated above it. The manufacturing process constructs the various layers to a predetermined sequence and configuration. Thus, the integrated circuit manufacturing process consists [5, 6] of a complete sequence of steps, summarized in the block diagram of Figure 2.8.

The operations listed on the left are technological and involve the use of clean rooms, furnaces, growth reactors, packaging equipment, and the like. The blocks on the right-hand side list a variety of heterogeneous steps to do with electronics, EDP, material processing, and reliability.

The principle that made possible highly complex, miniaturized ICs was to resort to lithographic techniques for the various fabrication processes. Once the type of circuit to be produced was defined, the necessary documentation (photomasks or magnetic tapes) for the application of lithographic techniques has to be prepared. The first step consists of converting the electrical diagram of the circuit into the topography and internal architecture of the integrated circuit. This operation, which results in the definition of the so-called circuit layout, is very com-

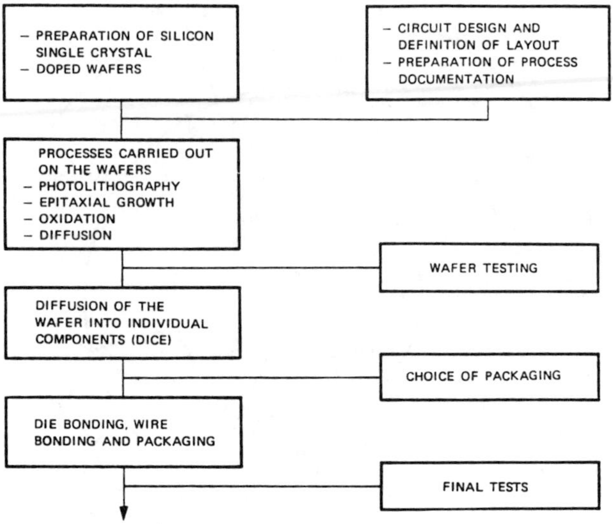

Figure 2.8 Schematic diagram of an IC manufacturing operation.

plicated. In order to determine the physical configuration of an integrated circuit, in fact, design engineers have to take into account the functional specifications of the product being developed, select suitable fabrication procedures, estimate its dimensions and establish its geometries, and approximate the positions of the various elements making up the circuit.

In the past (until about ten years ago), the job of working out circuit layouts from electrical diagrams was done manually. Today, designers can avail themselves of computers, graphic aids, and software packages that greatly facilitate this conversion through interactive human-machine communication. Layout design, however, may still be very lengthy when dealing with highly complex circuits and therefore much effort is currently being devoted to the development of design automation software.

Once IC layout has been defined, the next step consists of producing the documents necessary to the photolithographic manufacturing process. The computers used for layout design provide, at the end of the process, magnetic tapes from which, by photocomp machine, photographic masks can be derived, usually on a 10 : 1 scale. A different mask has to be produced for each of the various technological processes to be carried out on the wafer according to a predetermined sequence. The photographic masks are then reduced to a 1 : 1 scale, the scale necessary for contact lithography, while the so-called step and repeat operation is performed to allow a considerable number of identical circuits on each mask, arranged in lines and columns and extending over an area equal to the area of the wafer. Now, the masks are ready for wafer lithographic processing.

A wafer must possess some special features. First, it must have been cut out of a single silicon crystal, have a well-defined crystallographic orientation, and suitable physical and electrical characteristics. The typical photolithographic process, to be repeated in different circumstances for different type operations, is shown in Figure 2.9. Essentially, it uses a photosensitive material to cover a previously oxidized wafer. Through the photomask, the coating of light sensitive material is selectively exposed to ultraviolet light. Next, a light sensitive coating lays bare the oxidized layer of the wafer in predetermined areas. Regions of this coating, called photoresist, have either become more soluble due to exposure to light (positive photoresist), or remain soluble because they have been masked from light (negative photoresist). In the development process which follows, these regions of coating are etched away, and only in those areas the underlying oxide layer is exposed. Then the desired patterns are obtained by means of a selective chemical etching (the chemicals etch silicon oxide much faster than silicon and have no effect on unetched photoresist). Taking advantage of the protection afforded by the silicon oxide layer, unprotected areas are selectively doped, either by diffusion or by ion implantation.

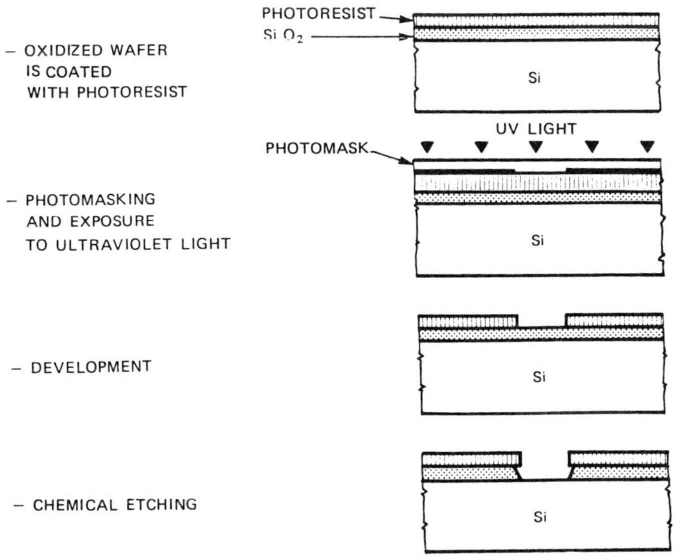

Figure 2.9 Photolithographic process.

Though conceptually similar, the technological processes performed on silicon wafers vary, depending on the type of circuit being developed or the performance desired. With increasingly complex and sophisticated circuits, the number of steps also increases, as does the number of photomasks. All processes, however, are essentially based on the two sequences shown in Figures 2.10 and 2.11 for bipolar and MOS ICs, respectively.

By analyzing the two series of steps, it can be seen that bipolar circuits call for a higher number of operations, reflected by the minimum quantity of photomasks required: six rather than four. Furthermore, the bipolar process involves an additional operation, as the individual circuit components all have to be electrically insulated from one another. This is not only an additional complication compared with the processing of MOS circuits, which are inherently self-insulated, but also consumes a portion of the wafer area and thereby reduces its possible integration density. Figure 2.12 shows a bipolar logic (TTL) integrated circuit, whereas in Figure 2.13 the die of a CMOS logic circuit is shown for comparison.

It should be pointed out that in actual practice the processes used to fabricate large-scale integration ICs, both for bipolar and MOS devices, generally involve a much higher number of operations and hence of photomasks, up to twice as many as the figures given earlier. These further complications in the fabrication processes arise from considerations of yield, packing density, and consequently multilevel interconnections, performance, and so on.

When all the processes have been completed, a wafer contains a certain number of integrated circuits (see, for instance Figure 2.14) that are perfectly finished and ready to operate as soon as the electrical connections to the bonding pads of each individual circuit are provided.

At this point, before the wafer is divided into separate ICs, it is necessary to perform electrical tests aimed at sorting out the circuits that work properly from the rest. This preliminary selection must be performed right away, as the yield of the processes described so far can be low (yield depends on circuit's complexity and the area it occupies; in some cases, even 10–20 percent values may be viewed as acceptable), and therefore, it is indispensable that subsequent individual operations (assembly, packaging), which are somewhat more expensive, be performed only on devices ascertained to be within specification limits.

Wafer testing is performed automatically by computer-controlled machines equipped with needlelike probes capable of producing by contact and pressure the electrical connection to the pads of the individual ICs. These machines can quickly test any circuit, regardless of its complexity: faulty circuits can be marked (for instance, with a drop of some magnetic material or ink) for elimination at a subsequent stage once the wafer cutting operation has been completed.

The number of tests to be performed to ascertain proper circuit operation can be very high (up to a few thousands in the case of highly complex circuits), hence, determining the test patterns to be applied to the circuit can be very critical and costly. As a rule, this procedure is defined at an early design stage, at a cost that may easily amount to a significant portion of overall design cost.

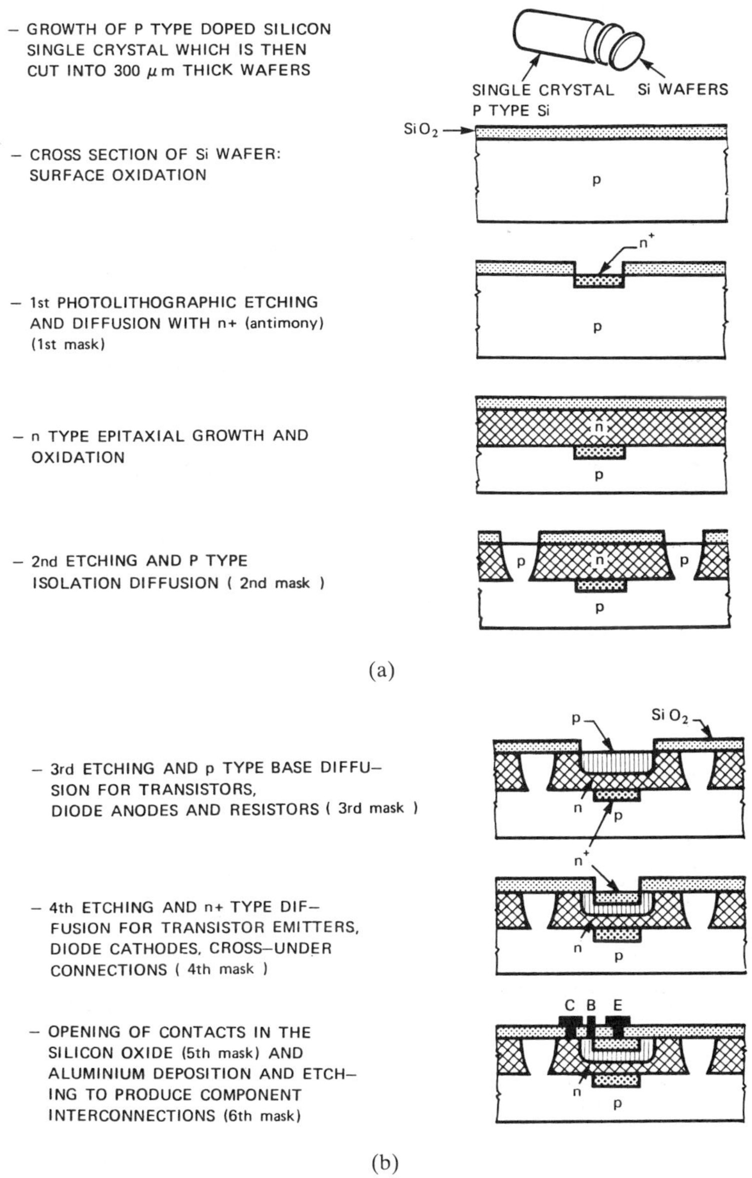

Figure 2.10 Basic planar process for bipolar IC manufacturing.

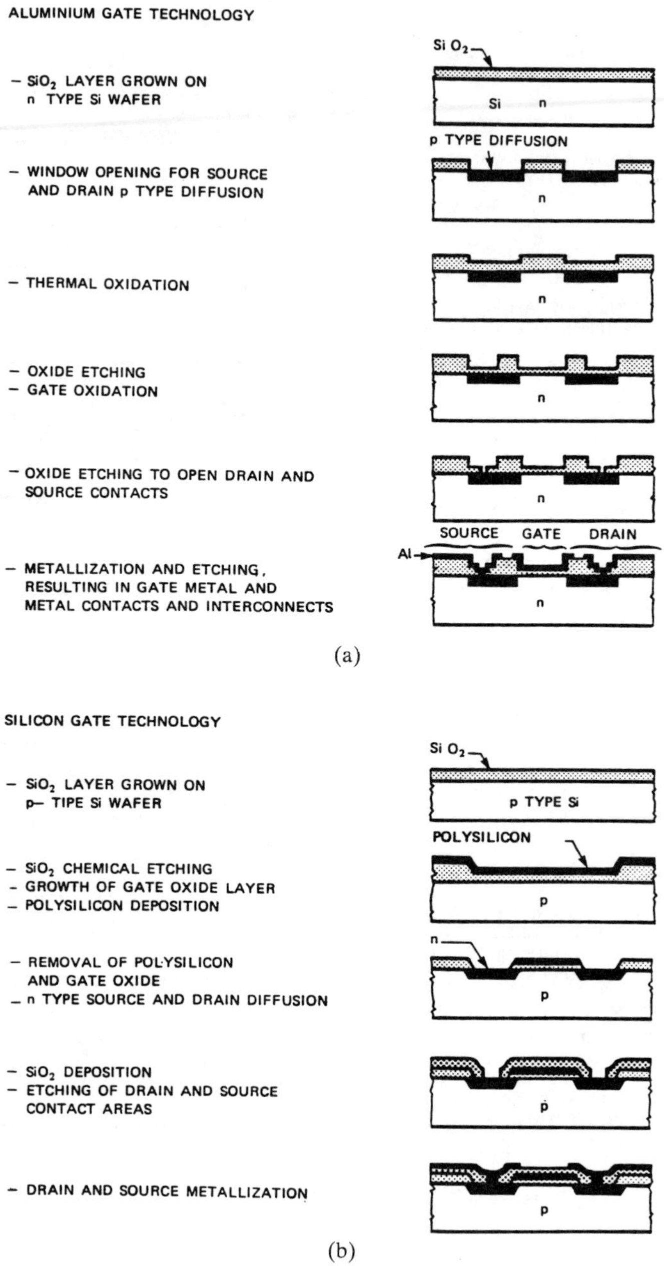

Figure 2.11 Basic planar process for (a) aluminium gate p-channel and (b) silicon gate n-channel MOS IC manufacturing.

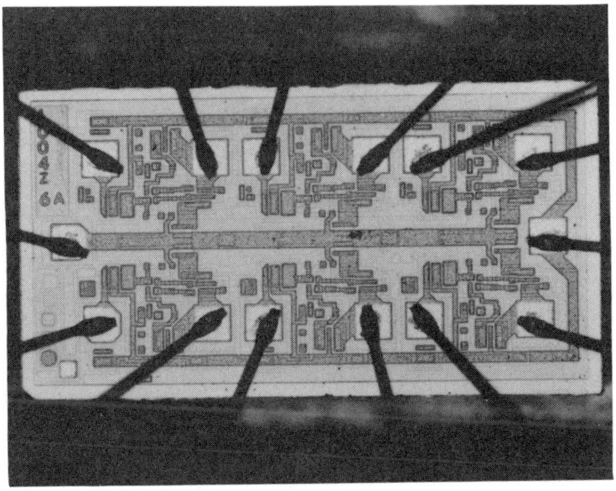

Figure 2.12 Bipolar digital (TTL) IC type 7404 (Hex inverter).

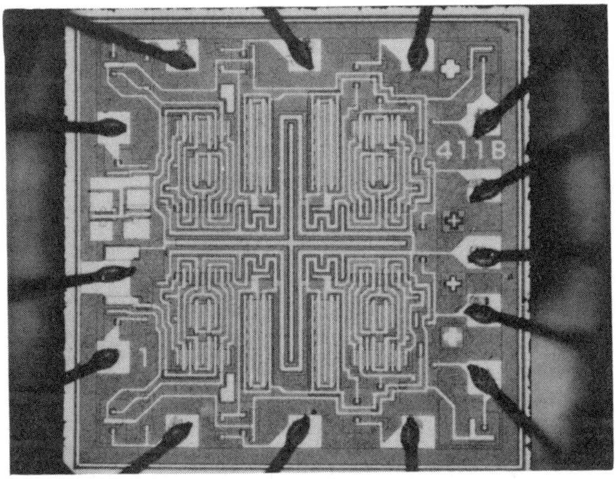

Figure 2.13 CMOS metal gate digital IC type 4011 (quad two-input NAND gate).

Following the location of faulty circuits, the wafers are scribed, mostly by means of very fine diamond saws, following the pattern of lines and columns to obtain individual ICs. At this point, faulty circuits, previously marked, can easily be removed by an electromagnet, while valid circuits stand ready for assembly and packaging.

Packaging is of the utmost importance from the viewpoint of finished product reliability, although it cannot be denied that the behavior of integrated circuits depends largely on the type and quality of the processes carried out on the wafer.

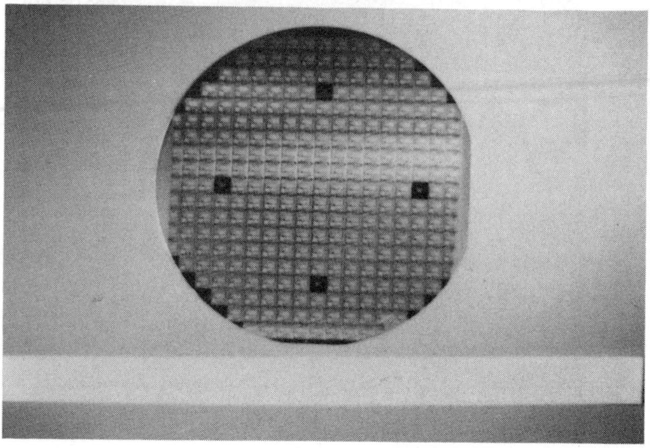

Figure 2.14 A wafer at the end of the manufacturing process.

IC packages must ensure circuit protection, heat dissipation, and efficient input and output signal interconnection between the inside and the outside (see Figure 2.15). Each of these tasks is associated with specific problems, especially in the case of very large scale integration.

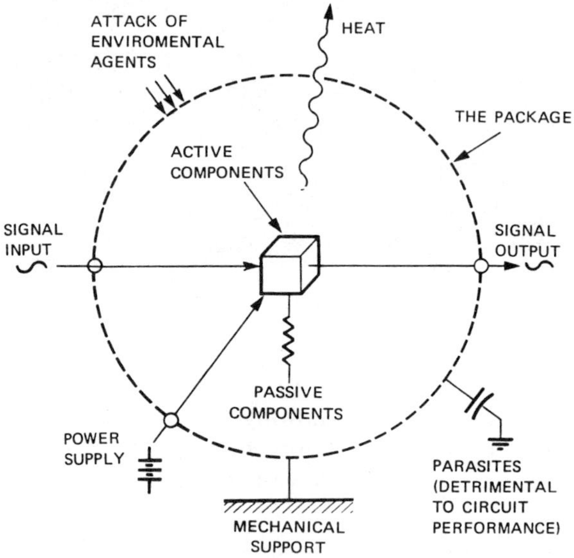

Figure 2.15 Semiconductor device package features.

As a rule, circuits are protected against external agents by a case, which should be as hermetic as possible. Choice of packaging materials, nowadays mostly ceramic and plastic compounds, is based on economic considerations to do with the great number of terminals required by modern circuit design [7]. Plastic compounds, albeit very economical, are not perfectly hermetic but permeable to moisture, a problem that requires additional surface protections. Ceramic packages, on the other hand, are perfectly hermetic and can be used to protect up to large scale integration circuits. When dealing with very large scale integration circuits, however, additional problems may arise. For instance, due to the reduced number of carriers (electrons) necessary to charge a stray capacitor to store the information in a memory cell of a DRAM, the circuit becomes sensitive to α particles emitted by radioactive elements present in parts per million in the ceramic material itself.

With hundreds of thousands of devices clustered to form a single circuit, heat dissipation may become a crucial issue, and because the average useful life of an integrated circuit depends strongly on its operating temperature, according to Arrhenius equation, great care must be taken to minimize the thermal resistance of the IC package.

Finally, in highly complex ICs, over 100 input and output pins can easily be required, and in actual practice, this is by no means easy to accomplish on miniature packages, which must retain the advantages of compact design afforded by integration.

All these problems have brought about a rapid evolution in packaging products intended for large scale integration circuits. At present, the most commonly used package is the so-called dual-in-line package, but new types are also being developed, such as the chip carrier and the pin grid arrays (see Figures 2.16 and 2.17) [8].

Once package style has been selected, the device is assembled and encapsulated through a series of steps that include, as a rule, die bonding (fastening the IC to the package) and wire bonding (providing electrical connections between the IC pads and the package terminals). These two operations are both crucial to reliability. Die bonding must minimize the thermal resistance between the active part of the IC, which produces heat, and the case, which must dissipate it. For instance, when a ceramic package with a gold plated cavity is used, this is accomplished either by heating the case until the eutectic temperature of the IC silicon and the gold (370°) is reached or by using a low-melting solder preform, such as a tin-lead alloy.

Wire bonding can be performed with either gold or aluminium wires, depending on the package and the outer surface finish of the IC. The wires (usually about 1 mil in diameter) are bonded to the IC pads either by thermal compression or ultrasonic bonding by automated microprocessor-controlled machines (see Figure 2.18). Mounted in its case (Figure 2.19) and encapsulated, the integrated circuit is now ready to undergo the final series of electrical tests.

Figure 2.16 Example of chip carrier (2 X).

Figure 2.17 Example of pin-grid aray (2 X).

As mentioned earlier, the technological processes carried out on the wafer differs for bipolar and MOS circuits. But even the different families of bipolar and MOS circuits admit a variety of processes. Thus, a logic bipolar device is subjected to considerably different fabrication processes than those for linear devices, and the differences are even greater when dealing with integrated power devices or those for high voltage applications. The various logic families (TTL, ECL, I^2L) also call for different processes.

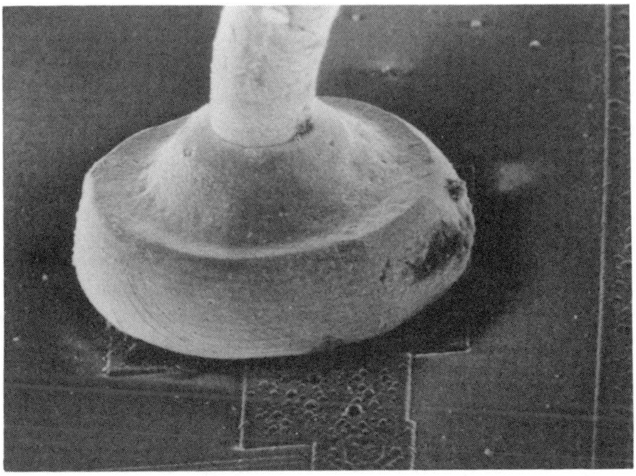

(a)

(b)

Figure 2.18 (a) Thermal-compression (550 X) and (b) ultrasonic (450 X) bonding.

The same applies to MOS integrated circuits, even though the differences are not as great where p-channel and n-channel MOS devices are concerned or between these and CMOS circuits. In this sector, processes differ mostly in relation to the scale of integration. Thus, we see that, for the same type of n-channel logic MOS circuit, the technology used a few years ago to process short- or medium-scale integration devices has been modified to a significant extent to deal with

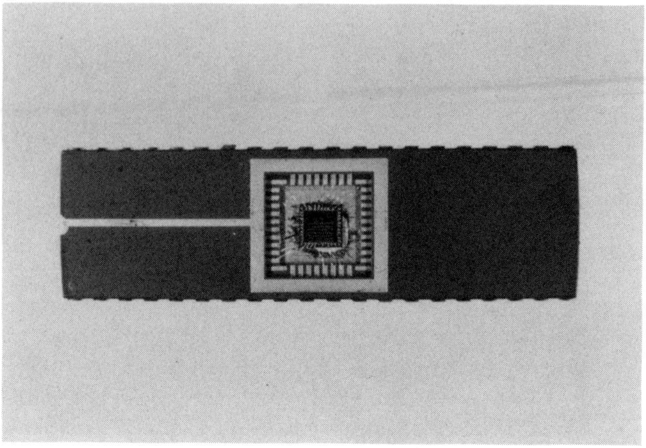

Figure 2.19 An integrated circuit die after mounting in its package (2 X).

today's LSI and VLSI circuits. This evolution was intially motivated by the desire to obtain increasingly high packing densities for increasingly complex ICs.

The first problem design engineers had to face was reducing circuit geometries. Thus, in a few years' time, circuit geometry was scaled down from 5–10 μm to micron size (see Figure 2.20). All of this necessarily entailed a corresponding evolution in photolithography. The traditional techniques, based on contact printing, were gradually replaced by photoresist impression by projection; and new techniques, such as X-ray, electron beam, and ultimately ion beam lithography, were developed, enabling semiconductor manufacturers to go to submicron geometries. Figure 2.21 illustrates in schematic form the evolution of lithographic techniques currently under way, and Figure 2.22 shows the limitations imposed on circuit geometry by the different techniques, as determined by yield and production constraints.

As soon as the reduction of the elementary devices got under way it became clear that a whole new series of problems were affecting both operation and reliability of ICs. A reduction in the length of MOS channels, in fact, led to punch-through phenomena: the edge effects of electric fields increased so that they were no longer negligible for normal source and drain doping profiles, power supply voltages proved too high for the new geometries, as a result carriers became so hot they could jump into dielectric layers altering threshold voltages, and so on.

Thus, it became necessary to reconsider the structure of ICs thoroughly and this led to the development of the notion of "scaling down" by which a reduction in any factor, *S*, of circuit geometry must be matched by a corresponding reduction in many other parameters (see Figure 2.23).

As it turned out, scaling down was beneficial in many respects: in addition to reducing the space occupied by circuit components and enhancing IC complexity,

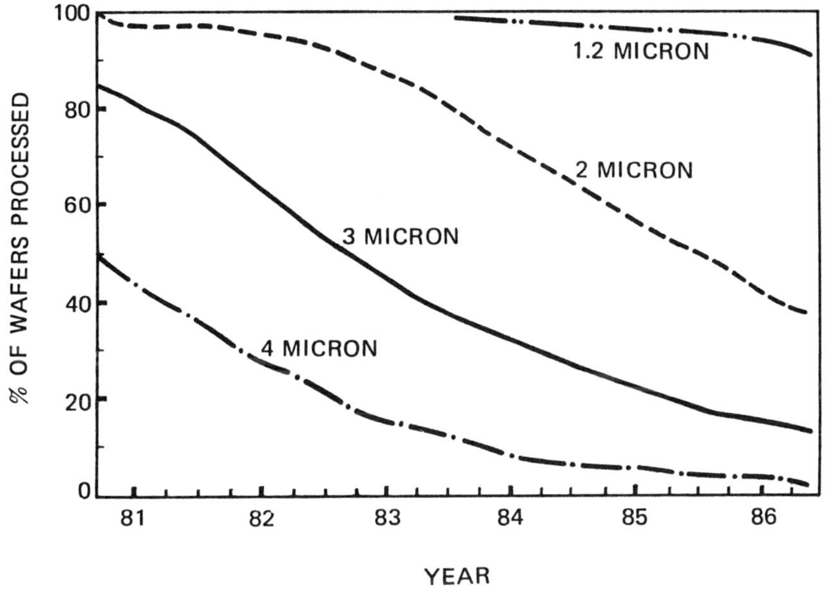

Figure 2.20 Production *versus* process standards in the US [from SGS].

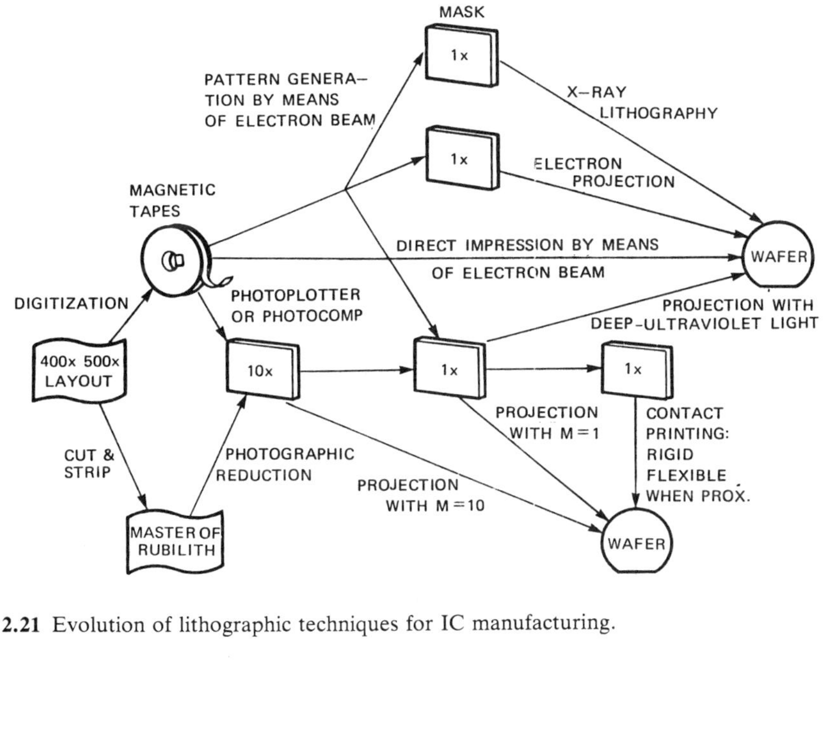

Figure 2.21 Evolution of lithographic techniques for IC manufacturing.

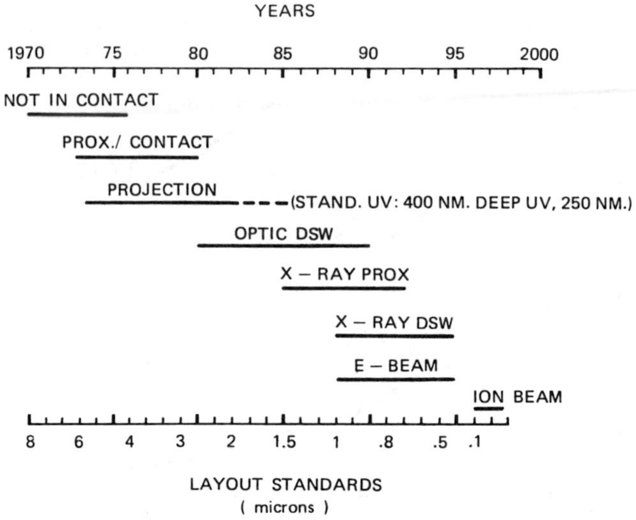

Figure 2.22 Evolution of the equipment used for lithography.

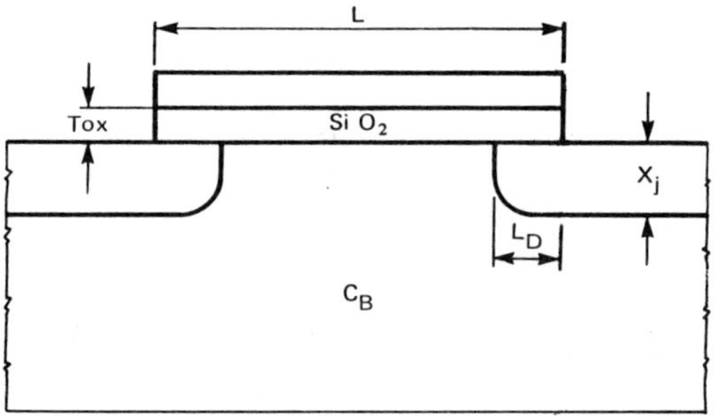

COMPONENT DIMENSIONS – PARAMETERS	SCALE FACTOR
– Tox , L, L_D, W, X_j	1/S
– SUBSTRATE DOPING C_B	1/S
– POWER SUPPLY VOLTAGE V	1/S
– POWER SUPPLY CURRENT I	1/S
– STRAY CAPACITANCE WL/Tox	1/S
– DISSIPATED POWER VI	$1/S^2$
– POWER – DELAY PRODUCT	$1/S^2$

Figure 2.23 Scaling down rules for MOS devices.

it also afforded considerable improvements in circuit performance speed, power dissipation, and above all Q factor (gate propagation delay multiplied by dissipated power), which consituted the true indicator of integration potential.

But scaling down also gave rise to a whole new series of technological problems. Lithographic methods had to be modified, as mentioned earlier, through the development of new exposure techniques, new equipment, new light-sensitive and electro-sensitive materials, and new etching techniques, which replaced wet chemical etchants (involving under-etching problems) with sophisticated anisotropic dry techniques such as plasma assisted etching.

Dielectric layer growth and deposition techniques had to be thoroughly investigated to produce extremely pure materials with a perfect structure, which made it possible to use progressively thinner layers down to the order of a few hundred Angströms. Doping by ion implantation was also improved to attain a much wider range of dopant concentrations and improve the control of doping profiles.

Multilevel interconnection techniques were studied afresh, experimenting with new conductive materials that might allow reduction of conducting tracks and contacts without entailing an exceedingly high increase in electrical resistance and might withstand high current densities without giving rise to the migration of metals over time.

Finally, in view of the incessant growth of IC areas, now exceeding 1000 mm^2 per device and of the concomitant need to maintain production yields within economically acceptable limits, it became necessary to improve the quality of silicon single crystals in order to have defect-free substrates. The diagram in Figure 2.24 shows how yield levels improved with decreasing silicon crystal defects: this improvement is particularly significant in large-size ICs.

Since it is economically advantageous to work on large-size wafers [9] single crystal growth techniques have also been improved, culminating in the production of silicon crystals up to 6″ (150 mm) in diameter. Figure 2.25 shows the evolution of wafer diameters in the US industry and expected future trends.

The evolution toward higher integration levels is expected to continue; and present day VLSI, capable of producing ICs with 10^5–10^7 transistors, should make room in a few years' time for *ultra large scale integration* (ULSI), with 10^7–10^9 transistors, to be supplanted in its turn by *giga scale integration* (GSI) with over 10^9 transistors. In order to meet these objectives, the miniaturization of IC devices will have to reach geometries on the order of 0.1 μm and solutions found to the tremendous problems in interconnections and IC active element isolation connected to progressive miniaturization. To cope with such complexity we shall probably have to resort to overlapping three-dimensional structures, and this will require further leaps to overcome numerous technological problems.

2.3 TECHNOLOGY OF GALLIUM ARSENIDE INTEGRATED CIRCUITS

As mentioned earlier, gallium arsenide ICs are taking the lead in many applications requiring very high operating speeds [10]. GaAs in fact, thanks to higher electron

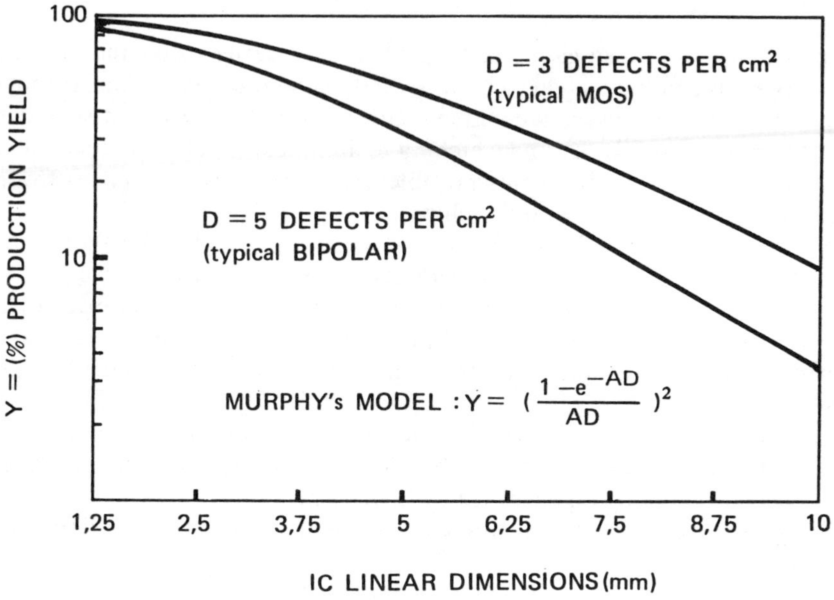

Figure 2.24 Production yield *versus* IC size on a silicon wafer.

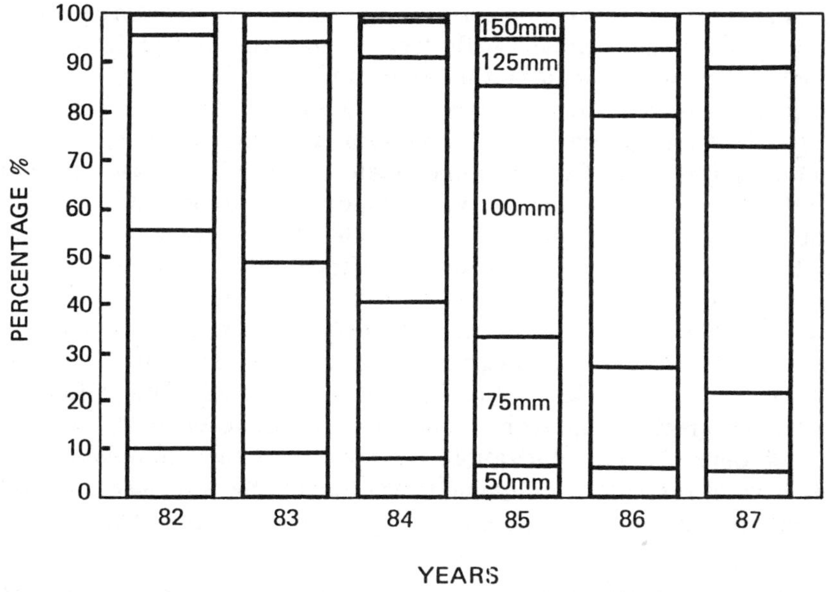

Figure 2.25 Diameter of silicon wafers processed in the US [from SGS].

mobility, offers theoretical operating speeds that compare favorably with those of corresponding silicon ICs by about 1 order of magnitude (see Figure 2.26).

The diagram in Figure 2.27 [11], taken from a paper by D.G. Fisher of ITT Gallium Arsenide Center, provides an example of application of GaAs ICs. We can see how, in this application, high-speed silicon circuits (CMOS and ECL) have to be flanked by digital and analog GaAs circuits for frequencies exceeding 1 GHz.

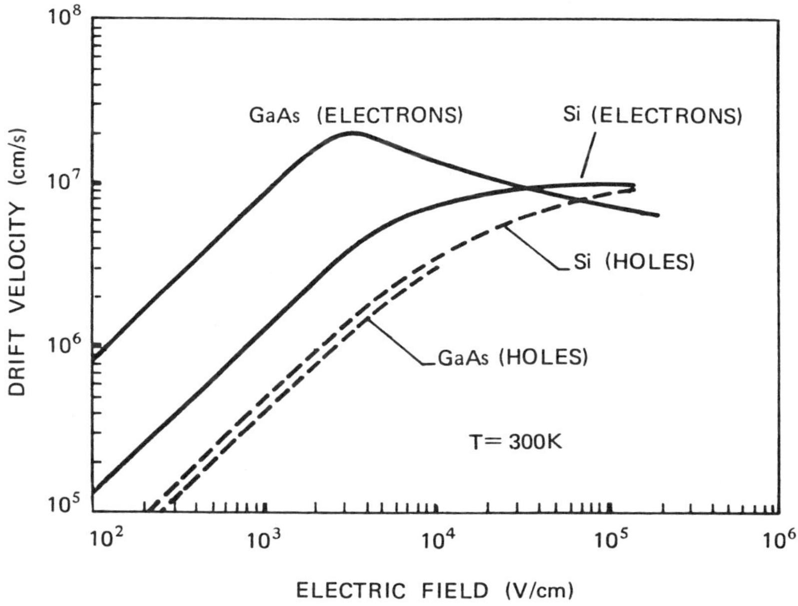

Figure 2.26 Carrier drift velocity *versus* electric field in GaAs and silicon. At low electric fields, electron velocity in GaAs is almost 1 order of magnitude higher than in Si.

Obviously silicon technology, which improved over the years, affords superior reliability and higher circuit densities at lower costs. GaAs technology, albeit very young, makes it possible to attain higher operating frequencies and to produce circuits that feature lower power dissipation rates, can work at higher temperatures, and possess greater radiation hardness. Figure 2.28 compares the *Q*-factor (the product of dissipated power times gate propagation delay) of GaAs ICs with that of the main silicon IC logic families. As can be seen, the *Q*-factor of GaAs circuits, roughly between 1 femtojoule and 1 picojoule, is definitely lower than the *Q*-factor of CMOS technology; whereas high speed silicon technologies, such as NMOS, TTL, and ECL, have even higher *Q*-factor ratings.

This factor, as pointed out earlier, is extremely significant from the viewpoint of large scale integration ICs, and we may safely conclude that GaAs circuit

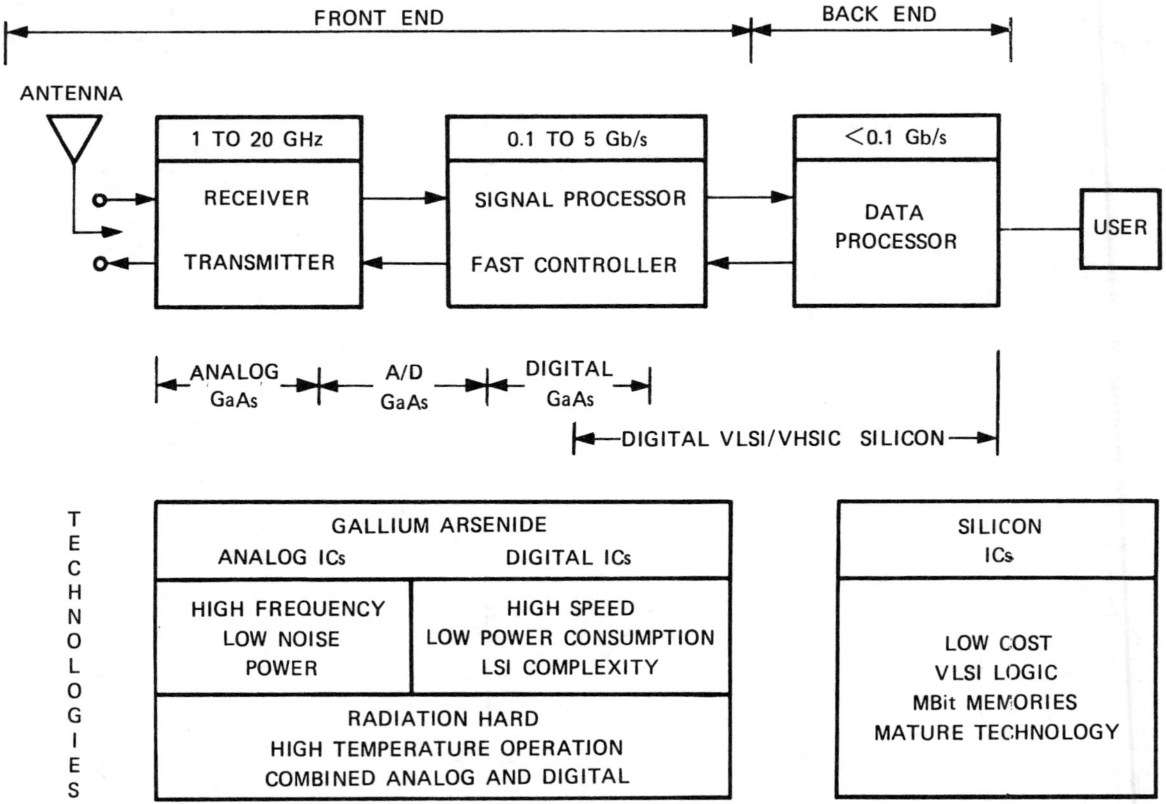

Figure 2.27 Communication system employing both silicon and GaAs ICs [from 11].

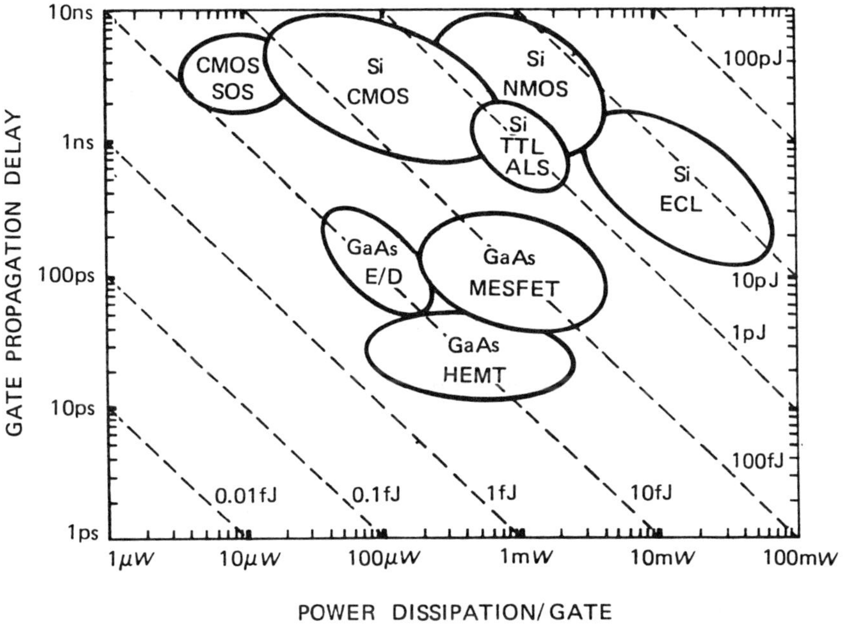

Figure 2.28 Operating characteristics of the main Si and GaAs logic families.

technology, despite its recent development and limitation to ICs of no more than a few thousand transistors, has the potential to surpass silicon technology in the future, affording even higher levels of complexity.

Since Gallium Arsenide does not have a thermal oxide with electrical properties similar to those of SiO_2, efficient MOS devices cannot yet be made with this material. At present, the basic element of GaAs circuits is the *metal electrode semiconductor field effect transistor* (MESFET) [12, 13].

Figure 2.29 illustrates the structure of a depletion mode MESFET that consists basically of two electrodes, source and drain, forming an ohmic contact with the semiconductor, and a gate electrode interposed between them that, through a Schottky contact with the semiconductor underneath, creates a carrier depletion area in the source-to-drain conductive channel, thus reducing the effective channel thickness.

Channel depletion may be increased by applying suitable polarization to the gate electrode. When a channel is completely depleted (pinched off) with no polarization, we have an E-MESFET (enhancement mode) device. In this case,

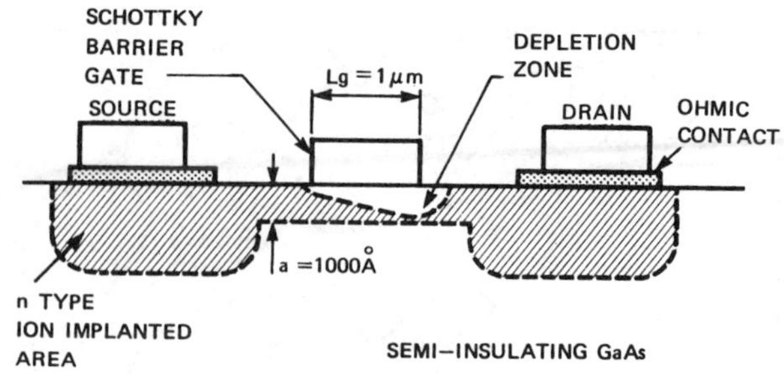

Figure 2.29 Depletion mode metal semiconductor field effect transistor (D-MESFET).

channel conduction is obtained by applying a slight positive voltage to the gate. Conversely, the channel of the depletion mode MESFET is conductive when the gate is unbiased (the FET is normally on); the off state can be reached by applying a negative bias to the gate.

Given the present state of GaAs technology, it is easier to produce D-MESFET integrated circuits, since they require less stringent process controls. In E-MESFET devices, in fact, the value of the so-called threshold voltage is particularly critical, and manufacturers find it hard to keep the tolerances of this parameter within sufficiently narrow bounds to ensure acceptable production yields when circuit complexity reaches a certain level. However, for application purposes, circuits containing E-MESFET devices are more interesting, because they require only one power supply (D-MESFETs require two) and exhibit lower power dissipation rates.

Two logic gate configurations have been developed that use only D-MESFETs, as shown in Figure 2.30; that is, buffered FET logic (BFL), mostly for SSI and MSI super-fast logic circuits, and Schottky diode FET logic (SDFL) used for high-speed LSI circuits (ALUs, multipliers, gate arrays, *et cetera*) [15].

Logic gates using enhancement mode FETs *direct coupled FET logic* (DCFL) have been developed by Hughes and McDonnell-Douglas (with junction FETs). This type of circuit features very low power consumption and is suitable for VLSI applications (see Figure 2.31). The table in Figure 2.32 compares the performance ratings of the different GaAs IC logic families discussed so far.

As we have seen GaAs circuit technology is still at an early stage of development and quite a number of problems still have to be solved before it can seriously threaten the predominance of silicon technology, if just for high speed applications. However GaAs technology has been able to benefit from the wealth of experience accumulated (and constantly being augmented) in the field of LSI silicon circuits: consider, for instance, lithography and interconnection techniques. But the problems that will prove harder to solve and to which manufacturing

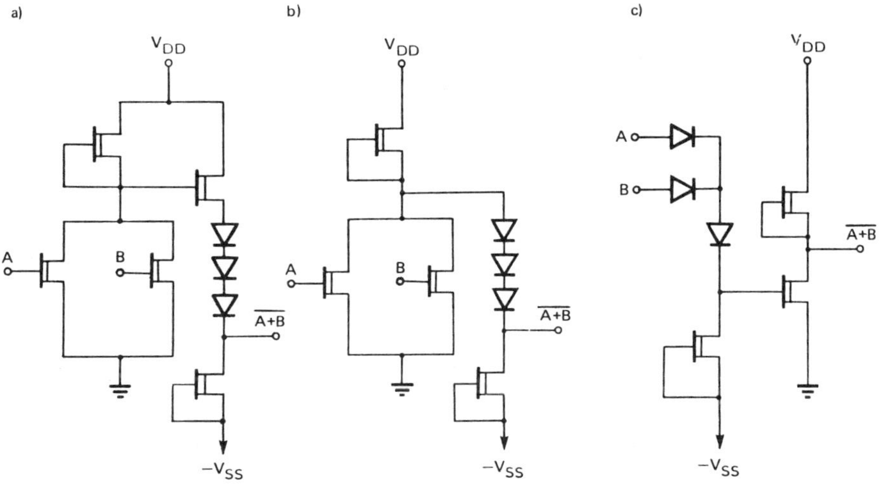

Figure 2.30 Logic gates based on D-MESFET technology: (a) and (b) buffered FET logic (BFL); (c) Schottky diode FET logic (SDFL).

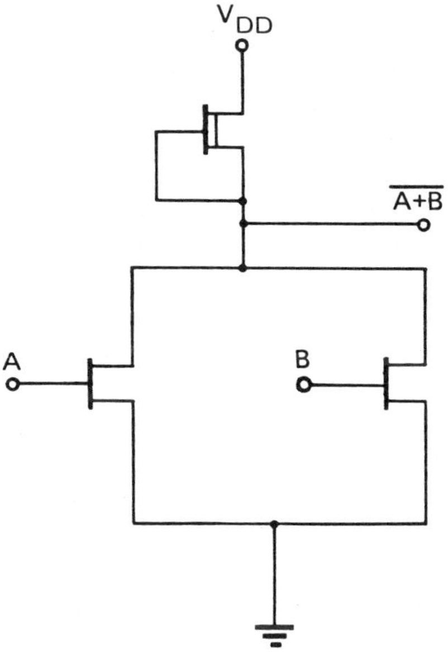

Figure 2.31 Direct coupled FET logic (DCFL) gate made with both enhancement and depletion mode MEFETs.

LOGIC GATE COMPARISON

BFL	SDFL	DCFL
HIGH SPEED WITH FAN–OUT	MEDIUM SPEED WITH LOW FAN–OUT	MEDIUM SPEED WITH LOW FAN–OUT
LARGE POWER DISSIPATION 5 mW/GATE	MEDIUM POWER DISSIPATION 1 mW/GATE	LOW POWER DISSIPATION $<50 \mu W$/GATE
GOOD DRIVE CAPABILITY FO $\geqslant 3$	POOR DRIVE CAPABILITY FO $\leqslant 3$	POOR DRIVE CAPABILITY FO $\leqslant 3$
POOR PACKING DENSITY $>1000\ \mu m^2$/GATE	MEDIUM PACKING DENSITY $>500\ \mu m^2$/GATE	GOOD PACKING DENSITY 200 μm^2/GATE
TWO POWER SUPPLIES	TWO POWER SUPPLIES	ONE POWER SUPPLY
POOR FAN–IN	GOOD FAN–IN	POOR FAN–IN
LIMITED TO MSI–LSI COMPLEXITY	LSI COMPLEXITY	VLSI COMPLEXITY

Figure 2.32 GaAs logic gate comparison.

companies are devoting their attention in trying to prepare GaAs products for large scale commercial diffusion, are the problems to do with the basic material of these circuits; that is with GaAs itself. At present, in fact; large sized GaAs wafers with good quality and uniformity characteristics that will ensure satisfactory production yields are hard to find.

The maximum diameter of GaAs wafers in current production is 3″ (costing from 30 to 40 times as much as silicon wafers), and to obtain the necessary quality and reproducibility, most IC manufacturers have been forced to produce the GaAs crystals on their own.

Finally, for an overall picture of the evolution of GaAs technology, we should point out that although manufacturers are still grappling with the technological problems stemming from the basic material and the shift from depletion to enhancement MESFETs (resembling in many respects the difficulties encountered several years ago in shifting from p-MOS to n-MOS technology), as well as with the numerous other obstacles to be overcome in the development of LSI gallium arsenide ICs, a second generation of devices has already begun to take shape, and it is expected to raise switching speeds to an unprecedented level (gate delay time of the order of 10 ps) by using other types of compound semiconductors (AlGaAs, InGaAs, InP) and heterojunctions.

The most promising devices in this new generation, requiring advanced layer growth techniques such as molecular beam epitaxy and *metaorganic chemical vapor deposition* (MOCVD), are based on the energy band discontinuity which originates at the junction with two materials having slightly different bandgaps; such as, for instance, GaAs and AlGaAs, as shown in Figure 2.33. In structures of this type, known as *high electron mobility transistors* (HEMTs) and as MODFETs, TEGFETs and SDHTs, carrier mobility can be considerably higher than in massive materials. This is because mobile carriers, which are separated from their dopant atoms, are then confined to the GaAs side of the junction and form a layer having quasi two-dimensional characteristics, which is generally referred to as *two-dimensional electron gas* (2-DEG). Discrete HEMTs for high-frequency, low-noise applications are already commercially available, and the practicality in LSI production has been demonstrated by various manufacturers. For example, a 4-kbit SRAM featuring 0.5 ns access time (at room temperature) has been recently developed by Fujitsu using 0.5 μm gate length devices [15].

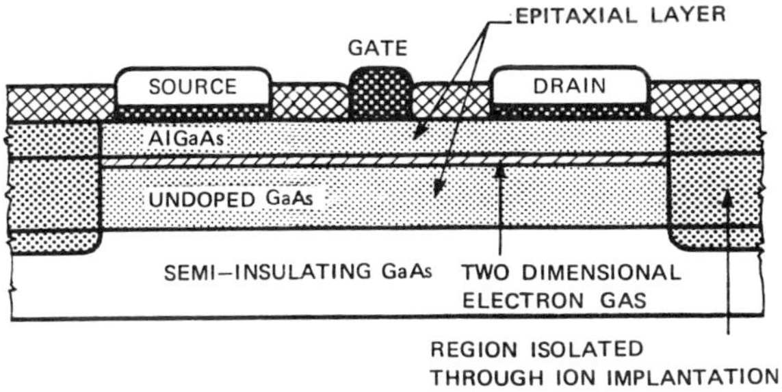

Figure 2.33 Basic structure of a high electron mobility transistor (HEMT).

REFERENCES

1. V. Ghergia: "Lo scenario della tecnologia microelettronica," SSGRR technical seminar.
2. P. Belforte et al., "New Switching Techniques for Wideband ISDN Environments," *CSELT Technical Reports,* June 1984.
3. P. Belforte et al., "Design and Development of an LSI Digital Switching Element (ECI)," *CSELT Technical Reports,* April 1984.
4. P.M. Solomon, "A Comparison of Semiconductor Devices for High Speed Logic," *Proc. of IEEE,* 70, May 1982.
5. G. Guarini and S. Iannazzo, *Circuiti Integrati,* Ed. Tamburini, 1971.
6. G. Zocchi, "Tecnologie VLSI," *Le Scienze,* Quaderni, No. 23, 1985.
7. R. Iscoff, "Exotic Materials for Integrated Circuits," *Semiconductor International,* July 1985.
8. V. Biancomano, "Semiconductor Technology," *Packaging Electronic Design,* June 14, 1984.
9. H. Huff: "Silicon Wafers Engineered for ULSI Circuits," *Semiconductor International,* July 1985.
10. "GaAs—Another Look," *ICECAP Report,* No. 4–5, 1984.
11. D.G. Fisher: "GaAs IC Applications in Electronic Warfare, Radar and Communications Systems," *Microwave J.,* May 1988.
12. R.C. Eden et al., "Integrated Circuits: The Case for Gallium Arsenide," *IEEE Spectrum,* December 1983.

13. A.G. Rode and J.G. Rope, "Gallium Arsenide Digital IC Processing. A Manufacturing Perspective," *Solid-State Tech.*, February 1985.
14. P.T. Greiling and C.F. Krumm, "The Future Impact of GaAs Digital Integrated Circuits," in *VLSI Electronics Microstructure Science*, vol. 11, Academic Press, New York, 1985.
15. *Fujitsu Facts*, 1988.

Chapter 3
DEGRADATION MECHANISMS IN INSULATING FILMS ON SILICON

Giorgio Baccarani and Fausto Fantini

3.1 INTRODUCTION

The physical properties of silicon dioxide (SiO_2) have played a key role in the development of silicon planar technology [1]. SiO_2 is in fact a nearly ideal insulating material, which, due to its chemical stability, has been widely used for both passivation of the silicon surface and for insulation of interconnection tracks. Also, it is largely impermeable to dopant diffusion, and for this reason it is used in device fabrication as a mask for the selective doping of the silicon surface. Finally, Si and SiO_2 are chemically dissolved by different etchants, thereby allowing selective processes that stop spontaneously after the SiO_2 has been locally dissolved.

With the advent of MOS technology, silicon dioxide has taken an active role within the device itself, which is based on the induction of an ohmic channel in the surface region of the semiconductor [2]. The channel conductance is controlled by the voltage applied to a gate on top of the oxide. In this case, another property of the Si-SiO_2 interface plays a fundamental role: the low surface-state density, which, by suitable annealing treatments, can be kept on the order of 10^{10} cm^{-2}, i.e., about 1 percent of the electron density within the channel under strong inversion conditions.

The small value of the ratio between interface-state density and number of Si atoms at the external surface ($\approx 10^{15}$ cm^{-2}) indicates that, at an ideal interface, there are no surface states in the silicon bandgap, and those states that are present are due to imperfections such as dangling bonds, vacancies, and so on.

The need to raise the device transconductance and the available voltage gain forces the process designer to reduce the thickness of the gate oxide to a few tens of nanometers. However, the supply voltage has been kept constant in the last few years, thereby increasing the value of the oxide field, which is nowadays often in excess of 2 $MVcm^{-1}$.

A further application of silicon dioxide is the antiscratch protection of the final device to prevent any possible damage during wafer handling and chip assembly. Similar protective functions are also performed by different dielectrics, such as silicon nitride (Si_3N_4), which is used both during the manufacturing process as a mask for field-oxide growth and as a contaminant-impermeable passivation layer, and polyimide for surface protection and insulation between conducting tracks. The main uses of dielectrics in devices is illustrated in Figure 3.1.

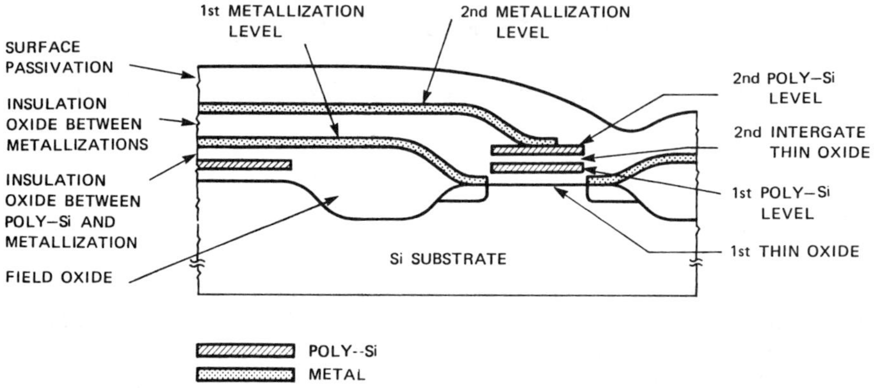

Figure 3.1 Schematic representation of the dielectrics employed in MOS technology with two levels of polycrystalline silicon and two metal levels.

The integrity of dielectrics is mainly affected by two factors that irreversibly degrade their properties: (1) current flow and (2) contamination induced by the fabrication process or by the operational environment. Faults in the dielectrics contribute up to 75 percent of the total number of faults occurring in MOSFETs. Hence, dielectric faults have great importance in the resulting device reliability.

Conduction across silicon dioxide can occur due to hot-electron injection from the semiconductor [3], or by tunnel effect under the influence of strong electric fields [4]. In the case of hot-electron injection, electrons reach an energy in excess of the barrier at the Si-SiO_2 interface and may therefore penetrate the oxide conduction band after a suitable redirecting collision. In the tunnel effect, however, the application of very large fields (on the order of 10 $MVcm^{-1}$) reduces the tunneling distance down to 3 nm, thus increasing the transmission probability across the barrier. In the former case, electrons can be trapped in the oxide whereas, in the latter case, more complex degradation phenomena take place, leading to the formation of positive charge and to a strong increase of the interface-state density.

Conduction phenomena across the oxide are in most cases undesirable effects kept under control by limiting the maximum values of the electric field within the oxide and the semiconductor. There are, however, specific devices (EPROM) that take advantage of the conduction properties of the oxide to store information in memory cells [5]. Such devices comprise two overlapping gates, the lower of which is electrically floating and fully immersed in SiO_2. By applying suitable voltages to the second-level gate and the drain, channel electrons are heated up and some of them are injected into the floating gate, where they lose energy and remain confined by the surrounding potential barrier. The resulting negative charge leads to a strong increase of the threshold voltage of the device, which remains off when the information is being read.

In *electrically erasable read-only memories* (E^2PROM) electron injection into the floating gate occurs by tunneling when information is both written and erased [6]. In order to avoid using too-large voltages, these devices are characterized by oxides having thicknesses below 10 nm or grown on textured polycrystalline silicon. The resulting memories are characterized by short writing and erasing times. They are, however, affected by reliability problems as the number of writing cycles increases, due to the degradation effects mentioned earlier.

Oxide contamination occurs primarily during the manufacturing process. The main contaminants are alkaline metals that can be easily ionized to produce a positive space charge, which is fairly mobile under the influence of oxide fields [1]. The resulting instability of the threshold voltage is of much concern for the device and circuit operation. Special attention must be paid to the cleanliness of the environment and to the procedures followed during the manufacturing process.

Finally, positive charge and interface states increase when the devices are subject to an intense radiation, as in space applications as well as during the high-energy process steps (e.g., ion implantation, plasma etching and deposition, *et cetera*).

This chapter deals with the reliability problems of silicon dioxide. We do not intend to examine the technology of material deposition or growth, nor do we want to address problems related to process quality and control that strongly influence production yield. Rather, attention will be paid to the degradation properties of insulators, and specifically SiO_2, and their effects on the reliability properties of devices.

After a brief survey of the types of charges detectable in silicon dioxide and the relative terminology, the next section discusses conduction phenomena in SiO_2, with special emphasis on the physical mechanisms of formation, injection, and capture of hot electrons in the oxide. The effect of electron tunneling is examined, followed by a short discussion on the possible phenomena responsible for dielectric breakdown. Finally, section 3.3 examines oxide contamination and the resulting effects on the performance of bipolar and MOS devices.

3.1.1 Charges in Silicon Dioxide

The standardization of the terminology used for silicon-dioxide charges is the result of a study carried out under the auspices of the Institute of Electrical and Electronics Engineers and by the Electrochemical Society [7]. The charges within the silicon dioxide and at the Si-SiO_2 interface have been classified in four groups (Figure 3.2):

1. Fixed charges in the oxide: these are positive and due to structural defects (ionized Si) in the vicinity ($\approx$ 2.5 nm) of the interface. Their density per unit area is indicated by Q_f.
2. Mobile ionic charges: these are mainly due to alkaline ions (Li^+, Na^+, K^+) and perhaps H^+. Their density per unit area is indicated by Q_m.
3. Charges trapped at the interface: these can be either positive or negative and are due to structural defects, metallic impurities, and dangling bonds. Their density per unit area is indicated by Q_{it}.
4. Holes or electrons trapped in the oxide: charge trapping can occur due to ionizing radiation (especially during the fabrication process), injection of charge from the semiconductor into the oxide, and other similar processes. Their density per unit area is indicated by Q_{ot}.

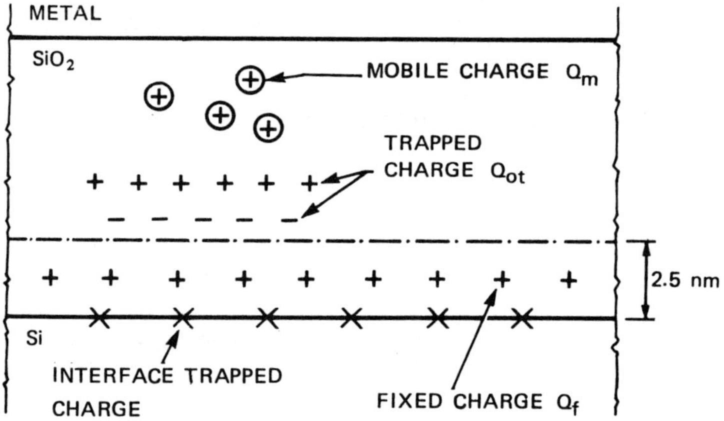

Figure 3.2 Space distribution of the different types of charges in SiO_2.

The presence of these charges may result in device instability: in particular, mobile charges due to alkaline ions can be particularly dangerous; for smaller MOS devices, however, the charge trapped by hot-electron injection becomes important from the reliability standpoint. Whereas Q_f and Q_{it} are spatially located in the vicinity of the interface, Q_m and Q_{ot} are distributed in the volume of the oxide.

Quantitatively, Q_m and Q_{ot} represent the effective charge densities per unit area at the Si-SiO_2 interface that are equivalent to the real charge distribution inside the oxide. The influence of a volume-distributed charge inside the oxide depends in fact on its density and position and increases if it is closer to the interface. The potential drop V_{ox} corresponding to an arbitrary distribution $\rho(x)$ inside the oxide is

$$V_{ox} = \frac{1}{\epsilon_{ox}} \int_0^{t_{ox}} \rho(x)x \, dx \tag{3.1}$$

where t_{ox} is the thickness of the oxide and ϵ_{ox} is its dielectric constant.

The effective charge at the Si-SiO_2 interface equivalent to the distribution $\rho(x)$ is

$$Q = C_{ox}V_{ox} = \int_0^{t_{ox}} \rho(x)(x/t_{ox}) \, dx \tag{3.2}$$

C_{ox} being the oxide capacitance per unit area. The total equivalent charge is obtained from the value of the flat-band voltage of the MOS capacitor:

$$V_{FB} = \phi_{MS} - \frac{1}{C_{ox}} (Q_f + Q_m + Q_{it} + Q_{ot}) \tag{3.3}$$

where ϕ_{MS} is the metal-semiconductor work-function difference. The flat-band voltage can be experimentally determined from the high-frequency capacitance-voltage characteristics.

3.2 CONDUCTION MECHANISMS IN THIN OXIDES

Over the past decade scaling device dimensions has resulted in gate-oxide thickness being reduced by a factor of 4. This process has occurred at constant voltage, due to the constraint of retaining the compatibility with existing systems; therefore, a large increase occurred in the maximum electric field applied to the oxide. Even assuming that supply voltages will reduce to 3–3.5 V (lower values cannot presently be assumed due to the difficulties in accurately controlling the threshold voltage and the nonscalability of transistor turn-on characteristics), a strong push still remains toward increasing the electric fields which have already reached about 2 MV/cm in thinner oxides.

The increase in semiconductor fields, together with scaling down device dimensions at constant supply voltage, has caused hot carrier effects that, due to the induced physical phenomena, became increasingly important for device reliability [8, 9].

This section examines the theoretical models of these effects and identifies the basic relationships linking the different physical quantities. The discussion will specifically refer to n-channel MOSFETs, even though positive charge carriers (holes) are not immune to similar effects. Hot holes in p-channel MOSFETs, however, become important at higher electric fields or, at the same field, have lower effects than the corresponding hot-electron phenomena.

3.2.1 Substrate Currents

The first major physical phenomenon exhibited in MOS transistors under the influence of large drain voltages is an impact ionization effect that occurs at the drain end of the channel. Such a process is governed by an energy-threshold mechanism, for the transition of an electron from the valence to the conduction band requires a transfer of energy at least equal to the bandgap. For reasonably uniform fields over a scale length of the order of the carrier mean-free path, a unique relationship between the average carrier energy and the electric field is established.

Since the ionization coefficient is of the type $\alpha_i = A_i \exp(-B_i/E_{\parallel})$, where $E_{\parallel}$ represents the component of the electric field parallel to the current flow, and A_i and B_i are two constants, it can be shown [10] that the substrate current I_{sub} is of the type:

$$I_{sub} = CI_{ds} \exp(-B_i/E_m) = 2I_{ds} \exp(-1.7 \times 10^6/E_m) \tag{3.4}$$

where E_m represents the maximum value of the longitudinal electric field in the channel (expressed in Vcm^{-1}); that is, the field at the drain end of the channel. Excessive values of the substrate currents can overload the bias-voltage generation circuit. In turn, any change of the substrate voltage modifies the threshold voltage of MOSFETs, which may then be affected by unacceptable leakage currents for preserving the information stored in dynamic RAMs and, in general, in all circuits where the information is stored as a charge in a floating node of the circuit.

In the most severe situations, a strong substrate current can forward bias the source-substrate junction, thus triggering a positive feedback that brings the transistor to a high-current state, corresponding to the upper branch of an S-type characteristic (snap back). In CMOS integrated circuits, the same mechanism can trigger a similar phenomenon by firing a parasitic thyristor (latch up).

3.2.2 Hot-Electron Injection into the Oxide

Hot-electron injection into the oxide, schematically illustrated in Figure 3.3, is exhibited when the longitudinal electric field in the channel of the transistor is so large that a number of electrons gain an energy larger than the oxide-semiconductor barrier. In this case, electrons are likely to overcome the potential barrier and to give rise to a detectable gate current. As the barrier height is of the order of 3 eV, the gate current will be smaller than the substrate current by a few orders of magnitude.

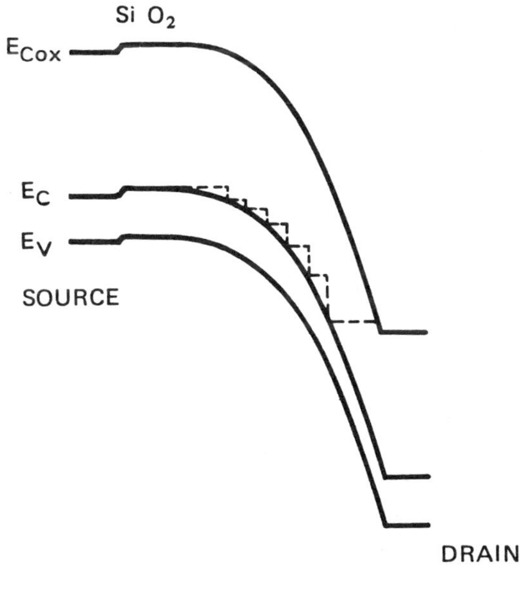

Figure 3.3 Schematic picture of the hot-electron injection mechanism into the oxide of a MOSFET.

Not only channel electrons but also electrons thermally generated in the substrate can be injected into the gate, so long as the substrate voltage is negative and sufficiently large as to allow them to overcome the potential barrier. Substrate hot-electron injection, although less important from a practical standpoint due to the small number of thermally generated electron-hole pairs, nevertheless has been thoroughly investigated in the past, since it is more easily susceptible to a quantitative interpretation due to the one-dimensional nature of the electric field in the test devices [11]. In any case, the influence of substrate hot electrons on the reliability of MOS devices is far less important than the effect of channel hot electrons, due both to the lower quantitative relevance of the effect and to the uniform trapping to which it gives rise. It should, however, be noted that the effect of the substrate electrons increases with temperature, since they are thermally generated, whereas the opposite holds for channel electrons.

Hot-electron injection into the gate oxide has been modeled according to distinct lines of thought that, however, lead to results in reasonable agreement. The first of them is the so-called *lucky electron* model, originally proposed by Shockley [12] for the study of transport phenomena in large electric fields. This approach was followed by Verwey, Kramer, and DeMaagt [13] and Ning, Osburn, and Yu [14] for the analysis of substrate hot-electron injection, and by Hu, Tam, and Ko [15, 16] for the study of channel hot-electron injection.

A second approach to the problem is the *equivalent temperature* approach [17]. This line of thought assumes that the electron distribution function is maxwellian at the equivalent temperature $T_e = T_e(x)$, where x represents the longitudinal coordinate along the channel. The injected current density is hence simply expressed by the Richardson formula of thermionic emission. The most difficult problem within this approach is that of properly determining the functional dependence of T_e on the x position.

In the theory of Hu et al. [15, 16], the injection mechanism is controlled by the following events:

1. The electron must gain from the longitudinal electric field energy in excess of the potential barrier at the semiconductor-oxide interface;
2. It must undergo an elastic collision, redirecting its momentum normal to the barrier;
3. It is supposed not to experience any anelastic collision when approaching the Si-SiO_2 interface and the top of the barrier within the oxide.

The use of suitable probability terms leads to the following expression for the gate current:

$$I_g = C(E_{ox})I_{ds} \exp[-\phi_b/(\lambda E_m)] \tag{3.5}$$

holding for $V_{gs} > V_{ds}$. In (2.2) I_{ds} is the source-drain current; $E_{ox} = (V_{gs} - V_{ds})/t_{ox}$ represents the oxide field in the vicinity of the drain, and $C(E_{ox})$ is a weak function of E_{ox} varying from 10^{-3} to 4×10^{-3} for a change of E_{ox} from 0 to 10^6 Vcm^{-1}; E_m is the maximum value of the longitudinal electric field and ϕ_b is the barrier height accounting for image force and tunnel reduction according to the expression:

$$\phi_b = \phi_{bo} - \beta E_{ox}^{1/2} - \theta E_{ox}^{2/3} \tag{3.6}$$

where $\beta = 2.59 \times 10^{-4} - (\mathrm{Vcm})^{1/2}$; $\theta = 4 \times 10^{-5}\,\mathrm{V}^{1/3}\mathrm{cm}^{2/3}$ [16], and λ represents the mean-free path of the energetic electron within the channel. The value of λ is influenced by optical phonon interaction and impact ionization. Its expression is

$$\lambda^{-1} = \lambda_p^{-1} + (1 - \gamma)\lambda_i^{-1} \tag{3.7}$$

where λ_p and λ_i represent the optical-phonon mean-free path and the impact-ionization mean-free path, respectively, and $\gamma = \phi_i/\phi_b$ is the ratio between the ionization threshold and the barrier height. The value of λ, from experimental data, turns out to be 9.2 nm at room temperature [16].

Figure 3.4 shows the theoretical and experimental behavior of the currents I_g and I_{sub}, normalized to the source current, against V_{gs}. It can be seen that the bell shape of the gate current is correctly predicted by the model, in spite of some quantitative discrepancies [16].

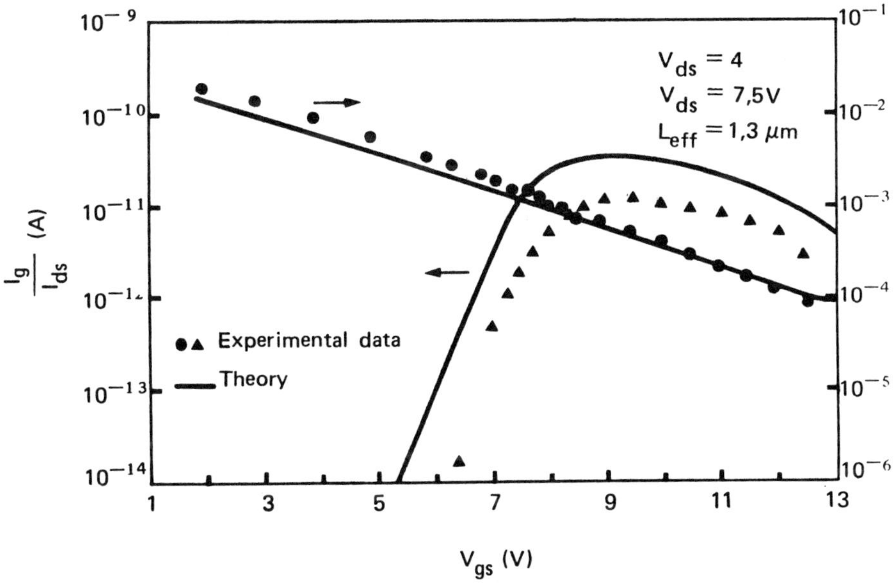

Figure 3.4 Theoretical and experimental behavior of the gate and substrate currents, normalized to the source current, against gate voltage. Theory refers to the lucky-electron model [16].

As already anticipated, in the equivalent-temperature model the gate current density is expressed by the Richardson thermionic-emission formula [17]:

$$J_g = qn_s(kT_e/2\pi m^*)^{1/2} \exp(-q\phi_b/kT_e) \tag{3.8}$$

where n_s represents the electron concentration at the Si-SiO_2 interface. Indicating by W the transistor width, the gate current is then expressed as follows:

$$I_g = W \int_0^L J_g(x)dx \tag{3.9}$$

where T_e is obtained by solving the energy balance equation:

$$qNvE(x) - \frac{d}{dx}\left(\frac{5}{2}Nk\Delta T_e v\right) = \frac{3}{2}Nk\Delta T_e/\tau_e \tag{3.10}$$

where $\Delta T_e = T_e - T_o$ represents the excess electron temperature, N is the electron density (per unit area), v is the electron velocity, and τ_e is the energy relaxation time. By considering the gate current as a small perturbation, the product Nv is constant along the channel. Furthermore, in the high-field region where injection takes place, electron drift velocity is nearly saturated, and (3.10) simplifies as follows:

$$\frac{d\Delta T_e}{dx} + \left(\frac{2}{5}\Delta T_e/\lambda\right) = \left(\frac{2}{5}q/\mathrm{k}\right)E(x) \tag{3.11}$$

having defined $\lambda = (2/3)v_s\tau_e$. The solution of (3.11) is

$$\Delta T_e(x) = \Delta T_e(0)\exp(-2x/5\lambda) + \frac{2q}{5k}\int_0^x E(x - x')\exp(-2x'/5\lambda)dx' \tag{3.12}$$

which expresses the excess electron temperature as the sum of two terms: the former depends on the initial condition and weakens quickly as x increases; the latter represents a convolution integral between the functions $E(x)$ and $\exp(-2x/5\lambda)$. If $E(x)$ along the channel is known, possibly by numerical integration of the semiconductor equations, (3.12) then provides the equivalent temperature as a function of the position x.

Equations (3.8), (3.9), and (3.12), in connection with a numerical evaluation of the electric field along the channel, provide the solid lines in Figure 3.5 under the assumption of a Gaussian impurity distribution at the source and drain end of the channel, in good agreement with experimental data. In Figure 3.5 it has been assumed $\tau_e = 8 \times 10^{-14}$ s [17].

Equation (3.12) accounts for the delay associated with a sudden variation of the electric field. By contrast, when $E(x)$ changes slowly over a scale length comparable with λ, (3.12) gives

$$\Delta T_e(x) = (q/k)\lambda E(x) \tag{3.13}$$

hence, assuming $\Delta T_e \approx T_e$, (3.8) becomes

$$J_g = qn_s(q\lambda E/2\pi m^*)^{1/2} \exp(-\phi_b/\lambda E) \tag{3.14}$$

Replacing (3.14) in (3.9) causes a dependence of I_g on $\exp(-\phi_b/\lambda E_m)$, in analogy with the lucky-electron model.

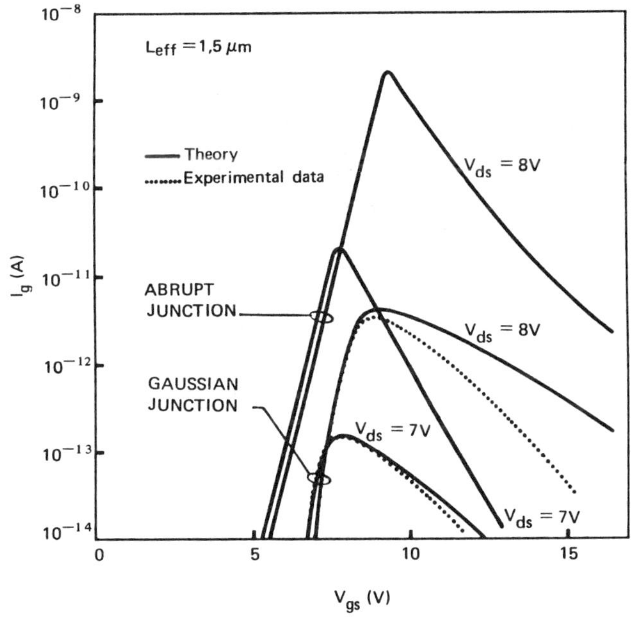

Figure 3.5 Theoretical and experimental behavior of the gate current I_g against V_{gs}, according to the thermionic emission model. Theoretical curves refer to the cases of abrupt and gaussian junctions. The latter provide results in reasonable agreement with experiments [17].

From (3.4) and (3.5) a correlation exists between substrate current and gate current [10]. Raising (I_{sub}/I_{ds}) to the exponent $(\phi_b/\lambda B_i)$, we get

$$I_g/I_{ds} = C(E_{ox})(I_{sub}/I_{ds})^{(\phi_b/\lambda B_i)} \tag{3.15}$$

Equation (3.15) is confirmed experimentally by Figure 3.6, where the ratio (I_g/I_{ds}) is reported on a logarithmic scale against (I_{sub}/I_{ds}) for constant values of $V_{gs} - V_{ds}$ and therefore of E_{ox} [16]. The straight lines corresponding to constant values of E_{ox} confirm the validity of (3.15), whereas the slopes of the straight lines provide the variation of ϕ_b *versus* E_{ox}. Such a variation is represented in Figure 3.7, from

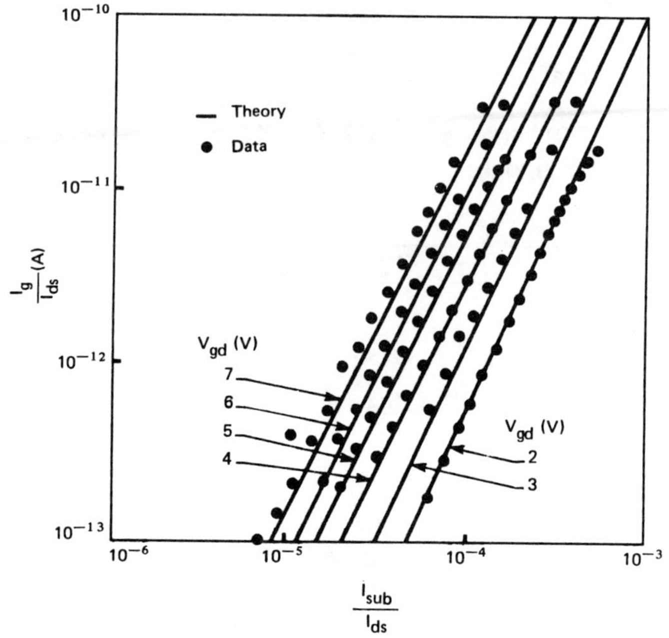

Figure 3.6 Gate current I_g against substrate current I_{sub} (both normalized to the source current) for constant values of $V_{gs} - V_{ds}$ and therefore of E_{ox} [16].

which θ can be extracted by minimizing the variance of the product λB_i [16]. With this procedure the values $\lambda B_i = 1.24$ V and $\theta = 4 \times 10^{-5}$ V$^{1/3}$ cm$^{2/3}$ were found.

A third mechanism induced by hot electrons is photon generation by *bremsstrahlung* [18]. These photons may then produce electron-hole pairs in the substrate, and the electron flow can be detected by a suitable collection junction. The emission probability for a photon with energy $h\nu$ is proportional to $\exp[h\nu/(q\lambda E_m)]$. Such a photon generation leads to the current:

$$I_n = 6 \times 10^{-5} I_{ds} \exp[-1.3/(\lambda E_m)] \tag{3.16}$$

where the factor 1.3 of the exponent of (3.16) may be interpreted as the average energy of the generated phonons.

3.2.3 Oxide Trapping

Hot-electron injection into the oxide normally represents an unwanted phenomenon due to the occurrence of charge trapping in the oxide itself. If an electron

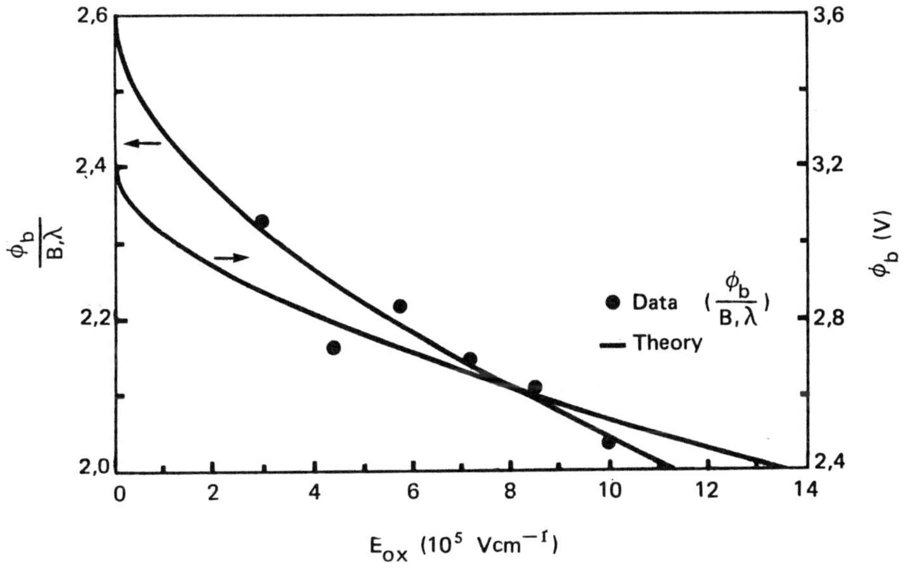

Figure 3.7 Si-SiO_2 barrier height ϕ_b, normalized to the factor ($B_i\lambda$) against the oxide field E_{ox} [16].

gets captured by a state in the oxide energy gap, the emission probability is often negligible, so that the latter will reside in such a state for a long time, contributing to the oxide space charge. Furthermore, injected electrons may be able to break dangling bonds at the interface, thus generating interface states.

The resulting effects are (1) a threshold-voltage shift and (2) mobility degradation of the carriers within the channel. The mobility degradation will of course result in transconductance degradation.

Different types of traps are generally contained within the oxide, each having its own energy level. Electrons and holes, however generated or injected into the oxide, can be captured by these centers through a cascade process of acoustic- and optical-phonon emission and will remain there almost indefinitely. The process of positive charge formation following the exposure of MOS devices to ionizing radiation is often interpreted as a generation of electron-hole pairs: in the presence of applied electric fields, electrons would be pulled by the field and then emitted into the metal or semiconductor; however, holes with their far-lesser mobility, would be captured, giving rise to the formation of positive charge. This charge can be removed only through annealing processes at 500°C in an inert ambient or in forming gas. Radiation may occur during the fabrication process, following ion implantation or sputter deposition of either metallic or dielectric films. Electrolithography on the wafer of course has enhanced the importance of the problem.

Indicating by n_t the concentration of trapped electrons in centers with energy E_t, and by N_t the concentration of such centers, the kinetic equation to be satisfied by n_t is [19]:

$$\frac{dn_t}{dt} = c_n n_c(N_t - n_t) - e_n n_t \tag{3.17}$$

where n_c represents the concentration of injected electrons in the oxide conduction band; c_n and e_n are the capture and emission probabilities, respectively. In (3.17) it is assumed that E_t resides in the upper half of the energy gap, so that interactions with valence electrons are statistically negligible. For any trap center, an equation like (3.17) may be written. A similar equation holds for holes.

If electron injection occurs at large electric fields, it is conceivable that $n_c = J_n/(qv_{th})$ [19], assimilating drift velocity to thermal velocity. Furthermore, the emission term in (3.17) is negligible compared with the capture term, due to the strong increase of n_c with respect to its equilibrium value n_{co}. In conclusion, (3.17) may be rewritten as

$$\frac{dn_t}{dt} = (J_n\sigma_n/q)(N_t - n_t) \tag{3.18}$$

where σ_n represents the electron capture cross section. The solution of (3.18) is

$$n_t = N_t[1 - \exp(-\sigma_n N_{inj})] \tag{3.19}$$

having indicated with N_{inj} the total number of electrons injected into the oxide per unit area.

The effective charge density, Q_{ot}, is then:

$$Q_{ot} = -q\int_0^{t_{ox}} n_t(x/t_{ox})dx = -qN_t(\bar{x}/t_{ox})[1 - \exp(-\sigma_n N_{inj})] \tag{3.20}$$

where $\bar{x}$ represents the distance of the charge centroid from the metal-oxide interface. Correspondingly, the threshold shift turns out to be:

$$\Delta V_T = -Q_{ot}/C_{ox} = (qN_t\bar{x}/\epsilon_{ox})[1 - \exp(-\sigma_n N_{inj})] \tag{3.21}$$

Equation (3.21) is compared with experimental data [19] in Figure 3.8. Each curve refers to a different value of the electric field and has been obtained by fitting experiments with a suitable capture cross section. The relationship between capture cross section and electric field [19] is shown in Figure 3.9. In the field range between 0.5 and 2.5 × 10^6 Vcm^{-1}, σ_n varies proportionally to E_{ox}^{-3}. This dependence is not fully accounted for by the Poole-Frenkel effect applied to Coulomb centers, which

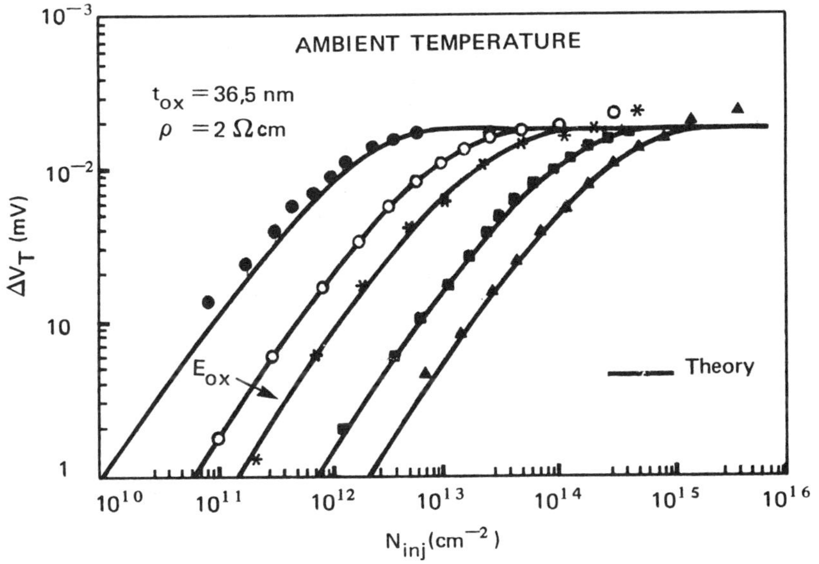

Figure 3.8 MOSFET threshold shift against injected charge for different values of the oxide field, E_{ox} [19].

would lead to a variation of σ_n with $E_{ox}^{-3/2}$, but is attributed to the combined action of the Poole-Frenkel effect and to the electron heating [19].

In addition to deep coulomb centers, there are also neutral shallow centers near the conduction band [20], which do not appreciably contribute to threshold shift, owing to the large emission probability of trapped electrons. At low temperatures, however, emission is severely reduced; hence, larger threshold shifts are to be expected for a given number of injected electrons.

Although device degradation appears to be correlated with the flow of hot carriers across the Si-SiO_2 interface, there is some controversy on the relative importance of charge trapping and interface-state formation [9, 21]. According to Tsuchiya [22], the initial degradation is mainly due to electron trapping, although interface state formation would become increasingly important as more electron traps become saturated.

In order to allow for reliability predictions, device lifetime must be correlated with some easily measurable parameters. A currently employed empirical model was first proposed by Takeda and Suzuki [23]. These authors observed that both threshold shift and transconductance degradation increase with t^n, with $n \approx$ 0.5–0.7, and both are proportional to $\exp(-1/V_{ds})$, which closely resembles the

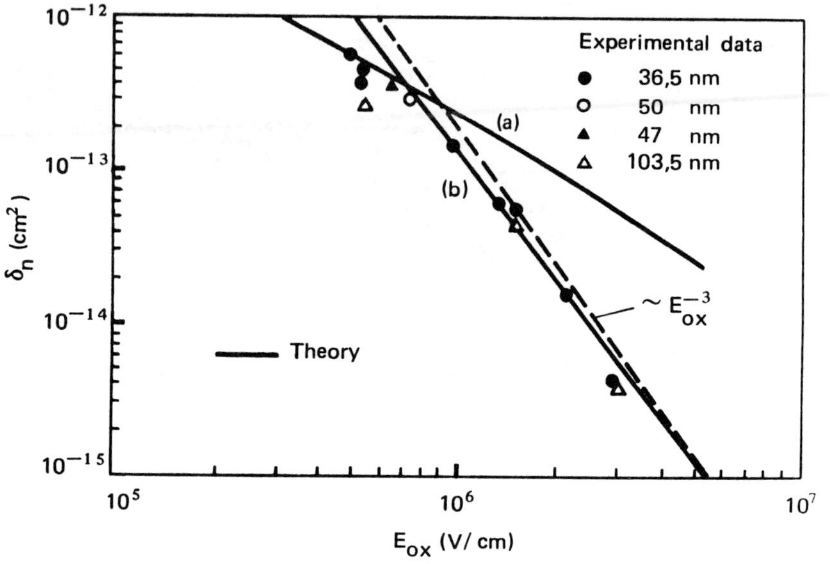

Figure 3.9 Electron capture cross section against the oxide average field, E_{ox} [19].

dependence of substrate currents upon impact ionization. The device lifetime t_f can be expressed as:

$$t_f = A(I_{sub})^{-m} \tag{3.22}$$

with $m \approx 2.5$–3.0 and A is a constant. A more complete, but similar expression, was derived by Hu et al. [9]:

$$t_f = B(I_{sub})^{-m}\,(I_{ds})^{(m-1)} \tag{3.23}$$

In dynamic conditions (3.23) can be generalized as [24]:

$$t_f = C\left[\frac{1}{T}\int_0^T (I_{sub})^m (I_{ds})^{-(m-1)}\,\mathrm{d}t\right]^{-1} \tag{3.24}$$

where T is the stress period. It is still a matter of controversy whether enhanced degradation results in ac conditions [25].

The robustness of MOSFETs against hot-carrier stress has been improved mainly by extending the source and drain with lightly doped regions, thus rendering the junctions more gradual and reducing the peak electric field within the channel. This technique, illustrated in Figure 3.10, usually designated *lightly doped drain* (LDD), is obtained with a double (n^- and n^+) implant, using the so-called oxide spacers at the side walls of the gate as a mask for the n^+ implantation. Unfortunately, a drawback of this structure is the increase of the source and drain parasitic resistances, which has deleterious effects in current-drive capability and speed performance of the transistors. The impurity concentration in the n^- region must

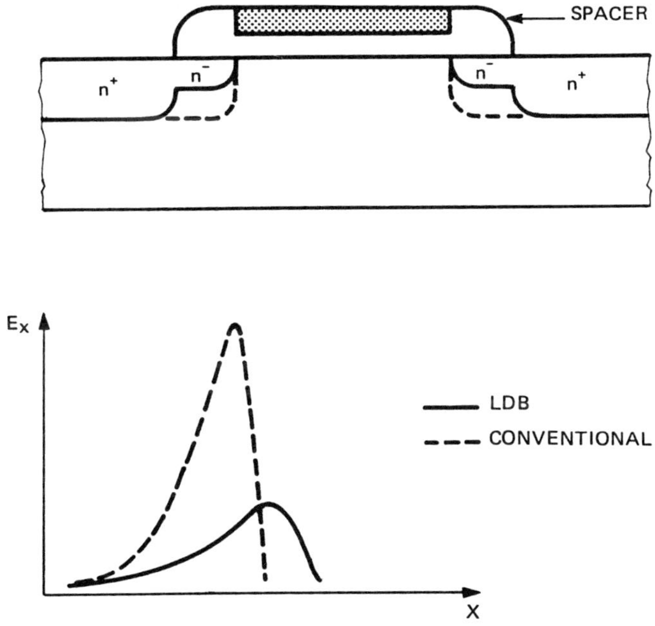

Figure 3.10 Schematic representation of the LDD MOSFET and its influence on the field profile along the channel.

be carefully designed in order to find the best trade-off between reliability and performance. Moreover, a lightly doped drain extension makes this region susceptible to a complete depletion induced by electrons trapped in the overlying oxide [26].

Current gain degradation has been observed in bipolar transistors, too, due to hot-carrier injection at the emitter-base junction near the surface [27]. In this case, interface state formation, rather than electron trapping, is responsible for the transistor gain degradation at low currents.

3.2.4 Conduction in Amorphous Materials

The study of conduction in amorphous materials is very challenging, due to the major difficulties associated with it. The lack of a periodicity condition for the inner potential does not allow using Bloch functions for the description of electronic states, and the usual results of the band theory no longer apply. Also, the effective-mass theorem, allowing for the consideration of external forces only, breaks down; that is, both internal and external forces ought to be considered simultaneously.

The transport properties of silicon dioxide have been investigated by disregarding its amorphous structure and using techniques suitable for crystalline materials. The latter procedure provided results in reasonable agreement with experiments. It may be worth pointing out that absorbtion measurements on amorphous oxide led to a state density similar to that of quartz, which proves that the short-range order of SiO_2 plays an important role as far as the electronic properties mentioned earlier are concerned.

Transport theory in SiO_2 [28, 29] is based largely on the assumption of a polar interaction with longitudinal optical phonons. The scattering probability associated with such an interaction is a rapidly decreasing function of electron energy; thus, the large breakdown field would imply a large coupling constant that limits the electron energy to about 0.1 eV. The electron mean-free path turns out to be 0.1–0.2 nm, and between two successive collisions, electron momentum is largely undefined, as opposed to materials where scattering events are not so frequent. The distribution function then becomes a questionable concept, which renders the Boltzmann equation of doubtful applicability.

A further difficulty related to the study of conduction in dielectrics is related to the use of the perturbation theory for evaluating the scattering probabilities. Due to the large fields necessary to accelerate electrons, the golden rule is unlikely to provide reasonably correct results.

These circumstances make the study of transport properties in SiO_2 very difficult, even disregarding its amorphous structure. An evaluation of the velocity-field relationship has been performed by Lynch [30] using a semiclassical method based on an energy balance equation. Ferry [31] came up with a similar relationship by numerically solving the Boltzmann equation. Thornber, finally, developed a fully quantum mechanical model, calculating the expectation value of electron momentum against electric field [32].

The results of these theories are compared with experimental data [33] in Figure 3.11, which shows the electron drift velocity versus field in fused quartz. The experimental low-field mobility turns out to be 21 $cm^2\ V^{-1}s^{-1}$, compared with a theoretical value of 32 $cm^2\ V^{-1}\ s^{-1}$. The function $v_d = v_d(E)$ is linear for fields smaller than 5×10^5 Vcm^{-1} and tends to saturate at higher values. However, when the electric field exceeds 10^7 Vcm^{-1}, the classical theory predicts that the energy-loss due to LO phonon emission cannot maintain a steady state, giving rise

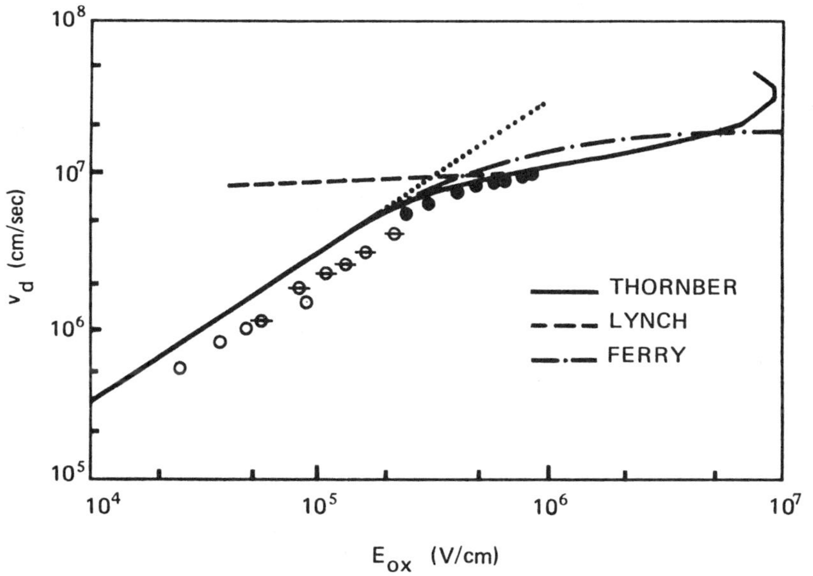

Figure 3.11 Theoretical and experimental drift velocity versus field in SiO_2 [26].

to an instability in momentum space called *runaway*. Thornber's quantum theory [32] predicts a nonzero probability for a tunneling effect in momentum space, leading to such a large electron energy to produce runaway even at fields below the threshold of 10^7 Vcm^{-1}.

This transport model has recently been contradicted by a number of experimental results. Photoemission measurements [34] and resonant tunneling across thin oxides [35] revealed carrier mean-free paths of about 2.5–4.0 nm, in disagreement with the predictions of the classical model. Furthermore, measurements of energy distribution of vacuum-emitted electrons [36] and measurements of the capture cross section of oxide traps [19] led researchers to believe that the average electron energy under the influence of strong fields may be as large as 1 eV or more, well beyond the predictions of the classical model. These experiments strongly support the hypothesis of an additional nonpolar interaction with high-momentum acoustic phonons [37]. Such an interaction is very effective in randomizing electron momentum with wide-angle events and leads to a scattering probability that rapidly increases with electron energy. Due to the wide-angle scattering events, the electron path across the oxide would be considerably longer, thus enhancing the LO phonon scattering rate. Monte Carlo simulations of electron transport in silicon dioxide, based on both scattering mechanisms, results in reasonable agreement with experiments [38].

3.2.5 Tunnel Injection into the Oxide

When the electric field applied to the oxide is of the order of 10 MV/cm, the tunneling distance for electrons impinging on the $Si\text{-}SiO_2$ interface is reduced down to 3 nm and less, and the transmission probability rapidly increases with the electric field. The situation is illustrated in Figure 3.12, which reports a band diagram of the $Si\text{-}SiO_2$ system. If the shape of the barrier is triangular, the current density across the oxide is expressed by the well-known Fowler-Nordheim formula:

$$J = AE_{ox}^2 \exp(-B/E_{ox}) \tag{3.25}$$

where the constant B depends upon the barrier height ϕ_B as follows:

$$B = \frac{4(2m^*)^{1/2}}{3\bar{h}q} \phi_B^{(3/2)} \tag{3.26}$$

Equations (3.25) and (3.26) were originally obtained for the case of injection from a metal into the conduction band of an insulator. In principle, these relationships can be used to determine the barrier height at the $Si\text{-}SiO_2$ interface, if the electron effective mass is known. For a semiconductor, however, the formation of quantized subbands within the inversion (or accumulation) layer suggests that some care must be exercised in the interpretation of the physical meaning of ϕ_B.

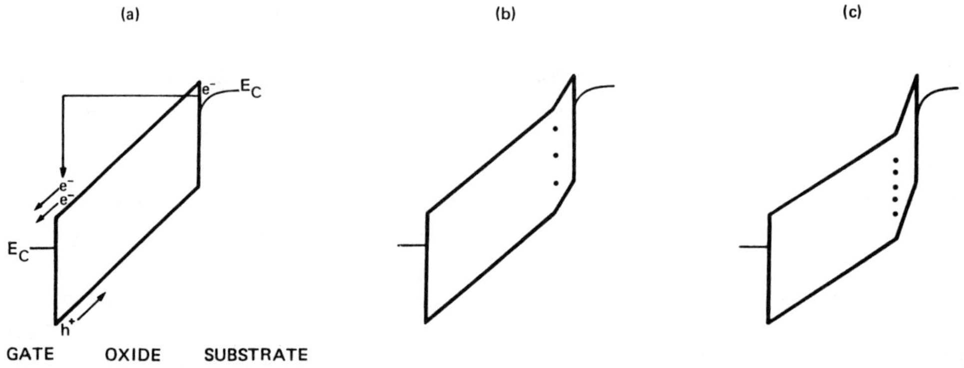

Figure 3.12 Band diagram of the $Si\text{-}SiO_2$ system.

From an experimental standpoint, though, the situation is even more complex. First, the current across the oxide is not constant for a given applied voltage. As illustrated in Figure 3.13, showing the injected current against time for a number of gate voltages, the current increases initially, reaches a broad maximum, and

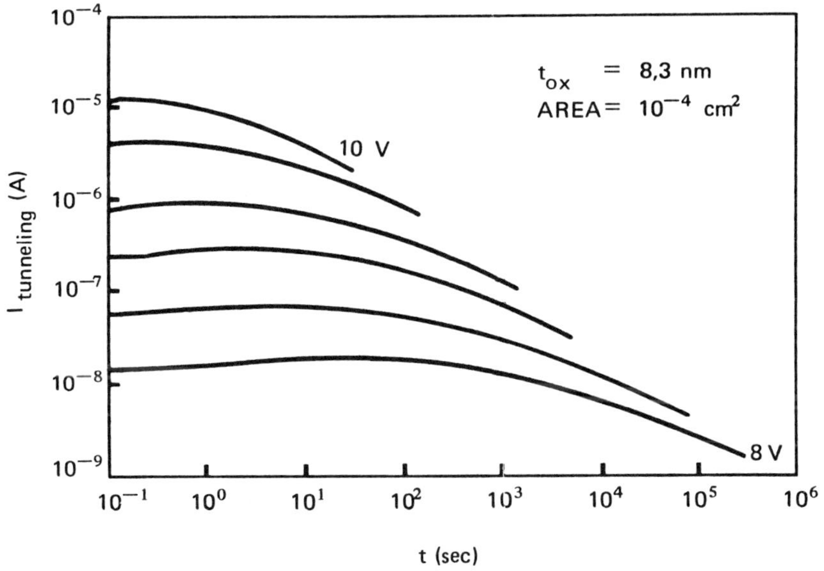

Figure 3.13 Time-dependent gate current for different values of the gate voltage. The end-point of the curves corresponds to the occurrence of the oxide breakdown [courtesy of P. Olivo, University of Bologna].

then decreases [39]. It should be noted that the log-scale in Figure 3.13 expands the short-term portion of the plot and further broadens the maximum. In a linear plot, the increasing portion of the curve would be very close to the vertical axis.

Dielectric breakdown of the oxide occurs at the point where the curve is interrupted. The breakdown time t_{BD} is represented as a function of the applied field in Figure 3.14. As shown, it is roughly exponential against field, although a law of the type $\exp(K/E_{ox})$ turns out to be in better agreement with experiments [40].

The interpretation of this data is very controversial. The rising portion of the curves in Figure 3.13 is associated with positive charge formation that, for a given applied voltage, distorts the bands, as shown in Figure 3.15, and increases the field at the emitting cathode. The formation of a positive charge is confirmed by the shift of the capacitance-voltage characteristics, but the physical origin of this charge is uncertain. Early assumptions were based on generation of electron-hole pairs due to impact ionization within the oxide [28, 29]. Electrons would be transferred to the anode by the field, whereas holes would be trapped in the oxide, thus generating a positive space charge.

An alternative assumption is that the energetic electrons impinging onto the anode can excite surface plasmons [41]. These decay very rapidly, generating

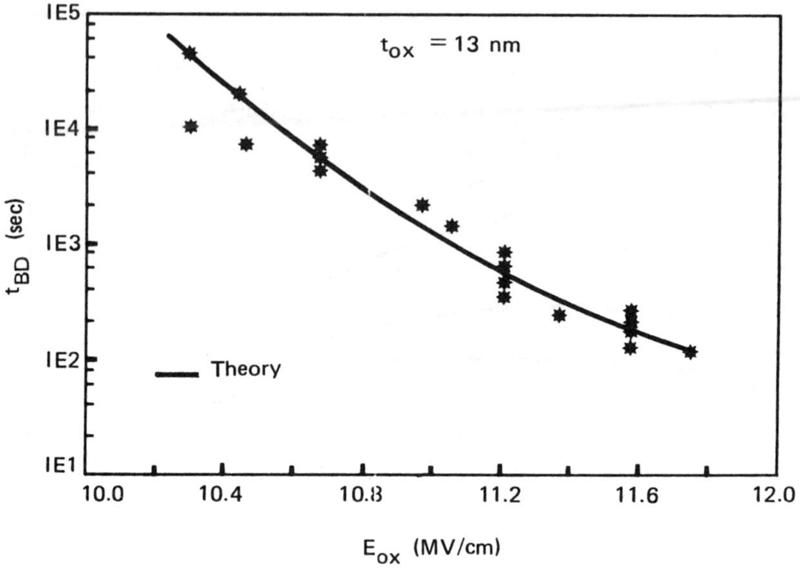

Figure 3.14 Time to breakdown t_{BD} against the oxide field E_{ox} [33].

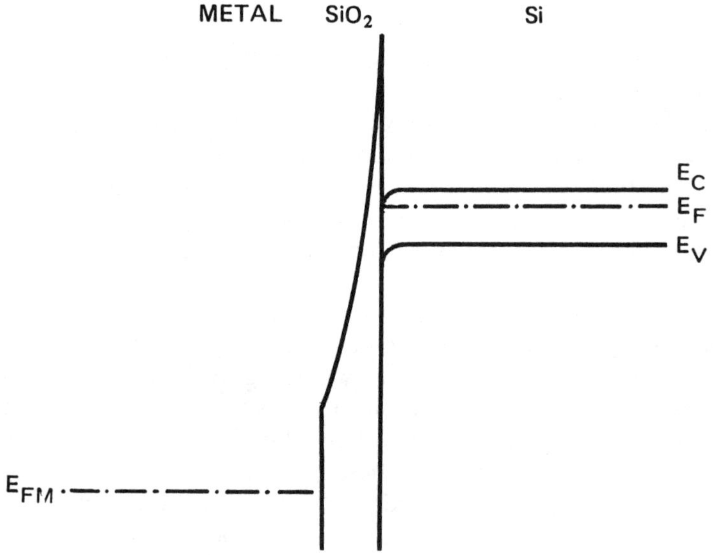

Figure 3.15 Band diagram of the Si-SiO_2 system in the presence of positive charge within the oxide. As shown, a positive charge leads to a field increase at the emitting electrode.

electron-hole pairs. Energetic holes could then overcome the potential barrier and be injected into the oxide.

The falling portion of the curves in Figure 3.13 is interpreted by assuming negative charge formation due to electron trapping, which would compensate for and subsequently overcome the positive space charge. Negative charge tends to bend the shape of the bands downward, thus reducing the field at the cathode for a given applied voltage; hence, the current is decreasing with time.

The interpretation of dielectric breakdown, which suddenly occurs in the negative-slope portion of the curves in Figure 3.13, was historically attributed to a positive feedback associated with formation of a positive charge [28, 29]. As a positive charge increases the cathode field, this would enhance the injected current, which, in turn, would generate a more positive charge, leading to the oxide runaway.

The classical model could predict fairly well the breakdown field in relatively thick oxides ($\sim$ 100 nm) [30, 42]. For thin oxides ($\sim$ 10 nm), however, the breakdown voltage turns out to be of the order of 10 V. If we assume that electrons are injected at an energy close to the bottom of the conduction band in silicon, the maximum kinetic energy they can reach in the oxide is about 7 eV, whereas the bandgap in SiO_2 is 9 eV. Consequently, impact ionization cannot justify the formation of positive charge [43].

Even assuming different mechanisms for positive charge formation, the weakness of this model is related to the lack of a logical relationship between the increase in negative charge, which must be taken for granted in order to justify the current decrease, and the sudden triggering of the positive feedback. Other assumptions have been suggested; among these, one of the most plausible seems to be that of resonant tunneling [64].

This effect occurs when an electron crosses by tunneling a double barrier, as shown in Figure 3.16, and its energy is identical to that of the state associated with the potential well between the two barriers. In such a condition, the transmission probability exhibits an acute peak. Furthermore, this maximum tends to be 1 when the transmission probabilities across the barriers are identical.

It is then plausible to assume that electron trapping and the subsequent band bending, by modifying the position of the energetic levels in the oxide, can produce the resonance condition with electrons sitting in the ground state. This condition could result in a current flow through the oxide that would quickly lead to severe thermal effects and, eventually, to the perforation of the dielectric.

An alternative model [44] is based on the observation that the total charge injected across the oxide at breakdown is largely independent of both current density and electric field. Thus, it is inferred that the critical parameter is the energy dissipated in the dielectric. When the electrons leave the conduction band of the oxide by trapping or by emission into the anode, their potential energy is dissipated over a short range, creating permanent damage. The cumulative effect of such a damage would be responsible for breakdown.

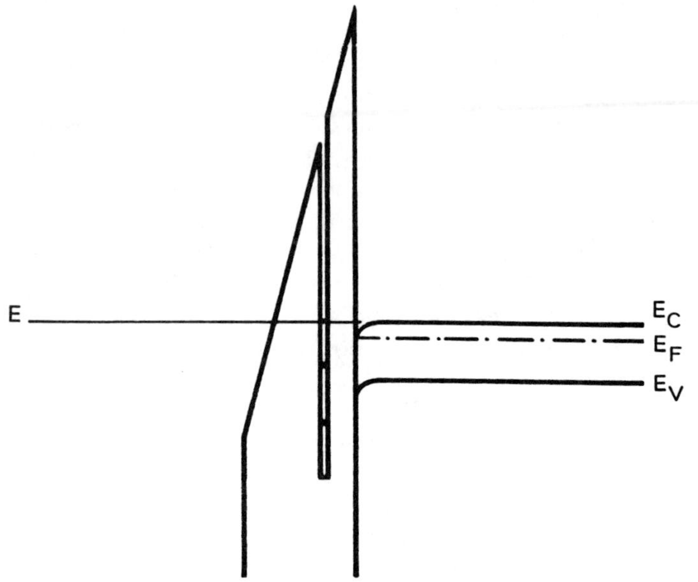

Figure 3.16 Band diagram of the Si-SiO_2 system in the presence of trap centers within the oxide. Resonant tunneling occurs when the energy of the impinging electron is the same as that of the trap center.

In general, it can be stated that the understanding of breakdown in silicon dioxide is still largely inadequate and several aspects in the picture are unclear. Further studies and research are currently underway to gain a better understanding of such an effect. Probably, still more sophisticated and diversified experimental techniques will be needed to gain a deeper insight in this problem.

As already stated in the introduction, tunneling across the oxide is used for the operation of E^2PROMs memories. Electron trapping within the oxide can occur in this case, too, negatively affecting the associated device reliability [45]. During the first writing and erasing cycles, positive-charge formation brings about a widening of the device operating window, defined as the difference between the erasing and writing thresholds. However, after repeated writing and erasing cycles ($\sim 10^4$) electron trapping causes this window to narrow, as shown in Figure 3.17. A negative charge within the oxide does indeed cause a smaller field at the interface, thus reducing electron tunneling. The effect does not tend to saturate, probably because of trap generation induced by the large fields, and eventually the memory becomes unusable. The number of allowed cycles depends on the cell technology: devices employing very thin oxides (FLOTOX) exhibit a longer endurance but

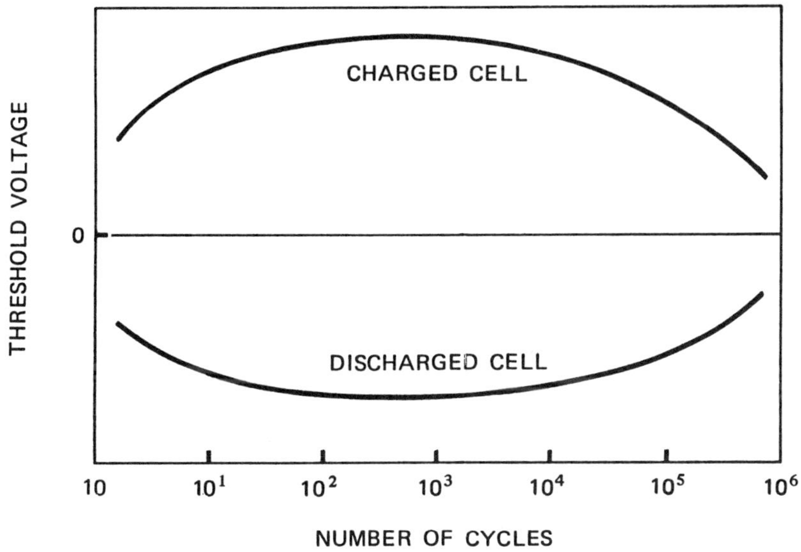

Figure 3.17 Behavior of the E^2PROM threshold voltages against the number of writing and erasing cycles.

suffer from defect-related breakdown; on the contrary, those with textured polysilicon are more prone to the long-term trapping effect but are less sensitive to premature breakdown thanks to the thicker oxide [46].

3.2.6 Acceleration Factors for Oxide Breakdown

From the standpoint of device reliability it is necessary that oxide breakdown does not occur during the lifetime of the device. For this reason, most attention has been paid to the time dependence of breakdown against electric field and temperature, with the aim of predicting the time of failure and identifying weaker devices. In this case, the definition of a suitable screening test is necessary [47–50]. The appearance of a failed oxide in an aluminum-gate MOS device is shown in Figure 3.18.

Most papers agree that the distribution of the time to failure is log normal. Metzler [52] has given a theoretical justification for that: if the failure probability is proportional to the injected current density (3.25), the log normal distribution results from the electric-field reduction at the injecting electrode as more and more electrons are trapped. This, in turn, implies that accelerated tests are possible using well-known techniques for reliability evaluation.

Figure 3.18 SEM picture of the failed oxide after removal of the gate metallization. Note that the failure occurred at the corner of the gate [courtesy of M. Vanzi, Telettra and LAMEL Institute].

The time to failure, t_f, is described by a relation that reduces to the Eyring one:

$$t_f = A \exp(E_a/kT - BE_{ox}) \tag{3.27}$$

where A is a constant, E_a is the activation energy, and B is the field acceleration parameter. The spread of values for E_a and B has suggested the existence of a mutual dependence between the parameters; from thermodynamic considerations [51], the following relationships have been obtained:

$$E_a = H - kCE_{ox} \tag{3.28a}$$

$$B = D + C/T \tag{3.28b}$$

where H is the enthalpy, C and D are constants.

These relationships provide a partial justification for the observed discrepancies among experimental results [49, 50] and allow for a greater confidence in the validity of the screening methods based on (3.27).

In accordance with the theoretical model [40], Chen and Hu proposed a relationship for the time to breakdown that follows the $1/E_{ox}$ dependence [52]:

$$t_f \propto \exp(F/E_{ox})$$

where F includes the Fowler-Nordheim and the hole-generation coefficients.

The dielectric used for insulation among the different conductor layers is especially critical. Chemically deposited and polycrystalline-silicon–grown oxides are indeed of a lower quality than thermal oxides [53], for they are affected by irregularities of the lower surface. There may be weak spots for breakdown caused by thinning of the oxide. An example of oxide nonuniformity among two polycrystalline silicon gates is shown in Figure 3.19.

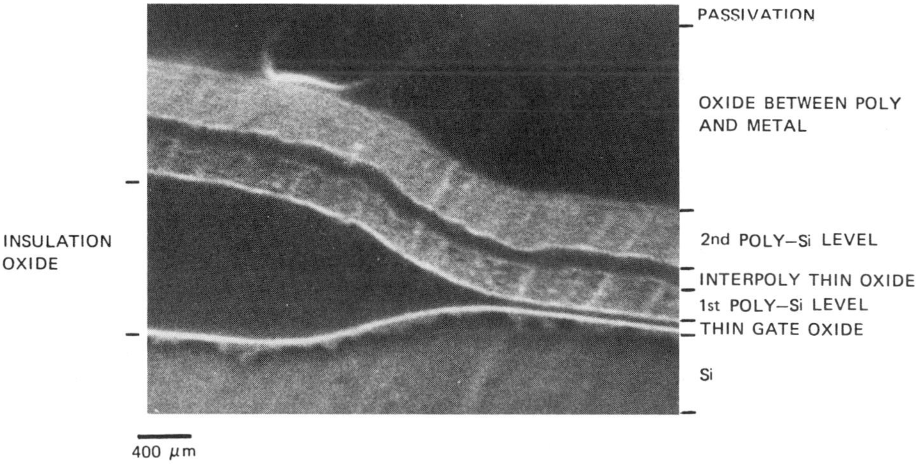

Figure 3.19 SEM picture of the cross section of a device with two levels of polycrystalline silicon [courtesy of M. Vanzi, Telettra and LAMEL Institute].

3.3 INSTABILITIES DUE TO CONTAMINATION

Reversible degradation phenomena, such as threshold-voltage shift and gain reduction, are due largely to the presence of a mobile charge within the oxide or at the interface [54–57]. The most dangerous contaminant is the ionic sodium (Na^+), which is extremely common in nature and can cause considerable effects even in quantities of 10^{11} cm^{-2}; the greater mobility of sodium, compared with other alkaline metal ions, is due to its narrower ionic radius ($\sim$ 0.095 nm) [58].

If the electric field within the oxide is such that a positive charge is pushed toward the interface, depletion and eventual surface inversion take place in the p-doped zones, while accumulation occurs in the n zones, as shown in Figure 3.20. Obviously, the effect of the contamination is more pronounced the closer the ions

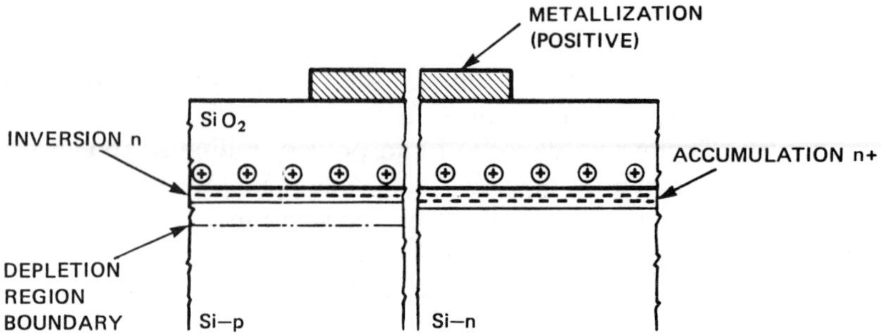

Figure 3.20 Schematic representation of the effect of a positive-charge contamination on *p* and *n* semiconductor surfaces.

are to the semiconductor surface and the lower the surface impurity concentration is.

As a consequence, bipolar transistors suffer mainly from an increase in the leakage current at the base-collector junction and a reduction of gain at low collector currents. In a reverse-biased p-n junction, the leakage current is caused primarily by the generation of electron-hole pairs in the depletion region. This current is made up by two components: I_{gm}, due to generation in the depletion layer of the metallurgical junction; and I_{gs}, due to surface generation. The lateral extension of the depletion region at the surface near the junction involves both the creation of an additional component of the generation current, I_{gi}, and the inclusion of a larger number of surface generation-recombination centers, with an increase of I_{gs}. Figure 3.21 shows this effect [1].

The degradation of current gain is more appreciable at low collector currents; in such a situation, the recombination current in the space charge region:

$$I_r = I_{ro} \exp(qV/2kT) \tag{3.29}$$

where $I_{ro} = I_{rm} + I_{rs} + I_{ri}$, again comprising three terms associated with bulk, surface, and lateral recombination, represents a large fraction of the total injected current. Thus, doubling I_r considerably increases the base current, whereas the collector current is largely independent of I_r.

If the ion source is finite, such as the metal-oxide interface, the degradation tends to saturate after a certain period of time, but if a virtually infinite ion source is available, such as the resin of the plastic package in contact with the chip, the gain degradation progresses indefinitely. The different speeds of this process, owing to the different plastic features, are illustrated in Figure 3.22.

The device most sensitive to ion contamination is the *n*-channel MOSFET [55]. Figure 3.23 shows how the presence of positive mobile charge in the gate

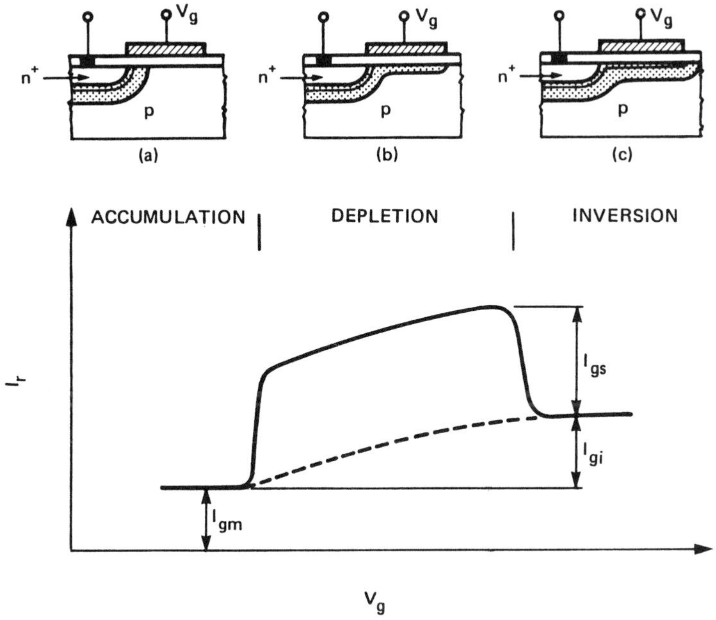

Figure 3.21 Influence of the lateral electric field on the reverse current of a gated diode. The field induced by the gated guard ring emulates the effect of ionic contaminants in the oxide and at the interface.

oxide tends to invert the silicon surface and lower the threshold voltage V_T. Thus, for a large ionic charge, the device can start conducting out of the gate control. For a p-channel MOSFET, instead, positive charge in the oxide tends to accumulate the silicon surface, so that the threshold voltage increases. Furthermore, the negatively biased gate attracts the mobile charge at the metal-oxide interface, thus reducing its effect on threshold. This effect is one of the main reasons for imposing a high level of cleanliness in the manufacturing areas. The presence of Na^+ contamination is currently checked during production by measuring the *capacitance-voltage* (C-V) characteristics on purposely designed MOS capacitors. Carrying out the measurements before and after a high-temperature stress test with positive bias, a parallel shift of the C-V characteristics is noted [59]. The latter is basically attributed to the mobile charge, for fixed charges and interface states are not affected by this stress. By measuring the maximum variation of the flat-band voltage, V_{FB}, as shown in Figure 3.24, the mobile charge is given by

$$Q_m = C_{ox} \Delta V_{FB} \tag{3.30}$$

where C_{ox} and Q_m represent the oxide capacitance and the mobile charge per unit area.

From the well-known relationship between flat-band voltage, V_{FB}, and threshold voltage, V_T:

$$V_T = V_{FB} + 2\phi_F + \gamma\sqrt{(2\phi_F)} \tag{3.31}$$

it is evident that V_T varies as V_{FB}.

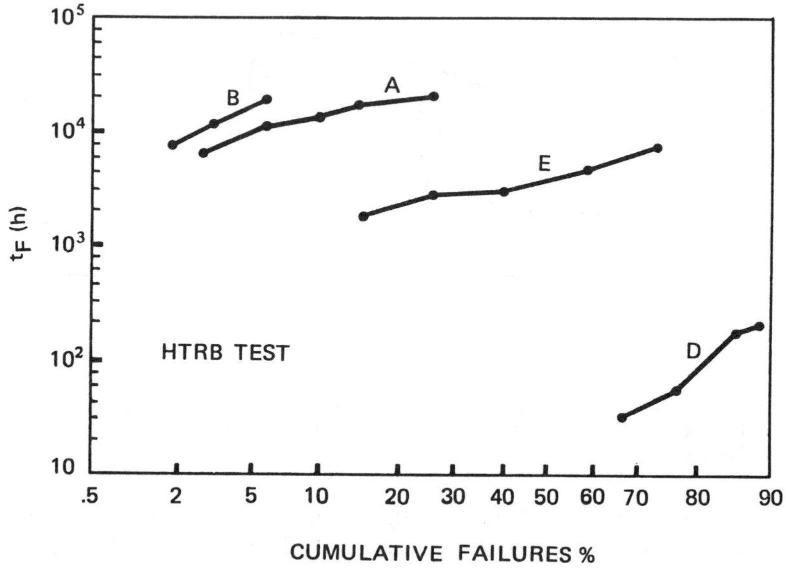

Figure 3.22 Log normal plot representing the results obtained on plastic packaged transistors in HTRB test at 125°C. Failure criterion was a 20 percent gain degradation.

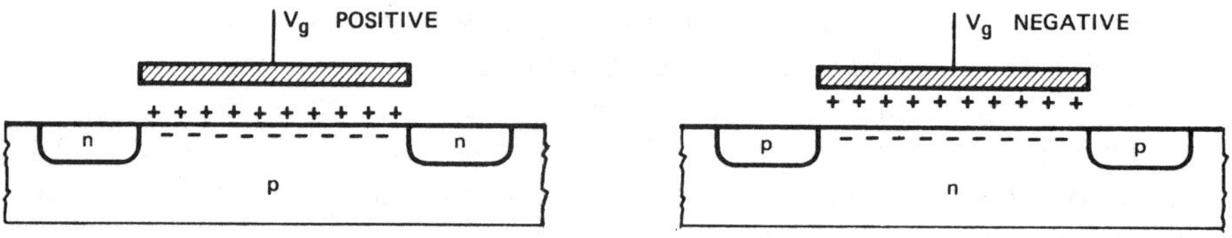

Figure 3.23 Effect of contaminants in the gate oxide in *n*- and *p*-channel MOSFETs.

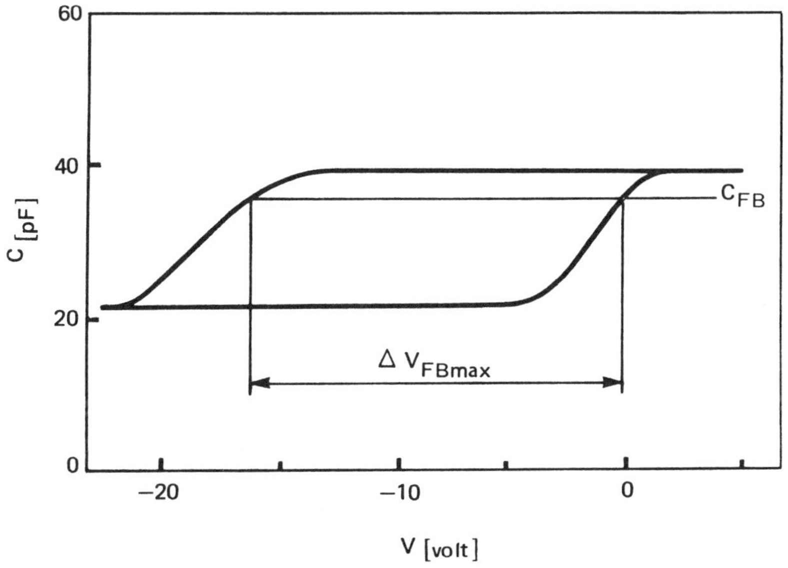

Figure 3.24 Flat-band voltage shift induced by positive charge in the oxide (courtesy of C. Morandi, University of Parma).

In addition to oxide contamination, mobile charge can cause parasitic channels among different transistors, as shown in Figure 3.25. The empirical law describing voltage variations due to mobile-charge contamination is [60]

$$\Delta V = At^{1/2}E^{1/2}\exp(-E_a/kT) \tag{3.32}$$

where A is a normalizing constant, E is the electric field, and E_a is the activation energy, which is generally found between 1.0 and 1.4 eV, even if values up to 2.0 eV, are reported [61], as it can be extracted from the data of Figure 3.26.

In this experiment, the contamination that appeared during high-temperature tests on n-channel EPROMs caused a special effect linked to the memory device structure. Within the EPROMs, information is stored as a negative charge injected through the oxide and trapped within the floating gate. The contamination due to positive ions shields the stored charge, as shown in Figure 3.27, causing information loss.

So far we have examined the effects of contaminants located within the active devices. However, the outer surface of the dielectric films is also of major importance for device stability. In fact, the surface conductance of dielectrics is strongly

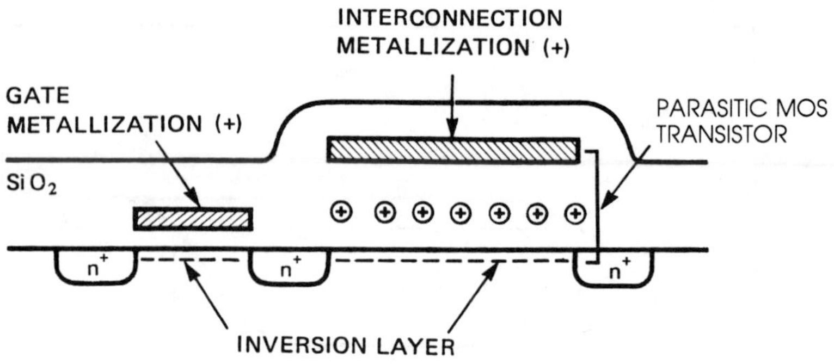

Figure 3.25 Contamination-induced parasitic *n*-channel MOSFET formation.

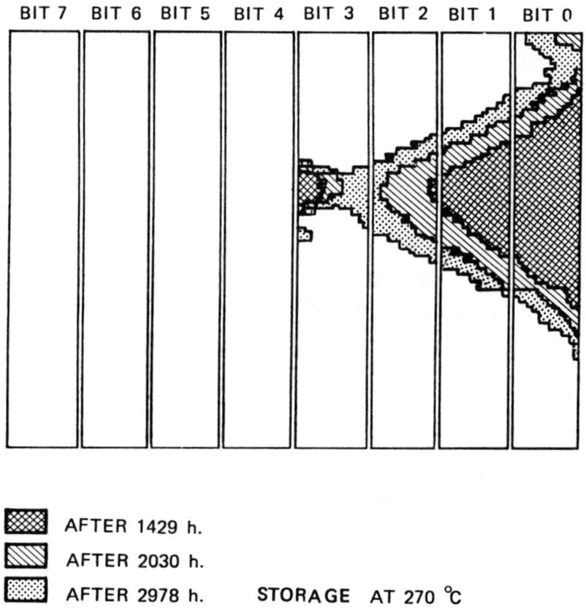

Figure 3.26 Map of information loss in an EPROM memory subjected to thermal stress at T = 270°C.

enhanced by surface contaminants and humidity. As the conductance increases the dielectric surfaces surrounding the metal layers are charged at a potential close to that of the metal itself (V_g in Figure 3.28) in accordance with a process described by a diffusionlike law. Here the diffusion coefficient is replaced by $1/(R_{sh}C_{ox})$ where R_{sh} is the sheet resistance of a surface square [1]. Hence, the potential on the dielectric surface fits a complementary error function:

$$V = V_g \operatorname{erfc}\left(\frac{x}{2\sqrt{(t/R_{sh}C_{ox})}}\right) \tag{3.33}$$

As shown in Figure 3.28, the main effect of this surface charge is again that of distorting the shape of the depletion layer, with results similar to those previously described. The effect of surface conduction is instead very limited, unless in the presence of very high humidity.

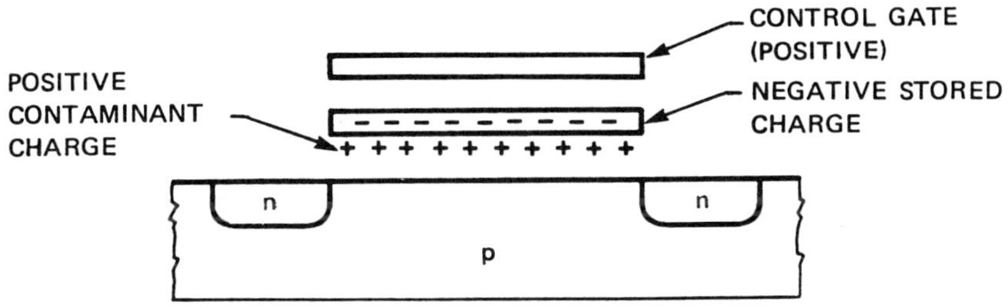

Figure 3.27 Effect of positive charge on an EPROM cell.

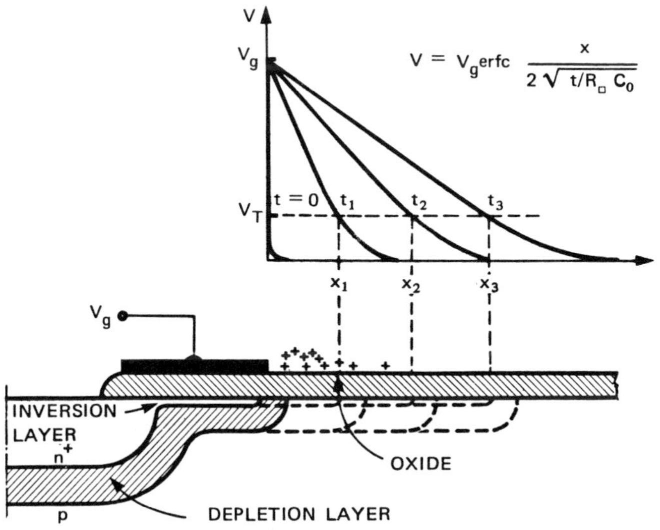

Figure 3.28 Depletion-layer deformation induced by the effect of charge at the insulator surface.

A common characteristic of the contamination-induced failures is their reversibility under thermal treatments. It is generally sufficient to store the contaminated devices for a few hours at high temperature (≈ 200°C) to fully or partially restore their nominal characteristics. This is due to the diffusion of contaminants, which tend to uniformly redistribute within the oxide. This phenomenon occurs even at room temperature in a longer time scale, causing trouble when the failure analysis is carried out long after the fault took place. This usually happens when devices fail during the field operation. In any case, it is normally sufficient to bias the device again in normal operating conditions for the fault to reappear.

In order to limit these effects, the practice has been universally adopted of protecting the device surface, including metalizations, with an additional passivation layer; this helps protect the device from mechanical damage during wafer handling and chip assembly. Both silicon dioxide and silicon nitride are used for this, as well as organic layers. Silicon dioxide is usually phosphorus doped (*phospho-silicate glass* or PSG), in order to trap contaminants far from the active region and reduce the mechanical stress associated with the different expansion coefficients of Si and SiO_2. Unfortunately, along with these beneficial effects, the addition of much phosphorus to the glass increases the risk of aluminum cathodic corrosion [62], so that a suitable phosphorus concentration must be selected. A detailed discussion of this effect can be found in Chapter 4. Figures 4.5, 4.6, and 4.7 show how the aluminum corrosion starts at the periphery of the bonding pad and then propagates along the metal line under the passivation. Since the formation of aluminum hydroxide is accompanied by a large increase in volume, as the corrosion progresses, the passivation is stressed and eventually breaks, exposing new metal to corrosion. This is one of the most common failure mechanisms for CMOS integrated circuits, due to their low dissipation and chip temperature, which helps water to reach the surface of the die.

More complex passivation systems have been developed, where multiple layers of Si_3N_4 or polyimide [63] are superimposed on the PSG. Nitride passivation is particularly effective against contamination, whereas the effectiveness of polyimide against humidity is still to be proved.

3.4 CONCLUSIONS

In this chapter we reviewed the main reliability problems induced by the dielectrics employed in semiconductor technology, with a special emphasis on silicon dioxide. Carrier injection from silicon into SiO_2 is the source of both catastrophic failures, due to thin-oxide breakdown and degradation of the electrical characteristics, resulting from carrier trapping into the oxide. The presence of contaminants inside the oxide, particularly alkaline ions at the Si-SiO_2 interface, causes device instability in terms of leakage currents and transistor gain.

These phenomena are closely related to the material and device technology. A significant improvement can be expected as the technology's learning curve progresses, and this has actually happened as regards oxide contamination and breakdown. However, much care must be exercised to ensure that any technological innovation that brings changes in a stabilized process does not result in the reappearance of undesired phenomena previously under control. For example, the replacement of polysilicon with silicides or polycides in MOSFETs may introduce increased contamination or a higher number of traps in the oxide. Furthermore, device scaling causes the increase of the electric fields in the devices, especially if the supply voltages are kept constant, as has been the case to date. This obviously induces an increasing concern for gate-oxide reliability, the research area where the most interesting studies are taking place, both for the improvement of technology and for a better understanding of the physics of carrier injection and charge trapping, and their effects on device reliability.

REFERENCES

1. A.S. Grove, *Physics and Technology of Semiconductor Devices,* John Wiley and Sons, New York, 1967.
2. W. Shockley and G.L. Pearson, *Phys. Rev.,* vol. 74, 1948, p. 232.
3. T.H. Ning, *Solid-State Electron.,* vol. 21, 1978, p. 273.
4. N. Klein, in L. Marton, ed., *Advances in Electronics and Electron Physics,* vol. 26, Academic Press, New York, 1969.
5. D. Frohman-Bentchkowsky, *Appl. Phys. Lett.,* vol. 18, 1971, p. 332.
6. W.S. Johnson, G. Perlegos, A. Renninger, G. Kuhn, and T.R. Raganath, *IEEE-ISSCC Tech. Digest,* 1980, p. 152.
7. B.E. Deal, *IEEE Trans. Electron Devices,* vol. ED-27, 1980, p. 606.
8. C. Hu, *IEEE-IEDM Tech. Digest,* 1983, p. 22.
9. C. Hu, S. Tam, F.C. Hsu, P.K. Ko, T.Y. Chan, and K.W. Terrill, *IEEE Trans. Electron Devices,* vol. ED-32, 1985, p. 375.
10. S. Tam, P.K. Ko, C. Hu, and R.S. Muller, *IEEE Trans. Electron Devices,* vol. ED-29, 1982, p. 1740.
11. R.R. Troutman, *Solid-State Electron.,* vol. 21, 1978, p. 283.
12. W. Shockley, *Solid-State Electron.,* vol. 2, 1961, p. 35.
13. J.F. Verwey, R.P. Kramer, and B.J. De Maagt, *J. Appl. Phys.,* vol. 46, 1975, p. 2612.
14. T.H. Ning, C.M. Osburn, and H.N. Yu, *J. Appl. Phys.,* vol. 48, 1977, p. 286.
15. C. Hu, *IEEE-IEDM Tech. Digest,* 1979, p. 22.
16. S. Tam, P.K. Ko, and C. Hu, *IEEE Trans. Electron Devices,* vol. ED-31, 1984, p. 1116.
17. E. Takeda, H. Kume, T. Toyabe, and S. Asai, *IEEE Trans. Electron Devices,* vol. ED-29, 1982, p. 611.
18. S. Tam and C. Hu, *IEEE Trans. Electron Devices,* vol. ED-31, 1984, p. 1264.
19. T.H. Ning, *J. Appl. Phys.,* vol. 47, 1976, p. 3203.
20. T.H. Ning, *J. Appl. Phys.,* vol. 49, 1978, p. 5997.
21. K.R. Hofmann, C. Werner, W. Weber, and C. Dorda, *IEEE Trans. Electron Devices,* vol. ED-32, 1985, p. 691.
22. T. Tsuchiya, *IEEE Trans. Electron Devices,* vol. ED-34, 1987, p. 2291.
23. E. Takeda and N. Suzuki, *IEEE Electron Device Lett.,* vol. EDL-4, 1983, p. 111.
24. W. Weber, C. Werner, and A.V. Schwerin, *IEEE-IEDM Tech. Digest,* 1986, p. 390.
25. J.Y. Choi, P.K. Ko, and C. Hu, *IEEE Electron Device Lett.,* vol. EDL-8, 1987, p. 333.

26. F.C. Hsu and K.Y. Chiu, *IEEE Electron Device Lett.*, vol. EDL-5, 1984, p. 162.
27. J.F. Verwey and B.J. De Maagt, *IEEE Trans. Electron Devices,* vol. ED-19, 1972, p. 245.
28. T.H. Di Stefano and M. Shatzkes, *Appl. Phys. Lett.*, vol. 25, 1974, p. 685.
29. P. Solomon and N. Klein, *Solid-State Comm.*, vol. 17, 1975, p. 1397.
30. W.T. Lynch, *J. Appl. Phys.*, vol. 43, 1972, p. 3274.
31. D.K. Ferry, *Appl. Phys. Lett.*, vol. 27, 1975, p. 689.
32. K.K. Thornber, *Solid-State Electron.*, vol. 21, 1978, p. 591.
33. R.C. Hughes, *Solid-State Electron.*, vol. 21, 1978, p. 251.
34. C.N. Berglund and R.J. Powell, *J. Appl. Phys.*, vol. 42, 1971, p. 573.
35. J. Maserjian and N. Zamani, *J. Appl. Phys.*, vol. 53, 1982, p. 559.
36. P. Solomon, in S.T. Pantelides, ed., *The Physics of SiO_2 and Its Interfaces,* Pergamon Press, New York, 1978.
37. B.K. Ridley, *J. Appl. Phys.*, vol. 46, 1975, p. 968.
38. M.V. Fischetti, private communication.
39. P. Olivo, private communication.
40. I.C. Chen, S.E. Holland, and C. Hu, *IEEE J. of Solid-State Circuits,* vol. SC-20, 1985, p. 333.
41. M.V. Fischetti, *Phys. Rev. B.*, vol. 31, 1985, p. 2099.
42. D.K. Ferry, *J. Appl. Phys.*, vol. 50, 1979, p. 1422.
43. E. Harari, *J. of Appl. Phys.*, vol. 49, 1978, p. 2478.
44. D.R. Wolters and J.J. van der Schoot, *Philips J. Research,* vol. 40, 1985, p. 137.
45. B. Euzent, N. Boruta, J. Lee, and C. Jenq, *IEEE Proc. Reliab. Phys.*, vol. 19, 1981, p. 11.
46. N. Mielke, A. Fazio, and H.C. Liou, *IEEE Proc. Reliab. Phys.*, vol. 25, 1987, p. 85.
47. N.J. Chou and J.M. Eldridge, *J. Electrochem. Soc.*, vol. 117, 1970, p. 1287.
48. C.M. Osburn and D.W. Ormond, *J. Electrochem. Soc.*, vol. 119, 1972, p. 591.
49. D.L. Crook, *IEEE-IEDM Tech. Digest,* 1979, p. 444.
50. J.W. McPherson and D.A. Baglee, *IEEE Proc. Reliab. Phys.*, vol. 23, 1985, p. 1.
51. R.A. Metzler, *IEEE-IEDM Tech. Digest,* 1979, p. 233.
52. I.C. Chen and C. Hu, *IEEE Electron Device Lett.*, vol. EDL-8, 1987, p. 140.
53. E.A. Irene, in S.T. Pantelides, ed., *The Physics of SiO_2 and Its Interfaces,* Pergamon Press, New York, 1978, p. 205.
54. E.H. Snow, A.S. Grove, B.E. Deal, and C.T. Sah, *J. Appl. Phys.*, vol. 36, 1965, p. 1664.
55. D.G. Edwards, *IEEE Trans. Reliability,* vol. R-31, 1982, p. 9.
56. B.A. McDonald, *Solid-State Electron.*, vol. 14, 1971, p. 17.
57. R.J. Kriegler, *IEEE Proc. Reliability Physics,* vol. 12, 1974, p. 250.
58. A.A. Bergh and G.H. Schneer, *IEEE Trans. Reliability,* vol. R-18, 1969, p. 34.
59. C. Morandi, private communication.
60. R.S. Hemmert, *IEEE Proc. Reliability Physics,* vol. 19, 1981, p. 172.
61. D. Bertotti, F. Fantini, and C. Morandi, *Microelectron. Reliability,* vol. 23, 1983, p. 717.
62. R.B. Comizzoli, *J. Electrochem. Soc.,* vol. 123, 1976, p. 386.
63. G.A. Brown, *IEEE Proc. Reliability Physics,* vol. 19, 1983, p. 282.
64. B. Riccò, M. YaAzbel and M.H. Brodsky, *Phys. Rev. Lett.*, vol. 51, 1983, p. 1795.

Chapter 4
CORROSION OF METALIZATIONS
A. Messina and E. Pollino

4.1 INTRODUCTION

Corrosion means the chemical or electrochemical reaction of a metal with the surrounding environment, which, if allowed to progress, will lead to the complete destruction of the metal [1]. In this context the metals affected by the phenomenon are basically those forming the semiconductor circuit metalizations (e.g., aluminum and gold) and the protective coatings of the encapsulations (e.g., tin and silver); the surrounding environment reacting with the metal consists of air, water, contaminant ions such as Cl, Na^+, and so on.

Metalization corrosion is the main reason for failure of plastic encapsulated semiconductors, and as we will see, it is closely linked to the presence of water on the chip surface [2].

4.2 GENERAL PRINCIPLES OF CORROSION

In general, mechanisms leading to corrosion can be divided into two types: namely, dry, such as the oxidation of aluminum in air, and wet, where the reaction requires the presence of an electrolyte, a moist environment, and an electromotive force.

Dry corrosion usually has minor importance for the reliability of semiconductor devices. For example, the oxidation of aluminum in air is considered self-passivating, forming an oxide thickness of less than 50 A. In a dry environment, such as inside an hermetically sealed container aluminum corrosion is virtually impossible and so this phenomenon will not be considered further.

A corrosion cell has to be formed to trigger the reaction. This is formed by a metallic anode and cathode connected externally (to the cell) by a circuit allowing electron flow and internally (to the cell) by a path for the ions (electrolyte).

The dynamics of the electrochemical corrosion are described by the Nernst equation:

$$E = E^{\circ} - \frac{RT}{2.3\ nF} \log \frac{A_P^p A_Q^q}{A_M^m A_L^l} \tag{4.1}$$

for a reaction of the type

$$lL + mM \rightarrow pP + qQ \tag{4.2}$$

where

E = the cell emf;

E° = standard oxidation potential;

A = chemical activities;

R = constant of gases (8.314 joule/mole · K);

T = absolute temperature;

n = chemical equivalents number;

F = Faraday constant (96,500 coulomb/eq);

l, m, p, q = moles of the L, M, P, Q species of reaction.

The free energy variation in the electrochemical reaction considered is $\Delta G = -nFE$, depending on the sign convention considered. If E is positive, G is negative and the reaction (4.2) can proceed at a rate dependent on its kinetics. If, instead, E is negative, ΔG is positive and the reaction cannot occur at all. The corrosion cell emfs are calculated from the Nernst equation using the standard oxidation potentials E° shown in Table 4.1.

Elements with a very positive E°, such as Li and K, are highly active whereas other elements exhibiting a very negative E°, such as Au and Pt, are inactive or noble compared to hydrogen.

Table 4.1 shows that Al has a high oxidation potential and hence will corrode easily. Electrochemical corrosion is a phenomenon consisting of two processes:

(1) Oxidation, due to a leakage of electrons from the electrode (anode) out of the cell and the formation of positive ions (cations) at the expense of the anode metal. The reaction

$$Al \rightarrow Al^{+++} + 3e^{-} \tag{4.3}$$

is hence an oxidation reaction which causes the erosion of the electrode.

Table 4.1 Standard Oxidation Potentials $E°$ Related to the Oxidation Potential of Hydrogen

Reaction	$E°$ (*Volts, 25°C*)	*Reaction*	$E°$ (*Volts, 25°C*)
$Li = Li^{+} + e^{-}$	3.05	$Cd = Cd^{++} + Ee^{-}$	0.403
$K = k^{+} + e^{-}$	2.93	$In = In^{+3} + 3e^{-}$	0.342
$Ca = Ca^{++} + 2e$	2.87	$Co = Co^{++} + 2e^{-}$	0.277
$Na = Na^{++} + e^{-}$	2.71	$Ni = Ni^{++} + 2e^{-}$	0.250
$Al = Al^{+3} + 3e^{-}$	1.66	$Pb = Pb^{++} + 2e^{-}$	0.126
$Be = Be^{==} + 2e^{-}$	1.85	$Mo = Mo^{+3} + 3e^{-}$	~0.2
$Ti = Ti^{++} + 2e^{-}$	1.63	$H_2 = 2H^{+} + 2e^{-}$	0.000
$Zr = Zr^{+4} + 4e^{-}$	1.53	$Cu = Cu^{++} + 2e^{-}$	−0.337
$Mn = Mn^{++} + 2e^{-}$	1.18	$Ag = Ag^{+} + e^{-}$	−0.800
$Zn = Zn^{++} + 2e^{-}$	0.763	$Pd = Pd^{++} + 2e^{-}$	−0.987
$Cr = Cr^{+3} + 3e^{-}$	0.74	$Pt = Pt^{++} + 2e^{-}$	~−1.2
$Fe = Fe^{++} + 2e^{-}$	0.440	$Au = Au^{+3} + 3e^{-}$	−1.50

(2) Reduction, which consists of a flow of electrons into the cell from the cathode and the formation of negative ions (or neutralization of positive ions) at the cathode. The reaction

$$2H^{+} + 2e^{-} \rightarrow H_2 \tag{4.4}$$

is a typical reduction reaction with gas development, and the reaction

$$Al^{+++} + 3e^{-} \rightarrow Al \tag{4.5}$$

is a reduction reaction with metal deposition at the cathode.

Therefore a complete corrosion cell consists of an anode and a cathode connected externally by a path for electrons (electrical circuit) and internally by a path for ions (electrolyte). Outside the cell the electrons flow between the anode (−) and the cathode (+), while inside the cell the charge is carried by cations migrating towards the cathode, where they are reduced, and by anions migrating towards the anode, where they are oxidized.

In their physical dimensions, the corrosion cells triggered in the semiconductor devices can also be very small (microscopic), and sometimes the anode and cathode are found on the same metalization. Three basic types of corrosion cells are important for reliability:

1. Galvanic cells (different metals)
2. Concentration cells
3. Electrolytic cells (with externally applied voltage)

In all cases triggering of the phenomenon requires the presence of two electrodes, an electrolyte and an emf between the electrodes. However, in the case of the electrolytic cell, the emf is applied externally, whereas in the case of the galvanic cell, the electrical potential is created by the presence of two different metals making up the electrodes (e.g., Cu and Zn or Al and Au). Finally, in the case of the concentration cell the emf is generated by the different electrolyte concentrations. An example of a galvanic cell composed of Cu and Zn is shown in Figure 4.1.

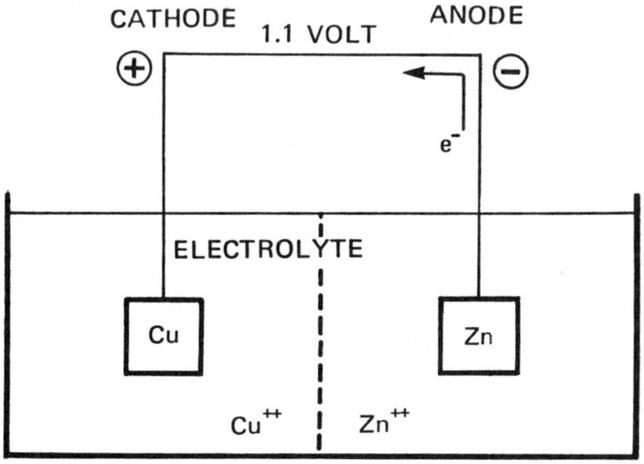

Figure 4.1 Cu-Zn corrosion cell.

The calculation of the emf generated in the Cu-Zn cell is carried out through the Nernst law, taking into account that the entire reaction

$$Zn + Cu^{++} \rightarrow Zn^{++} + Cu \tag{4.6}$$

can be divided into an oxidation reaction (the dissolution of the Zn at the anode):

$$Zn \rightarrow Zu^{++} + 2e^- \tag{4.7}$$

and a reduction reaction (the deposition of copper at the cathode):

$$Cu^{++} + 2e^- \rightarrow Cu \tag{4.8}$$

Applying the Nernst law using the tabulated values of E° and approximating the activities with the molar concentrations, we have

$$E(\mathrm{Zn}) = 0.763 - \frac{0.059}{2}\log\frac{[\mathrm{Zn}^{++}]}{[\mathrm{Zn}]} \tag{4.9}$$

$$F(\mathrm{Cu}) = 0.337 - 0.059\log\frac{[\mathrm{Cu}]}{[\mathrm{Cu}^{++}]} \tag{4.10}$$

In (4.10), $F(\mathrm{Cu})$ is a reduction potential for which the Nernst law with $F°\,(\mathrm{Cu}) = -E°\,(\mathrm{Cu})$ (Table 4.1) continues to apply.

The emf of cell $E = E\,(\mathrm{Zn}) + F(\mathrm{Cu})$ is hence

$$E = 1.1 - \frac{0.059}{2}\log\frac{[\mathrm{Zn}^{++}]}{[\mathrm{Cu}^{++}]} \tag{4.11}$$

which assumes the value of 1.1 volt when the ionic concentrations $[\mathrm{Zn}^{++}]$ and $[\mathrm{Cu}^{++}]$ are equal.

As an example of a concentration cell, we can consider the previous cell where the electrodes are equal (for example, Cu), but one electrode is in contact with a greater concentration of Cu^{++} ions than the other. The cell emf is hence

$$E = E(\mathrm{low}) - E(\mathrm{high}) = -\frac{0.059}{2}\log\frac{[\mathrm{Cu}^{++}]_{\mathrm{low}}}{[\mathrm{Cu}^{++}]_{\mathrm{high}}} \tag{4.11a}$$

The electrode in contact with the smaller ionic concentration acts as an anode toward the other electrode and, hence, will tend to corrode due to the increase in the $[\mathrm{Cu}^{++}]$ concentration.

Finally, in the electrolytic cell, the electrodes are of the same material and in contact with the same concentration of electrolyte: applying an external ddp causes one electrode to become anodic toward the other electrode and the first electrode starts to corrode. This situation is very common in biased integrated circuits.

Another approach to assessing the thermodynamics of a reaction is determining the probability that it will take place, represented by the potential-Ph diagrams or Pourbaix diagrams (Fig. 4.2) [3]. It can be deduced from Figure 4.2 that Al corrosion can take place for very low (less than 4) or very high (more than 8) pH levels.

Once the thermodynamic possibility of a corrosion reaction has been ascertained, the rate at which it progresses depends on system variables, including:

- Thermodynamic variables, such as cell type, applied voltage, and the pH of the zone in question.

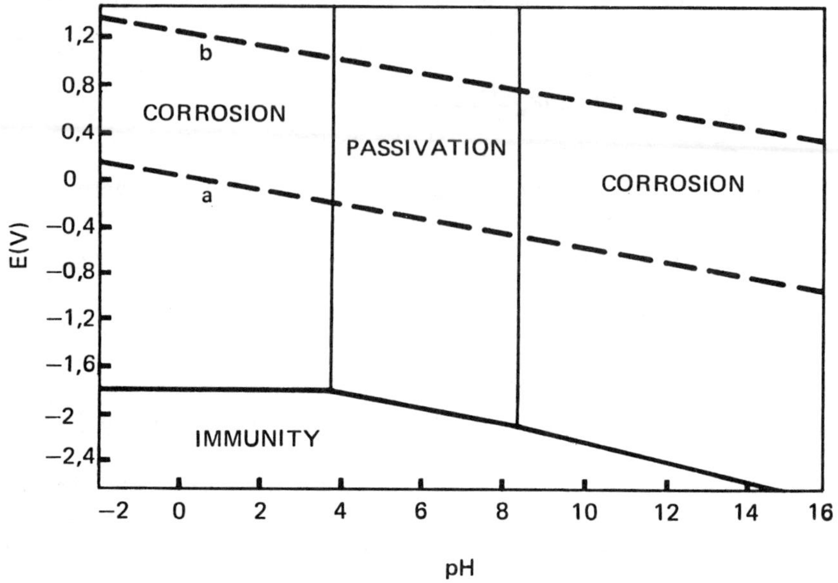

Figure 4.2 Pourbaix diagram for aluminum. In the configurations outside the band included between *a* and *b*, the reaction tends to passivate (*a* is the hydrogen curve and *b* the oxygen curve).

- Metallurgical variables, such as the composition and morphology of the electrode and so on.
- Environmental variables, such as temperature, humidity, contaminants present, and so on.

The different variables result in the corrosion phenomenon developing along one of the three typical slopes shown in Figure 4.3 [4]. Quantity A, which appears in the figure, can be any physically observable variable affecting the progress of the corrosion; for example, a change in dimensions, electrical characteristics, and so on. Curve a of Figure 4.3 is a function of the type $A = kt$. The corrosion rate is constant $dA/dt = k$. This trend occurs whenever the electrolytic reaction is not changed by secondary reactions originating with the corrosion products, such as the corrosion of Al or Cr metalization in a halogen-free (Cl^-) acid environment.

Curve b relates to a function of the type $A = kt^{\alpha}$, $\alpha > 1$. The corrosion rate increases with time: $dA/dt = \alpha Kt$. In this case, we say that the process is self-accelerating and it occurs when a secondary reaction takes place and stimulates further metal corrosion. An example of this is the corrosion of the Al metalization in the presence of chlorine ions: the hydrolysis of $AlCl_3$, which is a secondary reaction, leads to the formation of new Cl^- available for the corrosion reaction.

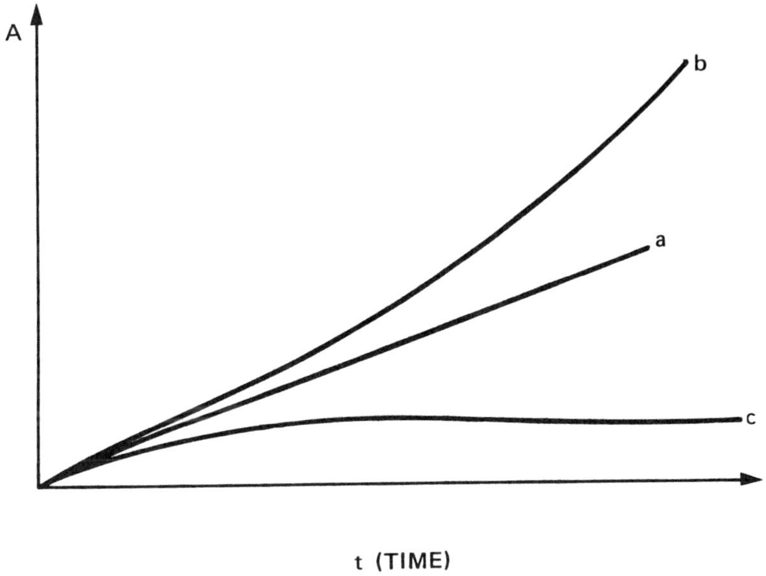

Figure 4.3 Corrosion dynamics: (a) normal reaction; (b) self-accelerating reaction; (c) self-passivating reaction. *A* is an arbitrary parameter.

Curve *c* relates to a function of the type $A = k \ln t$. The corrosion rate decreases in time: $dA/dt = kt^{-1}$. When this situation occurs we say that the process is self-slowing. It happens when the corrosion products are not very soluble and stick to the metal, performing a protective function; for example, Al corrosion in an average, halogen-free alkaline environment. The aluminum hydroxide, $Al(OH)_3$, formed is not very soluble and adheres to the metal. Since the hydroxide is a poor conductor, the metal tends to passivate and the electrochemical reaction slows down gradually.

The specific problems of corrosion in semiconductor devices inside the encapsulations are sometimes difficult to determine. In fact, corrosion starts at the microscopic level and often evolves rapidly until it completely destroys the device, even to the point of totally altering the initial situation that triggered the phenomenon. The kinetics of a corrosion reaction in an integrated circuit cannot, in practice, be evaluated on the basis of theoretical considerations. More pragmatic approaches for investigating the probability of corrosion-induced failure for a given device are through high-level accelerated tests; for example, tests of humidity, ionic contamination, and temperature.

Figure 4.4 gives an example of the type of information that can be obtained from these tests. It shows that the mean timing of failure due to aluminum corrosion

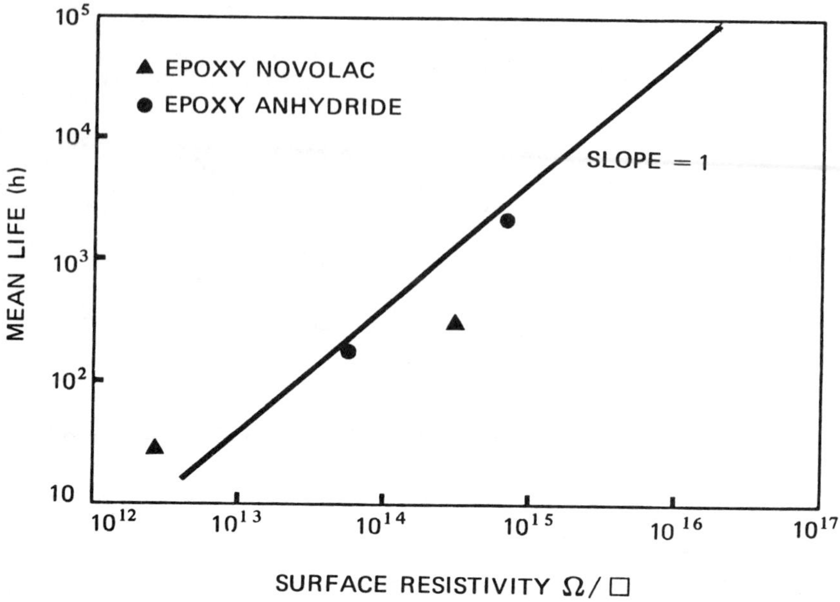

Figure 4.4 Relationship between average life due to aluminum corrosion at 110°C and 90 percent RH, and the surface resistivity of a humidity film at the interface between plastic and active plate.

depends on surface resistivity of devices incorporating various plastic resins. Surface resistivity is a particularly suitable quantity to be linked to corrosion since it is closely related to those variables (water, contamination, *et cetera*) on which corrosion depends most strongly. The climatic conditions used for this test are 110°C and 90 percent RH.

The ions most commonly found on the device surface and that, in the presence of water, give rise to the electrolytic solution required to trigger corrosion are

- Halogens (especially Cl) deriving both from the inadequate removal of fabrication process residues and from the container if it is plastic.
- Alkalines (especially Na) deriving from the diffusion ovens and more generally from glass containers and the hands of the operators.
- Phosphorus, which, unlike the above ions that are contaminants, is specially incorporated in the surface passivation glass to improve its mechanical characteristics and restrict the effects of the Na present.

Let us now examine a few of the most common examples of corrosion in semiconductor devices.

4.3 CORROSION OF ALUMINUM

Aluminum, as shown in the Pourbaix diagram in Figure 4.2, can be corroded in either acid or basic solution [2, 5]. In acid solutions aluminum corrodes as follows:

$$2Al + 6H^+ \rightarrow 2Al^{+++} + 3H_2$$
$$Al^{+++} + 3H_20 \rightarrow Al(OH)_3 + 6H^+ \tag{4.12}$$

whereas the corresponding reactions in a basic environment are

$$Al + 3(OH)^- \rightarrow Al(OH)_3 + 3e^-$$
$$O_2 + 2H_2O + 4e^- \rightarrow 4(OH)^- \tag{4.12a}$$

In both cases the product of the reaction is aluminum hydroxide, which tends to slow down the phenomenon, acting as a barrier to the penetration of humidity. However, the small quantities of ions present encourage the dissolution of $Al(OH)_3$ and, hence, also contribute to the corrosion of the aluminum.

Depending on the ions present, the erosion can take place either on the negative bias metalization (cathode) or on the positive bias metalization (anode). In the most common cases cathode corrosion can take place (1) if there is excess phosphorus in the passivation glass or (2) if there is a contamination from positive ions. In case 1, the excess P increases the absorption of humidity by the glass, the electrolyte formed is the acid type [6] since an excess of H^+ ions is formed and the reaction progresses in accordance with (4.12). The product of the reaction is $Al(OH)_3$, which takes up more volume than the initial aluminum. This causes the passivation glass containing the phosphorus to break, thereby allowing the corrosion to spread. Figure 4.5 illustrates this phenomenon. The figure shows how corrosion starts on the bonding pads where the passivation ends: that is, next to those areas featuring aluminum, phosphorus, humidity, and applied voltage all at the same time. Away from the pads, signs of corrosion on the chip surface are visible in the vicinity of microfractures in the passivation glass.

Figure 4.6 illustrates the kinetics of this phenomenon, showing the relationship of failure time, during humidity tests, to the quantity of phosphorus contained in the passivation glass. It can be seen that the optimum value is around 2 percent. Under this value the failure time decreases, since the passivation glass is more brittle, and it is therefore far more likely that humidity will reach the chip surface. For quantities of phosphorus above 2 percent, the failure time once again decreases due to the cathode corrosion phenomenon discussed earlier. Further discussions on this subject can be found in [8, 9, and 10].

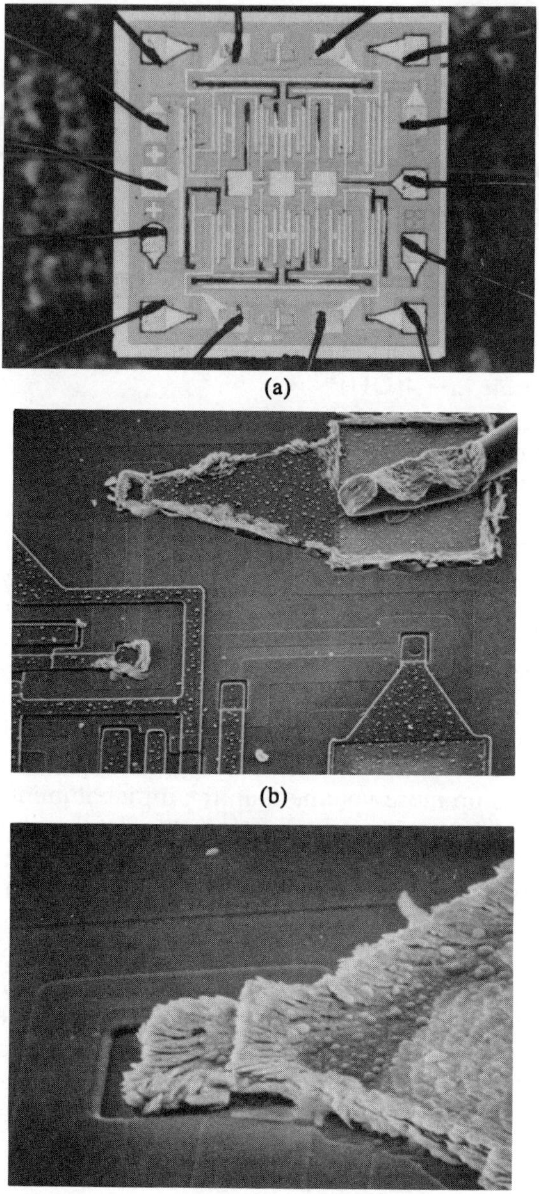

(a)

(b)

(c)

Figure 4.5 Cathodic corrosion of CMOS integrated circuits: (a) CMOS circuit (SEM 500x). (b) Detail of a corroded weld section (SEM 500x), the aluminum hydroxide takes up more space than the aluminum and causes the overlying passivation to break. (c) Detail of a metalization (SEM 1000x), where the corrosion has caused the track to break in the vicinity of the contact window.

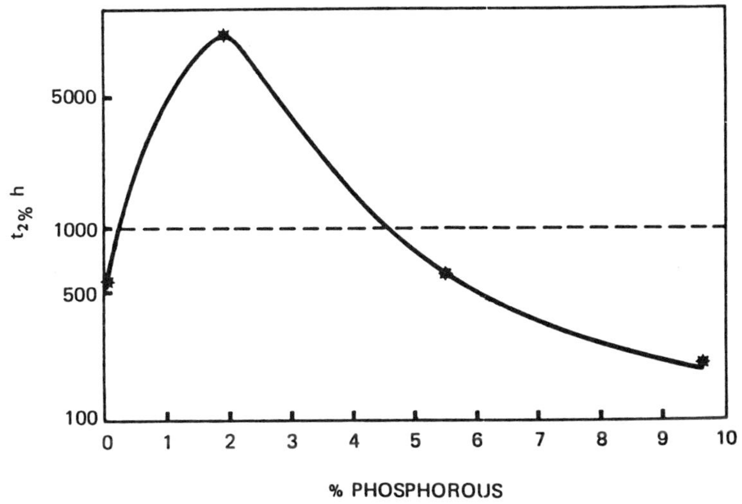

Figure 4.6 Relationship of failure time (2 percent of failures) to the percentage (by weight) of phosphorus in the passivation glass.

In case 2, when there is positive ion contamination, as from Na, the ions migrate under the effect of the electrical field toward the cathode [11] and the hydrolysis of water takes place in accordance with

$$2H_2O + 2e^- \rightarrow 2(OH)^- + H_2 \qquad (4.13)$$

This reaction causes a change in the pH, which becomes basic, and stimulates the corrosion of the aluminum in accordance with (4.12a), see Figure 4.7, curve *a*. Anodic corrosion, or corrosion of the positive bias metalizations, take place when negative ions are present (e.g., Cl^-) see Figure 4.7 curve *b*.

A certain number of hypotheses have been formulated to explain the phenomenon: the reactions that follow are based on the absorption of chlorine by the surface aluminum hydroxide [11]:

$$Al(OH)_3 + Cl^- \rightarrow Al(OH)_2\,Cl + OH^- \qquad (4.14)$$

This causes the oxide to dissolve and allows the direct reaction of the aluminum:

$$Al + 4Cl^- \rightarrow AlCl_4^- + 3e^- \qquad (4.15)$$

The ion $Al(Cl_4)^-$ reacts with the water present in accordance with (4.16):

$$AlCl_4^- + 3H_2O \rightarrow Al(OH)_3 + 3H^+ + 4Cl^- \qquad (4.16)$$

The Cl is once again available in greater quantities than at the start, and the corrosion progresses at the expense of the Al and the water. This failure mechanism is particularly important for plastic encapsulated semiconductors, since the Cl is used in the fabrication of epoxy resins.

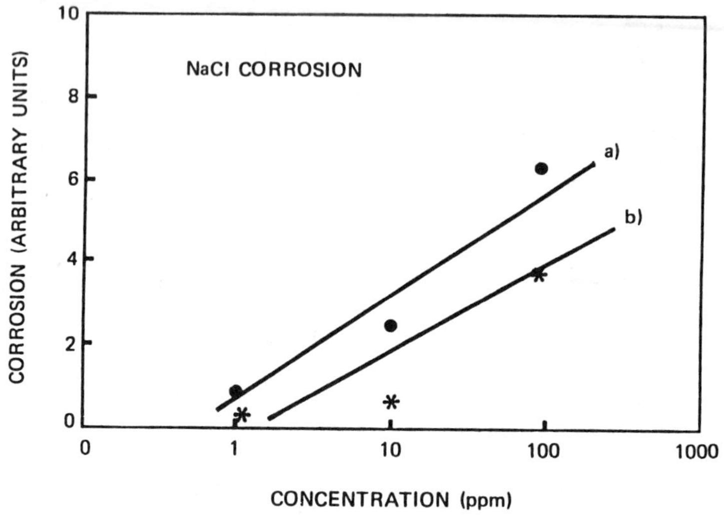

Figure 4.7 Corrosion due to positive ions (Na^+) curve *a* and negative ions (Cl^-) curve *b*.

Figure 4.8 illustrates the kinetics of chlorine-induced corrosion, showing the relationship between mean time to failure due to aluminum corrosion and the chlorine concentration. A detailed discussion of the effect of the presence of halogens on aluminum corrosion is given in [12]. This phenomenon can take place even in the absence of an external applied potential; in fact, in the case of devices with Al metalizations and bondings made with gold wire, an Au-Al galvanic cell is created that establishes emfs that can reach values up to 3.16 volts, as obtained applying the Nernst law. Figure 4.9 shows aluminum corrosion triggered by an Au-Al galvanic cell. A similar phenomenon has been observed on MESFET devices, which use gold metalization and aluminum gates [13].

4.4 CORROSION OF GOLD

Since Au is a noble metal, Au metalizations are much more resistant to corrosion than aluminum metalization, and this was one reason that led to the use of gold-based metalization systems (Ti-Pt-Au, Ti-W-Au, etc.). However, if Cl is present, anode Au dissolution may take place in accordance with the following formula [14]:

$$Au + 4Cl^- \rightarrow AuCl_4^- \tag{4.17}$$

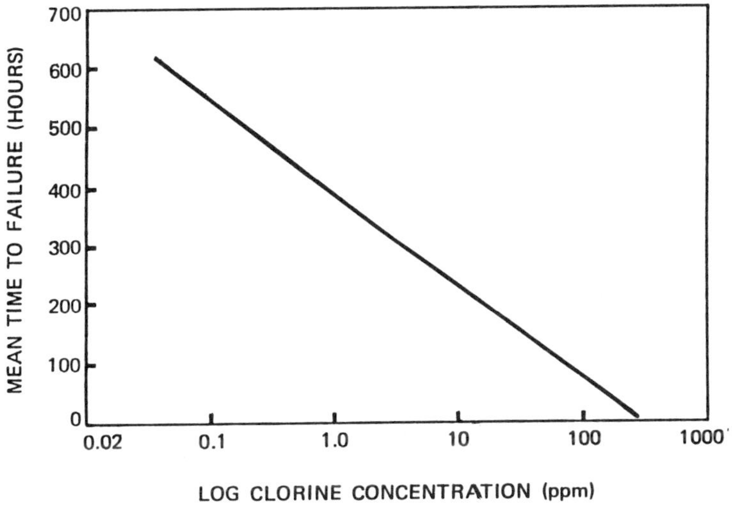

Figure 4.8 Mean time to failure for linear integrated circuits in relation to the concentration of Cl on the chip surface.

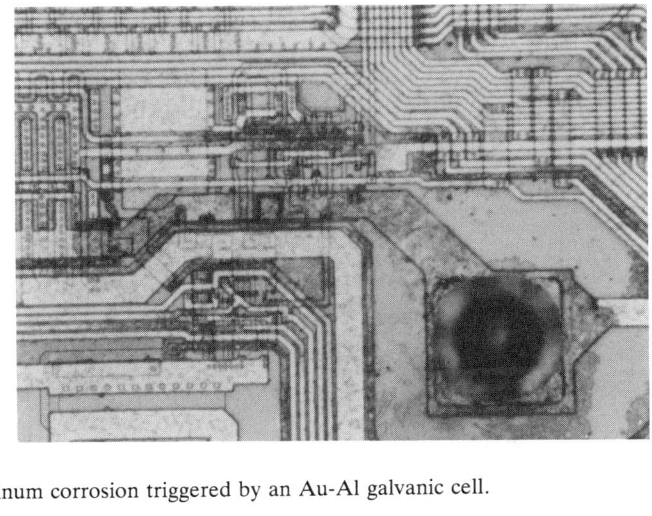

Figure 4.9 Aluminum corrosion triggered by an Au-Al galvanic cell.

Under the effect of the electrical field, the $AuCl_4^-$ ion is transferred to the cathode, where the Au is redeposited and the Cl is freed, in accordance with the equations:

$$AuCl_4^- + 2e^- \rightarrow AuCl_4^- + 2Cl^-$$

$$AuCl_2^- + e^- \rightarrow Au + 2Cl^- \tag{4.18}$$

The growth of Au at the cathode takes place with the formation of dendrite characteristics [15] that, at the end of the process, cause the metalizations to short circuit. This failure mechanism is known as *migratory gold resistive short* (MGRS). Figure 4.10 illustrates the formation of a gold dendrite starting from the bonding pad.

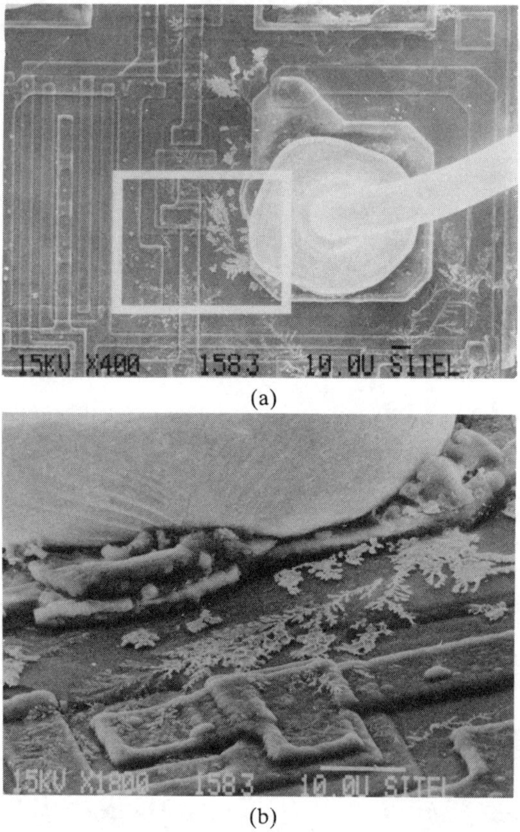

Figure 4.10 Formation of gold dendrites originating at the bonding pads: SEM (a) 400X and (b) 1800X.

The triggering of MGRS requires surface imperfections at the cathode, such as passivation spots or fractures, which initiate the deposition of the Au in dendrites. In their absence, the gold is deposited in a nonpreferential mode and does not give rise to short circuits. Figure 4.11 illustrates a growth of gold along a passivation fracture, which has shorted two metalizations.

The formation of gold dendrites (originating at the bonding pad) that shorted two aluminum metalizations is described in [16]. The phenomenon can be understood in the light of the earlier comments. In fact, the Cl dissolves a part of the gold, which, starting from the cathode, is redeposited in dendritic mode until the MGRS characteristics are formed. It appears to be more a growth than a corrosion phenomenon; in fact, the mechanism is so directional that approximately 10^{10} atoms

(a)

(b)

Figure 4.11 Deposition of gold along a passivation fracture: (a) optical microscope 800X; (b) detail magnified SEM 3000X.

are sufficient to connect two tracks 10μm apart without any sign of anode track erosion.

Gold corrosion has been observed even when there is no contamination but high humidity with very strong applied electrical fields [17]. The phenomenon is manifested as a blackening and swelling of the anode Au without the appearance of dendrites, while the negative pole metalization remains unchanged. It is maintained that the reaction leads to the formation of $Au(OH)_3$, which is sufficiently conductive to create short circuits due to the increase in volume (Figure 4.12).

4.5 CORROSION OF Ni – Cr_2

Other metals than Au and Al used in semiconductor technology, such as Cr and Ni – Cr, are also subject to corrosion. Again the presence of water, contaminant

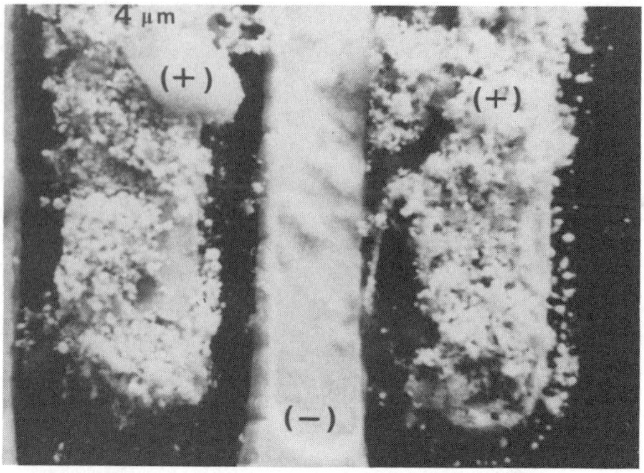

Figure 4.12 Corrosion of anodic gold without dendrite formation.

ions (particularly Cl^-), and applied voltage trigger an electrolytic cell with disastrous consequences for the circuit [7]. This is the main failure mechanism for plastic encapsulated thin film hybrid circuits; as they are produced in low volumes, their production lines are often inspected less regularly than monolithic integrated circuits.

Essentially, there are two critical points in the fabrication process in terms of corrosion sensitivity:

- The absence of a passivation glass on the surface of the device.
- The encapsulation with plastic resin; the subsequent polymerization may leave water and Cl on the surface of the device.

Apart from thin film hybrid circuits, Ni-Cr is used (even though increasingly less often) for the construction of resistors in PROM memories (fusible link) and in the provision of redundancies in VLSI RAM memories. In the case of PROM memories, anodic corrosion has been observed with applied voltages above 2.5 V [18]. This reaction can be predicted with the Pourbaix diagrams for Cr and Ni [3].

4.6 CONCLUSIONS

Apart from the environmental variables (temperature, humidity), which are discussed in Chapter 11 dealing with accelerated tests, metalization corrosion is highly dependent on certain aspects of the device construction process (package, passivation, assembly, *et cetera*) and the quality with which these processes have been carried out (e.g., presence of contaminations) [19, 20].

4.7 APPENDIX: CORROSION PHENOMENA AFFECTING ASSEMBLY

Considering the component–printed circuit or hybrid circuit board, the following cases of metalization corrosion can arise:

- Corrosion of printed circuit tracks and component rheofores.
- Corrosion of contacts.

4.7.1 Track and Rheofore Corrosion

This phenomenon is almost always due to residues of weld flux activated with highly ionizable components, such as ammonium halogens and ions. Again, it is an electrolytic corrosion model, where:

- The electrolyte consists of the humidity that has dissolved the flux residues (generally these residues are hydroscopic).
- The electrodes are the metalizations involved (printed circuit tracks and rheofores).
- The driving force is the applied potential, or in the case of dissimilar metalizations, such as a Sn-Pb track and gold-plated rheofore, the electrochemical potential difference between the two metals (in the case Au – Sn/Pb E = 1.5V).

Figure 4.13 illustrates an example of corrosion on Sn-Pb tracks after test cycles in a moist environment (IEC-50-1) supplied at 10 V dc. Note the formation of Pb dendrites between the two sections of the circuit.

4.7.2 Contact Corrosion

The coating most commonly used for the contact zones is gold plating, and a main cause of failure is the porosity of the Au etch. Porosity depends on thickness [21], according to

$$P = AS^{-n}$$

where P is the porosity index expressed in number of pores per cm^2; S is the thickness of Au expressed in microns; and A and n are constants that depend on the type of electrolyte, galvanization conditions, and the type of substrate.

The porosity is measured by counting the spots produced by a suitable chemical reagent, which is active toward the substrate and does not react on the gold; only in the presence of pores will the solution arrive at the substrate and give rise

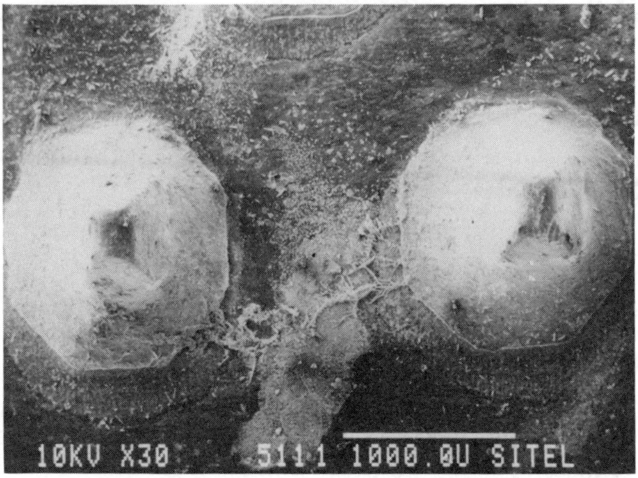

(a)

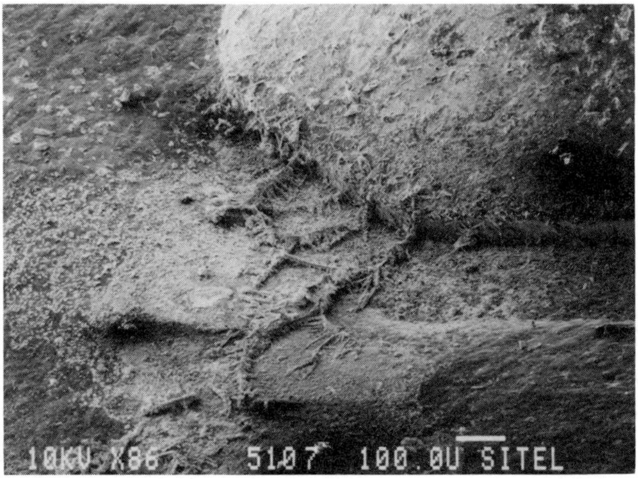

(b)

Figure 4.13 Dendritic growth of Pb, between two printed circuit sections, after humidity tests: SEM (a) 30X; (b) 86X.

to the spot. The typical selective reagents used for this purpose are wet sulphur dioxide and nitric acid.

Figure 4.14 [22] illustrates the relationship of thickness porosity for an acid gold-plating bath on a copper substrate (printed circuit laminate). The deposition of a nickel substrate (3–4 μm) improves the porosity curve considerably. Under

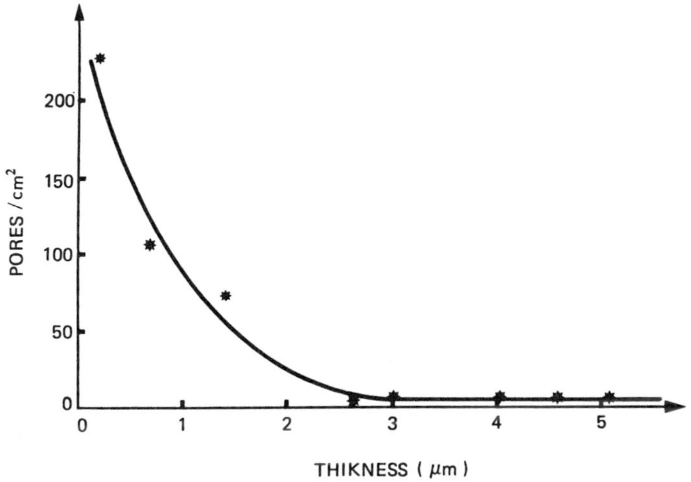

Figure 4.14 Porosity of a galvanic gold-plating on copper in relation to its thickness.

the same conditions, the increase in the porosity index starts at 1–1.5 μm Au, rather than at 2.5–3 μm, as shown in the figure. The corrosion model that can be triggered in the case of porosity is still galvanic where

- The electrolyte is the humidity, which may contain corrosive atmospheric agents (derived from the sulphur).
- The electrodes consist of metals overlaying the pore (gold and copper).
- The driving force is the electrochemical potential difference between the two metals.

The pores have an average diameter of 15 μm, and so the anodic surface is very small; this results in a high corrosion current density that speeds up the phenomenon. This mechanism is known as *pit corrosion* [4].

REFERENCES

1. S.C. Kolesar, "Principles of Corrosion," *RPS,* 1974, pp. 155–167.
2. N. Lycoudes, "The Reliability of Plastic Microcircuit in Moist Environments," *Solid-State Tech.,* October 1978, pp. 53–68.
3. M. Pourbaix, *Atlas of Electrochemical Equilibria in Aqueous Solutions,* Pergamon Press, Oxford, 1966.
4. G. Bianchi and F. Mazza, *Corrosione e protezione dei metalli,* Tamburini, 1979.
5. H.H. Uhlig, *Corrosion and Corrosion Control.* John Wiley and Sons, New York. 1963.
6. W.M. Paulson and R.W. Kirk, "The Effects of Phosphorous-Doped Passivation Glasses on the Corrosion of Aluminium," *RPS.* 1974, pp. 172–179.
7. A. Gallesio, E. Pollino, and D. Riva, *Rapporto Interno CSELT,* n. 79.04.103.

8. V. Bhide and J.M. Eldrige, "Aluminium Conductor Line Corrosion," *RPS,* 1983, pp. 44–51.
9. G.D. Giacomo, "Phosphorous Migration Kinetics from PSC to Glass Passivation Surface, *RPS,* 1984. pp. 223–228.
10. R.C. Olberg and J.L. Bozarth, "Factors Contributing to the Corrosion of Aluminum Metal on Semiconductor Devices Packages in Plastics," *Microelectronics and Reliabilty,* 15. 1976, p. 601.
11. W.M. Paulson and R.P. Lorigan, "The Effect of Impurities on the Corrosion of Aluminium Metallization," *RPS,* 1976, pp. 42–47.
12. M. Jannuzzi, "Bias Humidity Performance and Failure Mechanisms of Non-hermetic Aluminium SIC's in an Environment Contaminated with Cl," *RPS,* 1982, pp. 16–26.
13. J.C. Irvin and A. Lora, *Bell System Tech. J.* N.57. 1978, p. 2823.
14. F.J. Grunthaner, T.W. Griswols, and P.G. Clendening, "Migratory Gold Resistive Shorts, Chemical Aspects of a Failure Mechanism," *RPS,* 1975, pp. 99–106.
15. A. Shumka and R. Piety, "Migrated-Gold Resistive Short in Microcircuits," *RPS,* 1975, pp. 93–98.
16. I. Benedini, L. Diguida, and E. Pollino, "Reliability Evaluation of LSI Memories for Long Life Applications," 4e Colloque International Fiabilité & Maintenabilité, Perros-Guirec, 1984.
17. E. Brambilla, P. Brambilla, C. Canali, F. Fantini, and M. Vanzi, "Anodic Gold Corrosion in Plastic Encapsulated Devices," *Microel. Reliability,* 23, 1983, pp. 577–585.
18. W.M. Paulson, "Further Studies on the Reliability of Thin Film Ni-Cr Resistors," *RPS,* 1973. pp. 61–68.
19. R.W. Lawson, "A Review of the Status of Plastic Encapsulated Semiconductor Component Reliability," *British Telec. Tech. J.,* 2, April 1984, pp. 95–111.
20. S.P. Sim and R.W. Lawson, "The Influence of Plastic Encapsulants and Passivation Layers on the Corrosion of Thin Aluminium Film Subjected to Humidity Stress, *RPS,* 1979, pp. 103–112.
21. F.H. Reid and W. Goldie, *Gold Plating Technology,* Electrochemical Publication Ltd., 1974.
22. M. Perrone, "L'elettrodeposizione dei metalli preziosi," Engelhard Symposium. 1981.

Chapter 5
ELECTROMIGRATION IN THIN FILM CONDUCTORS

M. Finetti and E. Pollino

5.1 INTRODUCTION

Migration of electrons resulting from the application of a dc electric field is considered as one of the primary causes of failure in semiconductor integrated circuits.

More than a century ago, this phenomenon, known as *electromigration,* was discovered in liquid alloys [1]. Electromigration was first attributed to the electrostatic interaction between the electric field and metal ions. The importance of the collisions of atoms and mobile charges was suggested only considerably later [2]. Further experiments demonstrated that the direction of the atomic transport changes with the band structure and with the sign of the charge carriers [3].

Electromigration effects in thin films were extensively studied in the late 1960s, following the widespread use of thin film conductors in silicon integrated circuit technology. It was in fact observed that electromigration can be a significant cause of failure in metallic films used as interconnections in electronic circuits. This problem, which only a few years ago seemed to be completely solved, recently assumed a major relevance and extended to the area of thin films used in contact structures. In fact, the continuing trend toward scaling down device dimensions has led to a drastic reduction in the metal line sections, as well as in the contact areas, fabricated on increasingly shallower junctions. The consequent increase in current density, by a factor of K^2 for the interconnections and by a factor of K^3 for the contacts, where K is the scaling factor, requires the development of new materials. At the same time, new characterization techniques have been introduced, in order to evaluate the electromigration resistance with very short stress tests [4, 5].

The purpose of this chapter is to present some fundamental aspects of the electromigration in thin films, as well as to describe the main technological developments that have contributed to improving VLSI metalization reliability.

5.2 FUNDAMENTAL ASPECTS OF ELECTROMIGRATION

In a multicomponent solid, the action of driving forces induces an atomic flux that can be expressed as

$$\vec{J}_i = -D_i\vec{\nabla}\, c_i + \vec{F}(D_i/fkT)c_i \tag{5.1}$$

where c_i and D_i are the concentration and the diffusion coefficient of the ith component, and f is the correlation coefficient of the mechanism responsible for the atomic motion. The two terms on the right side represent, respectively, the contribution of the gradient concentration and the effect of a driving force of chemical, mechanical, thermal, or electrical type. In particular, electromigration is the phenomenon of mass transport resulting from the application of an electric field. At the same time, a vacancy flux proceeds in the opposite direction.

In general, the driving force for electromigration is composed of two contributions: one due to the direct action of the electrostatic field on the ions, the other derived from the effect of the mobile charge carriers on the atomic motion. At this point, it is convenient to introduce the concept of effective charge, Z^*, which represents the number of charges that the atom should have to undergo an electrostatic force equal to $\vec{F}$. The following expression can then be written:

$$\vec{F} = Z^*\, e\, \vec{E} \tag{5.2}$$

According to the previous considerations, (5.2) becomes

$$\vec{F} = (Z^*_{el} + Z^*_{wd})e\,\vec{E} \tag{5.3}$$

where the first term on the right side represents the effect of the electrostatic field, and the second term is the so-called electron wind.

In (5.3), the value of Z^*_{el} is assumed to be that of the ion's nominal valence, Z. The procedure typically followed to evaluate theoretically Z^*_{wd} is based on the exchange of momentum between scattered charge carriers and diffusing atoms [6]. Substantially identical results were also derived by applying a different mathematical approach, based on quantum mechanics principles [7]. By summing up the effects of the electrostatic field and the electron wind, the following relation is obtained in electronic conductors:

$$Z^* = Z(1 - 0.5\, N\, \rho_D/\rho N_D) \tag{5.4}$$

where ρ is the resistivity of the metal, ρ_D the excess resistivity contributed by N_D diffusing atoms, and N the total number of atoms.

In the case of $Z^* < 0$ the electron wind component prevails on the electrostatic force and the atoms diffuse in the same direction of the electron motion; that is, toward the anode. For $Z^* < 0$ mass transport proceeds toward the cathode. In electronic metals, such aluminum, atomic diffusion toward the positive electrode has been consistently observed, indicating the dominant effect of the electron wind force. We can also observe that, when the mobile carriers are holes instead of electrons, the sign in the expression of Z^*_{wd} becomes positive and the two components of the driving force are additive. In this case, mass transport is expected to occur toward the negative electrode, as experimentally observed in tungsten [8].

5.3 THE ATOMIC FLUX AND ITS DIVERGENCE

Following the description in the previous paragraph, the net atomic flux J_a in a lattice due to a current density j can be expressed as

$$\vec{J}_a = (ND/kT)\,Z^*e\vec{E} = (ND/kT)\,Z^*e\vec{j} \tag{5.5}$$

where the diffusion coefficient D is given by the relation

$$D = D_0 \exp(-E_a/kT) \tag{5.6}$$

and E_a is the activation energy of the diffusion process.

At the relatively high temperatures, close to the material melting point, T_M, typically used in the study of the electromigration in bulk samples, the diffusion coefficient and activation energy are well-defined quantities. As we will see in the following, the analysis of electromigration in thin films is much more complex, since in this case the experiments are typically carried out at lower temperatures (typically in the range of 200–300°C), corresponding to the conditions of interest for the use of these materials as conductors in the electronic circuits. At such temperatures, mass transport is mainly controlled by diffusion along grain boundaries. If we assume that, in all the grain boundaries of the film, mass transport proceeds with the same characteristics, the electromigration flux can be expressed as [9]

$$\vec{J}_b = N_b(\delta/d)\,[D_0 \exp(-E_a/kT)\,Z^*e\vec{j}\,]/kT \tag{5.7}$$

where N_b is the local density of ions at the grain boundary, d is the average grain size, δ is the grain boundary width (usually assumed equal to 10 Å) and E_a is the activation energy. D_0 and Z^* have their usual meaning, referred in this case to the grain boundaries.

The use of (5.7) allows us to predict the atomic transport only from a qualitative point of view, since in real films the transport properties vary as a function of grain size distribution and grain boundary orientation. Furthermore, besides the atomic diffusion along grain boundaries, other electromigration mechanisms can be present, each with its own activation energy. To this purpose, Figure 5.1 shows the different contributions, whereas the corresponding activation energies, as calculated in [10] for aluminum thin films, are reported in Table 5.1. Since the total flux of material is the sum of the various contributions, the effective activation energy, as measured in different electromigration tests, depends on the temperature range, grain size, and extent as well as structure of the defects inside the film.

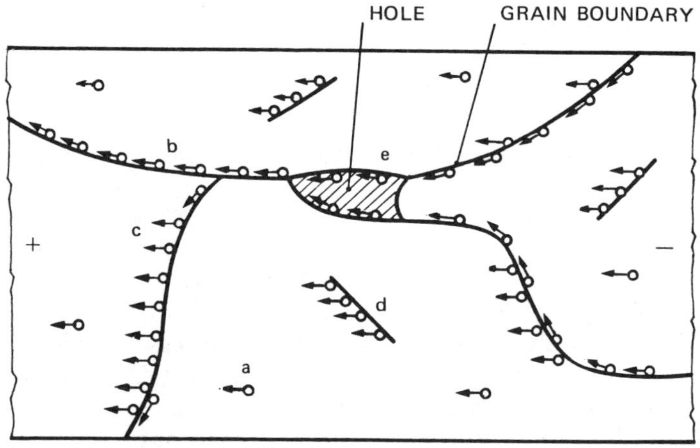

Figure 5.1 Mechanisms of eletromigration in polycrystalline Al (vacancy motion is indicated).

Table 5.1 Activation Energies of Aluminum Thin Films

Mechanism	*Activation energy (eV)*
Bulk EM	1.4
Grain boundary EM	0.4–0.5
Motion from grain boundaries into the bulk	0.62
Motion from defects in the bulk	>0.62
Surface EM	0.28

At this point it should be observed that electromigration alone cannot induce failures in a metal film, unless there is a nonvanishing divergence of the atomic flux. In fact, in the case of div $J_a = 0$, so much material would reach any section of the film, as it is the material that leaves the section itself. If on the contrary, div $J_a \neq 0$, voids occur in the areas where atomic transport results in mass depletion. As these voids grow in size, a crack develops that eventually leads to the rupture of the conductor line (Figure 5.2). Whiskers and hillocks are formed in the areas of mass accumulation (Figure 5.3). This second mode of failure may cause short circuits between adjacent or superimposed conductor lines, as well as breakage through passivating layers, with consequent corrosion problems.

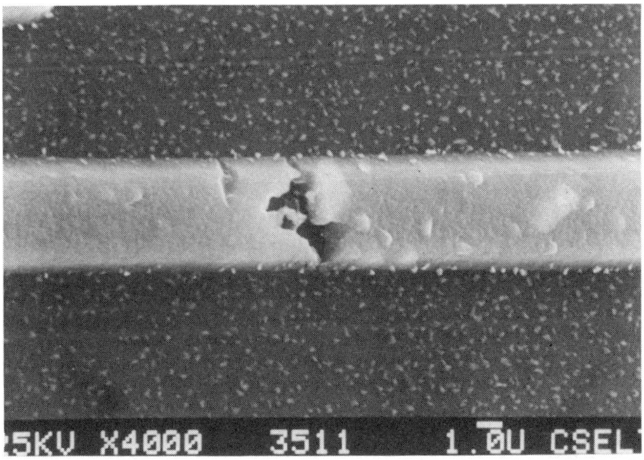

Figure 5.2 Eletromigration induced rupture of an Al interconnection.

By taking into account the expression of the atomic flux, we can see that several factors, like gradients in temperature, effective charge, diffusion coefficient and grain size, can be responsible for flux divergencies. In particular, the non-uniform grain structure is the most important factor for electromigration in thin films.

A well-known structural inhomogeneity can be found at the junction of three grains, the so-called triple point. The effect on electromigration has been studied in [11]. From Figure 5.4, it can be observed that, to avoid mass depletion or accumulation at the center of the triple point, the following equation should hold:

$$\vec{J}_1 = \vec{J}_2 + \vec{J}_3 \tag{5.8}$$

which can also be written as

$$D_1 \cos\Theta_1 = D_2 \cos\Theta_2 + D_3 \cos\Theta_3 \tag{5.9}$$

where D_i and Θ_i are diffusivity and angle between applied electric field and diffusion direction at the ith grain boundary, respectively. In the general case, where grain boundaries have reached an equilibrium configuration ($\Theta_1 = \Theta_2 = \Theta_3 = 120°$) as a consequence of a proper thermal treatment, the extent of the damage is determined by the difference in diffusivities, independent of the orientation of the grain boundary with respect to the electric field.

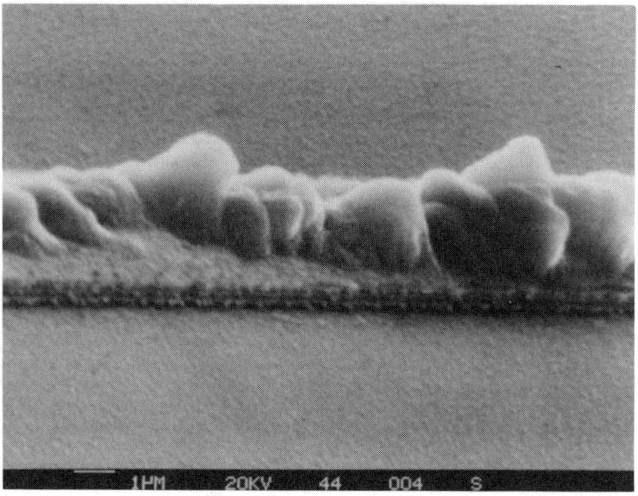

Figure 5.3 Example of extensive eletromigration induced hillock growth.

A few experimental studies, however, have shown that, in grains of highly preferential orientation, this electromigration mechanism becomes negligible [11, 12, 13]. In this case, a second source of damage, related to localized changes in grain size, is predominant. As schematically shown in Figure 5.5, significant mass build-up can occur at the boundary from small to large grains, whereas a corresponding mass depletion proceeds at the boundary from large to small grains.

It should also be observed that local variations in the conductor geometry, often associated with nonoptimized deposition parameters, can be responsible for accumulation or depletion effects, as shown in Figure 5.6.

In conclusion, a uniform grain size distribution is particularly desirable to reduce the electromigration damage. The increase in the average grain size, with respect to the line width, has also been shown to be particularly beneficial in

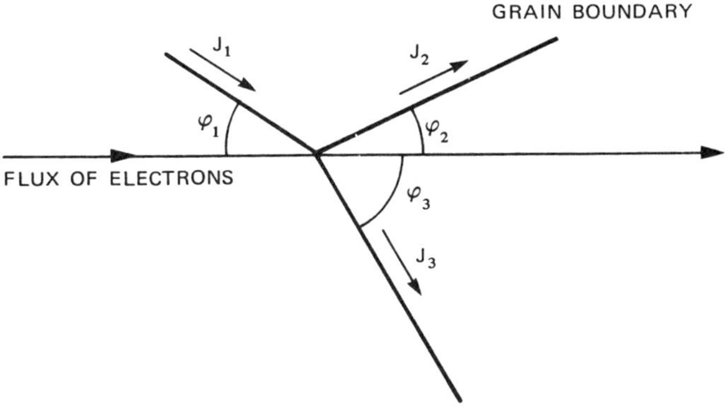

Figure 5.4 Schematic configuration of the atomic flux in a triple point.

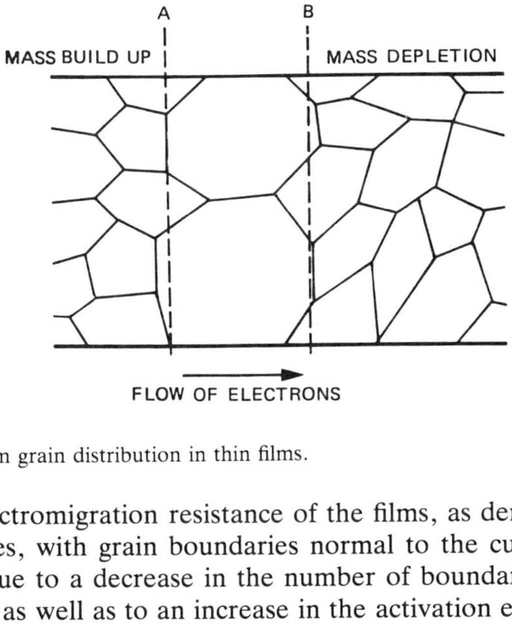

Figure 5.5 Nonuniform grain distribution in thin films.

improving the electromigration resistance of the films, as demonstrated in "bamboolike" structures, with grain boundaries normal to the current flow [14]. The improvement is due to a decrease in the number of boundaries available for the atomic transport, as well as to an increase in the activation energy, related to the larger bulk contribution.

5.4 ELECTRICAL TESTS TO EVALUATE ELECTROMIGRATION FAILURES

Several methods have been reported in the literature to characterize the electromigration resistance of metal thin films. The most used experimental technique is

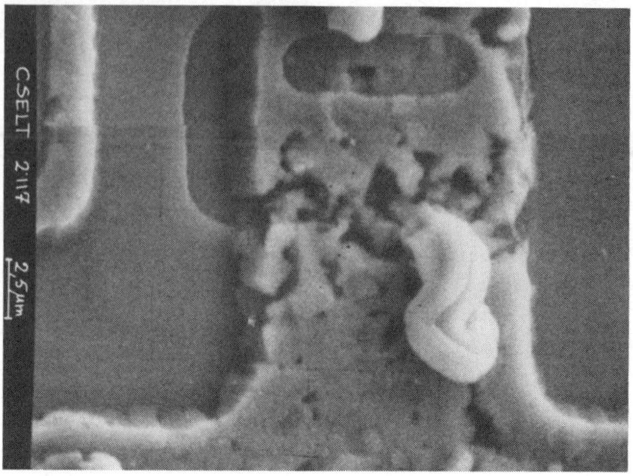

Figure 5.6 Example of failure caused by a local change in metal track geometry.

based on the determination of the average lifetime (*mean time to failure* or MTF). In this measurement the test samples are stressed at temperatures (200–300°C) and current densities ($1\text{–}5 \times 10^6$ A/cm^2), well above the values corresponding to normal operating conditions of the films. The electrical tests are typically performed at constant current density and temperature, on a sufficiently large number of conductors having the same characteristics, to determine the statistical distribution of the times to failure. Results have been also reported, for alternate [15] and pulsed [16] current conditions. Mathematical expressions have then been developed to relate the mean time to failure to current density and temperature. By these relations, values of MTF in any stress (T and J) condition could also be extrapolated.

In the model proposed by Black [17], the following equation is derived:

$$\text{MTF}^{-1} = Aj^n \exp(-Q/kT), \quad \text{with } n = 2 \tag{5.10}$$

where A includes the metal resistivity, the electron free path, a frequency factor for the atomic diffusion mechanism, and a geometrical factor that correlates the atomic transport rate to MTF. It is worthwhile to note the square dependence of MTF on the current density, whereas the diffusion theory predicts that the atomic flux, j_a, is directly proportional to j. In practice, tests performed on Al films at different current densities have shown that n depends on current density, temperature, and film geometry. In particular, values of n in the range 1–7 have been reported in [18–21].

Several models have also been suggested to explain the dependence of n on the current density, on the basis of current density changes and temperature gradients occurring in damaged areas and contributing to the acceleration of the electromigration phenomena [22–24].

By performing accelerated tests at constant current density and different temperatures, an Arrhenius plot of MTF *versus* $1/T$ is obtained, from which the activation energy Q can be determined. This energy does not indicate the activation energy of the diffusion process (E_a) but represents a quantitative measure of the cumulative effects on electromigration, related to structural dishomogeneities, average diffusion along grain boundaries, temperature gradients, substrate characteristics, and passivation layers.

On the basis of the previous considerations, it is evident that the extrapolation of the times to failure to real operating conditions of the films, in particular to lower current densities, can lead to appreciable errors [25]. In fact, by assuming $n \geq 2$, very optimistic evaluations are derived from (5.10). On the contrary, the use of $n = 1$ results in very conservative estimates of MTF.

The determination of MTF typically requires test times of the order of 10^2–10^4 hours. It is therefore particularly useful to be able to characterize the film electromigration resistance, by applying short stress tests, able to give information in times of the order of a few hours or even a few minutes. Two techniques of this type (the so-called TRACE and BEM) have been recently introduced [4, 5] and applied to the characterization of VLSI thin film metalizations [4, 5, 26–29].

The *temperature-ramp resistance analysis to characterize electromigration* (TRACE) has been used to study electromigration behavior in Al-Si, Al-Si-Cu, and multilayer Al-Si/Ti (or Ta) structures. In this method, the resistance change in a stripe under high current stress is monitored as a function of a ramping temperature. The measured conductor resistance is then separated in two components, R_T and R_{EM}, representing the temperature and the electromigration component, respectively. It has been shown that the data relative to the initial 5–10 percent resistance change can fit a thermally activated process described as

$$(1/R_0)\, dR_{EM}/dt = A \exp(-E_a/kT) \qquad (5.11)$$

where R_0 is the conductor resistance at the initial ramping temperature, and the preexponential A contains the dependence on the current density. Figure 5.7 shows, as an example, the fitting of the experimental data in Al-Si conductors [27]. The determined value of activation energy is typical for grain boundary diffusion in Al thin films.

Although there is interest for these new techniques, extensive work is still required to analyze more accurately the capabilities of both TRACE and BEM to provide reliable predictions of electromigration induced failure rates.

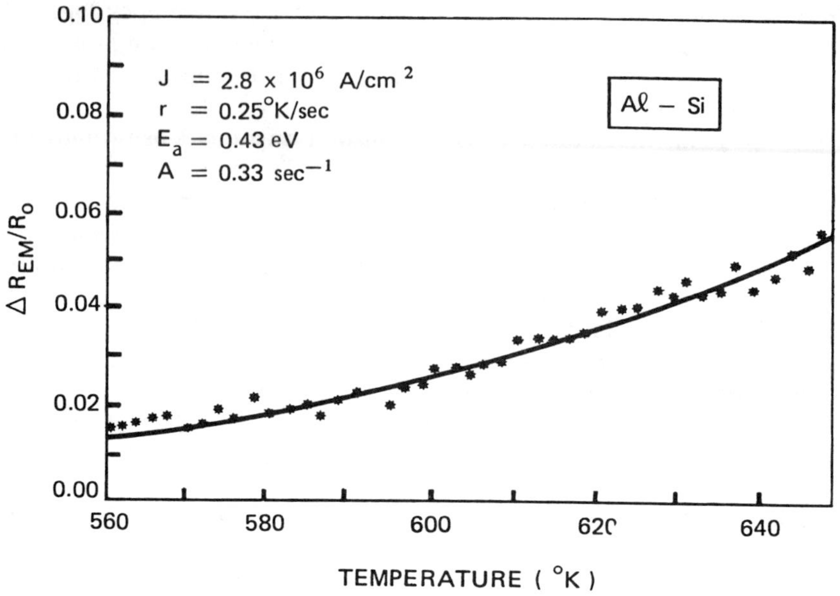

Figure 5.7 Application of TRACE to Al-Si films.

5.5 ELECTROMIGRATION PHENOMENA IN I.C. THIN FILMS

5.5.1 Aluminum

Aluminum is the material most widely used for metalization in integrated circuits because of its high thermal and electrical conductivity, good adhesion to the silicon oxide, easiness of deposition and patterning, good resistance to corrosion, and low cost.

Electromigration induced mass transport in Al films has been reported by several authors. A description of the most popular methods of investigation, together with the resulting measured values of activation energy for Al electromigration, has been presented in the literature [9], with E_a ranging from 0.4 to 1.3 eV, depending on technique of investigation and test conditions.

The methods typically used in these conductors to increase the electromigration resistance are based on

- The optimization of the grain structure, as previously described;
- The use of dielectric passivating layers, able to inhibit the growth of hillocks and, consequently, to retard the formation of voids [30–32];

- The deposition of Al films alloyed with small percentages of other metals or alternated with refractory metal films in sandwiched structures, both of which will be described in the following sections.

5.5.2 Al-Cu Alloys

Tests of electromigration resistance have shown that the addition of a second element to the conductor film significantly increases the MTF. In particular, the system Al-Cu has received a large interest from a technological point of view.

The study of the relation between Cu concentration and life time indicates that, at low Cu concentrations, the time to failure increases monotonously as the Cu content increases, up to a concentration, on the order of 3–4 percent by weight. At higher concentrations, the relation between Cu content and MTF is less obvious and depends on the film preparation [33]. A successive work [34] shows that the increase in MTF (~70) is directly related to the reduction of the atomic flux of Al (~80), mainly caused by the absorption of Cu atoms at the conductor grain boundaries. By assuming a reasonable value of binding energy for a solute atom at a grain boundary, it is possible to show that the number of defects available for diffusion can be effectively reduced (by more than 1 order of magnitude, for example, for an energy of 0.2 eV and at 120°C, at the presence of about 1 percent (at) of solute [12]) with a consequent reduction of the atomic flux. It is also clear that, since the Al diffusion occurs mainly at grain boundaries, if the positions available for diffusion are occupied by Cu atoms, an additional energy must be supplied for Al diffusion to occur, thus increasing the activation energy for electromigration.

The experimental results indicate that the Cu atoms, with respect to the 1 percent (at) required to saturate the grain boundaries, are used as a reservoir to replace the solute atoms diffusing in the direction of the electronic current, retarding in this way the Al electromigration process. The failure has then been observed mainly in the regions of Cu depletion.

5.5.3 Multilayer Structures

The electromigration resistance of Al, Al-Si, and Al-Cu conductors can be significantly improved by using multilayer structures, where the Al films are alternated with transition metal films [35–38]. From the comparison of MTF in homogeneous Al-Cu structures and in different types of multilayer structures (Table 5.2), we can note that the increase in MTF can be significant. However, the improvement in performance has been observed only in the films where, following a convenient

Table 5.2 Comparison among MTFs for Al, Al-Cu Alloys and Multilayer Structures [from 35]
Test Conditions 250°C, 1×10^6 Å/cm^2

	Linewidth (μm)	*Median failure (h)*	*Standard deviation* σ^a
Al	1.7	33	0.8
Al	1.5	9	0.97
Al–4%Cu	1.5	100	0.63
Al–4%Cu	1.5	170	0.67
Al–6%Cu	1.5	150	0.55
Al–4%Cu(Si)	2.2	120	0.55
Al–6%Cu(Si)	1.5	200	0.55
Al–4%Cu $CrAl_7$–Cr Al–4%Cu	1.9	11,000	0.8
Al–4%Cu $TiAl_3$–Ti Al–4%Cu	1.9	5,500	0.7
Al–4%Cu $TaAl_3$–Ta Al–4%Cu	1.7	4,500	0.6
Al–4%Cu $HfAl_3$–Hf Al–4%Cu	1.5	3,700	0.3

thermal treatment, an intermetallic compound could be formed (for example, $TiAl_3$ in the case of Al and Ti films or $CrAl_7$ for Al and Cr). The presence of this compound, which is an electrically continuous layer, improves the electromigration resistance, by blocking the propagation of the voids across the film, as shown in Figure 5.8. As discussed in [35], this improvement may be due to a decrease of effective charge or diffusivity in the intermetallic phase. Whereas measurements

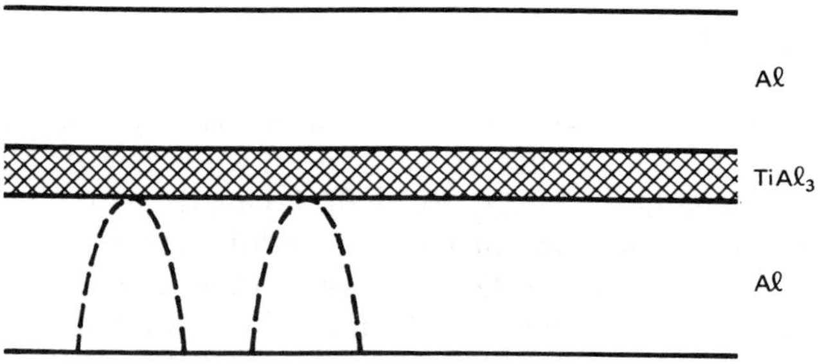

Figure 5.8 Effect of a $TiAl_3$ film in blocking voids propagation.

of effective charge in intermetallic compounds are not presently available, calculations based on the few known values of diffusivity indicate, at least for the intermetallic $CrAl_7$, a ratio of about 10^9 between the atomic diffusivities in the homogeneous films and in the intermetallic phase [35].

A secondary effect of the presence of the interfacial compound is the reduction in the number of damage sites in the Al films on top of the intermetallic phase, which is associated with the increase in the grain preferential orientation for similar films deposited over SiO_2.

In the case of transition metals like Zr and Hf, the resulting compounds ($ZrAl_3$ and $HfAl_3$) grow preferentially along the Al grain boundaries. The corresponding times to failure are shorter than the ones measured in planar compounds, like $TiAl_3$ and $CrAl_7$ (see Fig. 5.9). It is probable that when a compound is formed preferentially along grain boundaries, the resultant barrier is thinner or discontinuous and, therefore, of lower effectiveness than a planar compound.

Figure 5.9 TEM micrograph showing the planarity of a $TiAl_3$ compound, resulting from a thermally treated Al/Ti multilayer metallization [from 38].

5.6 CONTACT ELECTROMIGRATION

Contact electromigration has been observed in aluminum films deposited as contacts on shallow junctions in silicon [39–41]. Due to the momentum exchange between Si atoms and conduction electrons, in fact, Si atoms diffuse out of the contact region into the Al film, causing the formation of contact pits. As a consequence, Al spikes are then formed in the underlying junction. As the amount of mobile silicon increases, the spikes extend more deeply into the junction, which eventually becomes short-circuited. For very shallow junctions, as the ones required in VLSI, this type of failure is becoming more probable and should be taken into account in the design of the contact structures.

The accelerated tests of MTF to evaluate contact electromigration effects are generally performed at constant current density and temperature. The failure condition typically corresponds to a junction leakage current above the arbitrary value of 100 μA at 10 V. The mean time to failure can then be determined from the log normal distribution of the times to failure.

In Al-Si contacts the activation energy of the atomic flux has been found in the range 0.9–1.1 eV, consistent with the activation energy for silicon diffusion in the Al grain boundaries. As a consequence, the values of MTF are expected to depend on the metal microstructure; that is, on film thickness, composition, deposition technique, and successive thermal treatments.

Figure 5.10 shows schematically an Al-1%Si contact damaged by the electromigration. The electrons move from the left, crossing the contact region and flowing in the metal overlayer. It is important to note that, as a consequence of the current distribution underneath the contact, Si is removed only in the contact region; namely, close to the edge, where the current density is higher. The silicon atoms diffusing along the conductor then have been observed to pile up at the opposite end of the stripe [39]. This gives rise to a second mode of failure, in the dramatic increase of contact resistance.

The application of a diffusion barrier, like W or TiN, has been shown to be particularly beneficial in suppressing the leakage failure as well as any dependence on junction depth variations [39, 41].

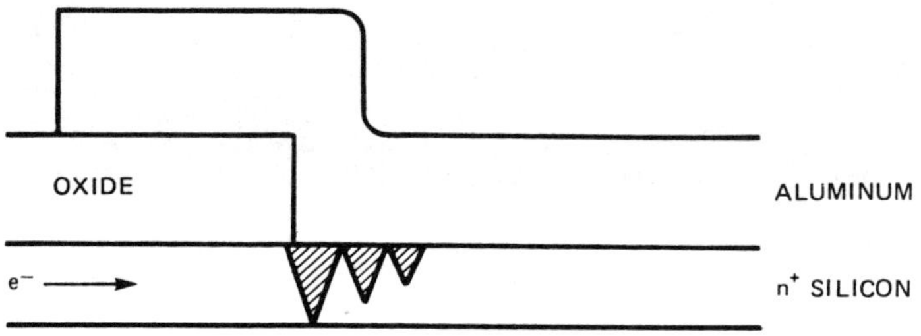

Figure 5.10 Schematic cross section of contact electromigration.

REFERENCES

1. M. Gerardin, "De l'action des piles sur les sels de potasse et de soude et sur les alliages soumis a la fusion igneé," *C.R. Acad. Sci.*, 53, 1861, p. 727.
2. F. Skampy, "Die elektrizitätsleitung in Metallen," *Verhandl Deut. Phys. Ges.*, 16, 1914, p. 156.
3. W. Seitz and H. Wever, "Uber einen neunen Effekt bei der elektrolytischen Überführung in Festen Legierungen," *Z. Elektrochem.*, 59, 1953, p. 942.
4. R.W. Pasco and J.A. Schwartz, "Temperature-Ramp Resistance Analysis to Characterize Electromigration," *Solid-State Electronics* 26, 1983, p. 445.

5. C.C. Hong and D.L. Crook, "Breakdown Energy of Metal (BEM)—A New Technique for Monitoring Electromigration Reliability at Wafer Level," *Proc. 23rd Int. Symp. on Reliability Physics,* 1985, p. 108.
6. H.B. Huntington and A.R. Grone, "Current Induced Marker Motion in Gold Wires," *J. Phys. Chem. Solids,* 20, 1961, p. 76.
7. C. Bosvieux and J. Friedel, "Sur l'ectrolyse des alliages metalliques," *J. Phys. Chem. Solides,* 23, 1962, p. 123.
8. G.M. Newmann and W. Hirschwald, "Electrotransport in Tungsten," *A. Naturforsch. A.,* 22, 1967, p. 388.
9. F. d'Heurle and P.S. Ho, "Electromigration in Thin Films," in *Thin Films—Interdiffusion and Reactions,* Wiley-Interscience, New York, 1978, p. 243.
10. H.U. Schreiber, "Activation Energies for the Different Electromigration Mechanisms in Aluminum," *Solid-State Electronics,* 24, 1981, p. 583.
11. M.J. Attardo and R. Rosenberg, "Electromigration Damage in Aluminum Film Conductors," *J. Appl. Phys.,* 41, 1970, p. 2381.
12. F. d'Heurle and R. Rosenberg, "Electromigration in Thin Films," in *Physics of Thin Films,"* vol. 7, Academic Press, New York, 1973 p. 257.
13. R. Rosenberg and M. Ohring, "Void Formation and Growth during Electromigration in Thin Films," *J. Appl. Phys.,* 42, 1971 p. 5671.
14. A. Gangulee and F.M. d'Heurle, "Anomalous Large Grains in Alloyed Al Thin Films. II. Electromigration and Diffusion in Thin Films with Very Large Grains," *Thin Solid Films,* 16, 1973, p. 227.
15. A. Bobbio, A. Ferro, and O. Saracco, "Electromigration Failure in Al Thin Films under Constant and Reverse DC Powering," *IEEE Trans. Reliablity,* R-23, 1974, p. 194.
16. J.M. Towner and E.P. Van de Ven, "Aluminum Electromigration under Pulsed D.C. Conditions," *Proc. 21st Int. Symp. on Reliability Physics,* 1983, p. 36.
17. J.R. Black, "Electromigration—A Brief Survey and Some Recent Results," *IEEE Trans. Electron Devices,* ED-16, 1969, p. 338; and "Electromigration Failure Modes in Aluminum Metallization for Semiconductor Devices," *Proc. IEEE,* 57, 1969, p. 1587.
18. D.S. Chabra and N.G. Ainslie, "Open Circuit Failures in Thin Film Conductors," IBM Tech. Rep. 22, 1967, p. 419.
19. J.C. Blair, P.B. Ghate, and C.T. Haywood, "Electromigration Induced Failures in Aluminum Film Conductors," *Appl. Phys. Lett.,* 17, 1970, p. 281.
20. M.J. Attardo, *op. cit.* [17].
21. K.A. Danso and T. Tullos, "Thin Film Metallization Studies and Device Lifetime Production Using Al-Si, Al-Cu-Si Conductor Test Bars," *Microelectronics Reliability,* 21, 1981, p. 513.
22. J.D. Venables and R.G. Lye, "A Statistical Model for Electromigration Induced Failures in Thin Film Conductors," *Proc. 10th Int. Symp. on Reliability Physics,* 1972, p. 159.
23. R.A. Sigsbee, "Electromigration and Metallization Lifetimes," *J. Appl. Phys.,* 44, 1973, p. 2533.
24. K. Nikawa, "Montecarlo Calculations Based on the Generalized Electromigration Failure Model," *Proc. 19th Int. Symp. on Reliability Physics,* 1981, p. 175.
25. P.B. Ghate, "Electromigration Induced Failures in VLSI Interconnects," *Solid-State Technology,* March 1983, p. 113.
26. J.A. Schwarz, "Effect of Temperature on the Variance of the Log-Normal Distribution of Failure Times due to Electromigration Damage," *J. Appl. Phys.,* 61 1987, p. 801.
27. M. Finetti, H. Ronkainen, M. Blomberg, and I. Suni, "Electromigration Resistance of Multilayer Aluminum/Titanium Interconnects," *Proc. MRS Meeting,* Boston, 1985, p. 812.
28. M. Finetti, I. Suni, G. De Santi, L. Bacci, and C. Caprile, "Comparison of Electromigration Induced Resistance Changes in Multilayer Al-Si/Ti and Al-Si/Ta Interconnects," *Proc. MRS Meeting,* Palo Alto, Calif., 1986, p. 297.

29. G. De Santi, "Valutazione di misure elettriche automatiche per la caratterizzazione di film metallici," *Rapporto Tecnico SGS,* N.01/86.
30. F. d'Heurle, A. Gangulee, and V. Ranieri, unpublished manuscript (see also [6]).
31. S.M. Spitzer and S. Schwartz, "The Effect of Dielectric Overcoating on Electromigration in Al Interconnection," *IEEE Trans. Electron. Device,* ED-16, 1969, p. 348.
32. N.G. Ainslie, F. d'Heurle, and O.C. Wells, "Coating, Mechanical Constraints and Pressure Effects on Electromigration," *Apply. Phys. Lett.,* 20, 1972, p. 173.
33. F. d'Heurle, "The Effect of Copper Addition on Electromigration in Aluminum Thin Films," *Met. Trans.,* 2, 1970, p. 681.
34. F. d'Heurle and A. Gangulee, "Solute Effects on Grain Boundary Eletromigration and Diffusion," in *Nature and Behaviour of Grain Boundaries,* 1972, p. 339.
35. J.K. Howard, J.F. White, and P.S. Ho, "Intermetallic Compounds of Al and Transition Metals: Effects of Eletromigration in 1–2 μm Wide Lines," *J. Appl. Phys.,* 49, 1978, p. 4083.
36. S.S. Iyer and C.Y. Ting, "Electromigration Study of the Al-Cu/Ti. Al-Cu System," *Proc. 22nd Int. Symp. on Reliability Physics,* Las Vegas, 1984, p. 273.
37. D.S. Gardner, T.L. Michalka, K.C. Saraswat, T.W. Barbee, J.P. McVitte, and J.D. Meindl, "Layered and Homogeneous Films of Aluminum and Aluminum/Silicon with Titanium and Tungsten for Multilayered Interconnects," *IEEE J. Solid State Circuits,* SC-20, 1985, p. 94.
38. M. Finetti, I. Suni, A. Armigliato, A. Garulli, and A. Scorzoni, "Correlation between Resistance Behaviour and Mass Transport in Al-Si/Ti Multilayer Interconnects," *J. Vacuum Science Tech.,* October 1987.
39. P.A. Gargini, C. Tsen, and M.H. Woode, "Elimination of Silicon Electromigration in Contacts by the Use of an Interposed Barrier Metal," *Proc. 20th Symp. on Reliability Physics,* 1982, p. 6.
40. J.J. Chern, W.G. Oldham, and N. Cheung, "Contact Eletromigration Induced Leakage Failure in Aluminum Silicon to Silicon Contacts," *IEEE Trans. Electronic Devices,* ED-32, p. 1341.
41. S. Vaidya, R.J. Schultz, and A.K. Sinha, "Shallow Junction Cobalt Silicide Contacts with Enhanced Eletromigration Resistance," *J. Appl. Phys.,* 55, 1984, p. 3514.

Chapter 6
METAL-SILICON INTERACTIONS
G. Ottaviani

6.1 INTRODUCTION

The diffusion of electronic circuits is mainly due to the availability of devices of increasing complexity and number of functions at decreasing price. An idea of the complexity of modern electronic devices can be obtained from Figure 6.1, which shows schematically a section of a modern electronic circuit containing about 50,000–100,000 active elements in an area of 0.5–0.8 cm^2, dissipating about 0.5–1.0 W and using 100–200 pins for connections. The interconnections between the active elements is obtained by a three-level metalization scheme.

Many factors form the basis of this technological achievement, the major improvements being the following: (1) increase of the active area of the devices, (2) innovative circuit design, and (3) shrinking of the physical dimensions of the elementary cells. Further significant developments in the first two factors are not expected. Much more promising is an improvement in the third one (i.e., miniaturization). An idea of the development in this area can be gained by considering the shrinking of the gate electrode in MOS devices. From 10 μm in 1971, it reached 2 μm after 10 years, and at present MOS devices are available with a 1 μm gate. In the research laboratories devices are in development with .25 μm gates.

The increase in complexity or performance, the shrinkage in dimensions, and the different technologies and materials used by manufacturers to build electronic circuits have pointed out the necessity for a better understanding of the factors affecting the reliability of whole devices.

Increasing requirements in both performance and reliability necessitate a better control of the construction technology and a deeper knowledge of the basic physical processes [1]. Epitaxy, ion implantation, oxidation, masking, etching, and metalization are only a few of the technologies involved in realizing an electronic circuit.

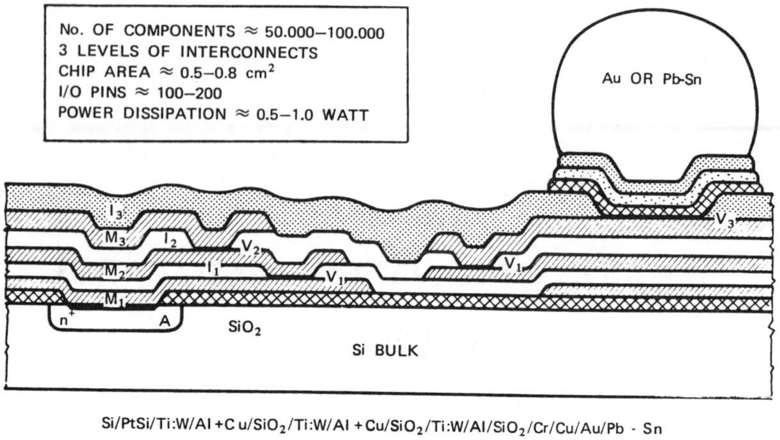

Figure 6.1 Cross section of an electronic device with three levels of interconnections. *M* represents metal layers; *I*, insulator; *V*, the routes between two metal levels.

In integrated circuits a design constraint that lies outside the semiconductor itself is the presence of metalization layers that provide a contact between the semiconductor and the external world. Processing techniques including diffusion, ion implantation, oxidation, and gettering are sufficiently well developed not to impose limitations in device degradation or performances. As devices are reduced to micron sizes, however, metalization becomes a major factor. The importance of this processing step can be understood by considering the case of aluminum. In the early days of circuit development, the contact regions were so thick that the aluminum used for metalization did not reach the junctions responsible for the electrical properties of the device. The annealing temperatures were high enough to melt the contacts. The necessity of increased performance led to a notable reduction of the physical dimensions and, consequently, liquid phases could no longer be permitted. This problem was easily solved by decreasing the alloying temperature. Further demands in circuit performance required a shallower junction depth, and for reasons that will be described in detail later, it became necessary to use aluminum containing a few atomic percent of silicon. In some actual devices, aluminum is approaching a new limitation, and it is necessary to place a diffusion barrier between silicon and aluminum to inhibit the reaction. Even this solution is not completely satisfactory, and new solutions must be devised for future devices. There is no doubt that any future remedy will be based on in-depth knowledge of the physical phenomena involved in the processes.

The purpose of this chapter is to discuss the principal physical aspects at the basis of the metalization process, taking into account both the techniques for film deposition and the requirements for better reliability. After a discussion of the peculiarities of the phase diagrams (which represent the condition for equilibrium and of the kinetic properties necessary to reach such equilibrium), we explore metal-semiconductor, and metal-metal contacts and problems involved in interconnections. Finally, some future developments in the metalization processes in integrated circuits will be presented.

6.2 PROPERTIES OF MATTER

Two or more metal films in contact with each other or in contact with a semiconductor material undergo over time, even at modest temperatures, changes in structure or composition. The prediction of such modifications is extremely difficult. However, a few simple criteria can be used as guidelines for determining the evolution of the structure.

It is worth noting that metal film structures generally used in microelectronics are not in thermodynamic equilibrium; the necessity to reach equilibrium is the driving force that often produces undesirable modifications. Equilibrium conditions, represented by the phase diagram, and kinetic properties of the system must both be considered.

6.2.1 Equilibrium Phase Diagram

The purpose of this section is to indicate in the phase diagrams of two elements the properties most relevant to the discussion of metal-semiconductor reactions. For a deeper analysis of phase diagrams, we suggest that you refer to more specialized texts.

Equilibrium phase diagrams are graphs that show which phases are present in a given system when it rests at equilibrium with its surroundings [2]. Temperature and composition are generally the two variables taken into consideration. Most of the data are taken at atmospheric pressure. Figure 6.2 shows an example of a phase diagram for isomorphous materials, here Ni and Pd, where the two elements can mix in any proportion. At high temperatures, above the line marked *liquidus*, only the liquid phase exists; at low temperatures, below the line labeled *solidus*, only the solid phase is present. In the region between the two lines the two phases coexist. The points where the two lines coincide indicates the conditions where solidification occurs with no change in composition. In order to have a high solubility, it is necessary that the crystal structure, the atomic volume, and the electronegativity are very similar between the two elements. A phase diagram with a eutectic point can be obtained if the two elements have chemical-physical properties slightly different and low solid solubility. As an example Figure 6.3 shows the Al-Si phase diagram. The eutectic point is characterized by a well-defined temperature

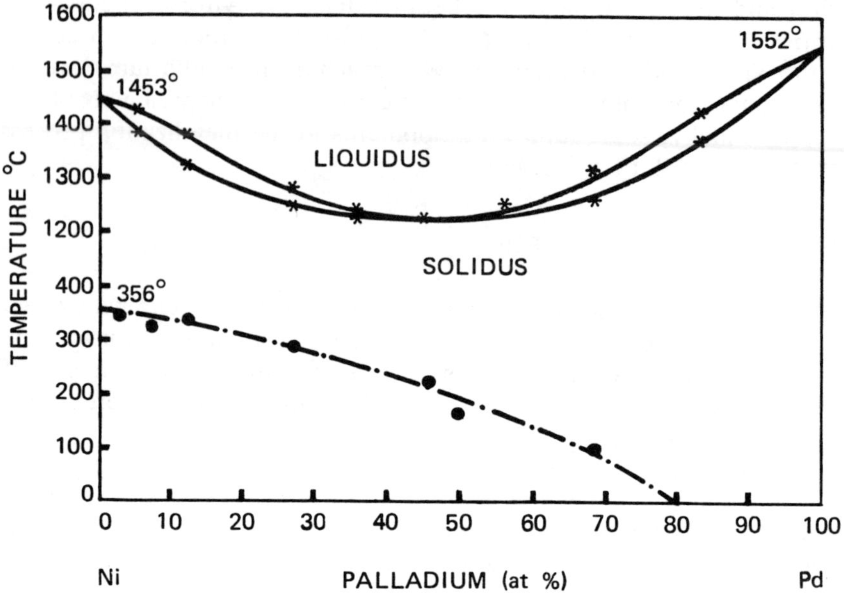

Figure 6.2 Nickel-palladium phase diagram.

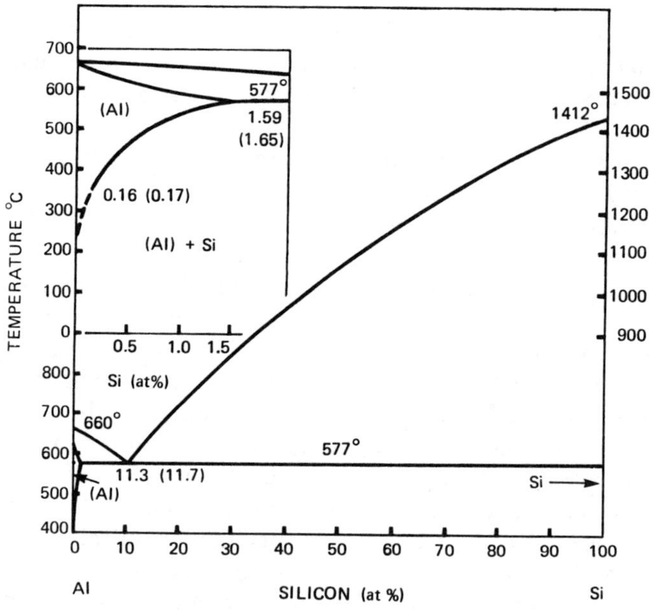

Figure 6.3 Silicon-aluminum phase diagram.

and composition. Such systems may or may not have some solid solubility; in the Si-Al system some small solid solubility of silicon in aluminum is evident (see the insert in Figure 6.3). Either the eutectic temperature or the shape of the solid solubility curve is a function of the hydrostatic pressure. Still referring to the Al-Si system, a hydrostatic pressure of 50 kBar produces an increase in the eutectic temperature of 100°C and a variation of the eutectic composition of 100 percent. The solid solubility of Si in Al changes from 1 atomic percent at atmospheric pressure to 7 atomic percent at 50 kBar.

When the two starting elements are very dissimilar, compounds can form with well-defined stoichiometries and crystal structures. The compounds act as extreme phases in a system with a eutectic point. A compound phase (see Figure 6.4) may exist at a specific composition, as in the cases of Pt-Si and Al-Al_2Au, or may have a large range of existence, as for PdAl-Pd_2Al and Pd_2Al-Pd. The compounds that form at the maximum of the solidus-liquidus separation curve, as Pt_2Si, PtSi, PdAl, $AuAl_2$, Au_2Al, *et cetera*, are congruently melting, since the compound present in the liquid phase remains during solidification. In the peritectic compounds, those not found at a maximum in the curve of the phase diagram, solidification occurs with a phase change, as for $PdAl_3$ or Au_4Al. The evolution of a structure from one configuration to another is controlled by the initial composition and by the stability of the various compounds.

The stability of the congruently melting compounds can be inferred by analyzing the phase diagram. The higher is the melting temperature, the more negative is the free energy of formation, and consequently the compound is the more stable. For instance, PdAl, having a melting temperature of 1645°C, is more stable than $AuAl_2$, which melts at 1060°C. The peritectic compounds are generally less stable than those that melt congruently.

6.2.2 Diffusion

Diffusion is the tendency of two materials to intermix and react chemically. In thin films the presence of defects, such as grain boundaries and dislocations, provide paths for efficient and fast transport of material [3–5]. Moreover, defects are located near surfaces and interfaces, leading to complex transport phenomena. A schematic view of the various processes occurring in thin films is displayed in Figure 6.5. The example shown is made up of three polycrystalline layers with different grain sizes. Element B interacts with element C, and a hypothetical B_xC compound is formed. Interdiffusion occurs between elements A and B, and the lengths of arrows J_A and J_B are proportional to the flux of the two materials. Any divergence of the flux results in preferential transport, producing a decrease of matter, mostly near the defects, and leading eventually to the formation of holes (Kirkendall effect). Diffusion may also result in increased porosity and may create stress.

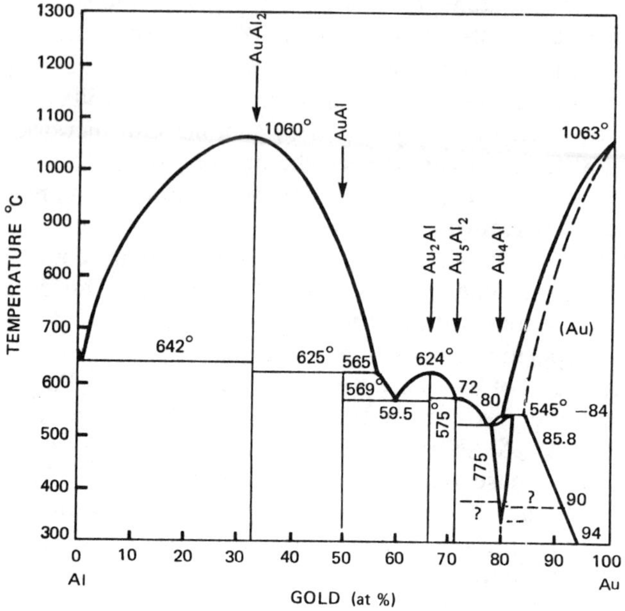

Figure 6.4a

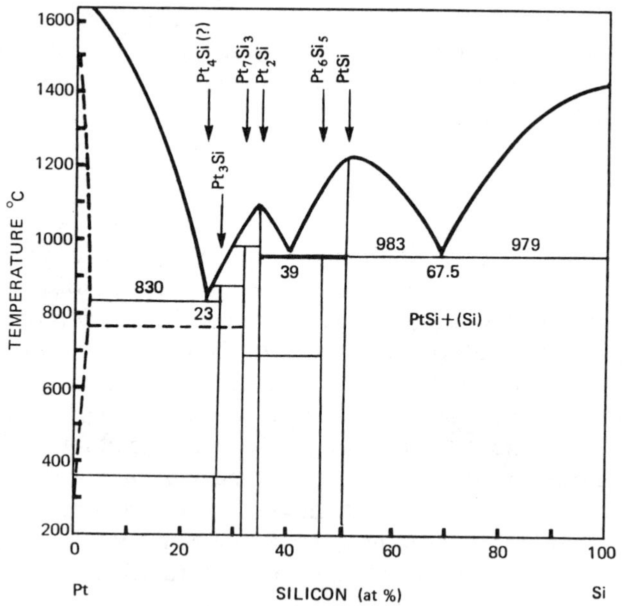

Figure 6.4b

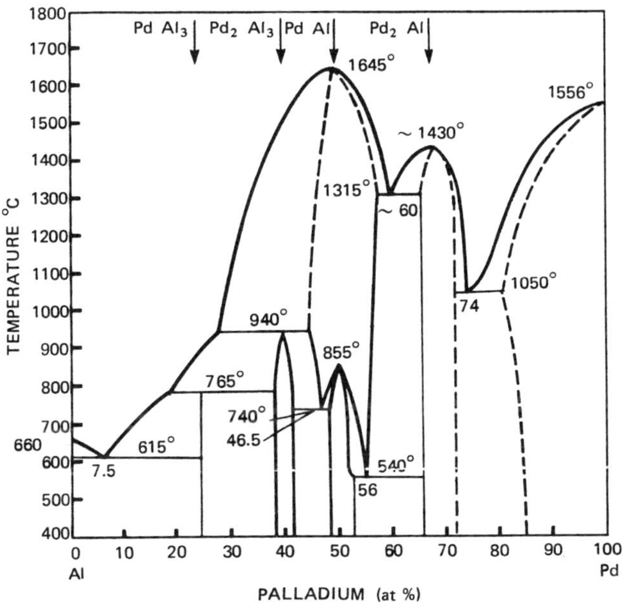

Figure 6.4c

Figure 6.4 Phase diagrams of the various alloy systems: (a) silicon-platinum; (b) gold-aluminum; (c) palladium-aluminum.

It is useful to analyze the phenomena correlated to diffusion in thin films considering the two basic parameters of mass transport: *diffusivity* and *driving force*. These factors are not only intrinsic properties of the materials but, in the case of thin films, depend on the structure, too. Because of this, thin film diffusion is considered separately from diffusion of material in bulk quantities. The driving force arises from free energy gradients, depending in turn on thermal, chemical, electrical, mechanical, *et cetera* potentials. Many combinations of potential gradients and structural defects are possible, making for a very complex situation in thin films. Generally, the easiest approach is to analyze each aspect singularly; consequently, in the following discussion, interactions between diffusion and driving forces will not be considered. In particular, structural aspects of diffusion will be examined with particular emphasis on grain boundaries.

The diffusivity, D, of many materials is found to obey the equation

$$D = D^{\circ} \exp(-Q/RT)$$

where D° is a temperature independent coefficient, Q is the activation energy, R is the gas constant, and T the absolute temperature. To distinguish among the

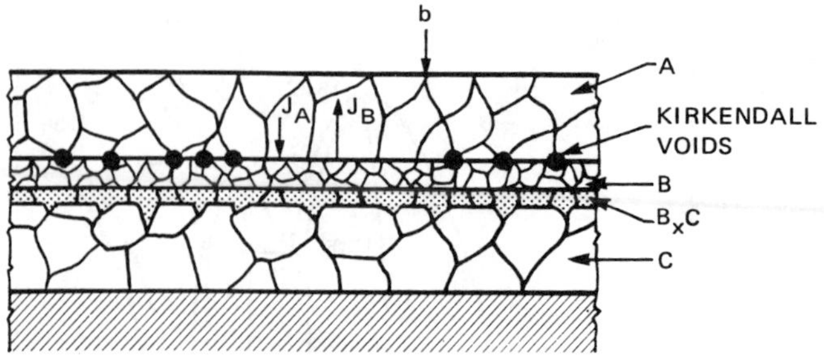

Figure 6.5 Interdiffusion and compound formation in thin films. The *A-B* couple interdiffuses, leaving holes due to the Kirkendall effect, since the flux of *B* in *A*, J_B, is higher than J_A. *B* and *C* interact, forming a hypothetical B_xC compound.

various mechanisms of diffusion the following notation will be used: the subscript *l* will identify lattice diffusion, *s* surface diffusion, *b* grain boundary diffusion, and *d* will indicate dislocation line diffusion.

The kinetics of the various mechanisms have been amply investigated, and results indicate that the activation energy is related closely to the melting temperature, T_m, of the host medium. The ratio T/T_m can be considered a useful parameter to characterize the diffusion process. The various mechanisms have different diffusion coefficients; the highest value is generally that for surface diffusion, and the lowest is the coefficient for bulk diffusion. The activation energies for surface, grain boundary, dislocation, and bulk diffusions are $13T_m$, $17T_m$, $25T_m$, and $34T_m$, respectively, with a preexponential term of 0.014, 0.3, 2.1, and 0.5 cm^2/s for each different mechanism. According to these figures, grain boundaries and dislocations contribute meaningfully to the diffusion process at temperatures lower than $0.5T_m$. More than one mechanism may be active at the same time, and generally they interact mutually, resulting in a very complicated process. In the approximation of linear combinations of the effects, the relationship relating the fluxes of matter due to the various mechanisms can be evaluated by assuming a film with cubic grains, distance δ between two adjacent grains, and a uniform density, p_d, of dislocations having A_d as active area. In the case of uniform and constant concentration of defects under steady state conditions, the ratios between the diffusion fluxes are given by

$$J_b/J_l = \delta D_b/LD_l$$

$$J_d/J_l = p_d A_d \, D_d/D_l$$

The combination of a slow diffusion mechanism, such as bulk, and a fast one, such as grain boundary or dislocation diffusion, produces various material profiles, depending on the kinetics involved. Three types of profiles, representing kinetic types A, B, and C, are drawn schematically in Figure 6.6. In case A, Figure 6.6a, bulk diffusion is comparable to that at the grain boundaries. The two mechanisms interact and a considerable quantity of matter diffuses in the grain, giving an almost flat impurity profile. In this case, the presence of grain boundary diffusion gives a profile substantially similar to pure bulk, only much deeper. In case B, Figure 6.6b, bulk diffusion is small compared to grain boundary diffusion. Considering the grains noninteracting, the diffusion occurs mainly along the grain boundary with a small amount of penetration into the grain. In the third case, Figure 6.6c, the bulk diffusion is negligible compared to that at the grain boundaries. Thus, most of the transport occurs along the grain boundary. In polycrystalline materials, as thin films, the most common kinetic situation is type B.

It is particularly important to predict, on the basis of the structural and morphological properties of the film and of the temperature involved, which of the various diffusion mechanisms is the most active. The data reported in Figure 6.7 may be used to reach a reasonable answer to this query. Each figure is divided into three regions in which diffusion prevails in the bulk (indicated by *l*), in dislocations (*d*), or at grain boundaries (*b*). The horizontal axis indicates the dislocation density, and the vertical axis shows the inverse of the linear grain dimensions. The ratio T/T_m is used as a parameter. The rectangular region in each diagram, given by broken lines, indicates the region of interest for thin films. It is clear that at low temperatures, $T/T_m \leqslant 0.3$ and grain boundary diffusion prevails, whereas at high temperatures, $T/T_m \geqslant 0.6$ and lattice diffusion is the most active.

6.3 METAL-SEMICONDUCTOR CONTACT

The electrical behavior of a metal-semiconductor contact is determined by the electronic properties of the metal, the doping at the surface of the semiconductor, and the contamination at the interface [6]. Generally, the contact is rectifying; the ohmic behavior is approached by heavily doping the surface of the semiconductor, producing current transport by a tunnel effect. An ohmic contact is characterized by a linear current-voltage dependence and a specific contact resistance so low that it does not affect the performance of the device. The specific contact resistance, defined as the reciprocal of the slope of the tangent to the curve of current density versus voltage, is expressed in Ω per cm^2.

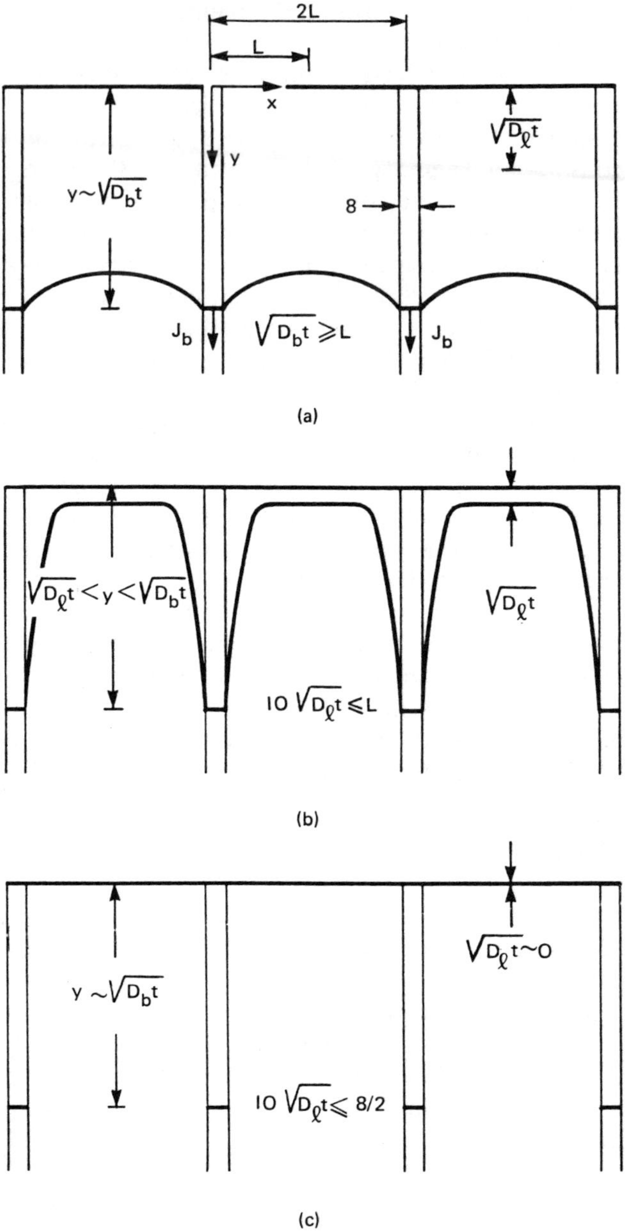

Figure 6.6 Schematic representation of the diffusion profiles obtained with three different cases of kinetics. The vertical lines indicate the grain boundaries and the curves represent regions of isoconcentration. The diffusion starts from the top straight line.

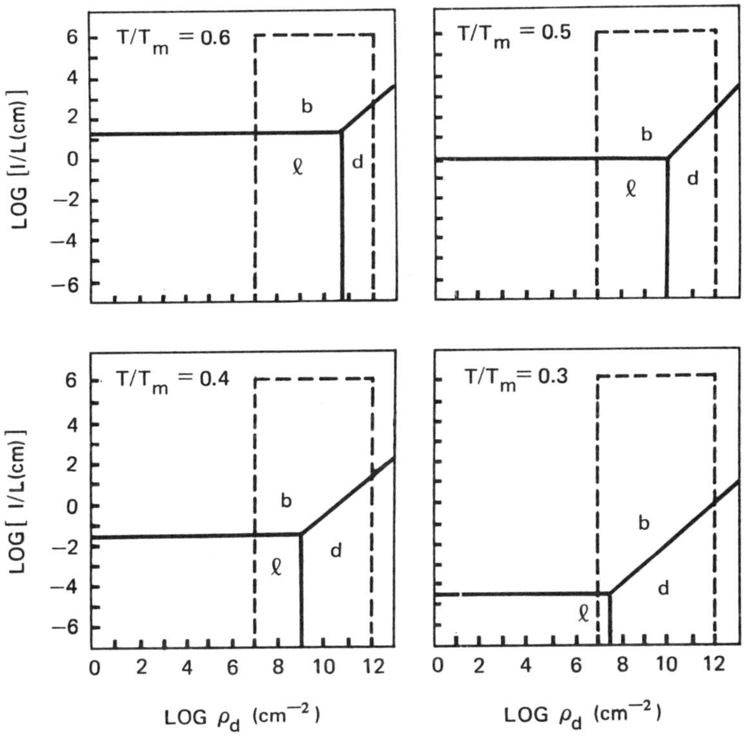

Figure 6.7 Schematic representation of the different regions where diffusion is dominant in the lattice, at the grain boundaries, or in dislocations. The parameter is the ratio T/T_m, where T_m is the melting temperature.

6.3.1 Si-Al Contacts

Aluminum is the most commonly used metal in electronic devices; it is used either as a contact on silicon or to make interconnections. Ohmic contacts are generally obtained by heating the metal-semiconductor system at temperatures on the order of 400–550°C for a few minutes, producing an intimate contact between the metal and the semiconductor. Heating reduces contaminants at the surface of the silicon [6,7]. Typical values for Al-Si contact resistance are on the order of $1 - 10 \times 10^{-6} \Omega cm^{-2}$.

The heat treatment that is absolutely necessary to establish the desired electrical properties produces a nonuniform erosion of the silicon and, in some cases, may lead to the destruction of the structure. The phase diagram in Figure 6.3 shows that at the temperatures considered the solid solubility of silicon in aluminum is about 1.6 atomic percent, and the times and temperatures involved in contact annealing are high enough to transport the silicon in aluminum for several microns, the silicon being provided by the substrate. In the presence of uniform erosion, there exists no problem with the interdiffusion; unfortunately the erosion occurs preferentially and deep holes are formed. Figure 6.8 shows the contact area after a typical heat treatment and aluminum removal. Deep holes are clearly identified, and their morphology is determined by the crystalline orientation of the substrate. A thin nonuniform oxide layer at the silicon-aluminum interface or the presence of stresses in the silicon can be responsible for the preferential erosion. In the case of nonuniform oxide, the aluminum dissolves silicon directly through the thinner regions of the barrier; in the areas of thicker oxide coverage the silicon is protected from aluminum dissolution. The result is that all the silicon dissolved in the aluminum comes from localized spots. The process is shown schematically in Figure 6.9. In the case of stress on the silicon or on the aluminum, it is necessary to consider that the solubility of Si in Al is strongly affected by pressure. Moreover, the system Si-SiO_2-Al is metallurgically heterogeneous so that during and after the heat treatment stresses can arise due to the different thermal expansion coefficients of the various layers. These stresses will be localized, mainly near the edges of the structures. Heat treatment relaxes the excess energy stored in the deformations with subsequent preferential dissolution. For this reason the holes are located mainly near the edges of the SiO_2. To minimize the phenomenon of erosion, silicon in the concentration of 1–2 atomic percent is added to the aluminum before it is deposited. This alloy allows the use of the aluminum metalization in devices with a contact area of $3 \times 3\ \mu m^2$ and junction depth of 3500 Å. However, for contacts smaller than $1 \times 1\ \mu m^2$ and junctions on the order of 1000–3000 Å, the Al-Si system gives rise to further problems requiring more studies.

Figure 6.8 Photograph of an aluminum-silicon contact area. The aluminum has been removed by chemical etching.

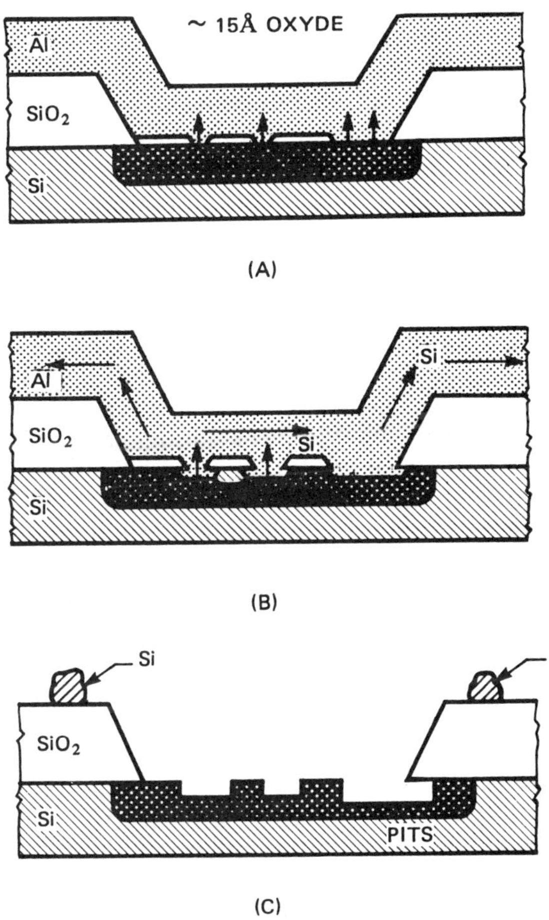

Figure 6.9 Schematic representation showing the progression of preferential silicon dissolution: (a) as deposited; (b) after heat treatment; (c) aluminum removed.

During cooling, silicon dissolved in the aluminum segregates and forms precipitates, on SiO_2 or directly on the contact area. The precipitates incorporate small amounts of aluminum, forming polycrystalline agglomerates of p-doped silicon. The presence of this silicon modifies the value of the contact resistance and produces several reliability problems.

Impurities in addition to silicon are added to the aluminum to solve particular problems and meet particular needs. For instance, copper is added to increase the resistance of aluminum to electromigration. Recently other materials, such as titanium and tantalum, have been added to achieve the same effect. The studies in this direction are in the initial stages only, and for several problems the solutions remain elusive. The tendency is to continue using aluminum as much as possible, trying to solve the problems as they appear.

The metallurgical stability of the contact is not preserved by the heat treatment. In fact, the segregation of silicon during cooling allows for the presence of "pure" aluminum at the onset of subsequent heat treatments. Dissolution then begins again from the substrate. Moreover, heating during device operation, although at relatively low temperatures, can produce a quick degradation of the contact, especially if the process occurs in presence of high current density.

6.3.2 Si-PtSi Contacts

The need for high performance transistors requires not only ohmic but also rectifying contacts. Aluminum cannot be used in this application because it does not meet the reliability requirements. Various kinds of materials have been considered in this regard ranging from near nobles, such as platinum, palladium, and nickel, to refractories, such as tantalum, titanium, and tungsten [8–12]. All of these materials react with silicon and form compounds. The major difference between the two classes of metals is the reaction temperature: it is relatively low for near nobles, around 200–300°C, and much higher for refractories, on the order of 500°C. Silicide formation is extremely important to obtain an intimate contact between metal and semiconductor; these contacts have the characteristics of reliability, reproducibility, and stability required in many devices.

Among the various possible metals, platinum is the most widely used. Silicide layers are formed by depositing a thin film of platinum, 200–1000 Å thick, onto a silicon wafer on which the contact area has been delimited. A heat treatment at temperatures around 400–600°C forces the formation of the silicide; the growth of the compound removes any impurities originating at the surface of the silicon, producing a highly reliable contact. The kinetics of formation of platinum silicide have been investigated by various authors. The first phase formed in the Pt-Si system is Pt_2Si, and during an isothermal treatment, the compound grows proportionally to the square root of the heating time. Figure 6.10 shows an example of kinetic data. The square root of time dependence implies that the growth is limited by diffusion and some particular investigations have shown that Pt is the diffusing species. When all the platinum has been converted to Pt_2Si, PtSi begins to grow, again at a rate proportional to the square root of time. The formation of new compounds in thin films occurs always when the previous compound or all the metal has been consumed.

The Pt-Si phase diagram shown in Figure 6.4a predicts five different stable compounds; only two can be formed by reacting a thin platinum film deposited on silicon. The other compounds can be obtained using suitable configuration, as shown in Figure 6.11. A careful analysis of the data reported in Figures 6.10 and 6.11 indicates that the formation of the various compounds is not determined by

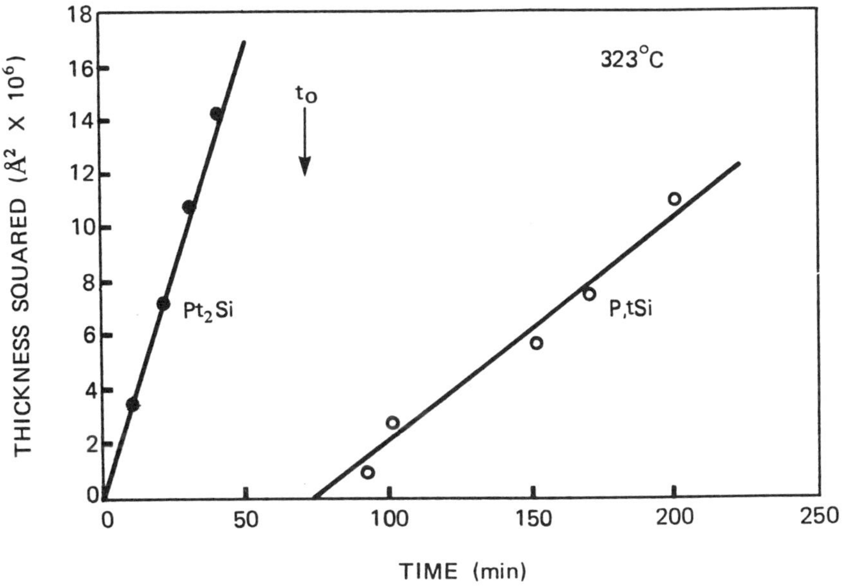

Figure 6.10 Formation kinetics of Pt_2Si and PtSi.

the speed of the growth but from the presence or absence of a particular material. In fact, with an excess of Pt, Pt $\gg$ Si, the last column at the right in Figure 6.11, the PtSi phase never forms and the reaction ends with Pt_3Si; whereas with an excess of Si, Pt $\ll$ Si, the last column at the left, PtSi is created.

At present it is not completely understood what physical processes are responsible for the preferential growth of some phases over others. Several empirical rules, based on the phase diagram, have been suggested to predict the sequence in which the various phases appear. One of these rules indicates that the first phase formed is the one that has a composition closest to the lowest temperature eutectic and that melts congruently at a maximum temperature. In the case of Pt-Si system, the minimum eutectic occurs at 23 atomic percent of silicon concentration, and the closest compound congruently melting is Pt_2Si. Experimental results indicate that Pt_2Si is indeed the first compound formed. To predict which phase forms sequentially, the phase diagram has been consulted again. If one element is in excess, for instance Si, the compound formed is the one having a composition between the first and pure Si; if more than one compound is present, the one formed is that which melts congruently and has the highest melting temperature. The compounds that are not congruently melting form later, in general. For instance, Pt_2Si_5 forms only after Pt_2Si and PtSi have been formed.

Impurities [7] play a fundamental role in compound formation. If the platinum film is contaminated with oxygen, the growth of the phase produces segregation

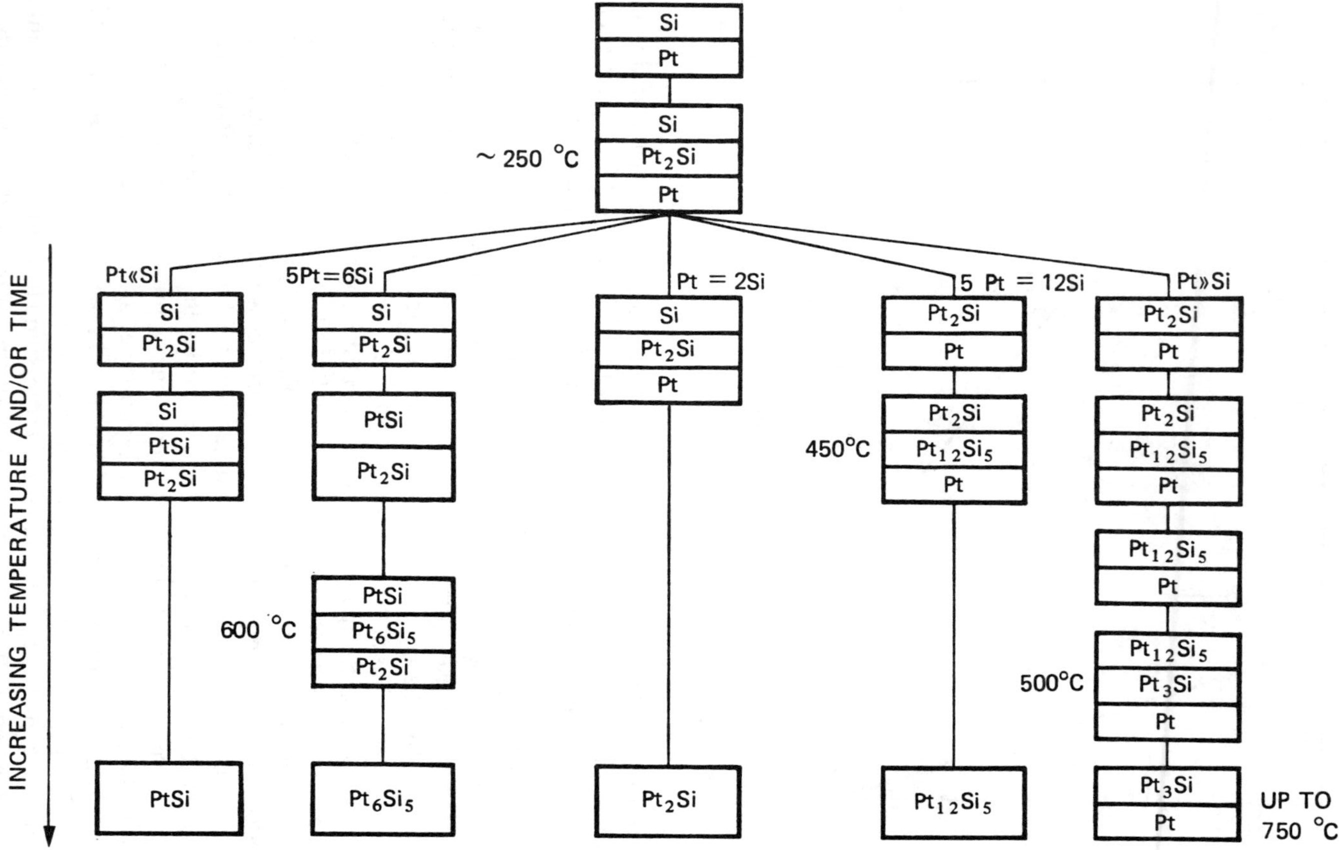

Figure 6.11 Diagram showing the compound formation sequence in the PT-Si system.

and consequently accumulation of the oxygen at the metal-silicide boundary. The presence of oxygen can change the formation mechanism of the compound and in some cases may cause the appearance of unexpected phases. Figure 6.12 summarizes the experimental results obtained in the Pt-Si system with small and large quantities of oxygen. Oxygen segregates to the surface of the sample, and with high oxygen content, the system's reaction is incomplete. In fact, with an oxygen content on the order of 3 atomic percent, some Pt is left unreacted, and for a particular combination of time and temperature, the simultaneous appearance of Pt, Pt_2Si, and PtSi is observed. The presence of oxygen also slows down the formation of the compounds themselves; for an oxygen content on the order of 3 atomic percent, the diffusion coefficient is reduced by more than 1 order of magnitude. A positive aspect of impurities segregation is that the impurities, including those at the silicon surface, are removed during the silicide formation, leaving a "clean" and reliable contact.

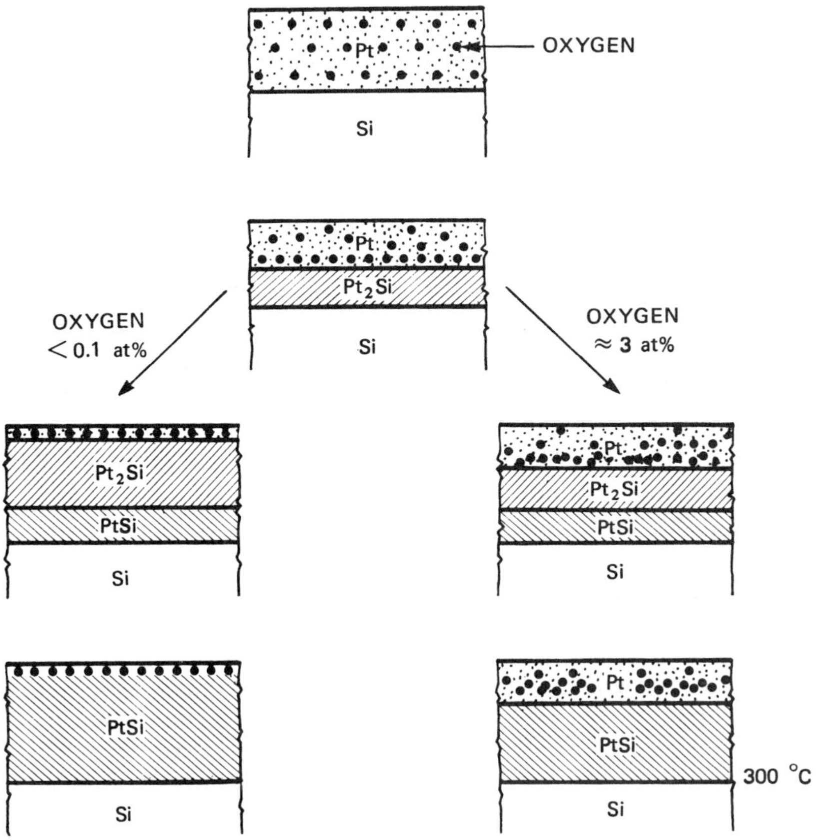

Figure 6.12 Schematic diagram showing the formation of platinum silicide in the presence of oxygen after heat treatment at 300°C. Oxygen segregates and accumulates at the Pt-Pt_2Si interface.

The oxygen in the annealing environment also effects the reaction kinetics. Although the silicide forms internally, oxygen diffuses in the platinum from the outer surface. When the silicide reaches the platinum region contaminated with oxygen, the reaction proceeds following the pattern previously described; a mixture of O, Pt, and Si segregates to the surface. Such a mixture, which seems to have a $PtSiO_4$ composition, tends to protect the structure from chemical etchings. This property is used in device fabrication to selectively remove any unreacted platinum.

Lateral growth of silicide, as shown in Figure 6.13, has been observed in structures where a continuous film of Pt covers bare silicon and SiO_2. The reaction starts in the region of pure silicon and continues after all the metal film directly overlying the silicon is reacted. The reaction proceeds laterally in the metal layer. Figure 6.13b shows a schematic lateral view of the sample and Figure 6.13a is a SEM image from a sample annealed at 700°C for 20 hours. The clearest squared parts show the regions where the contact occurs between silicon and metal; the irregular shapes at the corners indicate the thickness of the silicide formed. The lateral growth process is similar to the diffusion process of silicon in aluminum, therefore it could form pits. These phenomena start at temperatures around 500°C and the magnitude of the growth depends upon the time-temperature combination.

The formation of a stable and reproducible contact is an important step in the construction of electronic devices. The electrical properties of the contacts are determined by the kind of silicide used [13] and by the doping of the substrate. Table 6.1 reports the measured values of the Schottky barrier heights measured on n-type silicon for various silicides. The same value has been obtained for silicides of different compositions but from the same metal. In the case of heavily doped semiconductor material the effective barrier height is reduced by the presence of an electric field in the junction. The flux of the current can be controlled by tunnel effect; ohmic contacts are obtained when tunneling prevails. The contact resistance values obtained for PtSi deposited on n^+ and p^+ silicon are reported in Figure 6.14. It is shown that the contact resistance decreases with increasing dopant, and for platinum silicide, higher values are obtained on n-type than on p-type.

6.4 METAL-METAL CONTACTS

Once contacts are made to the silicon and all necessary interconnections are in place, electronic devices cannot be utilized until the chip is encased and connections are made to the outside world. External connections are built up on specifically selected areas of the devices. The two most widely used techniques for constructing contacts utilizes, in one case, external wires and, in the second method, direct soldering on the devices with alloys having low temperature melting points. In both cases it is necessary to combine various metals and consequently to verify

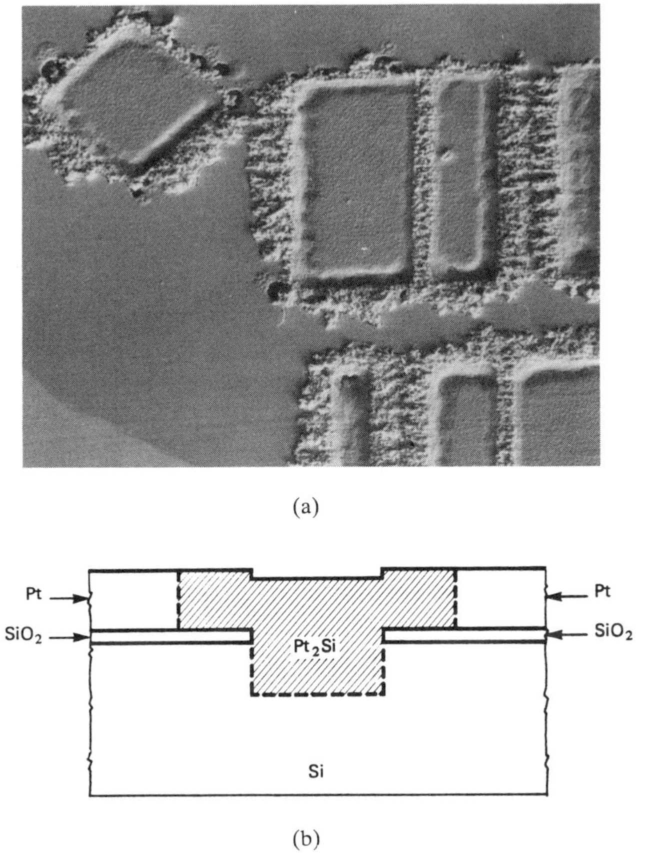

Figure 6.13 Lateral growth in the Pt-Si system observed after treatment at 700°C for 20 hours: (a) SEM image; (b) a schematic view of the structure.

their compatibility. Metal-metal interactions [7, 8, 14] are similar to the metal-semiconductor interactions previously considered: again the phase diagrams identify the equilibrium conditions that are reached by diffusion processes. At temperatures on the order of a few hundred degrees Celsius, interdiffusion processes as well as compound formation may be seen, depending upon the kinds of material used. In interdiffusion the minimum temperature at which mixing is observed depends upon the nature, poly- or single-crystalline, of the metals involved. For instance, polycrystalline metal films such as Pd and Au interdiffuse at 250°C, whereas much higher temperatures are necessary in the case of the single-crystal

Table 6.1 Barrier Heights of Various Silicides on n-Type silicon. Yttrium and the Rare Earths Show the Same Electrical Behavior

Group	*Element*	*Barrier height* (eV)
IV A	Ti	0.62–0.65
	Zr	0.55
	Hf	0.57–0.73
V A	V	0.59–0.64
	Nb	0.59
	Ta	0.59
VI A	Cr	0.57–0.59
	Mo	0.55–0.7
	W	0.65
VII A	Mn	0.65–0.67
	Tc	—
	Re	—
VIII A	Fe	0.67–0.68
	Ru	—
	Os	—
	Co	0.66–0.68
	Rh	0.7–0.74
	Ir	0.93
	Ni	0.63–0.67
	Pd	0.70–0.75
	Pt	0.82–0.87
II A (rare earths)	Y	0.39
	Gd	0.37
	Dy	0.37
	Ho	0.37
	Er	0.39

Ag-Pd couple. The presence of impurities strongly modifies the diffusion processes; heat treatment in an oxidizing atmosphere reduces the mixing in the Au-Pd system, whereas an increase is observed in the Au-Ti system.

One of the most thoroughly investigated metal couples that form compounds is the Au-Al system. This system has been studied both in the form of thin films and as a bulk material. the Au-Al equilibrium phase diagram contains five different stable compounds, all of which are formed in thin films reactions. The results obtained in thin film analyses are shown in Figure 6.15, where a strong similarity to the Pt-Si system behavior is apparent. At the onset of a reaction, the first phase formed is Au_2Al; it grows proportionally to the square root of time with an activation energy around 1 eV. In excess of Al, when all the gold has been consumed, $AuAl_2$ begins to form from the first compound, the growth continuing until all the

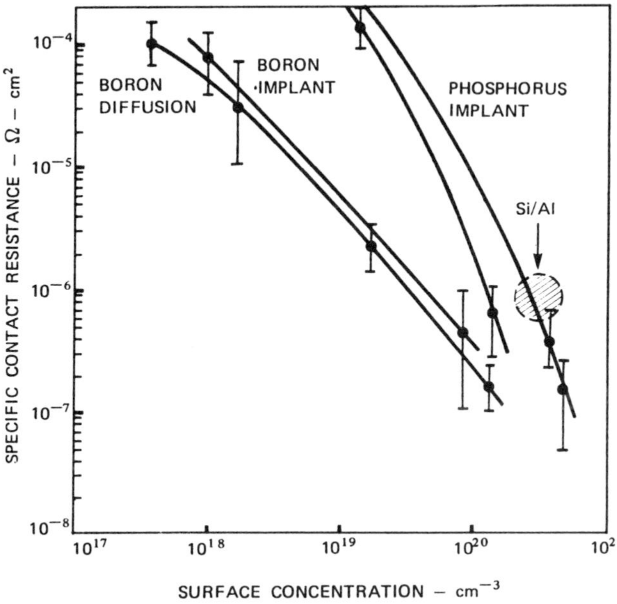

Figure 6.14 Specific contact resistance for PtSi on silicon. The surface dopings were obtained by ion implantation and thermal diffusion. As a reference, the values obtained for the Al-Si contact are reported.

Au_2Al has been consumed. In this transformation the mobile species has not been identified; however, it seems that one element moves faster than the other, producing Kirkendall voids. In bulk samples all the compounds seem to appear simultaneously; a schematic view of the results obtained are reported in Figure 6.16. Detailed considerations of these data suggest that the bulk experiences the same physical phenomena responsible for the phase appearances seen in Figure 6.15. When the thickness of the growing compound is so large that the quantity of material transported is strongly decreased, the formation of a new compound is triggered at an interface. These results are extremely important for device applications. In an electronic devices the interconnections are made by soldering a gold wire to an aluminum film. Thermal processes brought about by heating during normal operation are sufficient to provoke the formation of intermetallic compounds including $AuAl_2$, or "purple plague," the growth of which produces Kirkendall voids. Figure 6.17 shows a section of contact area between Au and Al; at the interface appear cracks due to the preferential transport of one element during the growth of purple plague. Careful investigations confirm the presence of $AuAl_2$. In the inner region, inside the gold wire, compounds rich in gold are present.

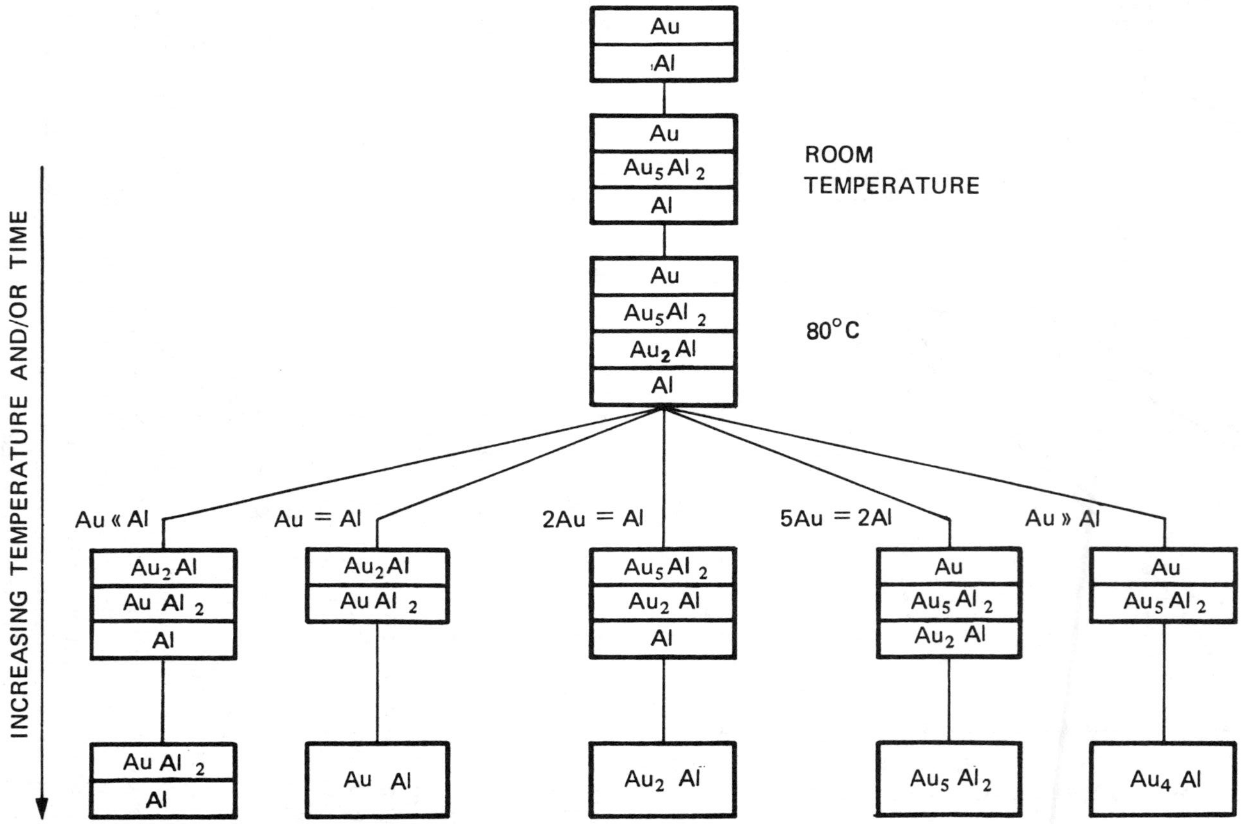

Figure 6.15 Schematic view of the interaction occurring in Au-Al thin film couples.

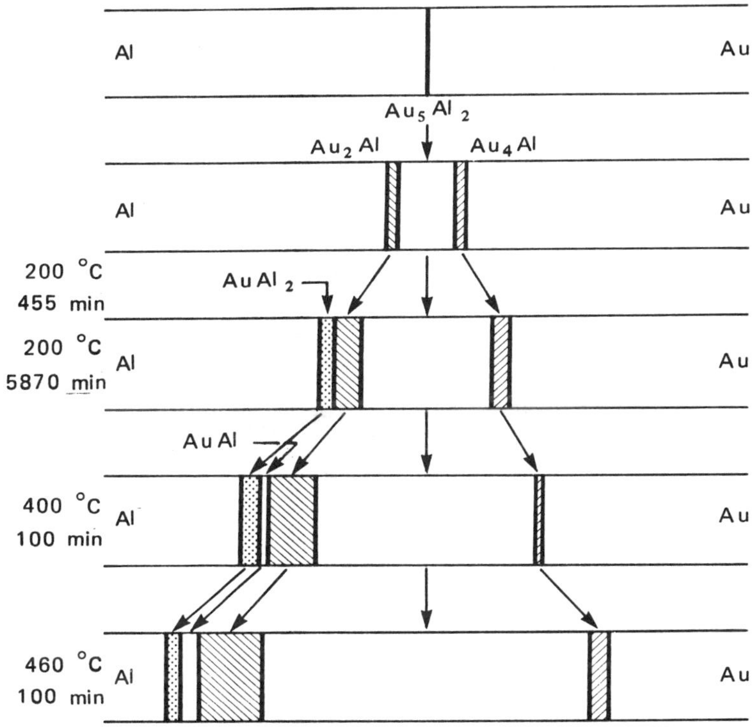

Figure 6.16 Schematic view of the reaction occurring in bulk Au-Al system.

6.5 DIFFUSION BARRIERS

The requirements on the metalizations in one integrated circuit are so many that one single material cannot satisfy all of them, and a combination of many different materials is needed [10–12]. For example, although Pt is a suitable metal for making the contact on silicon, it is not the best material for interconnections, where aluminum is preferred. Pt and Al must therefore coexist; there is a region in the device where platinum silicide and aluminum are in contact. The silicide-aluminum system is not stable, and during the life of the device, the silicide is probably decomposed, forming Pt-Al compounds. When aluminum reaches the silicon wafer, it causes an irreversible degradation of the electrical properties of the contact. As a consequence it is necessary to place a diffusion barrier between the two metals,

Figure 6.17 Picture of an Au-Al contact. The black spots at the bottom are Kirkendall voids.

as schematically indicated in Figure 6.18. In this case the diffusion barrier consists of a TiW alloy. From a thermodynamic point of view, such a contact is unstable since the system is not in the minimum possible free energy configuration. The introduction of a barrier slows down the processes necessary to reach such equilibrium. Generally the barriers are unable to produce true stability; their role is to increase the lifetime of the contacts only for a time comparable to the duration of the device. Since the introduction of a diffusion barrier adds to both the cost and complexity, it follows that the extension of the lifetime of the device is an economic question. Interdiffusion processes can occur by diverse causes: a concentration gradient, a negative value for the free energy of reaction, an electric field, a stress gradient, *et cetera*. In actual systems the first two are the most common, and consequently there are two principal barrier types: (1) barriers against diffusion and (2) chemical barriers. Recently attention has been focused on the importance of the electric field in structures having sizes on the order of a μm^2. However, on this subject reliable data are not available.

6.5.1 Barriers Against Diffusion

The process of atomic diffusion was considered in section 6.2.2, where we pointed out that atomic motion occurs by various fundamental physical phenomena, each one characterized by a different velocity. For reference we state again here that surface diffusion is the fastest of those types considered, bulk diffusion the slowest, with intermediate values obtained for grain boundary diffusion. The prevalence

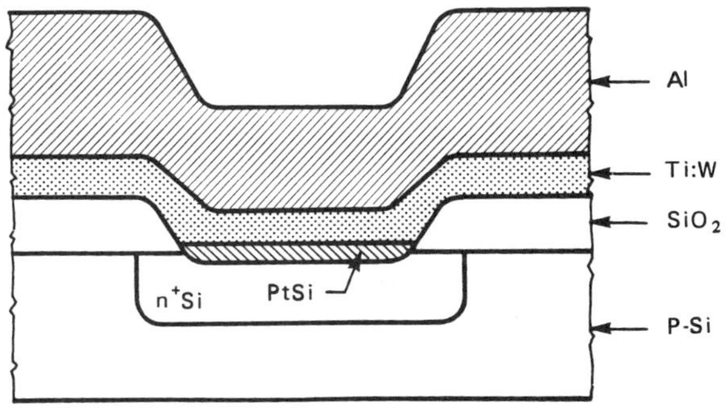

Figure 6.18 Schematic representation of a device using TiW alloy as a diffusion barrier.

of one of the mechanisms depends upon the quality of the film and temperature, as suggested by the data shown in Figures 6.6 and 6.7. At the temperatures that develop during device operation, generally no higher than 100°C, the diffusion is controlled by the grain boundaries. The flux of the material depends not only from the diffusion coefficient but also on the number of preferential paths present and consequently from the grain dimensions. For the best stability of the contacts, it is preferable to have films with grain sizes as large as possible and approaching, as a limit, single crystals. Grain boundary diffusion is strongly affected by solutes eventually segregated to grain boundaries. Although not always true, it can be assumed that the presence of impurities strongly reduces the atomic transport. It has been shown that impurities such as C, N, and O, generally incorporated into the film during deposition, slow down the diffusion. Numerous examples are reported in the literature of a better control during the deposition of the so-called diffusion barriers, leading to the loss of the barrier properties. The properties of the TiW alloy as a diffusion barrier are due particularly to the presence of oxygen and nitrogen. Even for TiN, the barrier properties seem to be due to oxygen contamination.

Another category of materials suitable for barriers is made up of compounds having amorphous structures. These materials have diffusion coefficients comparable to those of single crystals but are much easier to obtain than flawless structures. A difficulty with amorphous materials is that they generally have a relatively low crystallization temperature; therefore, a definitive decision on their use has not yet been reached. Also, the properties of amorphous materials are strongly affected by the presence of impurities.

6.5.2 Barriers Obtained by Chemical Reaction

To form a compound both nucleation and growth processes are necessary. Nucleation occurs only if the free energy is decreased by the formation of the compound; the diffusion coefficient gives the speed at which the compound grows. Thermodynamically, one criterion used to evaluate whether a reaction occurs at a given temperature is to consider the variation of Gibbs free energy:

$$\Delta G = \Delta H - T\Delta S$$

where ΔH is the variation in enthalpy and ΔS in entropy. In many practical cases, the contribution from ΔS is negligible; the reaction is controlled by ΔH and consequently by the variation in chemical energy.

Barrier layers can be used to reduce the total free energy of the system. Two films, A and B, in contact tend to react and to form a compound, AB, only if the energy is reduced in the process. An interlayer C, between A and B, can reduce the free energy of the system. The result of adding C is to increase the stability of the system; C acts as a diffusion barrier. Generally the presence of C does not bring the system to equilibrium; some reaction still occurs but at much slower rate. Figure 6.19 illustrates the behavior of this kind of barrier. In a practical case, C is a transition metal, and it reacts with A, aluminum, and B, silicide. The reaction occurs at both sides of the silicide, but the structure is not degraded until the two partially reacted films come into contact.

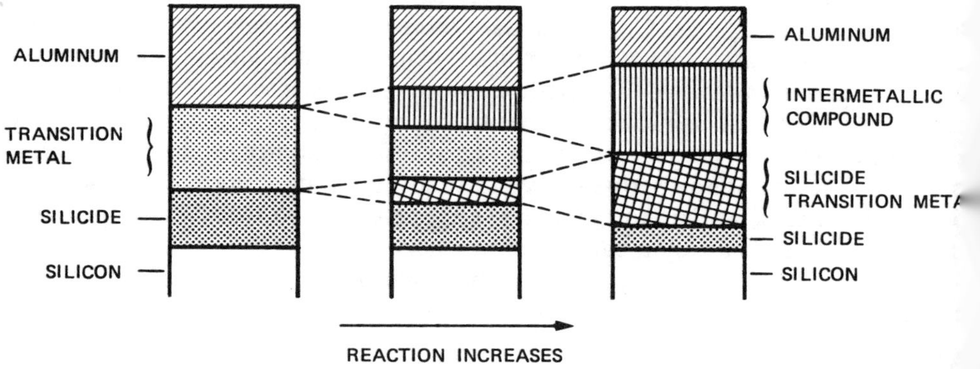

Figure 6.19 Schematic view of a diffusion barrier that reacts with two films, delaying the degradation of the contact.

6.5.3 Some Examples of Barriers

A large number of materials in various combinations have been used to produce barriers. Very often, diffusion barriers are necessary even when dealing with chemical barriers. Many data, available from the literature, are reported in Table

6.2–6.6, together with configurations used and failure mechanisms. The determination of the physical mechanisms responsible for the failure and the evaluation of the effective stability of the barrier are not always possible due to the sporadic presence of impurities.

Table 6.2 Stability of the Aluminum-Silicide-Silicon Contacts

Silicide	*Reaction temperatures*	*Reaction products*	*Failure mechanisms*
PtSi	350	Al_2Pt, Si	Compound formation
Pd_2Si	200	Al_3Pd_1Si, Si	Compound formation
	300	Al_3Pd_1Si, Al_3Pd_2, Si	Compound formation
	400	Al_3Pd, Si	Compound formation
NiSi	400	Al_3Ni, Si	Compound formation
$CoSi_2$	400	Al_9Co_2, Si	Compound formation
	450	Al_9Co_2, $Al_{13}Co_4$, Si	Compound formation
$TiSi_2$	550	$Al_5Ti_7Si_{12}$, Si	Diffusion (400°C)
$MoSi_2$	535	$Al_{12}Mo$, Si	Diffusion (400°C)

Table 6.3 Stability of the Aluminum-Refractory Metals-Silicon and Aluminum-Refractory Metal-Silicide-Silicon Contacts

Transition metal	*Silicide*	*Reaction temperature (°C)*	*Reaction products*	*Failure mechanisms*
Cr	—	300	Al-Cr	Compound formation
V	—	450	Al_3V, Al-V-Si	Compound formation
Ti	—	400	Al_3Ti, Al_5Ti-Si_{12}	Compound formation
Cr	Pd_2Si	450	$CrSi_2$, Al-Cr	Compound formation
	PtSi	450	Al-Cr	Compound formation
V	$GdSi_2$	450	$Al_{10}V$,	Compound formation
			VSi_2(525°C)	Compound formation
Ti	Pd_2Si	435	Al_3Ti	Compound formation
	PtSi	450	Al_3Ti	Compound formation
Mo	PtSi	500	$Al_{12}Mo$, Al_4Mo	Compound formation
W(250Å)	NiSi	500	—	Diffusion
W(2000Å)	$CoSi_2$	500	$Al_{12}W$	Compound formation
	$MoSi_2$	500	$Al_{12}W$	Compound formation

Table 6.4 Stability of the Aluminum-TiW-Silicon and Aluminum-TiW-Silicide-Silicon Contacts

Transition metal alloy	*Silicide*	*Reaction temperatures (°C)*	*Reaction products*	*Failure mechanisms*
Ti-W	—	500	—	Diffusion
Ti-W	PtSi	500	Al_2Pt, $Al_{12}W$	Diffusion
		550	$Ti_xW_{1-x}Si_2$	Diffusion
Ti-W	Pd_2Si	500	—	Diffusion

Table 6.5 Stability of the Aluminum-Transition Metal Compound-Silicon and Aluminum-Transition Metal Compound-Silicide-Silicon Contacts

Transition metal compound	*Silicide*	*Reaction temperatures (°C)*	*Reation products*	*Failure mechanisms*
ZrN	—	550	$Al_4Zr_3Si_5$	Compound formation
TiN	PtSi	600	AIN, Al_3Ti	Compound formation
	$CoSi_2$	550		Compound formation
	$TiSi_2$	550		Compound formation
TiC	PtSi	600	Al_4C_3, Al_3Ti	Compound formation
TaN	NiSi	600	AIN, Al_3Ta	Compound formation

Table 6.6 Resistivity of Various Silicides in Thin Film Form

Silicide	*Preparation*	*Annealing temperature (°C)*	*Resistivity* (μ Ω-cm)
$PtSi_2$	Metal on polysilicon	900	13–16
	Codeposited alloy	900	25
$ZrSi_2$	Metal on polysilicon	900	35–40
$HfSi_2$	Metal on polysilicon	900	45–50
	Codeposited alloy		60–70
VSi_2	Metal on polysilicon	900	50–55
$NbSi_2$	Metal on polysilicon	900	50
$TaSi_2$	Metal on polysilicon	1000	35–45
	Codeposited alloy	1000	50–55
$CrSi_2$	Metal on polysilicon	700	~ 600
$MoSi_2$	Metal on polysilicon	1100	~ 90
	Codeposited alloy	1000	~ 100
WSi_2	Codeposited alloy	1000	~ 70
FeSi	Metal on polysilicon	700	> 1000
$CoSi_2$	Metal on polysilicon	900	18–20
	Codeposited alloy	900	25
$NiSi_2$	Metal on polysilicon	900	~ 50
	Codeposited alloy	900	50–60

Table 6.6 continued

Silicide	*Preparation*	*Annealing temperature* (°C)	*Resistivity* (μ Ω-cm)
PtSi	Metal on polysilicon	600–800	28–35
	Codeposited alloy		
Pd_2Si	Metal on polysilicon	400	30–35

6.6 FUTURE DEVELOPMENTS

In an electronic device of the complexity displayed in Figure 6.1 the metalization should fulfill two main objectives: to make a direct contact to silicon and to interconnect the various active elements in the device. The problem of the contact is to deposit a metal layer onto n- and p-type silicon over junctions that, in the actual circuit, are located at a depth of 1000 Å. This contact must be applied while maintaining the structural integrity of the junction and with a contact resistance that is low, reproducible, and stable. No one material possesses all the desired characteristics; to obtain ohmic contacts, compromises must be made. If the contact on p-type junctions is the more critical, then Pt or Pd, with Schottky barriers of .3 eV, should be used; if n-type contact is the more critical, the rare earths, which have a barrier height of .3eV, constitute a possible solution. If both are critical, metals such as Ti and Cr, with intermediate values of Schottky barrier, should be used. When the suitable metal has been selected a heavily doped region can be produced using, for instance, ion implantation.

From the metallurgical point of view, the desire is to consume some silicon, to make an intimate and reliable contact, but to limit the amount of silicon reacted in order to avoid interfering with the junction. To avoid the problems associated with a nonuniform erosion, a general rule is to consume no more than one tenth of the total thickness, 1000 Å, of the silicon above the junction. It is not a solution to decrease continuously the thickness of the metal films. In fact, for thin films that are too thin the layer will become discontinuous. A possible solution includes the use of an alloy with two metals, such as Pt and W, that react with silicon at two different temperatures. By heating the system at temperatures between the two, the metal that reacts at the lowest temperature segregates to the interface and forms the silicide. Another solution includes the use of a metal with a certain quantity of silicon already dissolved. Thus, the quantity of silicon required to form the silicide is limited. Still another solution is to increase the thickness of the silicon above the junction by depositing a layer of semiconductor on the region of the contact.

A different discussion should be made for Schottky diodes. In the case of these contacts, it is necessary not only to stabilize the contact but also to maintain the same contact area to avoid altering the voltage drop for the same current. A nonuniform contact area can arise from imperfect cleaning and a nonplanar reaction. The reduction of the geometries only enhances these problems.

The field of interconnections on electronic devices requires considerable research efforts. As shown in Figure 6.20, the future requirements for VLSI devices can be recognized by considering two kinds of interconnection: those necessary for connecting the devices and those for connecting the gates. The actual needs are represented by the circle. It is clear that aluminum, with its low resistivity, is a suitable candidate. Polysilicon presents several problems, and for this reason, many studies in this area are trying to substitute, or complement, polysilicon with silicides or metals. Not all silicides have the same resistivity value; Table 6.6 reports the values measured in thin films in various laboratories. From the standpoint of resistivity the best choice is $TiSi_2$; however, the affinity of titanium silicide with oxygen suggests the use of other silicides, even if they have higher resistivities. Silicides are already used in production, and the devices have been made using tantalum and molibdenum silicide.

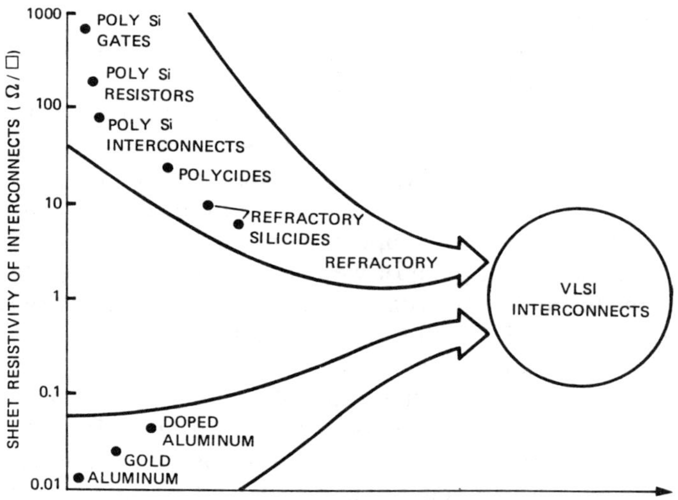

Figure 6.20 Schematic representation of the future needs in VLSI.

The films are deposited with the sputtering technique; there are studies using other techniques for deposition; as for instance, chemical vapor deposition, laser assisted chemical vapor deposition, and metal organic chemical vapor deposition. One of the most promising methods, used in pilot lines, is the chemical vapor deposition. This technique is hoped to provide better step coverages and a selective deposition.

A different line of approach to the connection problem is to investigate new materials, for instance polymers. Their resistivity may span from 10^{-15} to 100 $(\Omega\text{-cm})^{-1}$. Moreover such materials can be deposited and defined with techniques very similar to the ones actually used for photoresistance.

The trend in metalization is to utilize as much as possible those materials actually in use, trying to solve the problems as they appear, and to change the metalization only when it is absolutely necessary. Among the new materials, silicides are ready to enter into mass production.

REFERENCES

1. VLSI Electronics, *Microstructure Science,* Supplement A. *Material and Process Characterization,* G. Larrabee, Academic Press, New York, 1985.
2. W.G. Moffatt, *Binary Phase Diagrams Handbook,* General Electric Company, Schenectady, N.Y., 1976.
3. R.W. Balluffi and J.M. Blakely, "Special Aspects of Diffusion in Thin Films," *Thin Solid Films,* 25, 1975, pp. 363–392.
4. D. Gupta and P.S. Ho, "Diffusion Processes in Thin Films," *Thin Solid Films,* 72, 1980, pp. 299–418.
5. N.A. Gjostein, *Diffusion,* American Society for Metals, Metals Park, Ohio, 1973.
6. E.H. Rhoderick, *Metal-Semiconductor contacts,* Clarendon Press, Oxford, 1978.
7. J.M. Poate, K.N. Tu, and J.W. Mayer, eds., *Thin Films Interdiffusion and Reactions,* John Wiley and Sons, New York, 1978.
8. K.N. Tu and R. Rosenberg, eds., *Treatise on Materials Science and Technology,* vol. 24: *Preparation and Properties of Thin Films,* Academic Press, New York, 1982.
9. S.P. Murarka, *Silicides for VLSI Applications,* Academic Press, New York, 1983.
10. P.S. Ho and K.N. Tu, eds., *Thin Film and Interfaces,* Materials Research Society Symposia Proceedings, vol. 10, North Holland Press, Amsterdam, 1982.
11. R. Ludeke and K. Rose, eds., *Interfaces and Contacts,* Materials Research Society Symposia Proceedings, Vol. 18, North Holland Press, Amsterdam, 1983.
12. J.E.E. Baglin, D.R. Campbell, and W.K. Chu, eds., *Thin Films and Interfaces II,* Materials Research Society Symposia Proceeding, Vol. 25, North Holland Press, Amsterdam, 1984.
13. C. Calandra, O. Bisi, and G. Ottaviani, "Electronic Properties of Silicon-Transition Metal Interface Compounds," *Surface Science,* 4, 1985, pp. 271–364.
14. M.J. Howes and D.V. Morgan, eds., *Reliability and Degradation: Semiconductor Devices and Circuits,* John Wiley and Sons, New York, 1981.

Chapter 7

LATCH-UP IN CMOS INTEGRATED CIRCUITS

F. Fantini, M. Muschitiello, and E. Zanoni

7.1 THE PHYSICS OF LATCH-UP

A parasitic structure is inherent to CMOS integrated circuits, as shown in Figure 7.1a, where the cross section of a bulk, p-well, junction-insulated CMOS inverter is shown. The vertical npn transistor is formed by n^+ sources-drains (S/D) into the p-well (emitter), the p-well itself (base), and the n-substrate (collector). The lateral pnp structure is formed by p^+ S/D into the substrate (emitter), the substrate (base), and the p-well (collector). In the normal biasing condition, the substrate is connected to the positive bias and the p-well to the negative one. Therefore, the two transistors are connected in a SCR-like structure. Figure 7.1b reports the first order equivalent circuit of the parasitic SCR.

The two parasitic transistors Q_1 and Q_2, are normally in the off state; if one of the two transistors is brought into the on state and if the current gain product $\beta_1 \cdot \beta_2$ is sufficiently high, latch-up is fired, and both transistors are kept in the on state, leading to a short-circuit path between the two bias lines, until the device burns out, as shown in Figure 7.2, if enough current is supplied by the bias or the power supply is turned off.

Current gains β_1 and β_2 are functions of collector current; therefore, a minimum current I_{dd} should be supplied by the external current source and flow through the device to achieve regeneration. The latch-up path is characterized by the value of the *holding current*, I_h; that is, the minimum supply current that maintains the latched state; the corresponding voltage drop in the device is called the *holding voltage*, V_h. The current-voltage characteristics of the parasitic pnpn device are shown in Figure 7.3, where I_s and V_s are the dc switching current and voltage, respectively. Holding current (and voltage) and switching current (and voltage) are less ambiguously identified as the points on the I-V characteristics where differential resistance dV/dI is zero, respectively, in the low impedance (latched) state and high impedance (blocking) state of the parasitic structure [1].

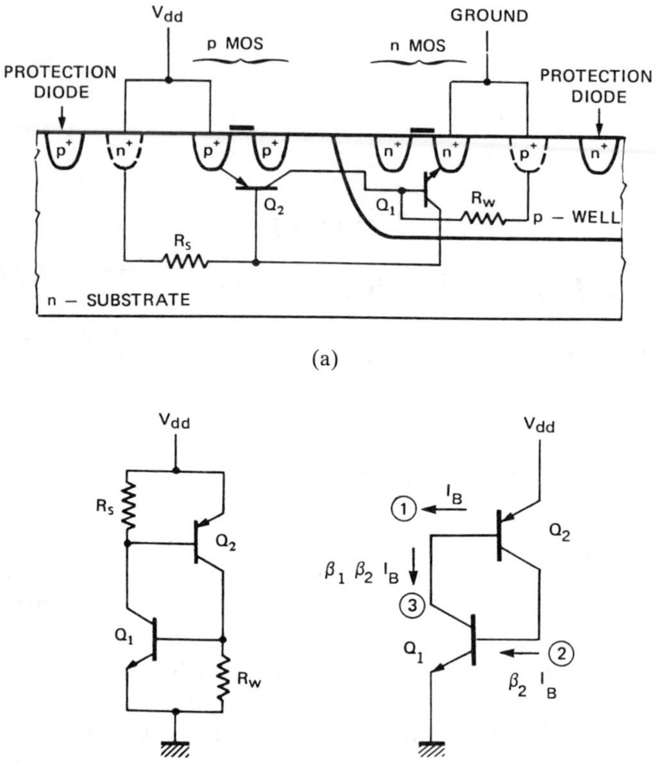

(a)

(b) (c)

Figure 7.1 (a) Schematic cross-section of a CMOS inverter showing bipolar parasitic transistors; (b) actual CMOS IC parasitic SCR structure with shunt resistances R_s and R_w; (c) parasitic SCR equivalent structure.

In a CMOS device, two parasitic resistances, R_s and R_w, respectively associated with the substrate and the well, shunt the base-emitter junctions of parasitic transistors. This introduces some differences between the parasitic pnpn structure in CMOS (Figure 7.1b), and the classical SCR, (Figure 7.1c). In particular, the holding current increases because, as R_s and R_w decrease, part of the current is diverted from the base of the parasitic transistors. However, lateral currents that flow in the n-substrate and p-well develop a voltage drop across R_s and R_w, and B-E junctions of parasitic transistors can become forward-biased, thus firing latch-up.

Many different causes can trigger latch-up in CMOS integrated circuits. They include [1,2] overshooting at the input-output pins, avalanching the well-substrate

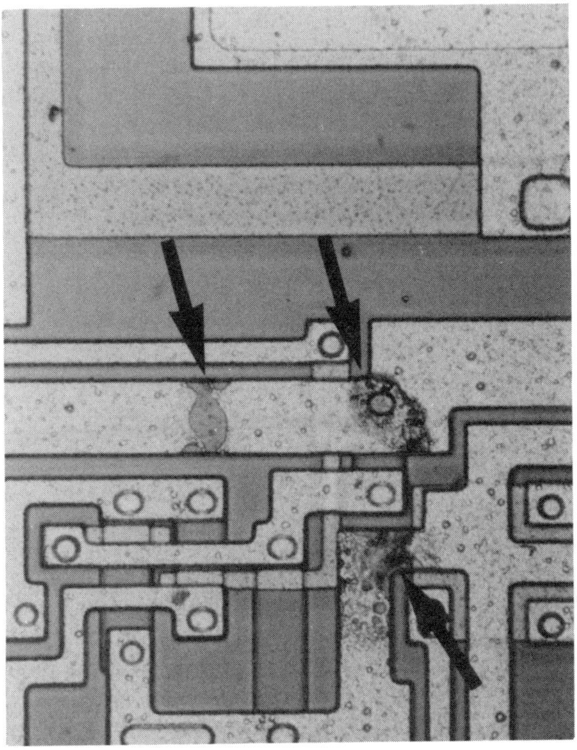

Figure 7.2 Optical micrograph of a custom CMOS IC showing the fused ground line.

junction by voltage overstress at the bias terminals, transient spikes exceeding bias voltages and introducing displacement currents through the well-substrate junction, and radiation-induced photocurrents.

The overshoots of input-output pins cause the S/D or protection diode junctions to turn on and inject a large amount of minority carriers in the substrate or well. These minority carriers are collected by the reverse-biased well-substrate junction and become majority carriers causing a voltage drop in the parasitic R_w/R_s resistance, thus turning on the B-E junction of one parasitic transistor. A common cause of this phenomenon is the signal reflection at the output of a device driving large capacitances (for instance, long wires).

The current in the parasitic resistances also can be induced by avalanching the well-substrate junction if a voltage exceeding the breakdown voltage is applied between the bias terminals. In this case, both parasitic transistors can be fired simultaneously, although, in general, the one with larger parasitic resistance turns on first.

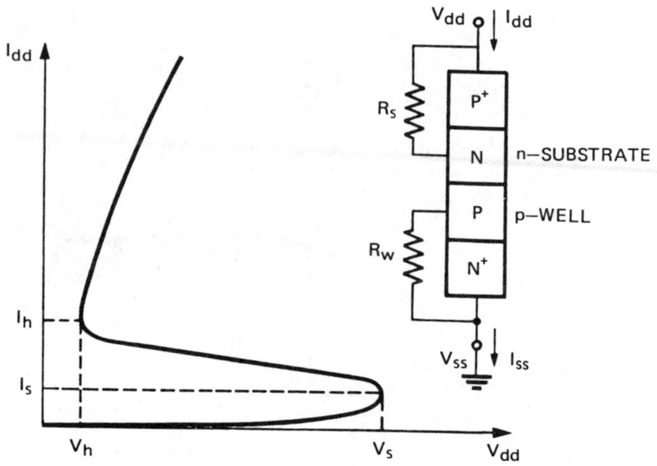

Figure 7.3 Current-voltage characteristics of the pnpn SCR present in CMOS ICs. I_h and V_h are holding current and voltage, whereas I_s and V_s are switching current and voltage, respectively.

Firing condition can be established before reaching the avalanche if *punch-through* of a parasitic transistor is induced by the spreading of the well-substrate depletion region, which reaches a close diffusion. The large punch-through current can be responsible for the voltage drop that fires the latch-up condition. A similar situation is caused by the presence of a parasitic FET in the region of the well-substrate junction. In this case, the parasitic MOSFET can be turned on by biased metal lines running over the surface or by oxide charges, which lower the threshold voltage of the parasitic device [3].

The spikes (ΔV) on the bias supply lines cause displacement currents through the well-substrate junction, which are proportional to junction capacitance (C_j):

$$I = C_j \Delta V / \Delta t$$

and can fire the latch-up condition, similar to discrete SCR devices.

Finally, a common cause of latch-up triggering is the photocurrent induced in the substrate by different kinds of radiation, such as x-rays or α-particles. The electron-hole pairs, produced by the radiation close to the well-substrate junction, are separated by the electric field and cause a majority current to flow in both the substrate and well, thus firing latch-up.

7.2 ELECTRICAL CHARACTERIZATION

7.2.1 Measurement of Basic Electrical Parameters of pnpn Parasitic Structures by Test Patterns

Characterization of the electrical sensitivity to latch-up of new CMOS technologies or structures can be accomplished by fabricating specially designed test patterns, which allow us to study the geometry and placement of supply contacts, to evaluate the effectiveness of guard rings, to evaluate the effects of device and layout scaling, and to compare different processes.

Using the specially prepared test patterns, all four terminals of the parasitic pnpn structure are individually accessible (Figure 7.4a) so that detailed characterization of both the blocking and latched state is possible.

Static characterization involves measurement of switching and holding points of the pnpn structures, previously defined as the I-V points that fulfill the condition $dV/dI = 0$, in the blocking and latched state, respectively (Figure 7.3). Latch-up can be triggered by connecting the npn emitter and its bypass resistor to ground, connecting the pnp emitter and its bypass resistor to V_{dd}, and then forcing current or voltage at one of the pnpn terminals. There are therefore eight different triggering modes, but only a few provide useful information.

The most straightforward way of triggering latch-up is to force a current IEP (or IEN) into (or out of) the emitter terminals, as shown in Figure 7.4a, where the pnp transistor is driven. The emitter voltage VEP and the currents INW and IEN are monitored. Supply currents IEN and IS are limited by external resistors.

Currents IEP, INW, and IEN are then reported as a function of VEP (Figure 7.4b). As we raise the level of the injected current IEP, the base-emitter junction of the pnp transistor is increasingly forward-biased and, because the npn is still turned off, the current INW is negative, as shown in Figure 7.4b. In this region, IEP, INW, and IS simply correspond respectively to emitter, base, and collector currents of the pnp transistor.

The collector current of the pnp develops a voltage drop on the RS resistor, which eventually turns on the npn. Consequently, the IEN current increases sharply (Figure 7.4b), rapidly reaching its maximum value. The collector current of the npn transistor drives the pnp base, so that current INW reverses (and it is no longer visible in the figure).

The switching point can be evaluated by plotting in linear scale the region where the differential resistance dVEP/dIEN changes its sign from positive to negative, identified in Figure 7.4b and reported in Figure 7.5. The switching point

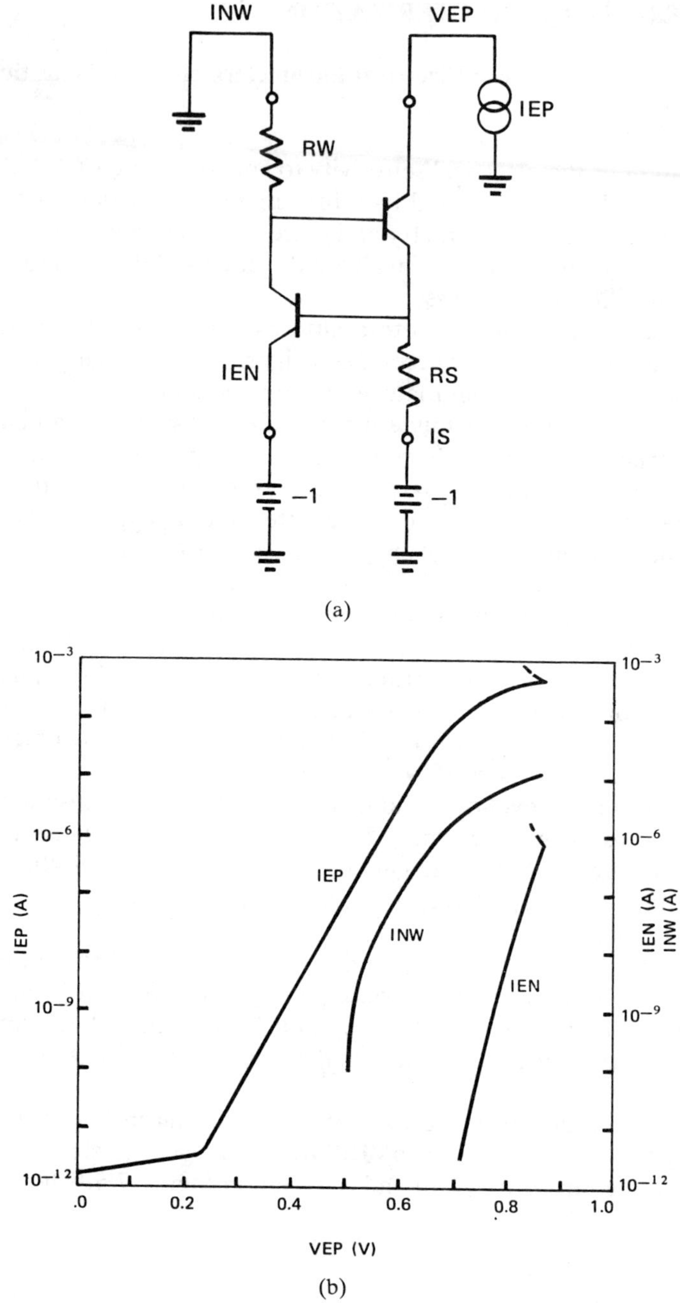

Figure 7.4 (a) Four terminal SCR structure, showing the pnp parasitic transistor turned on by emitter current injection; (b) pnp and npn emitter currents IEP and IEN, well current INW as a function of the pnp emitter voltage VEP.

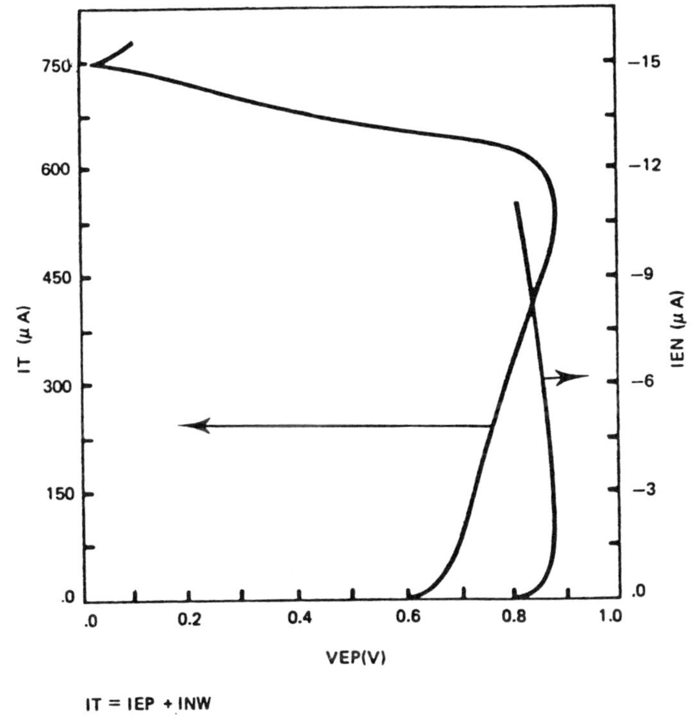

Figure 7.5 Linear plot of total current IT = IEP + INW *versus* VEP, identifying the switching point.

is then found by the condition dVEP/dIT = 0, as shown in Figure 7.5, where IT = IEP + INW. The value of IT at this point is the switching current that defines the latch-up resistance of the layout or technology under test.

To characterize the holding point (i.e. the point at which $dV/dI = 0$ in the latched state), the pnpn structure is driven into the low-impedance state; for instance, by injecting current into an emitter and then removing the triggering source. The pnpn structure is maintained in the on state, above the holding point, by an external current source. By lowering the total current IT = IEP + INW, the negative differential resistance portion of the curve can be observed (Figure 7.6), and the holding point (I_h, V_h) can be identified. Technologies characterized by a value of the holding voltage V_h that is higher than the maximum power supply voltage V_{dd} can never achieve a dc low-impedance condition and may be guaranteed as "latch-up free" [4]. Unfortunately, only greatly latch-up resistant structures show significant increase in V_h.

In absence of emitter or collector series resistances, as in the equivalent circuit considered in Fig. 7.1b, the holding voltage V_h is simply given by $V_{be} + V_{ce}$ (sat) = 0.8–0.9 V. Adding an emitter series resistance raises latch-up holding voltage and the SCR resistance in the on state, but the I-V characteristic for the latched

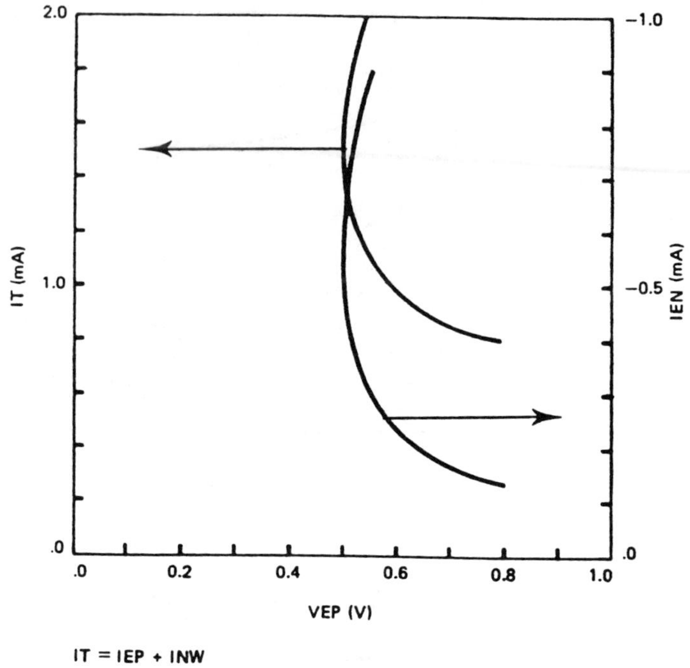

Figure 7.6 Linear plot of total current IT = IEP + INW *versus* VEP, with the device in the on state, identifying the holding point.

state, still extrapolates to the same value. Adding a collector series resistance not only raises the holding voltage, but also shifts the extrapolated latched state characteristic [5].

7.2.2 Characterization of Integrated Circuits

Direct measurement of switching and holding points of parasitic SCRs cannot be performed for completed CMOS integrated circuits, where pnpn structures are not directly accessible and a complex network of parasitic paths is unavoidably present. Moreover, latch-up paths can interact, causing anomalous electrical characteristics and possibly erroneous results in the evaluation of latch-up sensitivity [6].

Practical latch-up characterization of ICs considers the actual transient pulses to which the device may be subjected, and defines suitable electrical test methods to simulate them. An exhaustive characterization of every conceivable latch-up path is almost impossible for simple ICs. Moreover, precautions must be taken because the applied currents and voltages usually exceed the absolute maximum ratings of the IC such that they will certainly degrade its performance and may even destroy it, rendering the 100 percent inspection inapplicable.

Because of the large number of possible overstress phenomena, a complete latch-up characterization of a CMOS IC is difficult. The sequence of electrical tests may include:

(a) Supply overvoltage;
(b) Continuous current into inputs-outputs (I/O) with the supply voltage applied either before or after the input-output current has been established;
(c) Pulsed current into I/Os;
(d) Capacitor discharge into I/Os to simulate electrostatic discharge;
(e) Voltage transients on the supply line and power supply ramp-up.

In the following, these triggering mechanisms will be described in detail.

(a) Supply Overvoltage

A supply overvoltage can induce latch-up through different physical effects: breakdown of the well-substrate junction, punch-through of parasitic transistors, breakdown of source-drain diffusions, punch-through of one n^+ (or p^+) diffusion to another, parasitic FET triggering [3]. In any case, high currents flow in both substrate and well, thus initiating latch-up.

These mechanisms of latch-up initiation can be tested simply by increasing the supply voltage V_{dd} with a curve tracer [7] or a pulse generator [8], using the test configuration shown in Figure 7.7. An example of the former type of measurement is shown in Figure 7.8, where the I_{dd}-V_{dd} characteristics of a CMOS gate are reported. Switching current and voltage are 22 mA and 14.8 V, respectively.

Even if the minimum current sufficient to fire latch-up can be used as a rough estimation of the circuit latch-up sensitivity, no other information is forthcoming. Moreover, the high currents and voltages cause large power dissipation and device heating, and this may lead to erroneous results and damage to the device.

(b) Continuous (1 ms – 5 s) Current into Inputs-Outputs

In this case, the parasitic SCRs are triggered by forcing a current through the input protection diodes or through the parasitic bipolar structures, including MOS drain junctions at the output (Figure 7.9).

The test circuit shown in Figure 7.10a refers to injecting a current through the diode between an I/O and V_{dd}; the supply pin is at ground potential, while V_{ss} is negatively biased to avoid causing the supply voltage to increase during current injection.

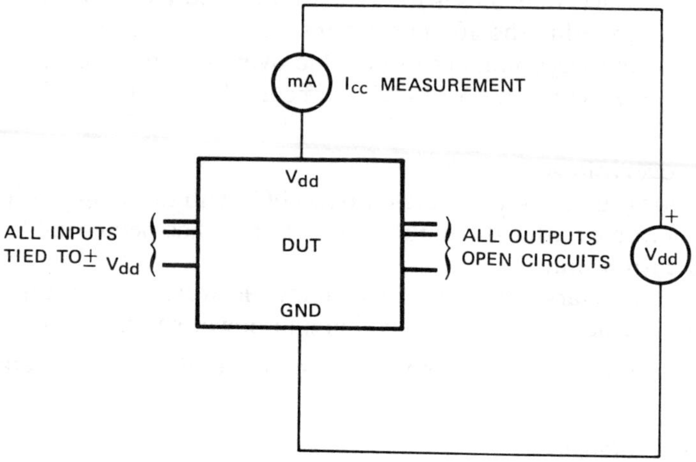

Figure 7.7 Measurement set-up for the characterization of CMOS IC susceptibility to supply overvoltages.

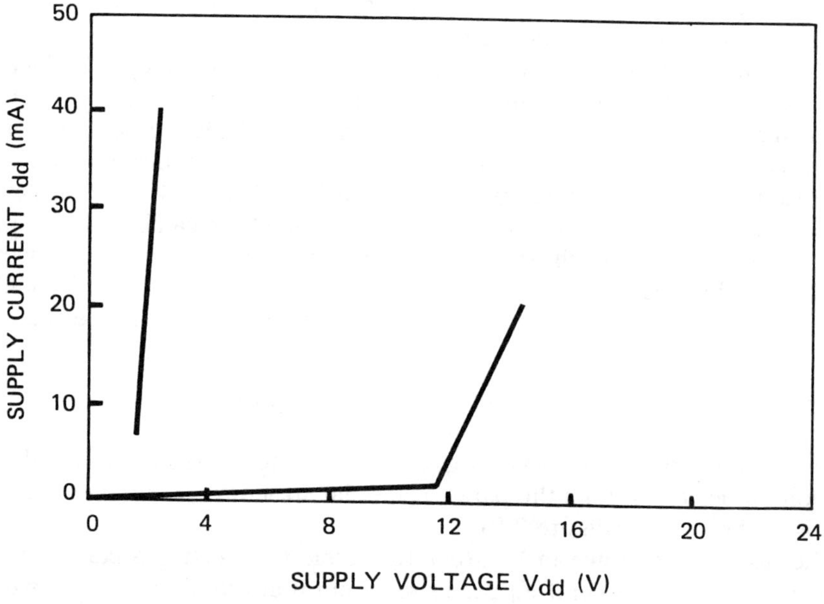

Figure 7.8 Latch-up triggering in a custom CMOS IC by supply overvoltage exceeding the breakdown voltage.

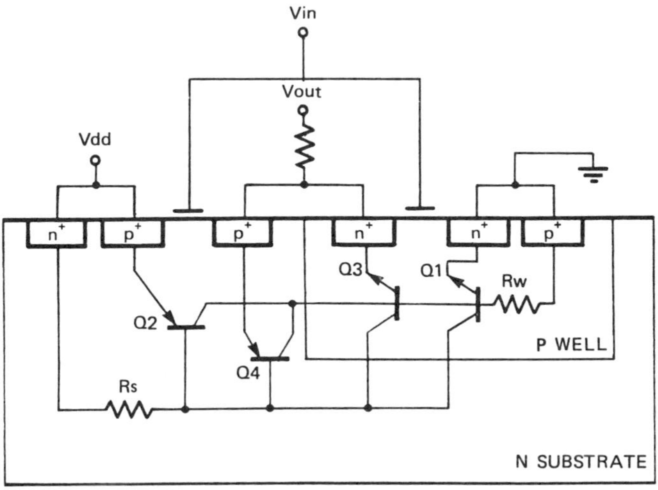

Figure 7.9 Schematic cross section of a CMOS inverter identifying parasitic transistors due to source and drain MOS regions.

The test circuit in Figure 7.10b sinks current from the diode between an I/O pin and V_{ss}. While testing an output, it must be put in three-state if possible; otherwise, its logic state must be 0 when current sinks from the output (negative overvoltage) and 1 when current is injected into the output (positive overvoltage).

The I/O current can also be established before applying V_{dd}. This is a more severe test, which simulates a condition in which one section of a logic system is powered down while its inputs remain active.

These tests can be easily performed by using a curve tracer to drive parasitic transistors connected to I/O (Figure 7.9) until latch-up occurs. To drive the npn transistor, for instance, the input or output pin is connected to the negative current step generator of the curve tracer (base drive inverted), while V_{dd} and V_{ss} are connected to the positive voltage (collector drive) and to ground, respectively. In this way, the npn transistor is driven in a common-base configuration, as depicted in Figure 7.11a, and the CMOS IC is correctly biased by a positive voltage V_{dd}. The base-step generator of the curve tracer sinks a current from the I/O of the device and drives the npn parasitic transistor, the characteristics of which are shown in Figure 7.11b.

When latch-up is triggered, current I_{dd} increases abruptly. The value of the sink current (i.e., the base current step of the curve tracer that causes latch-up, being 35 mA in Figure 7.11b), provides a measurement of the latch-up sensitivity of the device. The symmetric configuration (i.e., V_{dd} at the emitter drive of the curve tracer, V_{ss} at collector, and the inverted base signal at the I/O) can be used to inject positive current, thus biasing the I/O parasitic pnp.

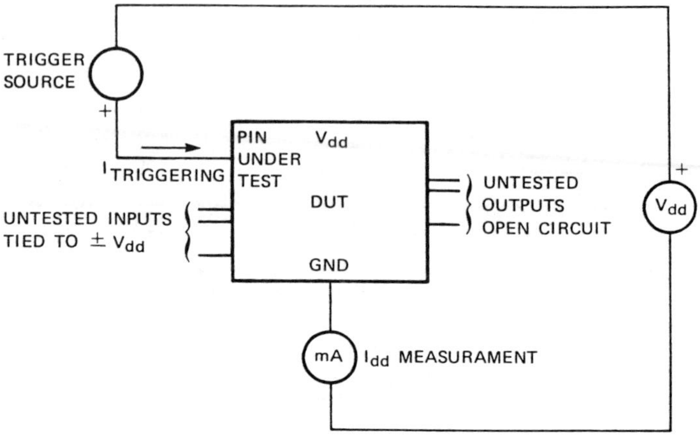

(a)

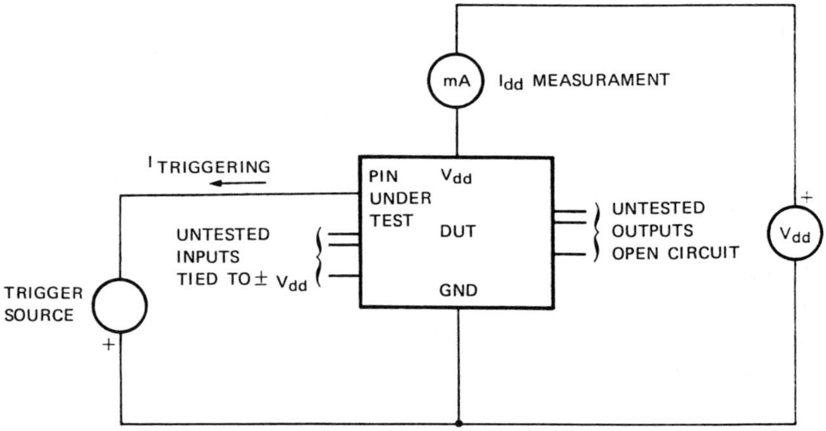

(b)

Figure 7.10 Measurement set-up for the characterization of CMOS IC to current injection into I/Os; (a) current injection (positive overvoltages); (b) current sinking (negative overvoltages).

(c) Pulsed Currents into Inputs-Outputs

Pulsed current tests provide a more realistic simulation of transients and reduce possibly erroneous results and damage that can arise in dc tests due to device heating. Moreover, they can be implemented in automatic test equipment.

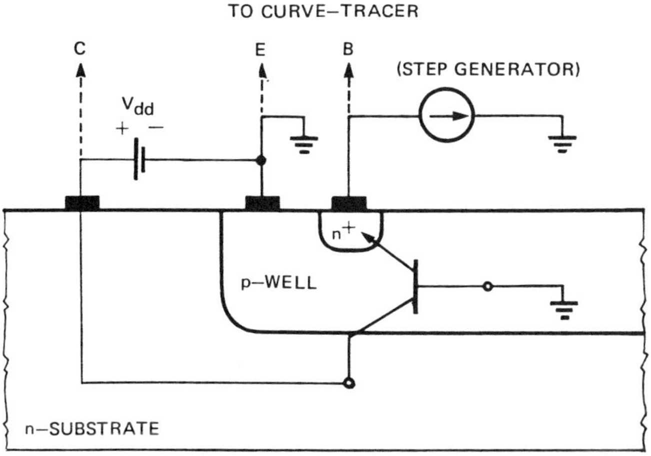

(a)

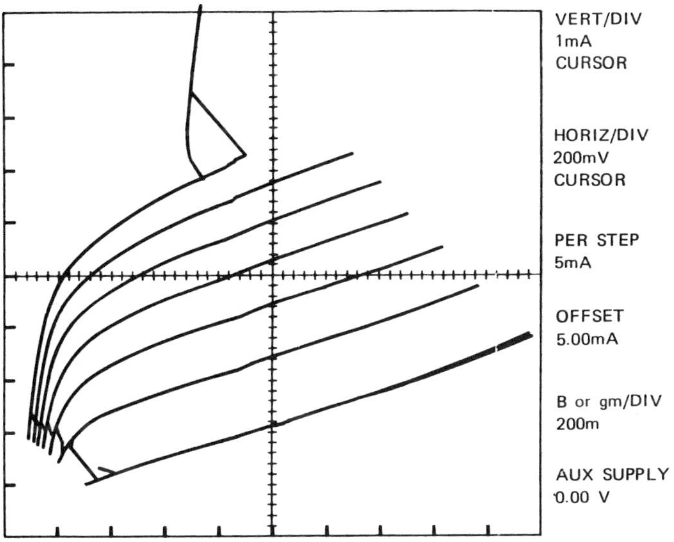

(b)

Figure 7.11 (a) Schematic cross section showing triggering of latch-up by driving an npn I/O parasitic transistor by the curve-tracer; (b) I_{dd}-V_{dd} characteristics of the output of a custom CMOS device, showing latch-up triggering at I_{out} = 35 mA.

Latch-up immunity is defined, as for dc tests, as the minimum I/O current able to cause latch-up, the *threshold current*. Threshold current increases with lowering pulse duration; below a defined pulse duration, latch-up cannot be triggered, regardless of how high the current pulse is, as shown in Figure 7.12a. Conversely, for sufficiently long pulses, the latch-up current threshold approaches the dc value. Note that this dc threshold current can be higher than the value obtained by means of a curve tracer because, in the curve tracer, the current at the I/O is applied by the step generator before increasing supply voltage V_{dd}, thus favoring latch-up.

Figure 7.12b shows the dependence of latch-up threshold current on pulse duration for a CMOS custom device. A similar dependence with two asymptotic limits is observed in test patterns driven by a current generator across shunt resistances [9]. The dependence on a minimum pulse duration for latch-up is attributed in this case to the base transit times of the parasitic transistors, which are longer than the RC times associated with the well-substrate junction.

Figure 7.12 also highlights the lowering of threshold current with increasing the temperature; in fact, bypass resistance values increase at increasing temperature, like the current gain of parasitic transistors, whereas base-emitter forward-bias necessary to support a given collector current decreases. All of these effects cause threshold and holding currents to decrease at higher temperatures. Electrical testing should therefore be performed at the maximum specified temperature in order to obtain actual "worst-case" values.

(d) Capacitor Discharge into Inputs to Simulate Electrostatic Discharge

I/O overvoltages and current injection can be also caused by *electrostatic discharge* (ESD). To test this triggering mechanism, a capacitor is charged to a specified voltage and then discharged into the I/O through a resistance, while the device is biased [7]. Alternatively, ESD simulators can be used for this purpose. This kind of test does not provide information different from continuous (b) and pulsed current (c) tests.

(e) Voltage Transients on the Supply Line and Power Supply Ramp-up

Voltage transients on the supply line can induce a lateral current in substrate and well. Referring to the cross section of Figure 7.1a, when supply voltage rapidly increases due to electrical noise or turn-on transients, the well-substrate depletion region widens, and a displacement current therefore appears between substrate and well. Induced lateral current depends on both junction capacitance and voltage ramp rate dV/dt.

To simulate this triggering mode, a voltage ramp can be applied to the supply line of the IC by a voltage pulse generator with variable rise time t_r. Keeping the

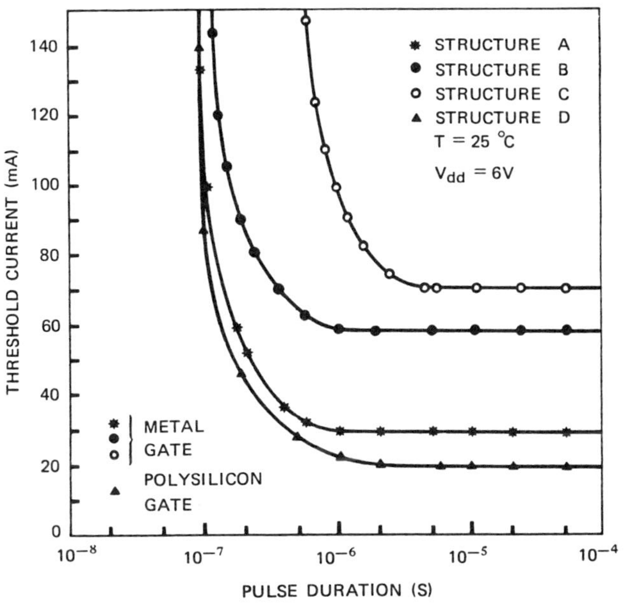

(a)

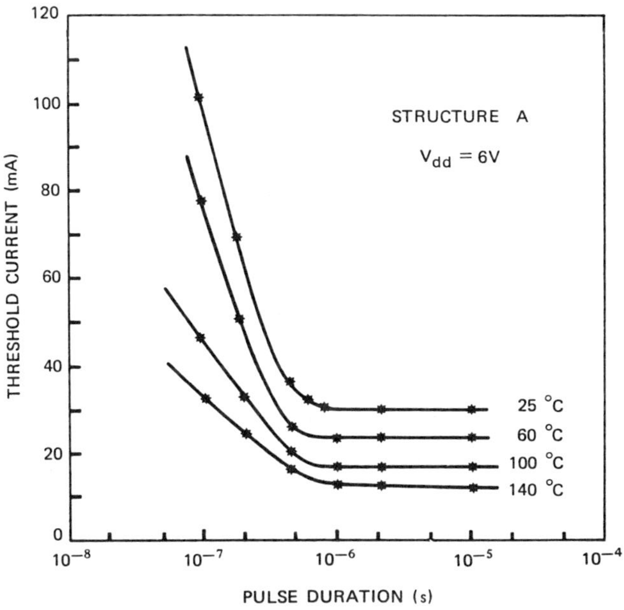

(b)

Figure 7.12 (a) Latch-up threshold current *versus* current pulse duration for various latch-up prone structures; V_{dd} = 6 V, temperature = 25°C; (b) effect of temperature on latch-up threshold current *versus* pulse duration characteristics.

maximum voltage value constant, equal to the maximum specified V_{dd}, t_r is decreased until the ramp rate dV/dt = V_{dd}/t_r becomes sufficiently large to induce latch-up. The minimum dV/dt sufficient to induce latch-up is an effective measure of latch-up susceptibility.

Figure 7.13 shows the effect of a voltage transient on the V_{dd} line of a custom 6μm polysilicon-gate CMOS IC. When the dV/dt value is too low (Figure 7.13a), latch-up does not occur and supply current returns to the extremely low normal values (Figure 7.13b). If the ramp rate is increased, the critical dV/dt value (10^8 V·s^{-1}) is reached (Figure 7.13c), the displacement current induces latch-up, and supply current increases sharply (Figure 7.13d). To avoid damaging the device, supply current was limited during the experiment.

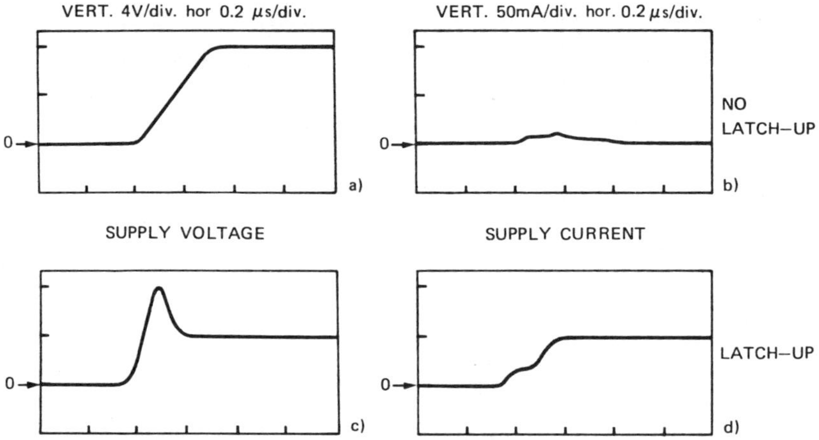

Figure 7.13 Effect of a supply voltage transient on a custom CMOS device: (a) when the ramp rate is insufficicent, latch-up does not occur, and supply current (b) returns to normal values. If the critical ramp rate 10^8 V · s^{-1} is exceeded, (c), the circuit latches and a large current flows through the device (d).

7.2.3 Standardization of Latch-up Electrical Testing on Finished Integrated Circuits

Despite testing for latch-up being a generalized industrial practice, standardization is still very poor [7,10]. A standard measurement procedure for the characterization of CMOS integrated circuit latch-up susceptibility or immunity, measured under the conditions (a), (b), and (c) of the previous section, is being developed by the JEDEC JC-40.2 Logic Standardization Committee [8] and by the International Electrotechnical Commission (IEC).* Only supply and input-output overvoltages are considered, following the test set-up of Figures 7.7 and 7.10, respectively.

These tests are considered necessary and sufficient for an exhaustive latch-up characterization.

Pulse durations longer than 10 μs are recommended. These durations allow us to introduce a cool-down time between repeated measurements and are sufficiently long to be considered static. Timing specifications on rise, fall, and delay times are also declared.

The specified sequence to be followed consists of applying power supply, setting up I/O conditions to place device in desired state, applying the trigger source for desired duration, measuring supply current to verify that latch-up has occurred, removing power supply, and inspecting the device to verify that it has been permanently damaged. The magnitude of the trigger current or voltage required to initiate latch-up is a measure of the device latch-up susceptibility. The test may be performed over the full range of temperatures, supply voltages, and input-output dc bias voltages specified in the manufacturer's data sheet. Also, all valid device logic states may be tested.

Because latch-up testing is potentially destructive, the JEDEC procedure is recommended only for device characterization and lot qualification purposes, and it is not a 100 percent inspection test.

7.3 ANALYTICAL TECHNIQUES

The electrical characterization of latch-up is of little help in correcting device technology or layout because, even if a high latch-up susceptibility is verified, no information is available concerning the reasons for poor resistance to latch-up. We therefore need to use analytical techniques, which enable us to identify active latch-up paths within the device under test. Analytical techniques can be divided into methods for the localization of latch-up sites and those for the study of the local latch-up sensitivity.

The localization methods enable active parasitic structures to be identified once the device has been brought to the latch-up condition by a suitable electrical stimulus. This implies that a detailed electrical characterization of the phenomenon must be performed beforehand.

Latch-up localization is based on detection of secondary effects induced by the large current flow:

(1) Localized heating, which can be detected by liquid crystal techniques [11] or by IR thermography [12, 13];

(2) Infrared (IR) radiation emission due to hole-electron recombination in forward-biased junctions, detected by IR microscopy [14, 15];

(3) Voltage drops in well and substrate, which can be detected by means of capacitively coupled voltage-contrast techniques in a scanning electron microscope (SEM) [16–19].

Other techniques, based on supply current modulation in the latched state by means of e-beam [16, 20] or laser [21], are characterized by reduced sensitivity and difficult interpretation of results, and so will not be considered here.

Techniques for the study of local latch-up sensitivity adopt a focused laser [19, 22–27] or electron beam [28] as a localized current generator, wherein latch-up is fired by the locally generated current. These techniques provide a direct indication of the local sensitivity to latch-up and prevent the designer from neglecting potentially dangerous latch-up sites due to an incomplete electrical characterization.

In the following, we will describe the different techniques referring to the same CMOS IC part, manufactured with a standard 7 μm metal-gate p-well junction-insulated technology. The large geometries employed enable an easier comparison among the different techniques. Fig. 7.14a shows the layout of this block; metal lines are omitted for the sake of simplicity. Fig. 7.14b shows the cross section of the block along the AA′ axis. The parasitic pnp transistor Q_1, formed by a p^+ source, the n-substrate, and the p-well, and the vertical npn transistor Q_2, composed by an n^+ source, the p-well, and the n-substrate, are identified. In Figure 7.14c the optical micrograph of the block is shown.

In this device latch-up can be initiated by means of a parasitic overvoltage applied to a specific input, which forward-biases the p^+/n input protection diode D_1 in Figure 7.14b.

7.3.1 Techniques for the Localization of Latch-up Sites

(1) Liquid Crystals and Infrared Thermography

This technique is based on the observation of "hot spots" in the neighborhood of parasitic SCR structures, where, due to the large current flowing, local temperature increases occur. These temperature increases can be detected either by infrared thermography [12, 13] or liquid crystal techniques [11], both of which are relatively simple to apply. In particular, whereas liquid crystals need tiresome operations, IR thermography offers good reproducibility, requires only IC package opening, and is completely nondestructive. New computer-controlled IR thermography microscopes [29] allow a temperature sensitivity better than 0.5°C and a spatial resolution of 15 μm. The traditional problem of IR thermography of integrated circuits, the correction of different surface emissivities due to the presence of different materials on the chip surface, is overcome by measuring the object radiance at two different temperatures and calculating emissivity for each image point [29]. The system computer automatically corrects the radiance map and then provides actual temperature maps at high magnification and high sensitivity.

Figure 7.15a shows the temperature map of the device in latch-up condition. To obtain this image, the device was mounted on a temperature-controlled heat

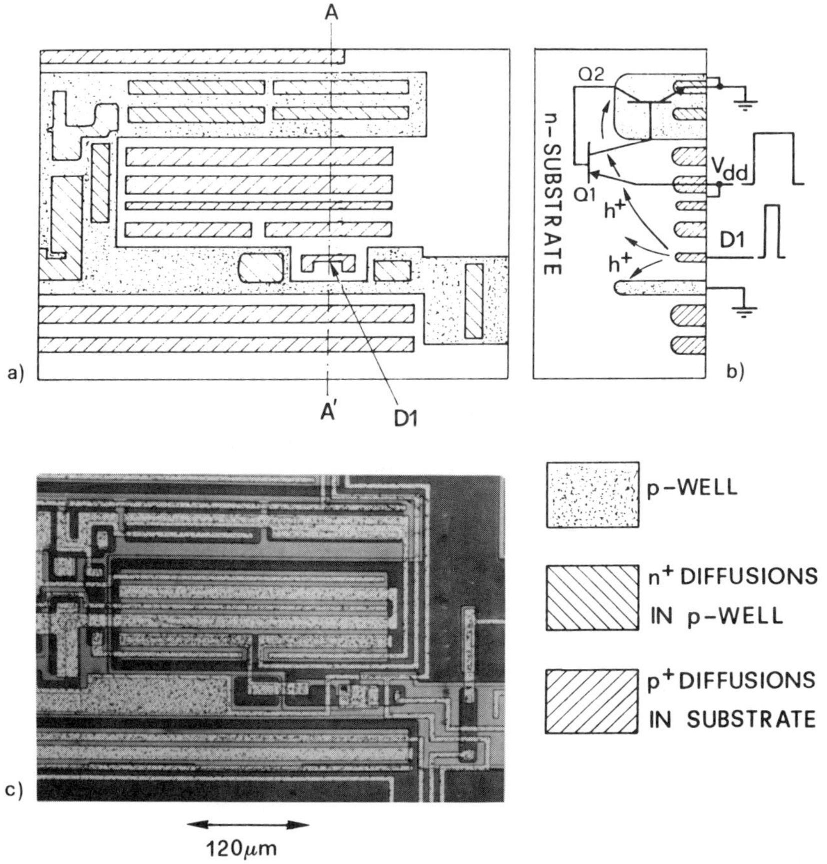

Figure 7.14 (a) Schematic layout of the CMOS analog block under test, metal lines are omitted; (b) cross section along the AA′ axis showing parasitic bipolar transistors $Q1$ and $Q2$; (c) optical micrograph of the device under test.

sink and was forced into latch-up by applying pulsed V_{dd} and V_{in} signals. The duty cycle of the V_{dd} signal was chosen to be low enough to minimize the device's heated area, which appears as the shaded area in the temperature map of Figure 7.15a. The heated area is also visible in the schematic of Figure 7.15b. As we can see, IR thermography provides a rough indication of the position of the latch-up site, but its effectiveness is limited by two main disadvantages:

(a) Lateral resolution is too poor to identify the pnpn structure, especially in densely integrated circuits. Moreover, device heating always spreads over a rather large area.

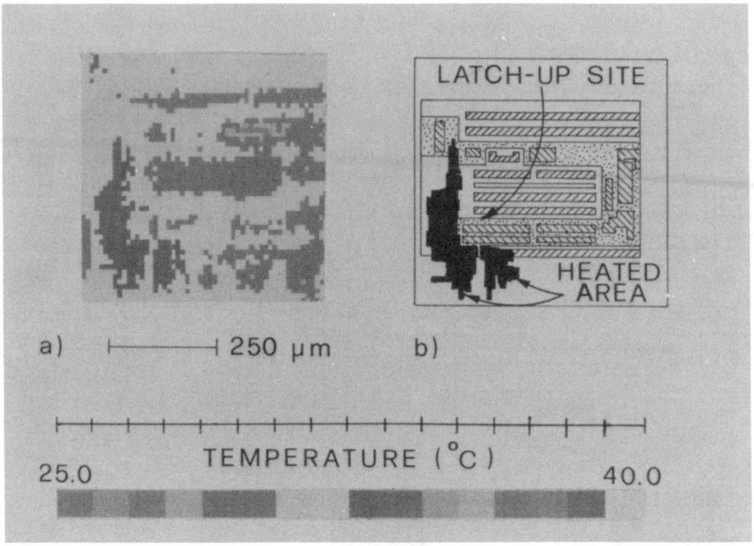

Figure 7.15 (a) Temperature map of the device under test in latch-up conditions obtained by IR thermography; (b) schematic layout showing the position of the hot spot.

(b) Local differences in device heat dissipation, for instance, because of the presence of metal lines, bonding pads, or defects in die bonding, can lead to erroneous results.

(2) Infrared Microscopy

Latched areas can also be identified by detecting the near-IR radiation ($\lambda = 1.1$ μm) due to electron-hole recombination in semiconductor regions crossed by large forward current [14,15].

To convert IR radiation into visible light, special IR-sensitive television systems are used. The televisions are modified and treated to minimize aberrations and glare and to optimize contrast and image quality in the IR spectrum. Vidicon cameras are usually adopted as detectors; their response covers the range between 450 and 1800 nm, but the visible part of the spectrum is usually filtered to avoid saturating the detector.

Because blackbody radiation of a sample at temperature $T < 300$°C has negligible intensity for wavelengths lower than 2 μm [29], the map of the IR radiation emitted by the device under test will show the actual distribution of recombination current, rather than areas of maximum heat dissipation.

The lateral resolution of the IR microscope is governed by the Rayleigh criterion $R = \lambda/NA$, where NA is the numerical aperture of the employed objective, which is here a factor of 2 to 3 worse than in conventional optical microscopes. Nevertheless, it is sufficient for latch-up analysis in densely integrated CMOS circuits.

Device topography can be observed by illuminating the device with an external IR radiation source and detecting the reflected light. Figure 7.16a shows the IR image (reflected light) of the CMOS analog block under test in the absence of latch-up. When latch-up is fired, bright spots appear, corresponding to semiconductor regions where recombination current is higher (Figure 7.16b). In Figure 7.16c, the external IR source was turned off, so that only the radiation emitted by the device would be detected. Due to the higher doping levels, recombination currents appear to be more intense in p-well, which therefore appears brighter than n-substrate.

As we can see, IR microscopy provides a clear and unambiguous detection of the latch-up site and current distribution within the pnpn path. When circuits are observed from the top, the presence of many conductor layers can mask the emission, as is the case with other techniques. A specific advantage of this technique is that, being silicon-transparent to near IR, recombination radiation can be observed from the back, through th silicon substrate. This can be simply, but exclusively, accomplished in plastic-packaged ICs by folding the terminals and lapping the package and the frame until the silicon substrate appears [30].

Figure 7.17a shows the backside IR image (emission and topography superimposed) of the device under test in the latched state; Figure 7.17b shows the IR radiation emitted by the device. Images are mirror-reflected with respect to the previous ones. Emission from regions that from the top side are covered by metal lines can now be revealed, as we can see by comparing Figures 7.16 and 7.17. This unique advantage of IR microscopy becomes more useful as the complexity of circuit layout and the density of metal tracks increase, including four or more conductor levels.

There are, however, also some disadvantages: (1) the Si emission due to latch-up may be very weak and sensitivity problems can arise; (2) no information is available on the logic state of internal IC nodes and switching of logic circuits, which may interfere with latch-up behavior; and (3) temporal resolution is poor and does not allow us to follow triggering and transient latch-up phenomena in detail. Despite these limitations, IR microscopy can be used effectively for the debugging of latch-up problems in CMOS ICs, even for advanced CMOS technologies with very narrow design rules.

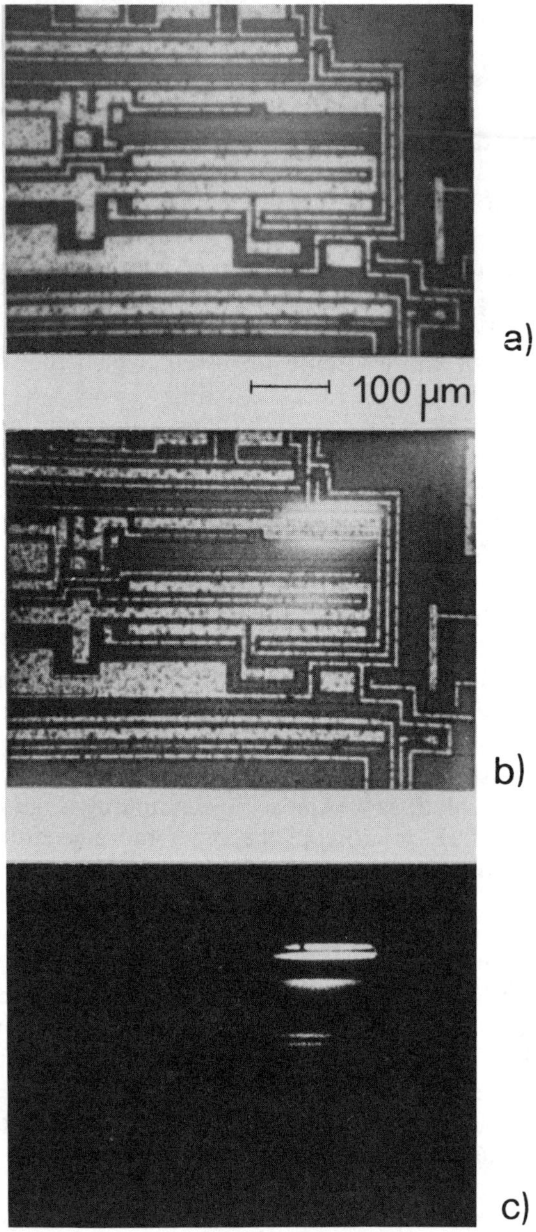

Figure 7.16 IR micrographs of the device under test observed from the top; (a) with topography superimposed, no latch-up; (b) with topography superimposed, in latch-up condition; (c) without topography, in latch-up condition.

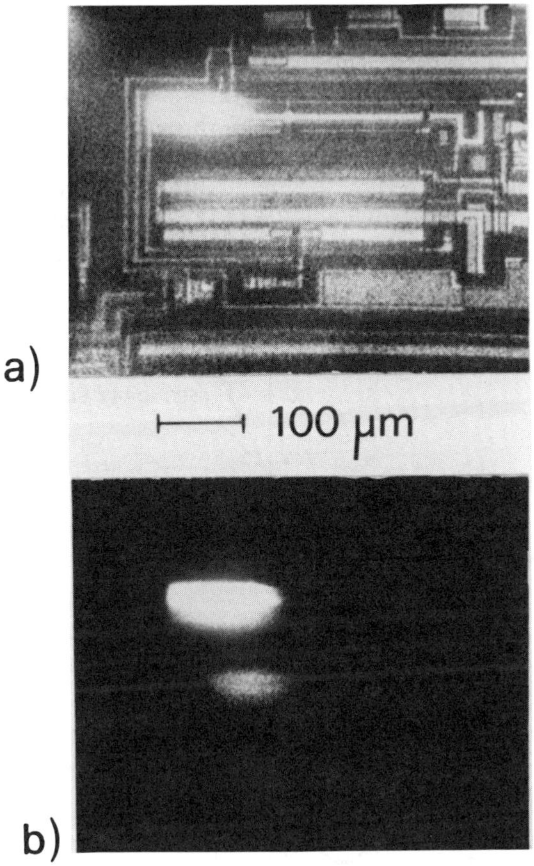

Figure 7.17 IR micrographs of the device under test observed backside; (a) with topography superimposed, in latch-up condition; (b) without topography, in latch-up condition.

(3) SEM Voltage Contrast Techniques

In the latch-up condition, both parasitic transistors are turned on, and a large current flows through the well and substrate areas involved in the parasitic pnpn paths. As a result, a large potential increase occurs in the p-well, which is normally at ground potential, while the potential of the n-substrate, normally at V_{dd}, decreases.

Latch-up current paths therefore can be effectively localized if a technique for revealing the voltages on the surface of p-well and substrate, which are covered by thermal oxide and passivation, is available [16–19].

Conventional voltage contrast cannot be used for this purpose because, to observe potentials on silicon surfaces, the primary electron energy ought to be high enough for complete penetration of overlying insulating layers. This is incompatible with the analysis of MOS structures, which must be observed at low energy levels (below 2 KeV) to avoid MOS threshold voltage alterations with a consequent change of logic states. Nevertheless, potentials on buried conductors and the silicon surface, covered by thermal oxide or passivation, can be detected at low beam energies by capacitive coupling [31, 32] between the insulator surface and the underlying conductor or semiconductor (Figure 7.18a).

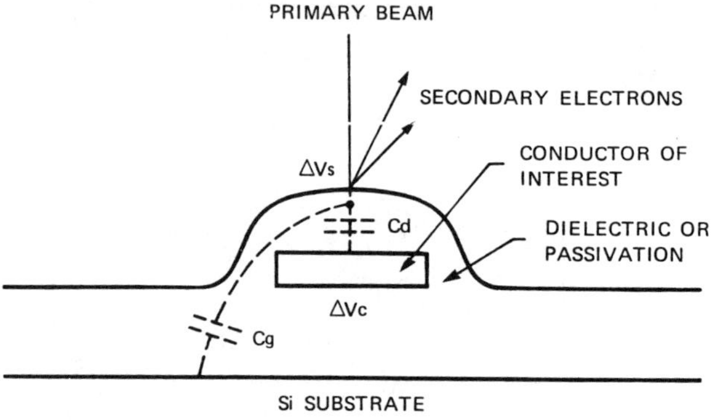

(a)

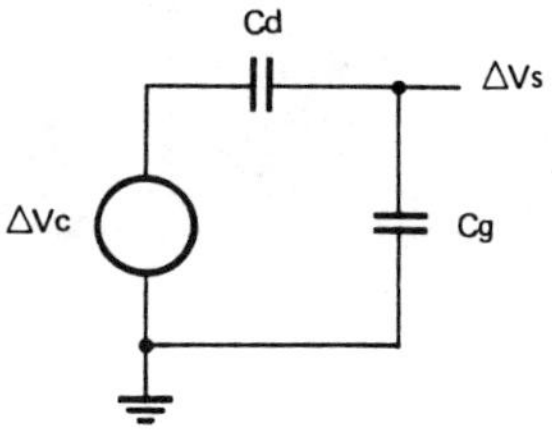

(b)

Figure 7.18 (a) Schematic cross section of a conductive track on an integrated circuit, illustrating the principle of capacitively coupled voltage contrast; (b) equivalent circuit. In general $C_d \gg C_g$, so that $V_s = V_c$.

The switching of a positive or negative voltage on passivated conductor tracks influences additional respective positive or negative charges on the insulator surface so that the variation of the surface potential V_s follows the variation of the conductor potential V_c. However, the equilibrium condition of the insulator surface is disrupted, and electron irradiation or surface conduction tends to compensate induced surface charging so that, after a time t, the surface potential returns to the equilibrium value and the information about the conductor voltage is lost.

Capacitively coupled voltage contrast therefore rapidly decays and cannot be used for a stationary image, which can be obtained only by switching the device bias on and off in synchrony with the beam scanning, faster than the decay time of the capacitively coupled voltage.

The effect can be viewed as the result of the connection of a capacitor C_d between the track of interest and the surface of the insulator (Figure 7.18a). As long as this capacitance is much greater than the stray capacitance C_g, the surface of the insulating layer accurately follows the potential of the buried track (Figure 7.18b). By use of capacitively coupled voltage contrast, it is possible to observe the potential present on metals covered by the passivation layer and semiconductor diffusion covered by the thermal oxide. Nevertheless, the superimposition of voltage and topographical contrast makes it difficult to identify potential drops along current paths [16].

Topographic and chemical contrast information may be effectively reduced by subtracting the signal detected when the device is unbiased from the signal obtained in the biased condition, thus enabling the resolution of small potential differences. A further improvement can be achieved by use of digital image acquisition and processing, which greatly improves the signal-to-noise ratio and enhances image contrast [17].

Figure 7.19 reports the capacitively coupled voltage contrast images of the device under test obtained by means of digital acquisition and topography subtraction. Figure 7.19a and 7.19b refer to the normal and latched state, respectively. In the latched state, a marked voltage drop is observed in the substrate and p-well regions. As a consequence of latch-up, this substrate zone becomes whiter as the p-well immediately beside it becomes darker. The central region, where the highest voltage drop is observed, appears to be crossed by the largest current flow.

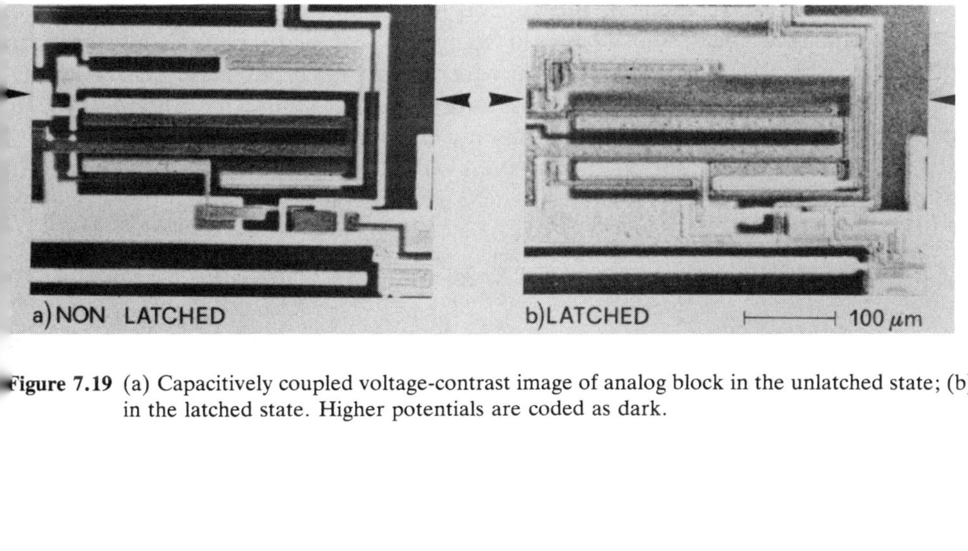

Figure 7.19 (a) Capacitively coupled voltage-contrast image of analog block in the unlatched state; (b) in the latched state. Higher potentials are coded as dark.

The capacitively coupled voltage-contrast technique reveals small voltage differences on the surfaces of semiconductor areas and identifies latch-up current paths in the steady state. In a large number of circuits, this can give sufficient information to find layout weaknesses. In complex CMOS ICs, however, the firing point can be different from the latch-up site in steady-state conditions, and a complete understanding of latch-up triggering requires a technique capable of following the temporal and spatial evolution of latch-up. Moreover, only a dynamic observation technique can give insight into transient phenomena [33], which, even if they do not lead to a sustained latch-up condition, can disrupt the correct behavior of the device and introduce logic errors.

Stroboscopic voltage-contrast techniques can observe temporal evolution of potentials on the surface of the ICs under test. Observation of potentials on the semiconductor surface can be obtained due to capacitive coupling of the surface of the thermal oxide covering the semiconductor by pulsing the device supply voltage V_{dd}.

As an example, Figure 7.20 shows stroboscopic voltage-contrast micrographs of the device under test taken at different times during the firing of the phenomenon. Figure 7.20a shows the potential distribution in normal conditions. When the input pulse is applied at $t = 2.2$ μs (Figure 7.20b), the p-well adjacent to the diode D1 reaches a high potential. Shortly after the input pulse is turned off (Figure 7.20c), the potential in the lower p-well decays, whereas the upper part of the p-well has not yet reached the latched condition and still holds a low potential. A voltage increase at the extreme right of the p-well occurs at $t = 4.2$ s, as shown in Figure 7.20d; if we increase the time, the area involved in latch-up widens. A marked increase in potential of the whole p-well area is observed when the steady state is reached (Figure 7.20e) along with maximum current absorption.

Identifying latch-up sites by using SEM voltage contrast usually requires comparison of voltage maps in the latched and unlatched states. Analysis of the results is therefore not straightforward, as in the case of IR microscopy, and can become nearly impossible for devices with a high metal density. Nevertheless, SEM voltage contrast provides other information that cannot be obtained by IR microscopy: (1) SEM enables triggering and transient phenomena to be analyzed with high temporal resolution (better than 1 ns); (2) SEM reveals the logic state of internal IC nodes and its changes, which may interfere with latch-up behavior.

7.3.2 Techniques for the Study of the Local Latch-up Sensitivity

These techniques provide a direct indication of the local sensitivity to latch-up and do not require in-depth electrical characterization. The device is normally biased and is scanned by an electron beam or a laser. Hole-electron pairs are generated in the semiconductor and collected by reverse-biased junctions. Depending on the

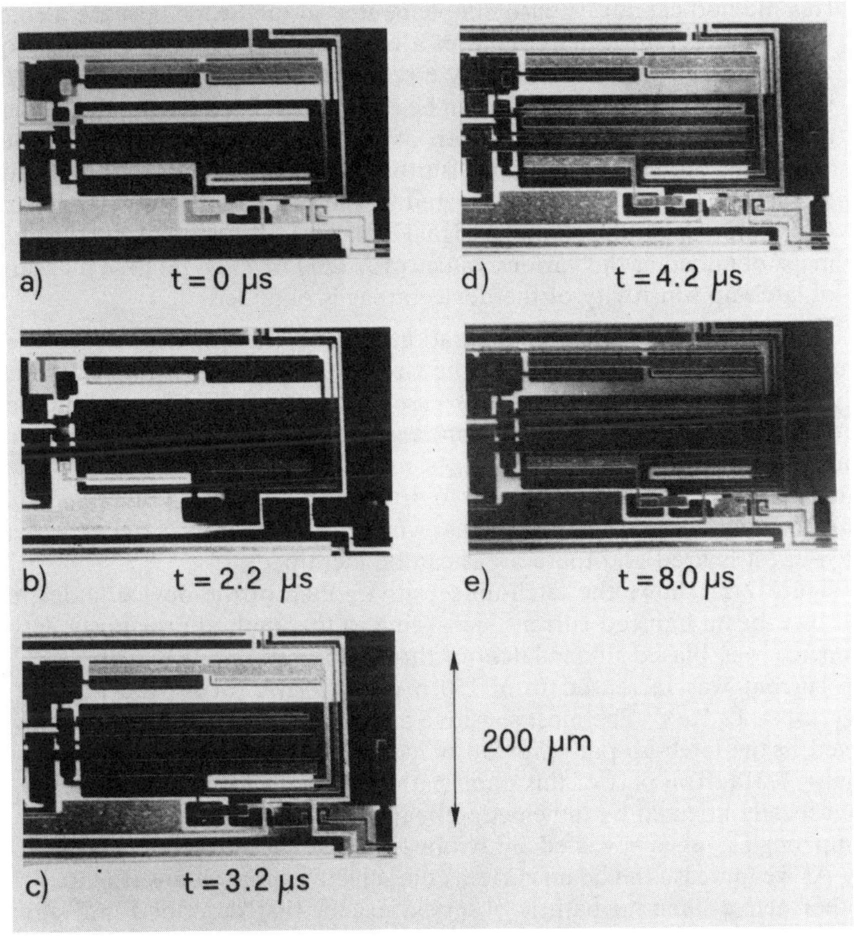

Figure 7.20 Stroboscopic voltage contrast micrographs of the analog block at different times during latch-up firing. Higher potentials are coded as dark.

scanning system available, the sensitivity analysis can be performed in two different ways:

(a) In computer-controlled laser scanning systems [24], for each position of the spot, the light intensity of the beam is increased until latch-up occurs, and the inverse of this intensity is taken as a measure of the local sensitivity to latch-up in that point. The resulting image is a gray level map or a y-mod graph of the sensitivity of the device under test.

This method cannot be easily implemented in the SEM, because a change in the beam current usually requires a readjustment of focus and position.

(b) Alternatively, the SEM primary e-beam current I_b (at fixed energy E_b), or the laser equivalent power P_0 can be maintained fixed during the frame scan, and different maps can be taken at different I_b (E_b) and P_0. By using a voltage as the video signal that is proportional to the device supply current, the areas where latch-up is triggered by the scanning beam will appear black (or white, depending on the signal polarity). By comparing the different maps obtained at the various values of I_b (E_b) or P_0, a relative measurement of latch-up sensitivity of the device areas is obtained.

Two different electrical configurations can be chosen. In the first one, only the supply voltage V_{dd} is applied to the circuit and latch-up is due to the current locally injected by the beam; in this case, a true "sensitivity measurement" is performed. In the second configuration, a voltage V_{in} is applied to a specific input (or output) in synchrony with the V_{dd} signal. The input pulse V_{in} alone will not initiate latch-up, but it is high enough to drive the circuit into a *marginal condition* [34] for the triggering of latch-up so that when the beam crosses the areas sensitive to latch-up, it is fired and these areas can be identified.

Figure 7.21 shows the latch-up sensitivity map of the device under test obtained by e-beam induced current, according to the method previously described. The device was biased under latch-up threshold at V_{dd} = 6V and V_{in} = 14 V. Beam current was increased from 150 nA to 970 nA, at a fixed primary beam energy E_b = 12 KeV. The most sensitive area is at the extreme right of the p-well involved in the latch-up path previously identified (Figure 7.14) and marked as *A* in Figure 7.21a. The part of this pnpn path that seems to be most sensitive to the current locally injected by the electron beam (Figure 7.21b) is also the area where latch-up begins, as is revealed by stroboscopic voltage-contrast analysis (Figure 7.10). As we increase the beam current, the sensitive area widens (Figure 7.21c–e). No other active latch-up path is observed except that described in Figure 7.14, thus confirming that the studied triggering mechanism is actually the most dangerous for the device.

Sensitivity analyses by means of e-beam scanning can be easily accomplished in the SEM by monitoring device supply current. The main problem of operating the SEM as a current injector is the damage of MOS devices that occurs if the high-energy electron beam is scanned in the neighborhood of MOS gates. As is well known [35], SEM observations at beam voltages greater than 2 kV can damage MOS devices, causing charge trapping in the gate oxide and shifts in threshold voltage. In effect, shifts in the threshold voltage of MOS transistors had been observed when conventional scanning systems were adopted [20]. Damage to MOS devices can be minimized by digital control of the beam position, which allows definition of areas that should be protected from charge injection and blanking the beam automatically when MOS gate areas are crossed. This technique can be

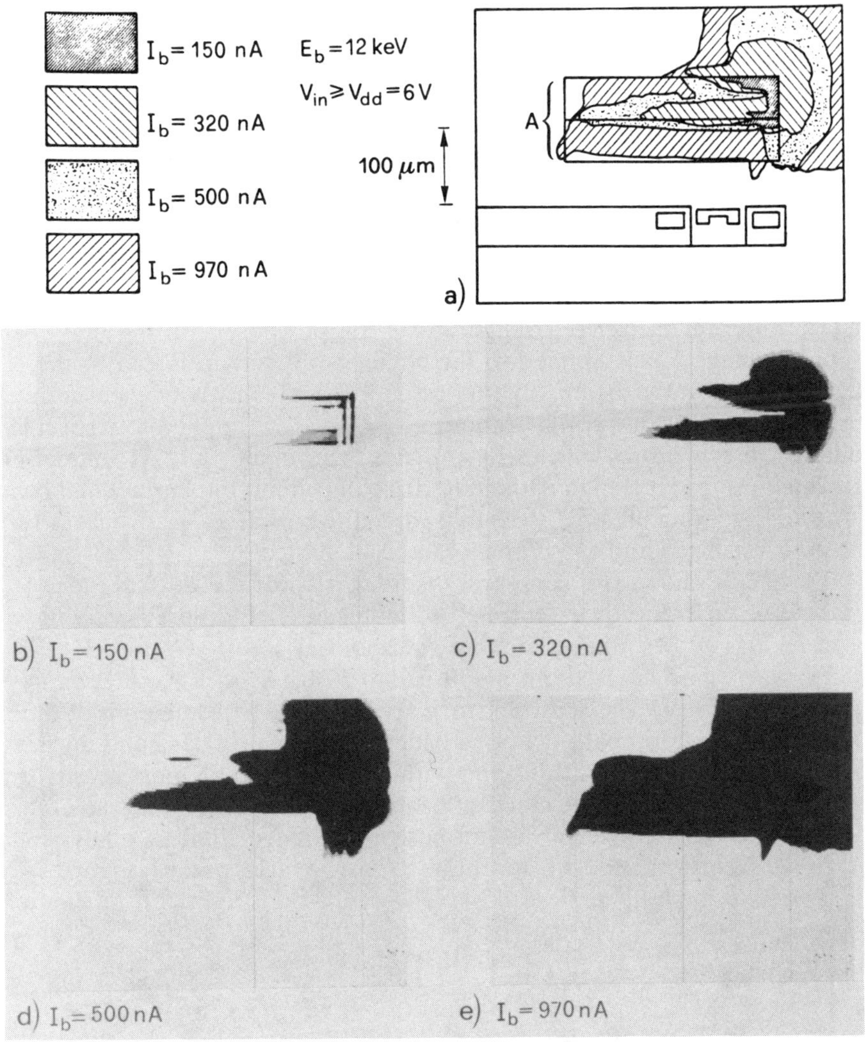

Figure 7.21 (a) E-beam sensitivity of to latch-up of the different areas of the analog block when a positive pulse $V_{in} = V_{dd} + 14$ V is applied at the input; $V_{dd} = 6$ V, $E_b = 12$ keV, (b) Gray level map of the supply current during the scan with $I_b = 150$ nA; (c) $I_b = 320$ nA; (e) $I_b = 970$ nA.

implemented in the same SEM apparatus that allows voltage-contrast analyses so that results of the different testing methods can be easily compared. Major drawbacks are (1) that accurate calibration of the primary beam current I_b must be repeated before each measurement, and (2) in densely integrated circuits, only spot or line measurements are possible.

Sensitivity analysis was repeated by using a scanning laser system (Biorad Lasersharp SOM 150), which employs two He-Ne sources, 5 mW and 15 mW, λ = 633 nm, with a series of calibrated attenuators. Figure 7.22 reports laser sensitivity maps under the same electrical conditions as Figure 7.21. In Figure 7.22a and 7.22c, the measured latch-up current is superimposed on the device layout. Figures 7.22b and 7.22d show only the areas where latch-up can be triggered by the laser beam, respectively with an equivalent power of 1 and 2 mW. As we can see, scanning laser sensitivity measurements perfectly reproduce results previously obtained by means of e-beam induced current.

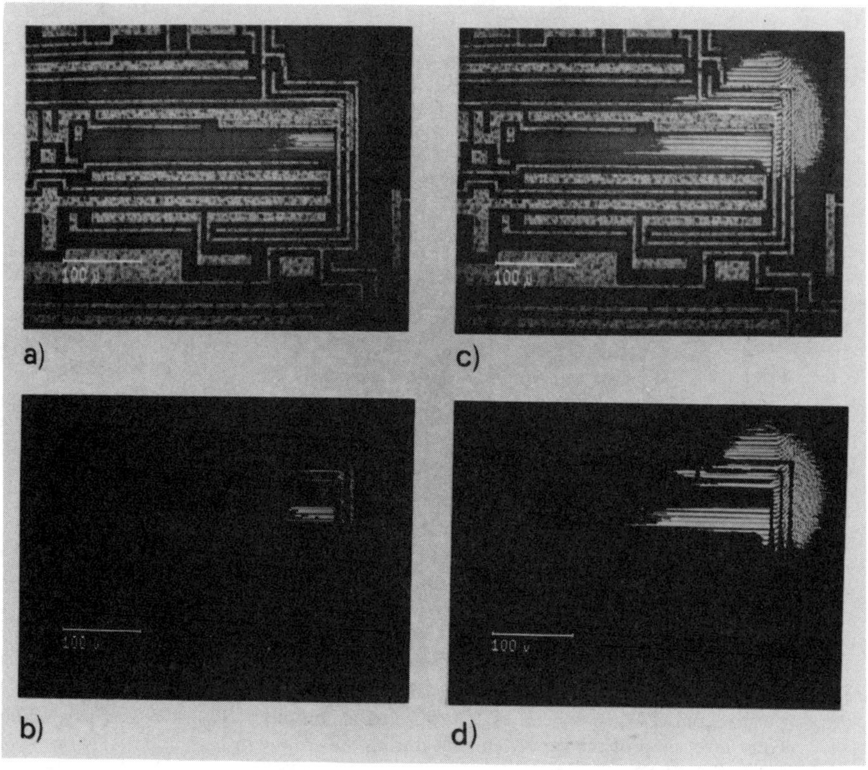

Figure 7.22 Scanning laser (He-Ne 15 mW, λ = 633 nm) sensitivity maps of the different areas of the analog block in the same electrical conditions as Figure 7.11; (a) and (b) equivalent laser power = 1 mW, with and without reflected light image of the layout, respectively; (c) and (d) equivalent laser power = 2 mW.

Moreover, laser scanning eliminates the problems of e-beam damage of MOS devices, thus providing a completely nondestructive latch-up evaluation. The disadvantage of this technique is that the depth of carrier generation can be varied only by changing the source wavelength, using a set of different sources from Ar (λ = 488 nm, depth of generation on the order of 1 μm) to Nd-glass (λ = 1060 nm, depth of generation greater than 500 μm). On the contrary, in the SEM technique, the generation depth can be easily increased by increasing the beam energy, but at the risk of damaging the device.

Tables 7.1 and 7.2 review the characteristics of analytical techniques available for the study of latch-up in CMOS.

Table 7.1 Analytical Techniques for Latch-up Localization

Technique	*Principle of operation*	*Disadvantages*	*Advantages*
Liquid crystals	Detect temperature increase	Poor lateral resolution	Easy application
IR thermography	Detect temperature increase	Poor lateral resolution	Easy application
IR microscopy	Detect IR radiation in Si (λ = 1.1 μm) due to carrier recombination associated with forward currents	Poor temporal resolution Sensitivity may be too low	Immediate interpretation Can detect latch-up under metalizations
Capacitively coupled SEM voltage contrast	Detects voltage changes in well and substrate due to latch-up currents	No observation of areas covered by metals	Completely nondestructive High lateral resolution and sensitivity
Stroboscopic voltage contrast analysis of latch-up triggering	Detects dynamic voltage changes in well and substrate due to latch-up currents and transient effects	Require depassivation No observation of areas covered by metals	High temporal resolution High lateral resolution
Supply current modulation in the latched state by electron beam induced current	Change in the supply current due to current magnification by transistor action or avalanche when e-beam scans latched areas	Unavoidable MOS damaging and possible misleading results	
Supply current modulation in the latched state by optical beam induced current	Change in the supply current due to current magnification by transistor action or avalanche when e-	Poor lateral resolution	Nondestructive

Table 7.2 Analytical Techniques for Latch-up Sensitivity Measurements

Technique	*Principle of operation*	*Disadvantages*	*Advantages*
Latch-up triggering by optical beam induced current; sensitivity analysis	Latch-up induced by the locally generated photocurrent Latch-up sensitivity measured by laser beam	Requires a dedicated scanning system	Completely nondestructive Electrical characterization not required Depth of generation variable with laser wavelength
Latch-up triggering by electron beam induced current; sensitivity analysis	Latch-up triggered by the locally induced current	Possible MOS damage	Electrical characterization not required Depth of generation variable with beam energy

7.4 LAYOUT AND TECHNOLOGICAL IMPROVEMENTS FOR AVOIDING LATCH-UP

Due to the high sensitivity of the standard CMOS technology to latch-up, increased by the scaling down of dimensions, the profitable use of CMOS in VLSI requires taking into account the latch-up problem and adopting the necessary precautions. Two ways have been followed: the first one is based on careful layout rules to avoid latch-up paths; the second applies to technological upgrades [1]. The first technique is obviously cheaper and easier to implement, at least in principle, because it does not impose change on the fabrication process. Unfortunately, it causes large waste of silicon area and may become unpractical in very large chips. Technological upgrades may be much more complicated and expensive, requiring a complete redesign of the whole fabrication process. In practice, both techniques have to be applied to VLSI.

7.4.1 Layout Guidelines

The layout optimization methods for increasing latch-up hardness are listed in Table 7.2 [36]. As already said, the main disadvantage for these techniques is the waste of silicon area. On the contrary, they are very easy to implement with a simple modification of the diffusion or implantation masks. In principle, the majority of these techniques try to reduce the R_s and R_w resistances, to avoid turning on the base-emitter junctions of parasitic transistors. One method is to provide many bias contacts to well and substrate to restrict the possible voltage drops between contacts to less than 0.5 V [1], for example. Contacts to V_{dd} and ground are placed into highly doped areas that can be regarded as "majority carrier guard rings," with the same type of doping of the background. The concept of guard rings dates back to the early MOS ICs, where it was used to decouple two active devices, avoiding the formation of the parasitic MOS because of contamination problems [37]. Here, as can be seen in Figure 7.23b, they act as a shunt for the parasitic transistor. A further improvement in latch-up resistance can be obtained by using adjacent contacts ("butted contacts") for guard ring and source regions, if it is allowed by the circuital design.

A different approach is that of minority carrier guard rings: in this case the guard ring has a doping opposite to its background, as shown in Figure 7.23a, and tries to collect the minority carriers injected by input protection diodes or by the drains of output buffers driving, for instance, long mismatched lines [38]. If these minority carriers are allowed to reach the well-substrate junction, they become majority carriers, are accelerated by the field, and may cause the latch-up to be fired. Minority carrier guard rings provide a closer reverse-biased junction to collect the minority carriers; in this sense they are more effective into the substrate than into the well, because the collecting diffusion may be made deeper into the substrate and because in the parasitic bipolar transistor associated with the well the carrier flux is mainly vertical, whereas guard rings are placed laterally and restricted to a thin sheet close to the surface.

7.4.2 Technological Improvements

By changing the fabrication technology, we can try to reduce the coupling between the parasitic transistors, to decrease R_s and R_w, and to degrade bipolar transistors characteristics, in particular their current gain [1]. Table 7.3 summarizes the techniques employed [36].

First, the techniques already known for lowering the gain of bipolar transistors were attempted: the lifetime of minority carriers injected into the base can be reduced by gold doping, as commonly done in TTL technology [39], by neutron irradiation [40], or even by internal gettering techniques, in which oxygen-induced defects are decorated by metal contaminants, resulting in very active generation-recombination centers. All these techniques cause a large increase in leakage currents, which may become unacceptably high in VLSI.

A second method to degrade the transistor gain is to use source-drain Schottky contacts, which greatly reduce the emitter efficiency of bipolar transistors [41]. Due to the greater ease of fabricating Schottky diodes on n-type silicon, mainly pMOS Schottky transistors have been attempted, (Figure 7.24a). However, the device characteristics of Schottky pMOS are much worse than the standard ones,

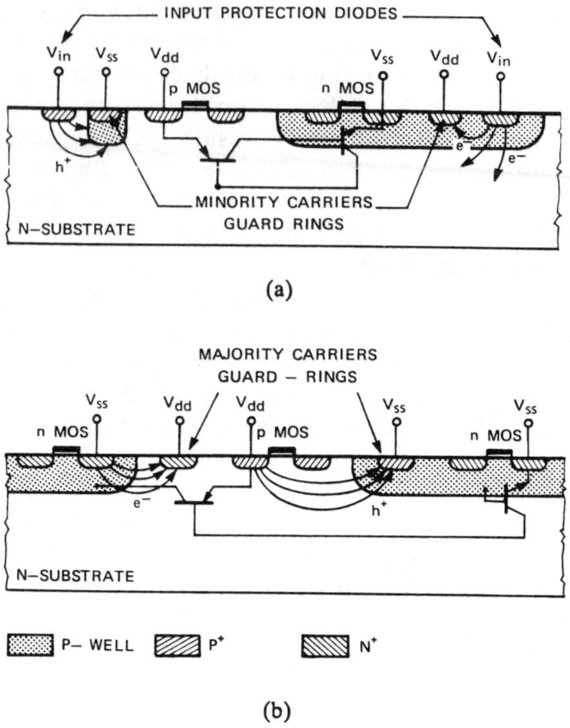

Figure 7.23 Guard rings employed to prevent (a) minority carriers and (b) majority carriers injection in well and substrate, thus avoiding latch-up.

in spite of possible technological improvements [42], and cause the process to be further complicated.

A simpler technique to reduce the parasitic transistor emitter efficiency is to reduce the doping of source and drain [43]. This reduction is limited by the degradation of MOS transconductance, due to the increase of source and drain resistance, so that a trade-off must be found between the two effects. Moreover, large improvements both in trigger current and holding voltage have been obtained simply by reducing the p-doping in an n-well process, in particular in connection with the use of thin epi-layers.

Instead of reducing the emitter efficiency, we may try to increase the minority carrier transit time through the base. To do this, we build a retarding field by counterdoping the well; only the vertical transistor gain is affected. This is now mainly performed by means of very high energy ($>$ 500 keV) ion implantation, which produces a doping that peaks close to the bottom of the well (Figure 7.25) and therefore increases in going from the emitter to the collector. The field induced

Table 7.3 Process Design Techniques to Avoid Latch-up in CMOS

Technique	*Principle*	*Effect*	*Complexity*	*Advantages-Disadvantages*
Gold doping	Lifetime reduction	Decrease β	Low	High leakage currents
Neutron irradiation	Lifetime reduction	Decrease β	Low	High leakage currents
Internal gettering	Lifetime reduction	Decrease β	Low	High leakage currents
Epitaxial layer on highly doped substrate	Increase in substrate conductivity Better carrier cofinement	Decrease R_s	Medium	Lowering of breakdown voltages Increased leakage currents Epilayer defects, reduced yield p^+/p less effective than n^+/n
p^+ buried layer in n-well process	Increase in substrate conductivity	Decrease R_s	Medium	Lowering of breakdown voltage Increased leakage currents
Retrograde well doping profile by	Electric field in the base reduces carrier transport	Decrease β of vertical transistor	High	Increased β of lateral transistor
counterdoping				Decreased mobility
high energy double implantation				Decreased mobility, defects
buried epi-layer	Decrease also R_W			Defects, autodoping
Schottky barrier pMOS	Reduced emitter efficiency of pnp	Decrease β	High	Decreased MOS transconductances Unstable threshold Higher leakage currents Compatible with shallow junctions S/D
Lightly doped source and drain pMOS	Reduced emitter efficiency of pnp	Decrease β	Medium	Decreased MOS transconductances Must be coupled with EPI-CMOS to be effective

Table 7.3 continued

Technique	*Principle*	*Effect*	*Complexity*	*Advantages-Disadvantages*
Trench isolation	Increases the path length of the SCR structure	Decouples parasitic transistors		
	Increases base width of lateral transistor	Decrease β of lateral transistor	High	Must be coupled with other techniques to be effective High parasitic currents Possible latch-up immunity lowering due to parasitic MOS
Silicon on insulator (SOI)	Eliminates parasitic transistors	Complete decoupling	Very high	High technological difficulties Very high cost Reduced MOS performances

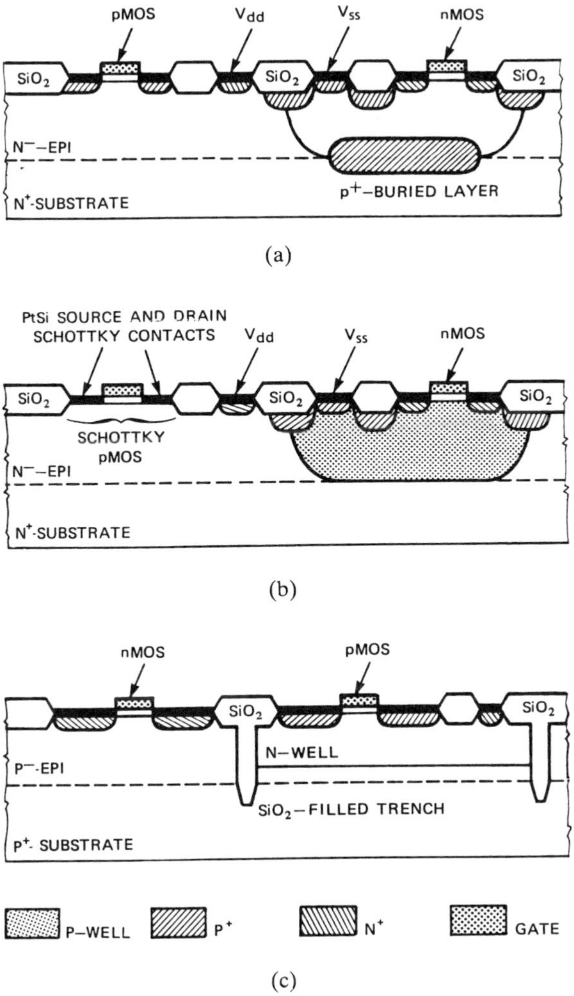

Figure 7.24 Process design techniques to avoid latch-up in CMOS: (a) n^- epi on n^+ substrate and p^+ buried layer under the p-well; (b) Schottky pMOS; (c) trench isolation.

by the doping delays the base transport of the minority carriers, thus reducing the gain of the parasitic bipolar transistors. The maximum levels of the well doping under the S/D help reduce the gain. The effectiveness of the retrograde well has been questioned, however, except in conjunction with a thin, low-doped epi-layer over a highly doped substrate [44].

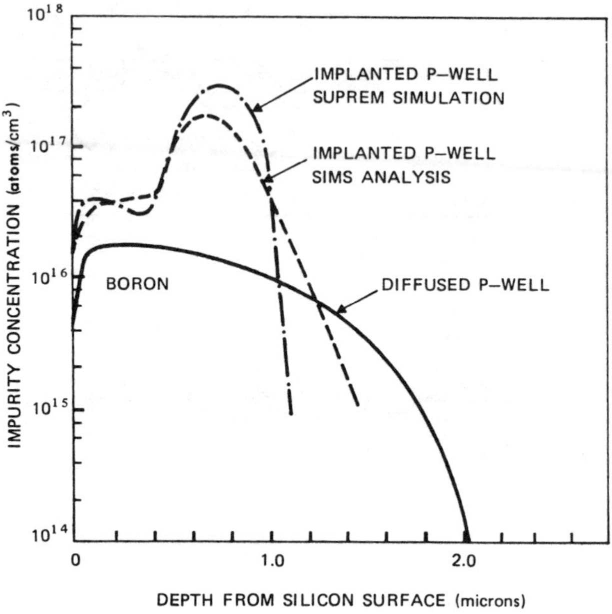

Figure 7.25 Retrograded p-well boron implanted impurity concentration profile [from 45].

Using a low-doped epi-layer over a high-doped substrate, shunt resistance R_S can be reduced without substantially changing the other steps of the MOS fabrication process. Reducing R_S decreases the effective current gain of the parasitic bipolar transistor. This is the more commonly employed technique to reduce latch-up, and it is shown in Figure 7.24b.

The effectiveness of the epi-layer increases with reduced thickness and increased doping of the substrate. A crucial point of this technology is the structural quality of the epi-layer, where active devices have to be built: the high doping could, in fact, be responsible for crystal defects, which will be transferred into the epitaxial layer during the growth and act as very efficient generation-recombination centers close to the surface, causing high leakage currents in the device.

A more complicated, drastic remedy to the latch-up problem is given by the physical separation of the two MOS transistors, which automatically leads to the elimination of the parasitic bipolar transistors. The use of a trench isolation, as shown in Figure 7.24c, is very effective in spoiling the lateral transistor. However, many difficulties are still present in this technology, so that surface current along the trench may cause serious problems. In the future this technology, coupled with selective epitaxial growth of two separate wells, may be a final solution to the problem.

Another drastic solution to latch-up already exists and has been known for many years: it employs the *silicon on insulator* (SOI) technology, long known as *silicon on sapphire* (SOS) [45]. In this case, the MOS transistors are fabricated into separate islands, grown or deposited over a dielectric substrate, so that no parasitic paths exist. However, this technology is still too expensive to be an industrial standard, and its use has been limited mainly to space and military electronics, with only few exceptions. Because of the effects of radiation, space electronics has always been very conscious of latch-up problems [46].

7.5 MODELING THE pnpn PARASITIC STRUCTURE

Modeling latch-up is a quite difficult task because of both the distributed nature of parasitic devices and the effects on high-level injection. The lumped-element two-transistor circuit model of Figure 7.1 can provide a simple description of the behavior of parasitic devices with negligible computational effort. However, it is affected by several heavy limitations [2]:

- Both substrate and well resistances are distributed in nature so that their value can not be easily computed.
- At high current densities, as occurs in the latched state, conductivity modulation effects tend to reduce the value of R_s and R_w.
- The lateral pnp transistor requires an appropriate model.
- The electric field in the base of the lateral pnp transistor, which takes place when transistors are connected in the SCR configuration, can increase the pnp current gain of more than 1 order of magnitude, so that measuring the current gain of individual transistors biased from external terminals generally leads to erroneous results;
- Current is a multivalued function of voltage and *vice versa*.

Several attempts have been made to improve the results obtained by means of the lumped-element model, including two-dimensional of substrate region to calculate the R_w value [47] and modeling of conductivity modulation and of field-aiding effects [48]. Lumped-element models, however, use empirical fitting parameters and therefore are more suitable for the characterization of a preexisting technology than for the study of new technological solutions. Two-dimensional simulators, like DESI [49], FIELDAY [50], and PISCES [51, 52] have been generally adopted in the evaluation of new technologies. Using these simulators, both the triggering and sustaining characteristics of the pnpn structure can be analyzed, and the influence of process and geometrical parameters on latch-up susceptibility can be evaluated. These programs calculate the solution of Poisson's equation and continuity equations for electron and holes in a two-dimensional cross section of the semiconductor device. Distributed device effects can be accounted for directly, no parameter fitting is required, and nonlinearities are implicitly included.

Several technological solutions have been checked using these simulation models; for instance, trenched Schottky barrier CMOS [53], epitaxial layers [54], silicided butted contacts [55], and lightly doped source and drain regions [43]. In [43], PISCES is used in conjunction with measured values of sheet and contact resistances to evaluate the contact resistance contribution to the holding voltage and to predict latch-up hardness of a low-doped, low-contact resistance source-drain process obtained with S-D silicide metalization.

Figure 7.26a refers to the two-dimensional potential contour map at the holding point obtained by means of PISCES for a 8 μm thick epi, standard p^+ S-D structure, whereas Figure 7.26b refers to the same structure with LDD source and drain; the contribution of contact resistance was taken into account in the simulation. The simulated values of the holding voltage were 2.45 V and 6 V, respectively.

Two-dimensional simulation programs have provided accurate prediction of device behavior, with the obvious exception of structures where three-dimensional effects become relevant due to the reduced device geometry [55] or to current redistribution within the pnpn structure. This last effect has been shown to cause hysteresis in the I-V characteristics of the latched structure [56].

The computational effort required for complete two-dimensional calculation of latch-up characteristics, however, is quite high, so that several authors [57,58] have tried to improve the lumped-element models, introducing corrections and device parameters suggested by two-dimensional simulations.

Troutman and Hargrove [56] have replaced the shunting resistances R_s and R_w of the conventional two-transistor model by transmission lines. In fact, in the substrate region are two highly conductive layers, separated by a low-doped layer: the field implant region under the field oxide (on top), and the outdiffused, highly doped region at the bottom of the epi-layer. The substrate therefore can be viewed as a transmission line, whose parameters can be computed easily, knowing process details. Results have been checked by means of the two-dimensional simulator FIELDAY and have been used to analyze the effect of different substrate contact placements. Finally, in [57], predictor-corrector continuation procedures have been employed to calculate both triggering and holding characteristics of a VLSI twin-tub, epitaxial CMOS process.

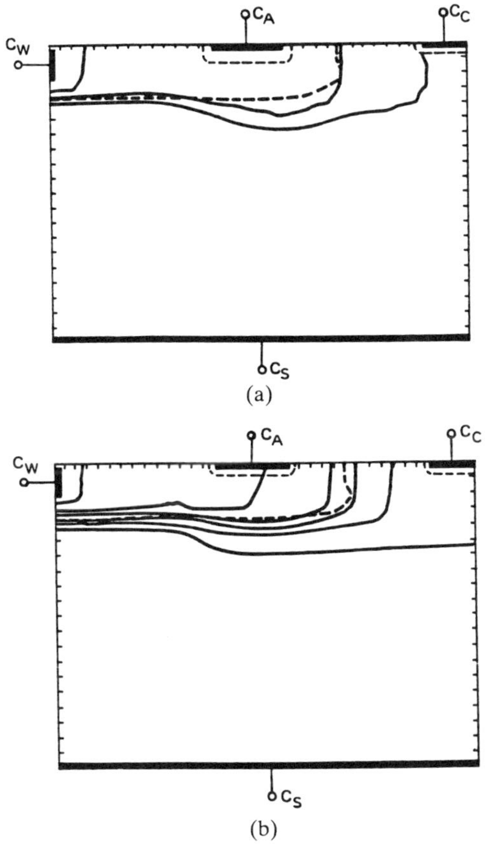

Figure 7.26 Potential contour lines at the latch-up holding point obtained by the PISCES II simulation of a pnpn structure. The difference between equipotential lines is 1 V: (a) 8 μm thick epitaxial layer and standard S/D; $VH = 2.5$ V, $IH = 30$–35 mA; (b) same epi with lightly doped S/D; $VH = 6$–7 V, $IH = 40$–50 mA (courtesy of B. Riccò and E. Sangiorgi, University of Bologna).

REFERENCES

1. R.R. Troutman, *Latch-up in CMOS Technologies,* Kluwer, Boston, 1986.
2. D.B. Eistreich, "The Physics and Modeling of Latch-up in CMOS Integrated Circuits," Stanford Electronics Laboratories Technical Report No. G–201–9, 1980.
3. D. Takacs, C. Werner, J. Harter, and U. Schwabe, "Surface Induced Latch-up in VLSI CMOS Circuits," *IEEE Trans. Electron. Devices,* ED-31 1984, pp. 279–286.
4. G.J. Hu and R.H. Bruce, "A CMOS Structure with High Latchup Holding Voltage," *IEEE Electron. Devices Lett.,* EDL-5, 1984, pp. 211–214.
5. P.V. Dressendorfer and A. Ochoair., "An Analysis of the Modes of Operation of Parasitic SCR's," *IEEE Trans. on Nuclear Science,* NS-28, 1981, 4288–4291.

6. C. Canali, F. Corsi, M. Muschitiello, and E. Zanoni, "Correlation between Anomalous Latch-up I-V Characteristics and Observation of Current Distribution by IR Microscopy in CMOS IC's," *Electron. Lett.*, 24, 1988, pp. 204–205.
7. R. Croes and P. Hendriks, "Standardizing Latch-up Immunity Tests for HCMOS Logic ICs," *Electronic Components and Applications,* 7, 1986, pp. 187–197.
8. EIA JEDEC JC–40.2 CMOS Logic Standardization Committee, "A Standardized Test Procedure for the Characterization of Latch-up in CMOS Integrated Circuits," Draft 3.4, August 19, 1987.
9. R.D. Rung, and H. Momose, "DC Holding and Dynamic Triggering Characteristics of Bulk CMOS Latchup," *IEEE Trans. Electron. Devices* ED-30, 1983, pp. 1647–1655.
10. R. Rocchelli and E. Zanoni, "Latch-up Analysis of CMOS Integrated Circuits for Automotive Applications," *Alta Frequenza,* 54, 1985, pp. 292–301.
11. J. Hiatt, "A Method of Detecting Hot Spots on Semiconductors Using Liquid Crystals," *IEEE 19th Ann. Proc. Int. Reliability Physics Symp.,* 1981, pp. 122–127.
12. R.S. Payne, W.R. Grant, and W.J. Bertram, "Elimination of Latch-up in Bulk CMOS," *1980 IEDM Tech. Digest*, pp. 248–251.
13. M. Muschitiello, F. Crosi, M. Stucchi, E. Zanoni, and C. Canali, "Analytical Techniques for Localization and Sensitivity Analysis of Latch-Up in CMOS ICs," to be published in *Microelectronic Engineering,* North-Holland, 1989.
14. N. Khurana, "Pulsed Infra-red Microscopy for Debugging Latch-up on CMOS Products," *IEEE 22nd Ann. Proc. IEEE Int. Reliability Phys. Symp.*, 1984, pp. 122–127.
15. C. Canali, F. Corsi, M. Muschitiello, M. Stucchi, and E. Zanoni, "Characterization of Anomalous Latch-up Effects by Means of IR Microscopy and SPICE Simulation," *Journal de Physique,* 49, 1988, pp. C4–809–802.
16. S.M. Davidson, "Latch-up and Timing Failure Analysis of CMOS VLSI Using Electron Beam Techniques," *IEEE 21st Ann. Proc. Int. Reliability Physics Symp.,* 1983, pp. 130–137.
17. F. Fantini, M. Vanzi, C. Morandi, and E. Zanoni, "Observation of Latch-up Phenomena in CMOS IC's by Means of Digital Differential Voltage Contrast," *IEEE J. Solid-State Circuits,* SC–21, 1986, pp. 169–174.
18. E. Zanoni, M. Giannini, A. Senin, G. Simeone, and C. Canali, "Observation of Latch-up Time Evolution in CMOS IC's by Means of SEM Stroboscopic Voltage Contrast Techniques," *IEEE J. Solid-State Circuits,* SC–22, 1987, pp. 65–70.
19. Kendall Scott Wills, C. J. Pilch, Jr., and A. Hylsop, "A Study of CMOS latch-up by Laser Scanning and Voltage Contrast Techniques," *IEEE Int. Rel. Physics Symp.* 1986, pp. 115–119.
20. P.V. Dressendorfer and M. G. Armendariz, "A SEM Technique for Experimentally Locating Latch-up Paths in Integrated Circuits," *IEEE Trans. Nuclear Science,* NS–27, 1980, pp. 1688–1693.
21. D.J. Burns and J.M. Kendall, "Imaging Latch-up Sites in LSI CMOS with a Laser Photoscanner," *IEEE 21st Ann. Proc. Int. Reliability Physics Symp.,* 1983, pp. 118–121.
22. A.H. Johnston and M.P. Baze, "Experimental Methods for Determining Latch-up Paths in Integrated Circuits," *IEEE Trans. Nuclear Science,* NS-32, 1985, pp. 4260–4265.
23. M.P. Baze and A.H. Johnston, "Latch-up Paths in Bipolar Integrated Circuits," *IEEE Trans. Nuclear Science,* NS–33, 1986, pp. 1499–1504.
24. F.J. Henley and M.H. Chi, "CMOS Latch-up Characterization Using a Laser Scanner," *IEEE Int. Reliability Physics Symp.*, 1983, pp. 125–129.
25. T. Shiragasawa, H. Shimura, K. Kagawa, T. Yonezawa, and M. Noyori, "Latch-up Analysis on a 64 kbit Full CMOS Static RAM Using a Laser Scanner," *IEEE Int. Reliability Physics Symp.*, 1984, pp. 63–68.
26. M.H. Wechold, and D.L. Parker, "Imaging Latch-up Sites in CMOS Integrated Circuits Using Laser Scanning," *IEEE Electron. Components Conf.*, 1985, pp. 408–410.
27. E. Lange and J. Winnerl, "The Laser Scanning Microscope as an Analytical Tool for Characterization of the Operation of Integrated Circuits," *ESSDERC* 1983, pp. 166–167.

28. C. Canali, M. Giannini, A. Scorzoni, M. Vanzi, and E. Zanoni, "Measurement of the Local Latch-up Sensitivity by Means of Computer-Controlled Scanning Electron Microscopy," *IEEE Solid-State Circuits,* SC–23, 1988, pp. 597–603.
29. R.W. Astheimer, *Handbook of Infrared Radiation Measurement,* Barnes Engineering Co., 1985.
30. P.K. Footner, B.P. Richards, C.E. Stephens, and C.T. Amos, "A Study of Gold Ball Bond Intermetallic Formation in PEDs Using Infrared Microscopy," *IEEE Proc. of Int. Reliability Physics Symp.,* 1986, pp. 102–108.
31. S. Gorlich, K.D. Herrmann, and E. Kubalek, "Basic Investigations of Capacitive Coupling Voltage Contrast," in *Microcircuit Engineering 1985*, Academic Press, London, 1985, pp. 451–460.
32. S.C.J. Garth, "Electron Beam Testing of Ultra Large Scale Integrated Circuits," *Microelectronic Engineering,* 4, 1986, pp. 121–138.
33. W.-H. Chang and M.D. Rodriguez, "Transient Switching of the Parasitic Bipolar Device of an Epitaxial CMOS Transistor," *IEEE Electron. Devices Lett.*, EDL-8, 1987, pp. 275–276.
34. D.J. Ager and J.C. Henderson, "The Use of Marginal Voltage Measurements to Detect and Locate Defects in Digital Microcircuits," *Proc. of the IEEE Int. Reliability Phys. Symp.*, 1981, pp. 139–148.
35. W. Reiners, S. Gorlich, and E. Kubalek, "On the Primary Electron Energy Dependance of Radiation Damage in Passivated n-MOS Transistors," *Inst. Phys. Conf. Ser. No. 76: Section 12, Microsc. Semicond. Mater. Conf.*, Oxford, March 25–27, 1985, pp. 507–512.
36. C. Canali, M. Giannini, and E. Zanoni, "Techniques for Latch-up Analysis in CMOS IC's Based on Scanning Electron Microscopy," *Microelectron. Reliability*, 28, 1988, pp. 119–161.
37. F. Fantini and C. Morandi, "Failure Modes and Mechanisms for VLSI IC's—A Review," *Proc. IEE,* 132 Pt. G, 1985, pp. 74–81.
38. P. Brambilla, F. Fantini, and G. Mattana, "Updating of CMOS Reliability," *Microelectron. Reliability,* 23, 1983, pp. 761–769.
39. W.R. Dawes, Jr., and G. F. Derbenwick, "Prevention of CMOS Latch-up in MOS Integrated Circuits," *IEEE Trans. Nuclear Science,* NS-23, 1976, pp. 2027–2030.
40. J.R. Adams and J.R. Sokel, "Neutron Irradiation for Prevention of Latch-up in MOS Integrated Circuits," *IEEE Trans. Nuclear Science,* NS-26, 1979, pp. 5069–5073.
41. M. Sugino, L.A. Akers, and M.E. Rebeschini, "CMOS Latch-up Elimination Using Schottky Barrier PMOS," *IEEE Trans. Electron. Devices,* ED-30, 1983, pp. 110–118.
42. E. Sangiorgi and S. Swirhun, "Trenched Schottky Barrier PMOS for Latch-up Resistance," *IEEE Electron. Devices Lett.* EDL-5, 1984, pp. 293–295.
43. B. Ricco, E. Sangiorgi, and G. Ferriani, "High Holding Voltage CMOS Technology with Lightly Doped Source and Drain Regions," *IEEE Trans. Electron. Devices,* ED-34, 1987, pp. 810–816.
44. A.G. Lewis, R.A. Martin, T.Y. Huang, J.Y. Chen, and M. Koyanagi, "Latchup Performance of Retrograde and Conventional n-well CMOS Technologies," *IEEE Trans. Electron. Devices,* ED-34, 1987, pp. 2156–2164.
45. D.M. Brown, M. Ghezzo, and J.M. Pimbley, "Trends in Advanced Process Technology—Submicrometer CMOS Device Desing and Process Requirements," *Proc. of the IEEE,* 74, No. 12, December 1986, pp. 1678–1702.
46. J.A. Criscuolo, J.S. Cable, and D.C. Lee, "Validation of Latch-up Mitigation in Complex VLSI Circuits," *IEEE Trans. Nuclear Science,* NS-33, 1986, pp. 1510–1514.
47. M.-J. Chen and C.-Y. Wu, "A Structure Oriented Model for Determining the Substrate Spreading Resistance in Bulk CMOS Latch-up Paths and Its Application in Holding Current Prediction," *Solid-State Electronics,* 28, 1985, pp. 855–866.
48. A.W. Wieder, C. Werner, and J. Harter, "Design Model for Bulk CMOS Scaling Enabling Accurate Latch-up Prediction," *IEEE Trans. Electron. Devices,* ED-30, 1983, pp. 240–245.
49. G.J. Hu, "A Better Understanding of CMOS Latch-up," *IEEE Trans. Electron Devices,* ED-31, 1984, pp. 62–67.

50. C.S. Rafferty, M.R. Pinto, and R.W. Dutton, "Iterative Methods in Semiconductor Device Simulation," *IEEE Trans. Electron. Devices,* ED-32, 1985, pp. 2018–2027.
51. E.C. Sangiorgi, M.R. Pinto, S.E. Swirhun, and R.W. Dutton, "Two-Dimensional Numerical Analysis of Latchup in a VLSI CMOS Technology," *IEEE Trans. Electron. Devices,* ED-32, 1985, pp. 2117–2130.
52. J.A. Seitchik, J.-H. Chern, and P. Yang, "Latchup Modeling Using PISCES," *IEEE Circuits and Devices,* November 1987, pp. 35–38.
53. E. Sangiorgi, B. Ricco, and L. Selmi, "Three Dimensional Distribution of CMOS Latch-up Current," *IEEE Electron. Device Lett.* EDL-8, 1987, pp. 154–156.
54. L. Selmi, F. Venturi, E. Sangiorgi, and B. Ricco, "Three Dimensional Distribution of Latch-up Current in Scaled CMOS Structures," in P.U. Calzolari and G. Soncini, eds., *Proc of ESSDERC '87,* Bologna, 1987, pp. 783–786.
55. M. Strzempa-Depré, J. Harter, C. Werner, H. Skapa, and R. Kassing, "Static and Transient Latchup Simulation of VLSI CMOS with an Improved Physical Design Model," *IEEE Trans. Electron. Devices,* ED-34, 1987, pp. 1290–1295.
56. R.R. Troutman and M.J. Hargrove, "Transmission Line Modeling of Substrate Resistances and CMOS Latchup," *IEEE Trans. Electron. Devices,* ED-33, 1986, pp. 945–954.
57. W.M. Coughran, Jr., M.R. Pinto, and R. Kent Smith, "Computation of Steady-State CMOS Latchup Characteristics," *IEEE Trans. CAD,* 7, 1988, pp. 307–323.

Chapter 8
RELIABILITY OF COMPOUND SEMICONDUCTOR MICROWAVE FIELD EFFECT DEVICES

D.E. Riva and G. Clerico Titinet

8.1 INTRODUCTION

From the beginning of solid-state microelectronics, materials such as germanium (Ge), and later silicon (Si), have predominated in terms of use. Silicon in particular has been enormously successful, chiefly because its relative ease of processing and capacity to form stable dielectrics first encouraged the planar process and then opened the way to *very large scale integration* (VLSI). Growing interest in optical communication has spurred researchers to investigate new materials and structures, thus leading to the development of new processing methods.

Improvements in these methods, which were developed for both Si technology and optoelectronics, have helped to overcome many of the difficulties associated with the production of devices capable of making full use of the superior characteristics of compound semiconductor materials such as GaAs. In the case of GaAs, for example, physical properties show the following advantages over Si:

- For low electric field values, charge carrier drift velocity is up to five times higher, which makes very high frequency operation possible.
- The energy "gap" is larger (1.42 eV compared to 1.21 eV), which guarantees good performance, even at relatively high temperatures. The gap is also direct, thus permitting integration with optoelectronic devices.
- There is greater radiation resistance because of both the wider gap and the shorter carrier life.

- High-resistivity substrates can be prepared, which are excellent for use as dielectrics in high frequency applications and facilitate insulation of active areas in digital circuits.

The major processing problems are related to the following factors:

- GaAs is a binary compound and tends to decompose at fairly low temperatures (somewhat above 600°C).
- GaAs is very fragile, which complicates processing.
- GaAs has no stable native oxide.

These problems have gradually been overcome. Current crystal growth techniques, of which Bridgman and *liquid encapsulated Czochralsky* (LEC) are the major types, permit commercial production of good quality substrates with diameters up to 4 inches and minimum guaranteed defect densities of 5000–10,000/cm^2. The influence of these defects (mostly dislocations) on the electrical performance of devices that use these substrates is not yet fully understood. However because of the introduction of variations and refinements (vertical magnetic fields and In doping in the Czochralsky method [1], and computerized continuously variable temperature controls for ovens divided into multiple heat zones for Bridgman growth techniques), these substrate defect densities can be further reduced.

The development of the ion implantation process (much used in Si technology) makes possible the production of selective and preferential doping with optimal control over junction depth and carrier concentration, whereas crystalline lattice damage reduction and activation of implanted species are achieved by annealing techniques developed to prevent semiconductor decomposition induced by heat treatment. These techniques, which employ a protective dielectric layer or are applied with As overpressure, are now considered routine.

Metalization techniques have been identified that are sufficiently stable for production of rectifying contacts (for FET gate electrodes) and ohmic contacts, their function differing according to the type of application (analog and digital).

Finally, the development and improvement of epitaxial layer growth techniques, providing close control over the characteristics of the materials grown (such as MBE and MOCVD), make possible new structures with alternating binary and ternary compounds and the exploitation of band discontinuity effects, which provide even higher performance, at least as far as speed is concerned. Although the reliability of devices based on these new technologies has improved with time and the experience gained by manufacturers, much remains to be done, especially because the most promising applications are in the military, aerospace, and telecommunication sectors, where reliability is imperative.

This chapter will thus deal with the technological and reliability problems relating to components of greatest interest in the field of microwaves; that is, MESFETs and *microwave monolithic integrated circuits* (MMICs), as well as the promising new technologies employing heterostructure field-effect transistors.

8.2 MICROWAVE GaAs MESFETs

The *metal semiconductor field effect transistor* (MESFET) is a three-port device where the current flow inside the active zone, or channel, between two electrodes called the *source* and *drain,* is modulated by the voltage applied to a third electrode called the *gate.* In this device conduction takes place through the flow of majority carriers and hence is unipolar. It shares many features with the vacuum triode, and its use as an amplifier has a considerable theoretical advantage over the bipolar transistor. This is because it is a square law device, as shown by the equation

$$I_d = I_{dss} [1 - (V_{gs}/V_p)]^2 \tag{8.1}$$

where V_p is the pinch-off voltage; that is, the voltage applied to the gate electrode (for a given voltage applied to the drain, V_{ds}) that reduces the value of the current flowing through the channel, I_d, to a value close to zero. This voltage is such that for $|V_{gs}| > |V_p|$ there is no significant change in the electrical behavior of the FET, and I_{dss} is the saturation current; that is, the current flowing through the channel for a given drain voltage and $V_{gs} = 0$. This basically means that a GaAs FET amplifier generates little odd order distortion, and even order distortion can be eliminated by suitably designing the amplifier with balanced stages [2].

From the structural point of view the MESFET is not a new device, but only recently has improvement in process techniques expanded its applications. Since the maximum working frequency is closely linked to the transit time of the carriers in the active zone, reducing the lateral dimension of the gate electrode to the minimum and using a very high-charge mobility material with good thermal dissipation produces devices supplying several Watts in very high-frequency bands (tens of GHz). It is equally true that the latest generations of FET amplifiers are capable of replacing conventional *traveling wave tubes* (TWT) in many applications. Among the considerable advantages that this brings is reduced space and weight (factors of basic importance in equipment on board spacecraft and vehicles), the possibility of reduced supply voltages (from several thousand volts to a few volts) with a consequent reduction in the criticality of the power supplies, and increased reliability. In fact, although many problems are still to be resolved, the average life of a power TWT is of the order of several hundereds of hours, whereas the average life of a MESFET in the same application is greater by at least 1 order of magnitude.

8.2.1 Fabrication Technology

The starting material is a GaAs semiinsulating substrate characterized by very high resistivity ($> 10^7$ Ωcm) obtained by adding impurities such as oxygen or chromium into the melt, which should be thermally stable and have a very low density of crystal defects. To ensure that the interface with the active layer is in its best possible structural, chemical, and morphological condition so that the conduction mechanisms are unaffected, the layer is extended by a thick epitaxial growth, called a *buffer*. The active layer is then grown on the buffer using the same procedure but different doping conditions. In this way, it is possible to obtain a sharp profile of carrier concentration and high electron mobility up to the interface with the substrate. Under these conditions, the noise figure and gain are markedly improved.

A further improvement in these parameters has been achieved using two high-doping areas (n^+) as "pedestals" for the source and drain contacts. Apart from achieving a lower contact resistance, the source-gate and gate-drain resistance of the recessed channel structures are also reduced. This technology, which consists in growing a thicker active layer and reducing the thickness in the channel zone by chemical etching, reduces the tolerance required for epitaxial growth and helps improve the performance and reliability. Alternatively, the n^+ contact regions can be obtained by implantation.

The source and drain metalizations generally consist of several layers made up of different materials, the choice of which is a compromise between the lowest contact resistance and the highest resistance to degradation from interactions with the semiconductor or the device working conditions. The contact is generally obtained through alloy with Au-Ge, on which are deposited layers of Ni, Pt, Ti, Cr, or combinations of these and other elements, and finally a surface layer of Au. Other metallurgical combinations, however, are feasible and currently being studied.

Various solutions also have centered on the gate. The dimensional control of this electrode is of fundamental importance to the device's performance. Gates up to 0.5 μm long can be obtained using conventional photolithography. A lower value can be achieved using electron beam lithography. The gate material should create a Schottky junction with the semiconductor.

Many metals feature suitable characteristics for forming a rectifying contact with the GaAs; Al has been, and continues to be, widely used due to its simple processing and good thermal stability characteristics. However, despite its significant thermal and electrical conductivity characteristics, Au cannot be deposited in direct contact with the semiconductor, with which it reacts even at relatively low temperatures (250°C). Hence, a barrier is necessary to prevent the interdiffusion between GaAs and Au but that barrier must not impede formation of a stable Schottky contact on the semiconductor side. It should also have low resistivity and sufficient reactivity with the gold to ensure minimum resistance associated with the gate. The types of barriers most commonly used are TiPt, TiW, and TiPd. Hence, even gate electrode metallurgy can vary considerably, and it is still difficult to indicate the best compromise between performance and reliability.

The amount of current that can flow through a MESFET, which is linked to the power of the device, depends on the total "width" of the gate. Different approaches have been developed to increase this parameter without introducing parasitic elements, and attempts have been made to minimize the difficulties of assembly inside the package. Some of these approaches include the following:

- Crossover with dielectric: The gate electrode consists of several fingers, interconnected through a metalization that overlaps the drain and source fingers, from which it is insulated by the dielectric (SiO_2). This solution, illustrated in Figure 8.1, provides a compact structure, but there is an increase in stray capacitance at the crossover points.
- Crossover without dielectric, or air bridge: The source metalization overlaps the other electrodes and makes contact only with the source fingers, as shown in Figure 8.2. The capacitance per unit area is smaller.
- Wraparound: The source metalization surrounds the chip allowing a significant reduction in the associated inductance, but crossover is still needed to connect the drain and gate fingers.

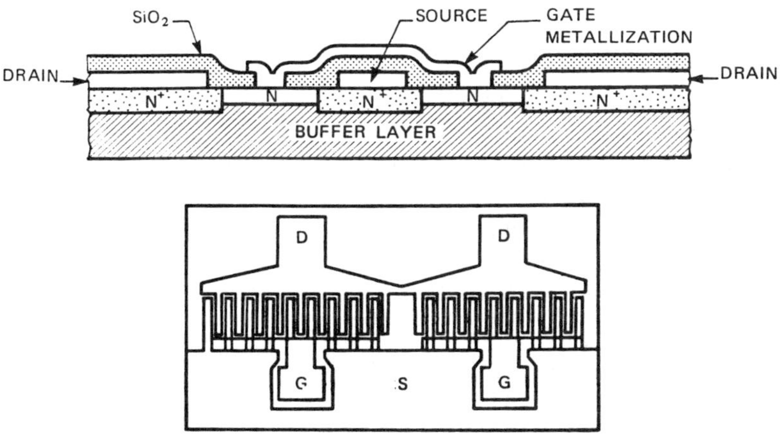

Figure 8.1 Power FET with gate crossover [105].

Two further approaches connect the source electrode to the package ground plane to further reduce the parasitic elements, particularly those associated with the wires connecting the chip and the package, and enhance heat dissipation.

- Via holes: Holes are made on the back of the chip in the areas requiring contact with the source metalization previously deposited on the upper section, which stops the chemical etch. The holes are filled at the same time as the metalization of the device's lower face (see Figure 8.2).

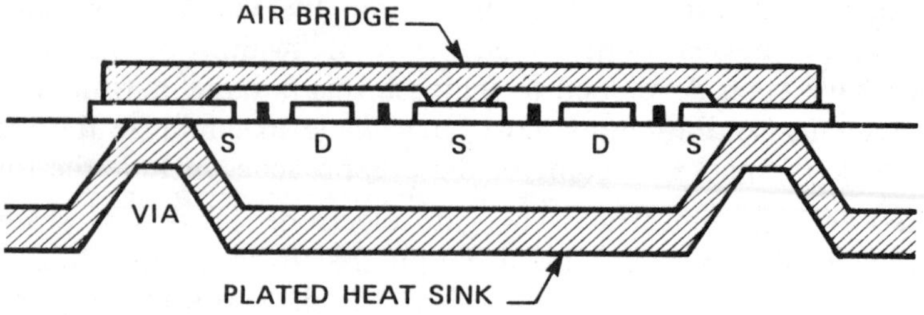

Figure 8.2 Power FET with source connection via holes on the lower face and air bridge on the upper face.

- Flip-chip: A galvanic gold layer is grown on the source fingers and the external gate and drain sections, then the chip is bonded face down by thermocompression. The structure described and some design details are shown in Figures 8.3 and 8.4.

8.2.2 Determination of MESFET Reliability

The literature to date on the reliability of GaAs MESFETs is relatively substantial but equally disparate. The results available generally have been obtained from tests, as no valid field data exist. This is due to the specific nature of the applications and their recent development, as well as the technological developments in the search for an acceptable settling level. The literature reports mainly accelerated tests that, with few exceptions, were conducted on fairly small samples and are often difficult to compare because of the wide range of characteristics and technological solutions adopted. A test program clearly has several objectives:

- To provide information allowing a sufficiently approximate estimate of the expected useful life of the devices for a particular application, and to provide this information in a reasonably short time.
- To lead to the identification of the most significant potential failure mechanisms. An understanding of these mechanisms can indicate corrective actions or at least suggest operating precautions.
- To direct the preparation of the screening procedures and hence the purchasing specifications.

It is clear that the type of stress applied should be chosen to emphasize the typical failure mechanisms of the technology. These are the mechanisms most likely to occur under normal operating conditions, although not all together and suddenly but in a diversified, gradual, and controlled manner.

The metalization system seems to represent one of the most critical aspects in terms of reliability. Aging due to high temperatures is an important cause of

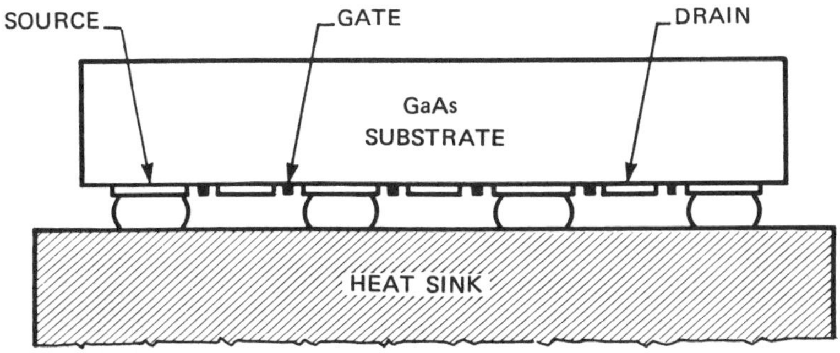

(a)

(b)

Figure 8.3 Power FET with flip-chip mounting: (a) illustrative diagram; (b) actual device (MGF2148P).

failure. It would be useful in studying this metalization thermal stability to test samples of the semiconductor substrate on which the various systems to be examined have been deposited. However, this is not feasible because the bias, type, and amplitude of signal applied play an important role and are particularly significant, especially for power MESFETs.

For some time, the validity of direct current as opposed to *radio frequency* (RF) tests for power MESFETs has been discussed. The former may be less critical

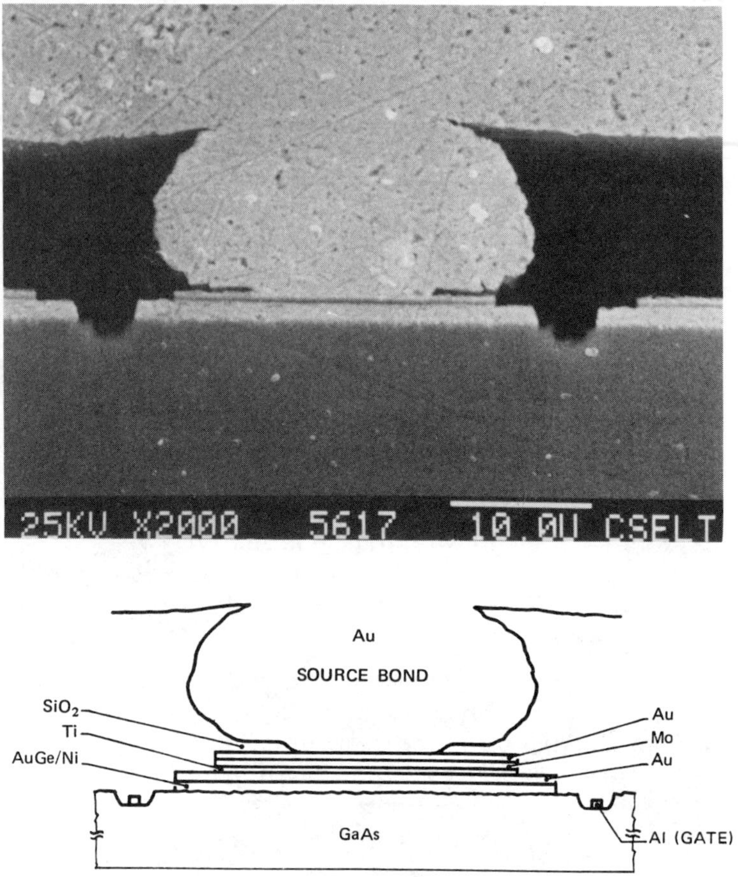

Figure 8.4 Cross section of MESFET of Figure 8.3b with design details.

in terms of the necessary structures, although they can have uncontrolled self-oscillations that may lead to the rapid destruction of the devices. The suppression of these unwanted oscillations is far from easy and requires individual treatment for each component under test. Moreover, the RF tests require the use of suitable signal sources and the assembly of the components in amplification modules to be individually tuned, connected to instruments measuring the parameters being tested. In addition to all these factors, the test structures should be capable of withstanding high temperatures (up to more than 250°C), and the components themselves are not cheap. Therefore, very real problems are involved in the development of an exhaustive test program.

Although radio frequency tests are more representative of actual operating conditions, they do not appear to detect failure mechanisms substantially different from those found by simply applying the *direct current* (dc) bias. However, they provide a more accurate estimate of the useful life of devices, which is less than that extrapolated from the dc test results. Some failure mechanisms (for example, corrosion) are not particularly enhanced by the type of bias applied, and their study does not require any particularly sophisticated technique.

Another rather controversial point is the correlation between the degradation of dc parameters with the degradation of the RF parameters, and even the choice of failure criteria is still generally subjective. One example of this correlation, used within a qualification program incorporating such procedures as accelerated tests, was proposed by Irie et al. [3], which is shown here for illustration purposes.

In this case, the component being examined is a low-noise MESFET whose technological characteristics are as follows:

- Semiinsulating substrate, $r = 10^8$ Ωcm
- Buffer layer of several μm, $n = 10^{13}$–10^{14} cm^{-3}
- Active layer 0.2 μm thick, $n = 10^{17}$ cm^{-3}
- Both layers have been grown through $Ga/AsCl_3/H_2$ VPE
- Al V-shaped gate, L = 1μm, W = 300μm, thickness = 0.6μm
- Source and drain in Au (88 percent) + Ge (12 percent)/Pt, thickness = 0.19μm

The small signal equivalent circuit is shown in Figure 8.5. In relation to the specific needs the failure criteria adopted for the frequency parameters is based on a variation of more than 0.5 dB in the noise figure and 2dB in the *maximum gain* (MAG) measured at 4 GHz. The corresponding failure criteria for the direct current parameters obtained through the computer is shown in Table 8.1.

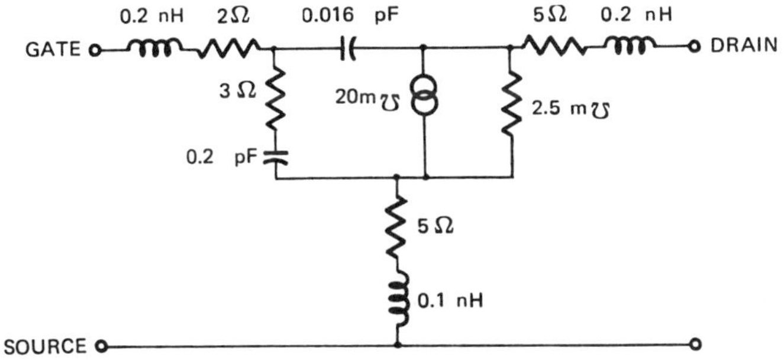

Figure 8.5 Equivalent small signal circuit of MESFETs tested by Irie et al. [3].

Table 8.1 Failure Criteria for Specific dc Parameters

Parameter	*Symbol*	*Conditions*	*Failure criteria*
Saturation drain current	I_{dss}	$V_{ds} = 3$ V $V_{gs} = 0$ V	±10%
Specific drain current	I_{ds}	$V_{ds} = 0.5$ V $V_{gs} = 0$ V	±25%
Pinch-off voltage	V_p	$V_{ds} = 3$ V $I_{ds} = 0$ mA	±10%
Transconductance	g_m	$V_{ds} = 3$ V $I_{ds} = 30$ mA	±10%
Direct gate voltage	V_{gf}	$I_{gf} = 20$ mA	±20%
Reverse gate current	I_{gs}	$V_{gs} = -5$ V $V_{ds} = 0$ V	Ten times the initial value or 1 μA

The processing was carried out using the model described and incorporated an increase in the parasitic resistance corresponding to the degradation of the ohmic contacts encountered during the life tests. However, since, even for MESFETs, temperature is a main accelerating factor in degradation, it is essential to know the precise channel temperature for correct valuation of the mean time to failure (MTF). Channel temperature can be derived only through accurate measurement of the thermal resistance of the devices. The most suitable method is undoubtedly that which uses the thermal drift of the direct I-V characteristic of the Schottky gate diode, which has been developed in different versions [4, 5, 6]. Unlike methods based on the use of liquid crystals or infrared microscopy, this technique is applicable to encapsulated and integral devices and can be performed on relatively simple, cheap instruments.

8.2.3 Degradation Mechanisms

Some of the degradation mechanisms encountered are peculiar to the technology; others are more familiar because they have already been studied in Si devices and are related to the use of the same materials. Some of these mechanisms, such as corrosion and electromigration, are catastrophic; others are manifested with a gradual degradation of some of the electrical parameters. The main mechanisms will be considered in this section.

8.2.3.1 Burnout

This catastrophic phenomenon can occur at different times but with similar results; in fact a distinction is made between short-term, or instantaneous, burnout and long-term burnout. Short-term burnout occurs when the bias is applied and can be induced by electrostatic charges applied to the device during handling or assembly, by poor supply circuit design or impedance mismatch, or by local defects in the device structure. Hence, some potential causes can be eliminated through precautions in the design and assembly stages and thorough inspection.

This is particularly true for low-noise FETs, in which the mechanism is induced mainly by transients applied by operators or equipment [7]. Some studies [8, 9] conducted with the aim of verifying the sensitivity of the devices to energy and polarity have shown that positive impulses applied to the gate tend to damage the active region and give rise to spherical eruptions, or holes, due to localized overheating. The application of negative impulses, however, produces effects that are similar but related to process defects, where the gate breakdown voltage is reduced due to lack of uniformity in the material or misalignment. The spherical forms are made up of semiconductor material and gate electrode metal, and failure may be due to either a short circuit between the electrodes or an open circuit in the gate. Since low-noise FETs generally consist of a single gate, or at most a few gates in parallel, the degradation of the characteristics due to this phenomenon is catastrophic. One way of increasing burnout resistance may be the use of refractory materials having a high fusion point for the Schottky barrier.

The power devices are biased with much higher drain voltages than low-noise devices. Hence, for the purpose of obtaining increasingly higher powers, significant effort has been concentrated on the development of structures capable of good burnout resistance and a high breakdown voltage between source and drain.

It should be specified that the burnout triggering mechanism is still not fully understood; however, many explanations have been given based on simulations [10, 11, 12]. Because this is an extremely important phenomenon, we will attempt to describe it at a sufficiently detailed level. The phenomenon is essentially of thermal origin: local heating or overheating of the crystal lattice, especially next to the ohmic drain contact, causes the material to melt. This gives rise to eruptions on the surface and below the surface to the substrate level and in the direction of the source contact, which in any case cannot be reached.

The presence of radiation in the areas where eruptions occurred prior to failure brings to mind the presence of a large number of hot electrons and material nonuniformities caused or accentuated by heat. Light emission due to high drain voltages, which can be seen under the microscope [13] seems to be the only precursor to burnout, which otherwise happens suddenly and without warning. Yamamoto, Higashisaka, and Hasegawa [13] have studied the characteristics of the light emission, localized at the edges of the drain contact for $V_{gs} = 0$ and increasing in intensity with the increase of V_{ds}.

Applying a reverse bias to the Schottky gate diode causes the phenomenon to weaken, once again increasing at high gate voltages but now located at the edge of the gate contact on the drain side. The dependence of emission on drain and gate bias is shown in Figure 8.6. The device structure seems to affect the intensity

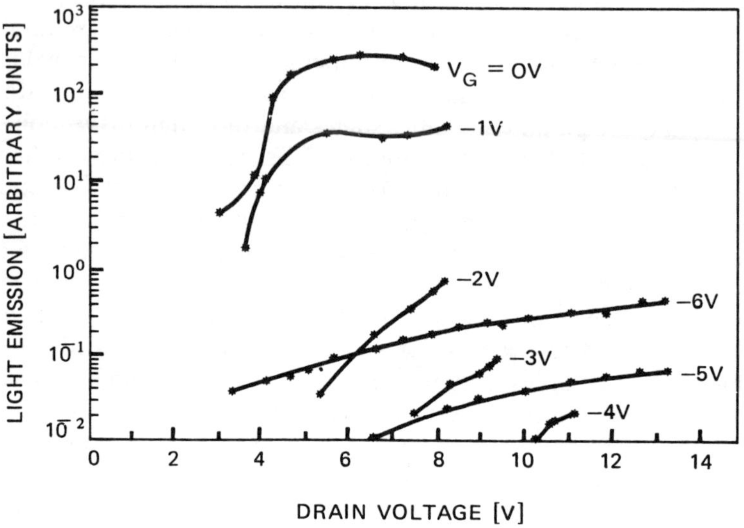

Figure 8.6 Light emission dependence on drain and gate bias in a MESFET.

of radiation whereas the precise location of the zones from which the radiation originates is uncertain; it is indicative of "weak points" and apparently originates at the interface between the various device layers. However, the intensity observed is closely correlated to the drain breakdown voltage, in the sense that the higher the voltage the lower is the radiative emission.

The analysis conducted by Wemple et al. [10] on devices having the structure shown in Figure 8.7a suggests that burnout starts when a lattice temperature of $> 400°C$ is reached at the edge of the drain electrode; in some regions inside the structure, the temperature rises to 500–550°C. This is illustrated by the experimental results given in Figure 8.7b, where the burnout power intersects with zero at precisely that temperature range. Having observed experimentally a sharp change in the temperature coefficient from negative to positive immediately before burnout with an instantaneous increase of I_{dss}, Wemple et al. concluded that this region represents the normally nonconductive device zone (that is, the buffer layer and the substrate) in which a conduction mechanism is set up.

Therefore, dc burnout could be considered simply a consequence of the thermal runaway associated with the positive temperature coefficient of the buffer layer and the substrate when the semiconductor becomes intrinsic. Other significant data have been obtained from studying the behavior of devices when impulses are applied to the drain electrode; these led to the conclusion that burnout voltages

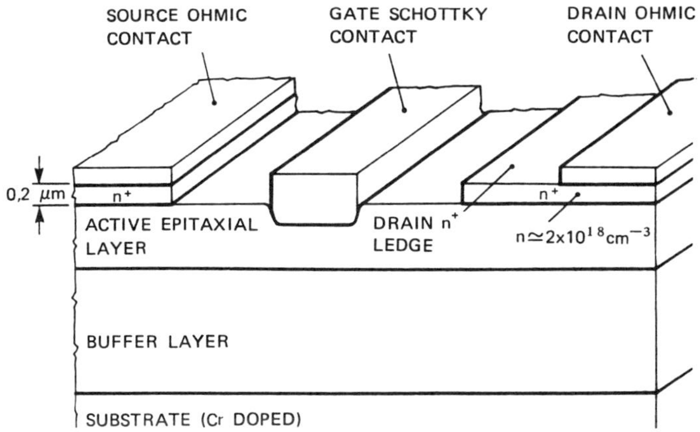

(a)

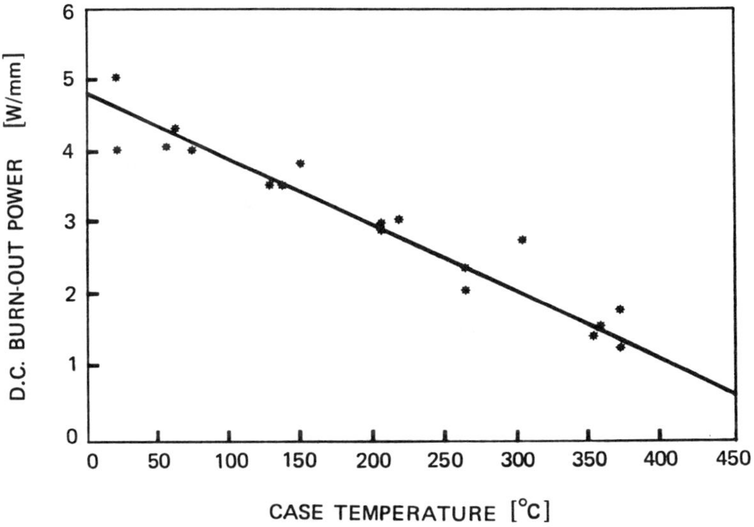

(b)

Figure 8.7 (a) Diagram of channel geometry with n^+ ohmic contacts (Au-Ge/Ag) and the Schottky gate; (b) burnout power dependence (W/mm) on case temperature for these devices with a gate periphery of 6 mm and 30–50 μm substrate.

in pulsed condition are independent of temperature up to 400°C (ambient) and that the mechanism is triggered by the instantaneous power applied in peaks of 4–6 W/mm, which is similar to the dc case but with different implications. The phenomenon always starts at the drain contact, set off by high temperature processes with high current densities and electrical field. In fact, if we remember that ohmic contacts are obtained by alloy at temperatures of the order of 400°C and that the temperature reached by the lattice next to the drain can be much higher, we can appreciate how interdiffusion between semiconductor and metal induces high current density filaments, which extend to the substrate and cause thermal instability or even localized GaAs microscopic fusion (T_f = 1238°C).

Computer-aided simulation [11] has shown that a step voltage pulse applied to the drain leaves a highly energized, or hot, electrical field region and electron domain in the gate region and any discontinuities of the structures toward the drain. It also shows that a distributional peak of these electrons is located at the edge of the drain electrode for a fraction of the switching time, before the high electrical field region dissipates to the drain. The temperature of the hot electrons in a MESFET biased in nominal conditions can be of the order of thousands of degrees C. Figures 8.8 and 8.9 illustrate typical examples of instantaneous burnout.

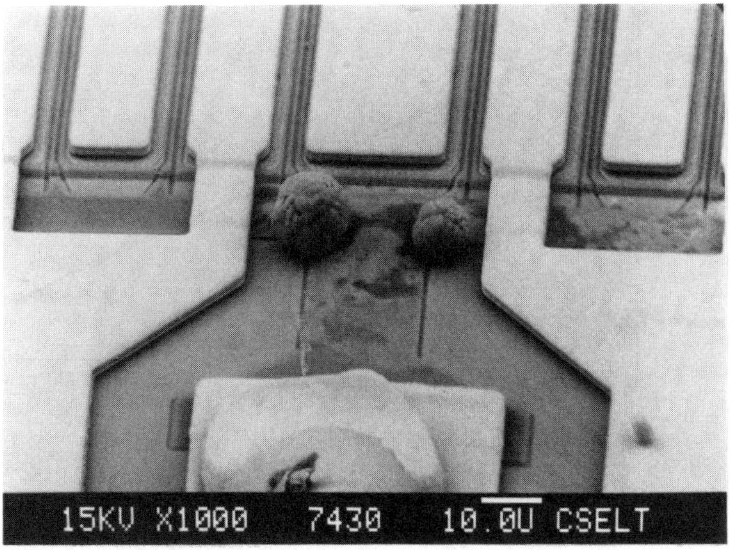

Figure 8.8 Eruptions of GaAs derived from instantaneous burnout caused by local overheating next to the gate metalization.

Figure 8.10 shows the V_{ds} - I_{ds} characteristics of the device of Figure 8.9 taken before and after burnout. Since power MESFETs consist of several electrodes in

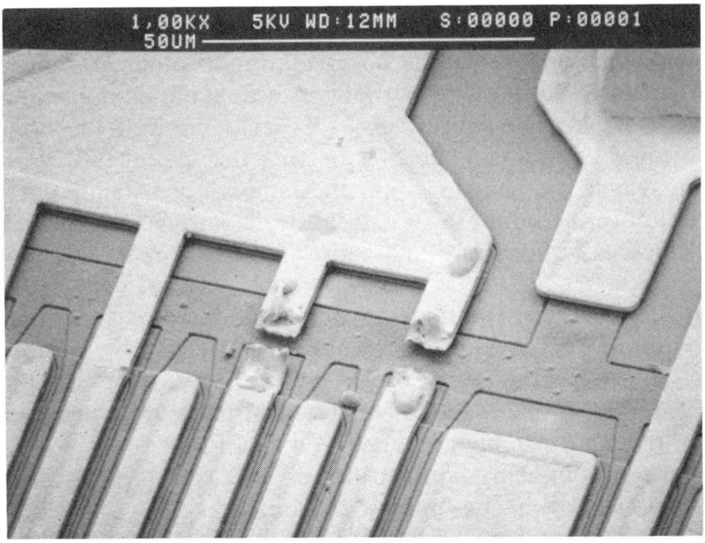

Figure 8.9 Consequences of burnout on a power MESFET; interruption of two drain fingers and spherical eruptions along the gate electrodes.

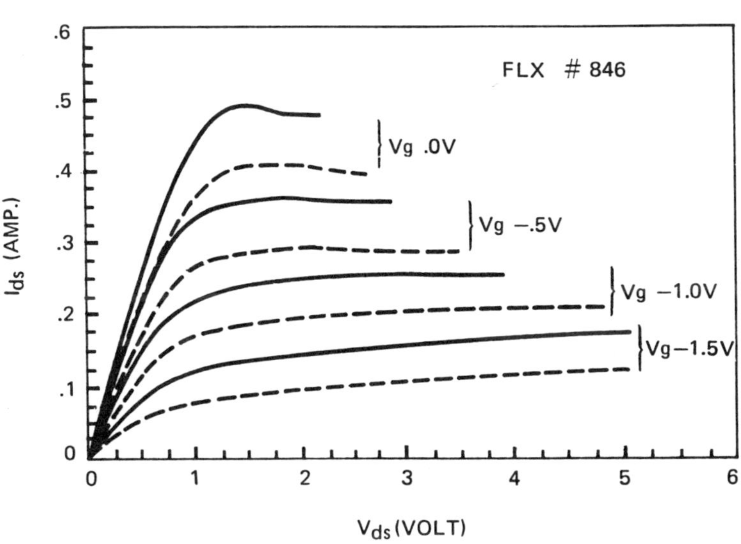

Figure 8.10 Characteristics of the MESFET of Figure 8.9 measured before (unbroken curves) and after (dashed curves) burnout.

parallel (40 in the case described), the breakdown of only some does not affect operation, although it will reduce efficiency in the long term. However, a gradual deterioration in functionality will be observed due to other mechanisms induced by the changes caused by the first burnout. A schematic model summarizing this is shown in Figure 8.11. The figure shows a basically unstable reaction system and examines the most typical structures. Rejecting the hypothesis of an avalanche multiplication, which is stabilizing from the temperature standpoint, we assume that the condition of the buffer layer and substrate is triggered when high current-density fused filaments penetrate and cause instability in the reaction circuit of Figure 8.11.

Various technological solutions have been adopted to allow the application of high bias without the resultant electrical field triggering the burnout condition. The planar structure has the lowest breakdown voltages due to the high electrical fields at the gate and drain edges; the characteristics of the ohmic contacts were first improved by using two highly doped, low-resistance zones at the source and drain, either through implantation, further n^+ epitaxial growth, or reduction of the active layer thickness at the "recessed channel."

Furutsuka, Tsuji, and Hasegawa [14] have shown the effectiveness of simply thickening the epitaxial layer under the drain contract to relax the electrical field in this critical region, without having to use n^+ contact layers. In this way, they obtained devices capable of operating with an applied drain voltage of 24 V. Wemple et al. [10] have obtained devices with burnout voltages in the 40–50 V range, adopting the structure schematized in Figure 8.7a, which has the n^+ layer ledge outside the drain metalization toward the gate. Tiwari, Eastman, and Rathbun [15] propose increasing the space between drain and gate while keeping the gap between gate and source to a minimum. Moreover, they confirm the validity of a recessed channel and gate stucture and propose that the gate recess have the same depth as the surface depletion zone. They measured burnout voltages of more than 60 V at what is considered to be an ideal active layer doping level equal to $8 \times 10^{16}/\text{cm}^3$.

Apart from the consequences of energy intensive transients perhaps the most worrisome factor affecting reliability is time: essentially, during its lifetime, a reduction occurs of a device's instantaneous burnout voltage, for reasons that still are not clear. As already mentioned, light emission seems to be the only warning of burnout; studies by Wemple et al. [10] suggest this at both high (T_a = 350°C) and low (T_a = 190°C) aging temperatures. Furthermore, the time required for light emission to take place is a clear indication of burnout time and, hence, burnout voltage. Although light emission could be eliminated through surface chemical etching, it will reappear albeit more slowly after further aging. This led to the conclusion that degradation of the surface properties caused such breakdown. In particular, the changed distribution of the surface electrical fields, probably due to structural modifications, could be explained by a gradual development

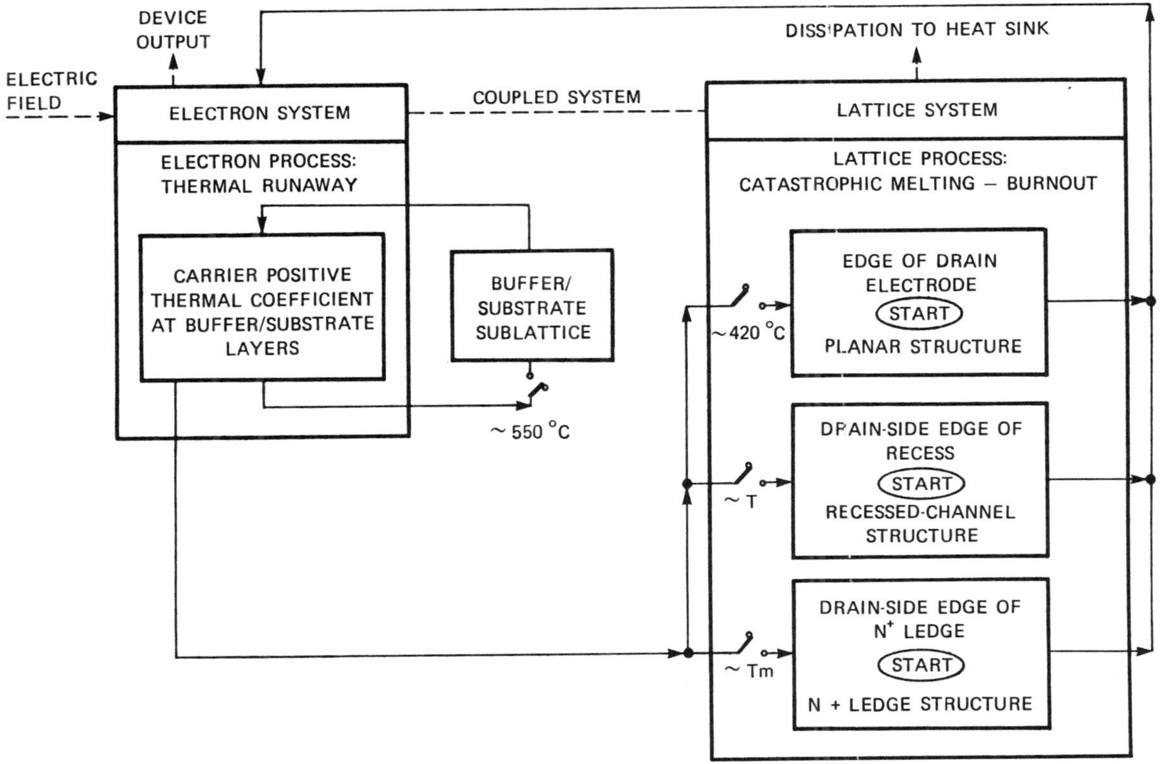

Figure 8.11 Schematic diagram of a feedback model to explain instantaneous burnout. The reaction circuit (continuous lines) leads to catastrophic fusion with unstable or exponential growth. The triggering temperatures are in the lattice, above which the unstable reaction circuit becomes operational.

of elemental arsenic, according to the following chemical reaction:

$$Ga_2 + As_2O_3 + 2GaAs \rightarrow 2Ga_2O_3 + 4As \tag{8.2}$$

which involves the reduction of the As_2O_3 by the GaAs [16].

The presence of crystalline As on the surface may also cause the creation of conductive paths between the gate and drain, as hypothesized by Dumas et al. [17] to explain the high leakage current and associated reduction in breakdown voltage detected after a sample of nonpassivated commercial devices were aged. Surface erosion can lead to a reduction in current by reducing the effective thickness of the channel, as illustrated by Igi et al. [18]. Rapid erosion may take place when the gate of non-passivated FETs is biased above the breakdown voltage and light emission is observed, due to photo-enhanced oxidation in the channel region [19]; the same phenomenon was reported as an explanation for facet degradation in AlGaAs lasers [20]. In the absence of precise data, the cause of the final catastrophic fusion is thought to come from a slow shifting of the high electrical field region toward the drain contact, where the phenomenon is triggered. Since the presence of native arsenic oxide is undesirable, methods have been developed to avoid its degrading effects. Kretschmer and Hartnag [21] have tested various chemical etches for cleaning the active surface, and report excellent results using KOH:H_2O in the ratio 5:1 and NaOH(25 percent): H_2O_2 (30 percent):H_2O in the ratio 2:1:300.

Wemple et al. [10] propose passivating the device with a layer of silicon nitride deposited by vapor phase with hydrogen-rich plasma-assisted reaction. The hydrogen would combine with the residual surface oxygen to block the reaction described. Fukui et al. [22, 23] illustrate that dc and frequency life tests on devices having aluminum gates and Si_3N_4 passivation, at channel temperatures of 250, 210, and 175°C, resulted in an estimate of failure rate less than 100 FIT at 110°C. It is important to stress that these tests incorporated about 1 million device hours and that no failure could be attributed to burnout. Previously, a study conducted by Jordan, Irvin, and Schlosser [24] on similar devices passivated with SiO_2 resulted in a failure rate between 4500 and 20,000 FIT —(calculated for $I_{ds} = 0.7$ A and $T_a = 52$°C), already giving an appreciation of the comparable validity of the Si_3N_4 passivation. However, Dumas et al. [25] show that a SiO_2 surface passivation can cause Ga to escape during operation, with a consequent change in the surface properties. The Auger analysis of the degraded devices has detected the presence of oxidized Ga and an excess of As in the form of material accumulated in the active region. This situation can be related to the formation of layers of surface spatial charge in the regions between source and gate and drain and gate, with the consequent drift of electrical parameters. Hence, it is like the previous case. The dc and RF test results on some 250 commercial devices from six different manufacturers, the technological characteristics of which are shown in Table 8.2 [26], are given in Figure 8.12.

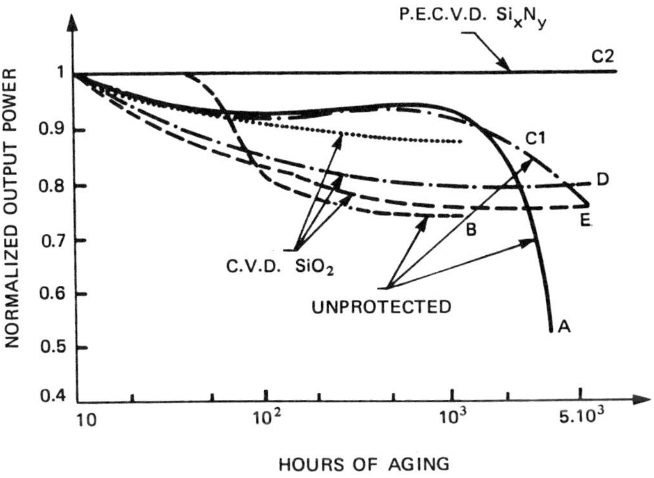

Figure 8.12 Output power normalized at a gain compression of 1 dB measured at 8 GHz for the test with reverse gate bias [25, 26].

The devices protected with silicon nitride (C_2) show an excellent stability over time, whereas those protected with vapor-phase deposited silicon oxide (D, E, and F) are degraded due to surface erosion and the resultant reduction in channel thickness. The devices without protection (A, B, and C_1) are degraded by a reduction in breakdown voltage (BV_{dgo}) and an increase in leakage current, attributable to the formation of a surface conductive layer between the gate and drain due to the release of elemental As from the semiconductor [17]. An excessive reduction in breakdown voltage resulted in burnout in the case of the A devices.

8.2.3.2 Electromigration

The combined effect of temperature and high current density can cause transport phenomena in the material composing the semiconductor device metalization. The physical appearance and aspects of this subject are discussed fully in another chapter of the book. We will limit ourselves here to a discussion of the effects on MESFET reliability, separately examining ohmic contacts and gate metalization.

Abbot and Turner [9] when discussing a study of devices with ohmic contacts from four metalization systems (all Au-Ge based, however), conclude that metal migration has a major effect only when junction temperatures are at least 250° C. Under these conditions, and for current densities of about 10^5 Å/cm^2, all of the MESFETs tested showed a net degradation of the characteristics within the first 1000 hours. Figure 8.13 shows schematically the effects of electromigration; namely, material accumulation around the edges of source contacts, depletion

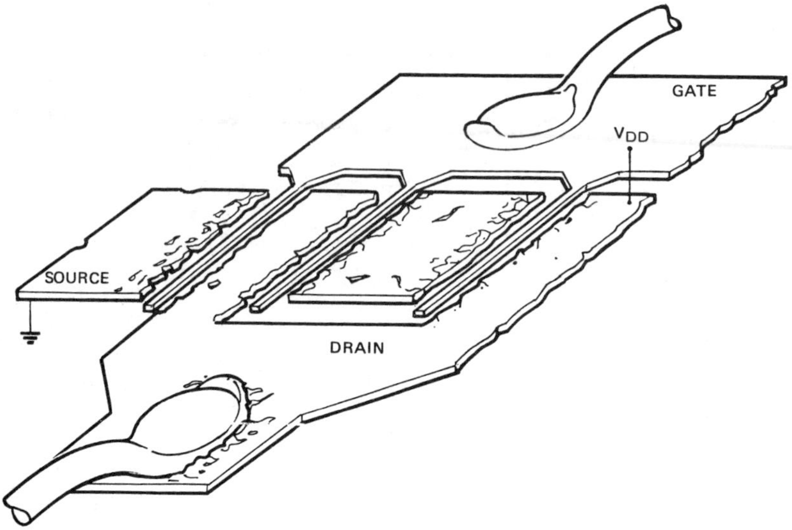

Figure 8.13 Schematic representation of metal electromigration in a MESFET.

around the edge of drain contacts, and accumulation around the wire bondings on this electrode. Metal atoms are carried along the direction of electron flow, toward the edges at higher positive voltage, thus leaving holes where they are removed. Using 1 eV as the activated energy value, they obtained an MTF value, extrapolated to ambient temperature, of 10^7 hours.

It is interesting to note how a subsequent study of the influence of contact form on metal transfer rate indicated a marked superiority of rectangular contacts over other shapes; specifically, triangle and T-shaped. However, this can be easily explained by the fact that the rectangular form allows the current to be distributed more evenly. Christou, Cohen, and Macpherson [27] have identified the main cause of power MESFET degradation in the electromigration of Au-Ge/Ni ohmic contacts tested at radio frequency at a channel temperature of 200°C. Microanalysis of the extrusions formed at the edges of the source contact, causing failure at several points of the SiO_2 passivation, has shown that the accumulated material consisted basically of Au with traces of As and an appreciable quantity of Ga, which is a sign of a reaction between semiconductor and metal. This was confirmed by the presence of Ga depleted zones in all the source contacts examined. The MTF at 200°C is 2200 hours, with an associated activation energy of 1.1 eV.

A fairly similar value (2100 hours at T_{ch} = 228°C) was found by White, Rogers, and Hewitt [28], who observed the growth of metallic extrusions at the drain bonding pad and depletion at the edges of the relevant ohmic contact. However, the increase in contact resistance measured in this case is too low to justify the reduction in gain encountered during the tests, which has been attributed to a deterioration of the active layer, in agreement with Fukui et al. [22].

Generally, however, the activation energy for the electromigration of gold-based metalizations, starting from a value of 2 eV [29], is strongly affected by dissolution effects, and so it is likely that the diffusion of Ga in the top Au layer of the ohmic contact will cause it to fall by even more than 50 percent. This problem also requires the provision of an effective barrier layer for its resolution.

Electromigration represents a very real danger to gate metalization, especially for power MESFETs. However, since the gate lateral dimension (and hence the cross section of the electrode itself) is constantly being miniaturized to increase operating frequencies, it is generally necessary to adopt appropriate technological solutions to maintain acceptable reliability levels. In a power MESFET, gate forward current densities of 10^4–10^5 Å/cm^2 are fairly common; and during the RF cycles, peak values greater than even 1 order of magnitude can be easily reached. Furthermore, a local increase of current density can be caused by the depletion of the metalization following the formation of intermetallic compounds (see section 2.3.5) or the reduction of the useful section next to any steps [30] due to poor process control (or poor step coverage).

It is also known that Al has a lower electromigration threshold than Au. This, together with other factors such as semiconductor adhesion problems caused by the inability of Al to reduce the native oxides on the GaAs [31] during the evaporation phase, stimulated the search for alternative metalizations.

Furutsuka, Tsuji, and Hasegawa [14] have obtained an MTF of 10^6 hours at a channel temperature of 130°C for power devices with a 0.5 μm-long gate subjected to direct current life tests (T_{ch} = 240, 270 and 300°C). Although observing only formation of holes along the electrodes and not accumulation of material in the direction of current flow, they attribute the cause of the failures detected to electromigration [33].

The same researchers were able to obtain an MTF higher by at least 1 order of magnitude using a Ti/Al system, where they did not encounter electromigration but final failures were due to burnout. Figure 8.14 provides the Arrhenius diagrams obtained for three different metalization systems. Similar tests performed on devices with Ti-W/Au gate metalization have given an MTF at T_{ch} = 125°C equal to 5×10^6 hours [34, 35] with an activation energy of 1.85 eV. This is illustrated in the diagram of Figure 8.15, which shows that one cannot speak of a marked superiority of one system over the other. It can be stated nonetheless that a good Si_3N_4 passivation layer is effective in reducing the risk of electromigration even for current density values above the threshold, as illustrated by the data provided by Fukui et al. [23].

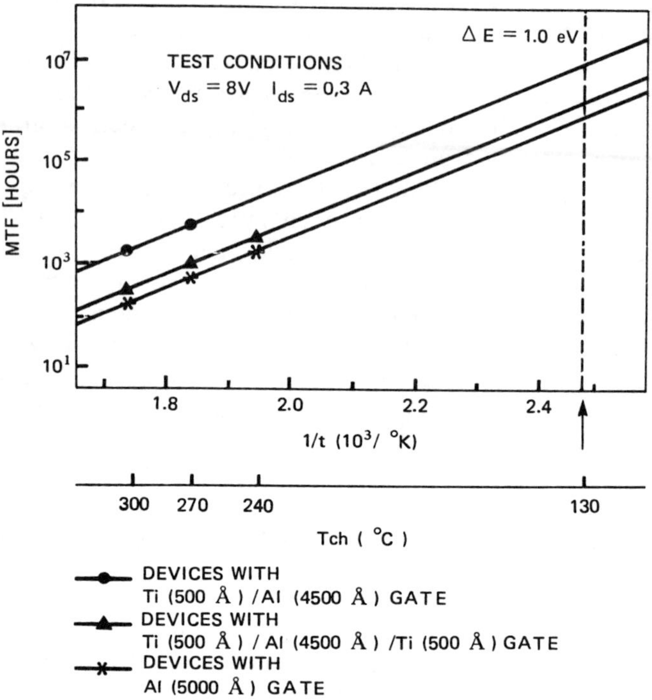

Figure 8.14 Arrhenius diagrams derived from direct current life test results on power MESFETs with 0.5 μm-long gates consisting of three different metalization systems.

8.2.3.3 Stability of Ohmic Contacts

The main factors of good ohmic contact are a linear voltage-current characteristic, stability in time, and low associated parasitic resistance. For high doping levels (at least $10^{19}/cm^3$), which are not typical of actual devices since they are difficult to obtain without sophisticated techniques such as *molecular beam epitaxy* (MBE), the deposition of a metal on the semiconductor would result in the formation of an ohmic contact. This avoids the need for the alloy operation, involving fusion and subsequent recrystallization of the metal and hence high temperatures, which can affect surface morphology and uniformity [36, 37].

The method most widely used for ohmic source and drain contacts is the deposition (by evaporation, for example) of a layer of gold-germanium in proportions that promote the formation of a eutectic (88 percent Au, 12 percent Ge by weight; the eutectic temperature is 363°C). The deposited material is then covered by a layer of Ni (about 280 Å of Ni for every 1000 Å of AuGe) or Pt, which improves the wettability of the alloy. The Ni, apparently, promotes the

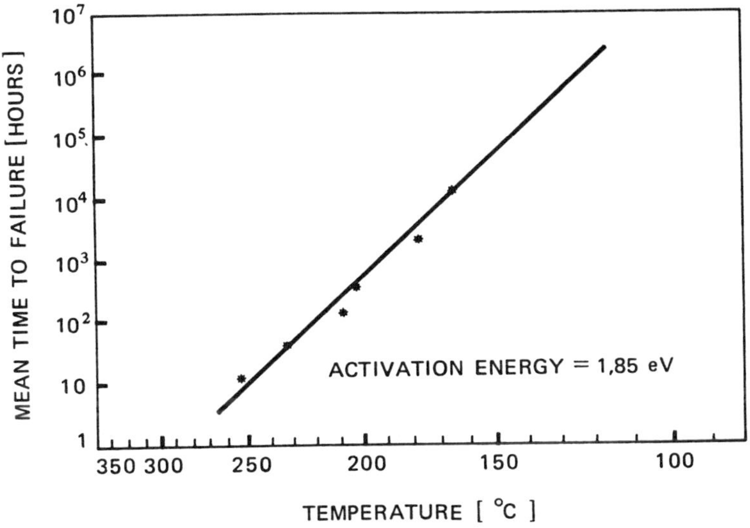

Figure 8.15 Arrhenius diagram derived from the direct current life test results on devices with Ti-W/Au gate metalization.

diffusion of Ge in the semiconductor and inhibits balling up during alloying [38]. During the alloy process, the Ge replaces the atoms of Ga diffused in the Au [39, 40]. The ideal alloy temperature is 425–450°C, which corresponds to a specific contact resistance of 10^{-6} Ωcm^2 [40] due to the formation of extensive solid phase epitaxial regions having the same substrate orientation. The distribution of elements after the alloy process is illustrated in Figure 8.16 [41a].

Ohata and Ogawa [39] have studied the behavior of ohmic contacts composed of Au-Ge, Pt/Au-Ge and Ni/Au-Ge (Au 88 percent-Ge 12 percent by weight) deposited in different thicknesses on n-type GaAs with a carrier concentration equal to $2 \times 10^{16}/cm^3$. Unlike the first system, the other two systems have a very regular alloy interface; the structures of these contacts before and after aging at 330°C, as observed earlier, are given in Figure 8.17. Figure 8.18 shows the variation in conductance in time during the test, indicating greater stability in contacts having a lower Pt and Ni thickness, the conductance of which is held virtually constant.

The enhanced contact with Au-Ge/Ni as opposed to the Au-Ge/Pt system are, for example, described by Lundgren and Ladd [41], who recorded in the former system an increase in resistivity of 10 percent after a heat treatment for 80 hours at 300°C. The second system, apart from about a 50 percent higher initial resistivity when subjected to the same treatment, showed an increase of 75 percent. This comparison, however, does not include the thickness data, which, as shown in [39], is of major importance. The contacts, nonetheless, tend to degrade in time

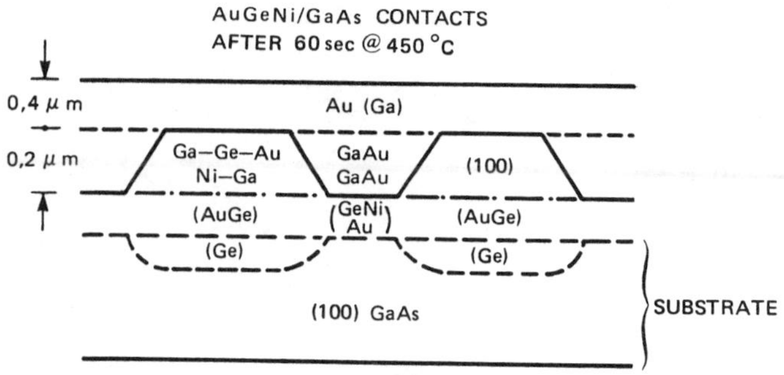

Figure 8.16 Diagram of interdiffusion at Au-Ge/Ni contacts with GaAs and distribution of consistent materials.

due to the temperature effect; namely, the diffusion of Ga in the metalization and Au and Ni in the semiconductor [42].

To prevent the formation of high-resistivity regions due to the excess diffusion of Ga in Au, a barrier layer, consisting of Ti-W, for instance, is often interposed between the Au-Ge/Ni layer and the top Au layer deposited at the end of the process to aid the bonding of the wires.

After subjecting 400 low-noise devices to high-temperature (175–265°C) aging tests, Mizuishi et al. [43] attributed the degradation of the conductance with zero drain bias (g_{d0}) to the increase in resistance of the ohmic contacts. The metallurgy used was slightly different to that studied by Ohata and Ogawa, comprising one layer of Au-Ge (7 percent by weight) 1500 Å thick, one Ni layer 300–500 Å thick and one Au layer 1500 Å thick. The alloy was obtained at 400°C. In contrast with [39], Ga was observed in the Au surface layer, even though the Ni layer was relatively thin and a high-resistance layer was formed under the interface due to the diffusion of Ni in the crystal. The associated activation energy was 1 eV for the majority of the devices. Using a 10 percent reduction in g_{d0} as the failure criteria—according to the derived model, this corresponds to an increase of 0.5 dB in the minimum noise figure F_{min} (at 7.4 GHz), an MTF has been estimated of 10^7–10^8 hours at a channel temperature of 80°C.

Katsukawa et al. [44] provide more optimistic data taken from accelerated aging tests at ambient temperatures of 259, 295, and 337°C performed on conventional MESFETs for use in X-band having the ohmic contacts made of Au/Ti/Ni/Au-Ge.

For a 10 percent reduction of I_{ds} attributed to an increase in the contact resistance an MTF of 3×10^7 hours has been calculated at a channel temperature of 130°C with an activation energy of about 1.8 eV.

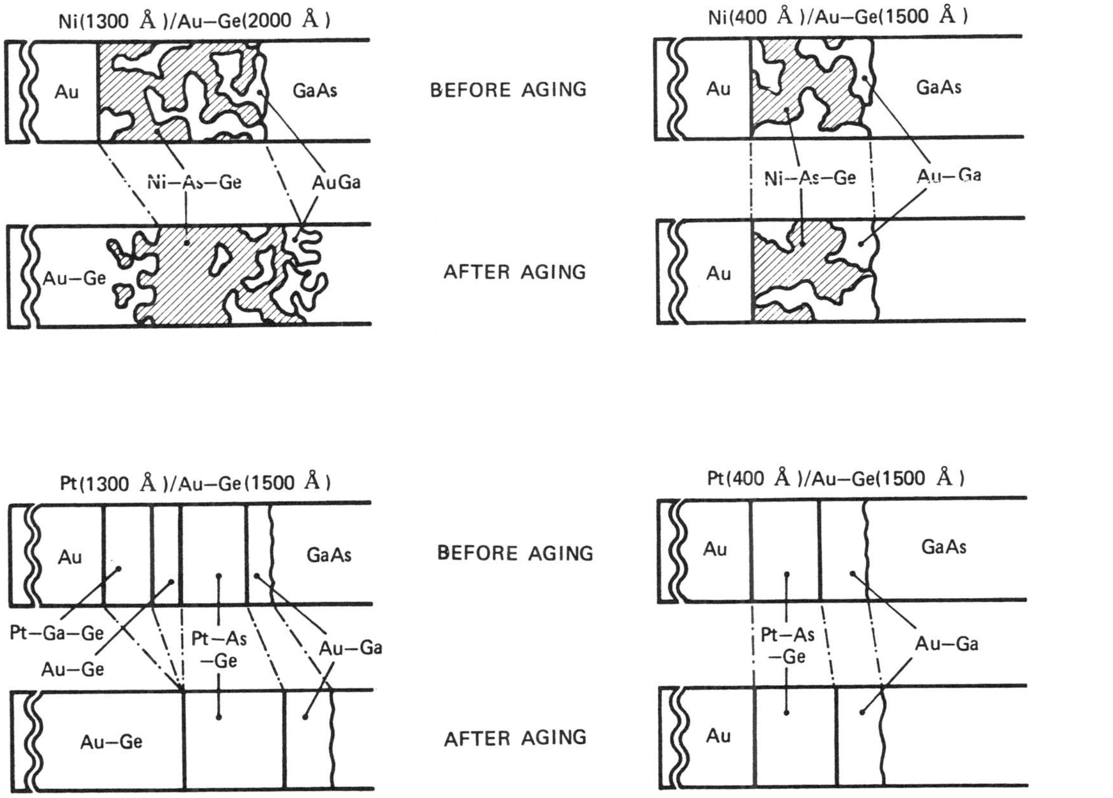

Figure 8.17 Initial structure and structure after aging of contacts by alloy in accordance with [39].

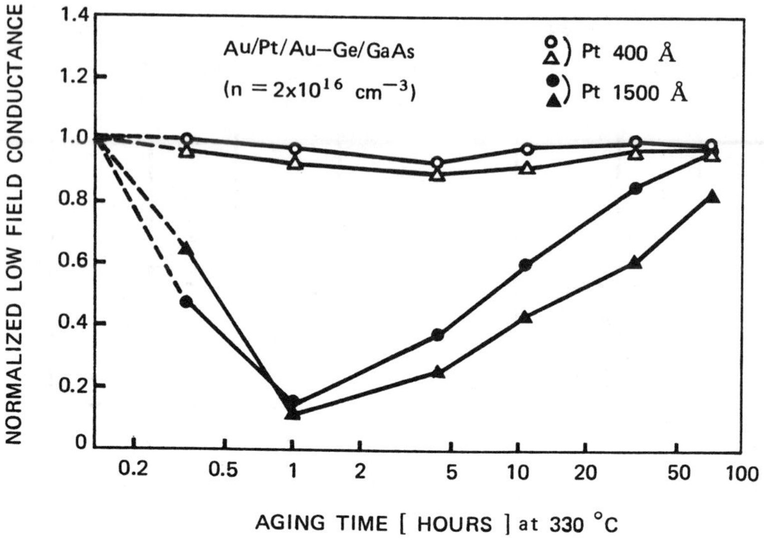

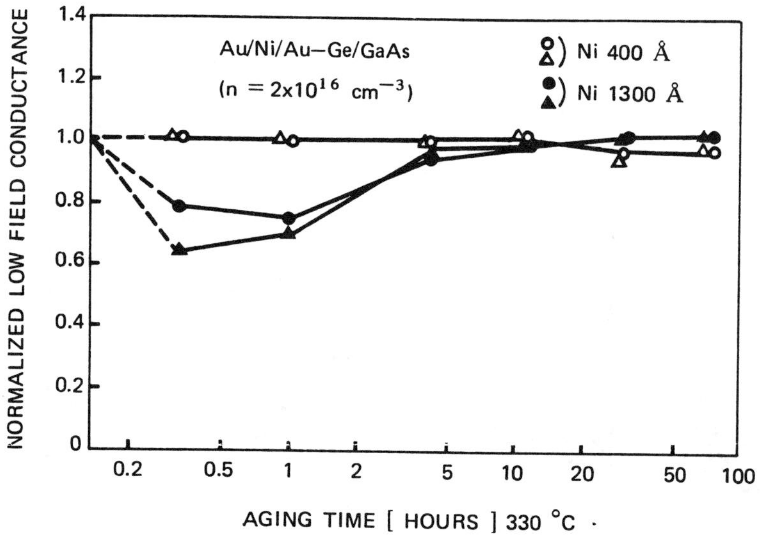

Figure 8.18 Variation of conductance normalized for low field values during the aging of the structures of Figure 8.16.

Irie et al. [3] consider the degradation of the Au-Ge/Pt ohmic contacts to be a major influence on the average life of low-noise devices tested at high temperatures (227, 259, and 295°C for 1500 hours) without applied bias. They observed a gradual reduction in I_{ds} (drain current in nonsaturation conditions) attributed to the increase in contact resistance. The physical explanation for the phenomenon (not verified analytically) is a slight penetration of the alloy interface into the active layer or the formation of point defects and dislocations near the contact. The activation energy determined was 1.8 eV and the MTF value extrapolated to a channel temperature of 100°C is greater than 10^8 hours.

New techniques for obtaining ohmic contacts are being researched and assessed. The possibility of growing epitaxial layers with high carrier concentration ($n = 5 \times 10^{19}/\text{cm}^3$) using MBE has resulted in contacts with Au-Ge or Sn without reheating at the eutectic temperature, as well as the direct deposition of metallic layers [36, 37]. Although no data are available on their long-term performance the fact that the contact obtained is uniform, that the probability of crystallographic defects induced by surface tensions is reduced to a minimum, and the devices produced by this method exhibit operating characteristics at least equal to conventional devices means that this new system can be considered more than just promising.

8.2.3.4 Schottky Contact Stability

The characteristics and the integrity of the gate-channel interface are of major importance in the operation of MESFETs. The deposition of a metal on the active GaAs layer by conventional techniques results in a contact with rectifying characteristics, and the barrier height is virtually independent of the metal used. The value falls on average between 0.8 and 0.9 eV for n-type semiconductors and between 0.5 and 0.6 eV for p-type ones. These values are maintained even if the metal is deposited in an ultra-high vacuum, and at ambient temperatures after thermal surface cleaning. This was demonstrated by Waldrop [45] on ideal contacts, which were free of interface oxide as shown in Figure 8.19. Hence, the choice is related to technological factors, including metallurgical stability, compatibility with the other metals in the fabrication process, and adhesion to the semiconductor.

In Waldrop's study of transition metals, such as Ti, Cr and Al, which are those used most widely in the fabrication of current devices, the interface is not abrupt but chemically reactive, and the products of the disassociation of the GaAs are Ga and an arsenide containing the contact metal. In the case of Au, there is Ga and As diffusion in the metal and the interface becomes confused and deteriorates. However, the Au-GaAs system is not stable in temperature, and does not represent an acceptable alternative.

The Auger analysis, with an in-depth profile of structures consisting of a 2000 Å thick Al film deposited with an electron beam at ambient temperature on an n-type GaAs epitaxial layer, has shown a very extensive interface region low in arsenic [46]. After only an hour's annealing at 250°C in a vacuum, Ga was detected

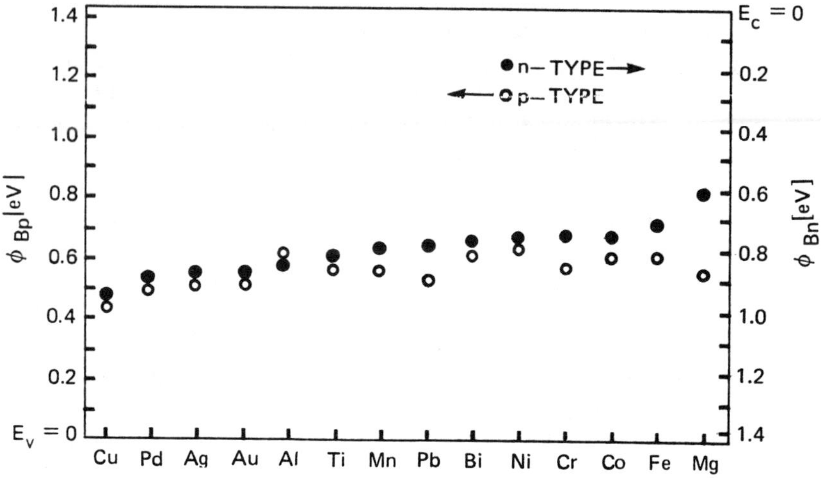

Figure 8.19 Barrier height of ideal contacts with the GaAs (100) obtained via I-V measurements. The full dots and the scale on the right correspond to n-type contacts, the empty dots and the scale on the left to p-type contacts.

throughout the Al thickness and the tendency of O to migrate toward the surface was seen. After 24 hours, the Ga had reached the surface, on which Al oxide had also formed. The situation is shown in Figure 8.20.

Similar results have been obtained after a 24-hour annealing at 275°C on FETs specially fabricated by Sleger and Christou [47]. It can be deduced that the surface stoichiometry of the active region under the gate is significantly changed, and the interface shifts inside the channel to reduce its actual thickness and hence decrease the drain current. Sleger and Christou [47] have also detected an increase in resistivity (from 2.5×10^{-6} to 10.6×10^{-6} Ωcm) explained by the probable inclusion of oxygen into the Al due to interdiffusion. This highlights the importance of fabricating gate electrodes sufficiently thick to reduce the associated resistance and prevent an excessive increase in the long term; at the same time, an increase in the conductor's useful section results in a reduction in the risk of electromigration.

This problem, in particular, and the need for compatibility with the bonding wires to avoid the formation of intermetallic Au-Al compounds have favored the choice of a refractory metalization. One metal receiving attention is Ti, which is widely used today, but, as with Si technology, its use involves building multilayer structures, since its low electrical conductivity and tendency to oxidate makes wire bonding difficult. Moreover, coating it with a layer of gold is insufficient, because of barrier instability due to the interdiffusion between the two metals, which can occur even at temperatures like 200°C [48], with the subsequent rapid formation

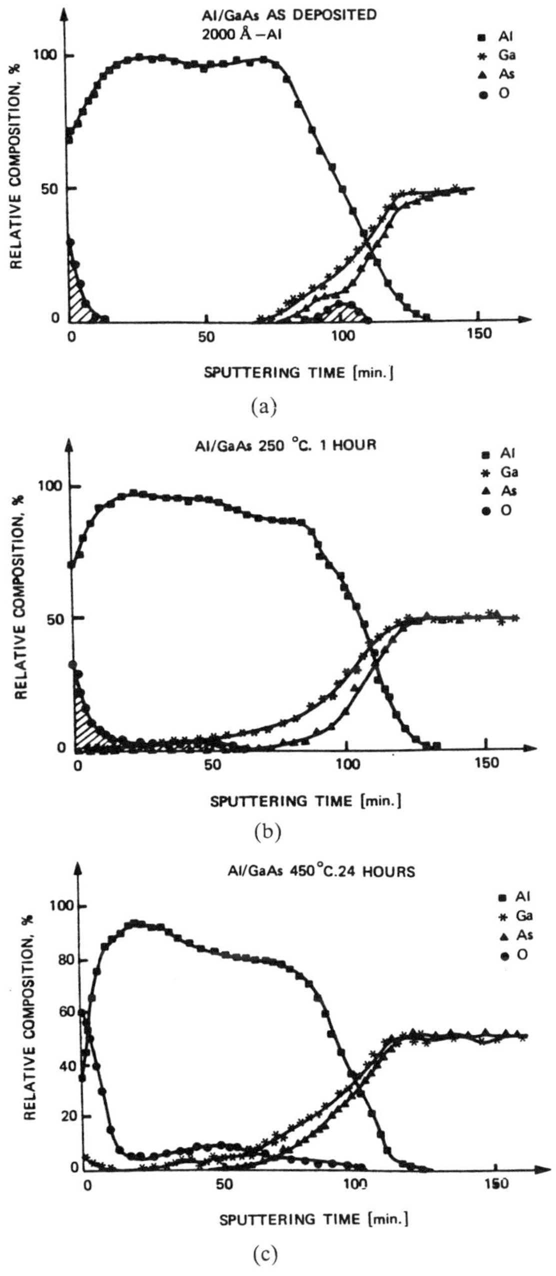

Figure 8.20 Results of the Auger analysis on a film of Al deposited on GaAs: (a) at time of deposit; (b) after 1 hour; (c) after 24 hours.

of Au-Ga phases. The systems most widely used at the moment are TiW-Au and TiPt-Au.

Day, Christou, and Macpherson [49] have shown the stability of the Au-TiW/n-GaAs system up to temperatures above 500°C, recording a slight increase in the barrier height but noting that integrity is maintained. Baker et al. [50] conclude that the presence of impurities, such as oxygen in quantities between 3 and 7 atomic percent, drastically improves the quality of the barrier, blocking the diffusion mechanism along the grain edges both of the gold and species coming from the semiconductor.

Studying the reliability of MESFETs with a TiW-Au gate metalization, Canali et al. [51, 52] have measured a parametric degradation (reduction in I_{dss} and V_p and increase in R_0) common to all the accelerated tests carried out. The cause of the degradation, induced just by the temperature, has been attributed to the sinking of the gate metalization in the active layer due to penetration of the Au in the semiconductor through the TiW barrier (see Figure 8.21). In agreement with Baker et al. [50], this phenomenon was not found when the oxygen content was higher than 4 percent in weight.

The compensating effect of channel doping due to Au diffusion cannot be excluded. However, this phenomenon can also lead to catastrophic burnout due to hot spots induced by disuniformities created in the channel and the consequent triggering of thermal runaway. The shift of the interface inside the channel is the explanation also given by Ogbonnah and Fraser [53] to account for degradations of FETs with TiPd/Au gates fabricated by implantation and investigated for use on monolithic integrated circuits. The activation energy associated with the reduction in effective channel thickness due to interdiffusion found in this study is 1.6 eV and the mean projected life at a channel temperature of 100°C is 1.6×10^9 hours.

The stability of other systems has been studied by Lundgren and Ladd [41] who did not observe degradation in metalization of the Ti/Pt/Au and Mo/Pt/Au type after 80 hours heat treatment at 300 and 350°C, respectively, whereas the barrier represented by Cr/Pt/Au after 85 hours at 350°C changed to a nonlinear, nonrectifying ohmic contact.

8.2.3.5 Formation of Intermetallic Compounds

As indicated in the previous section, almost any metal could be used for the gate Schottky contact; for this reason, the choice is made mainly on the basis of system stability and process complexity. Al has been widely used from the start, but its use has involved a search for solutions to some quite major problems in reliability. It is well-known that, apart from a few circumstances, it is not feasible to bring Au and Al into close contact. In previous silicon devices, gold bonds on aluminum

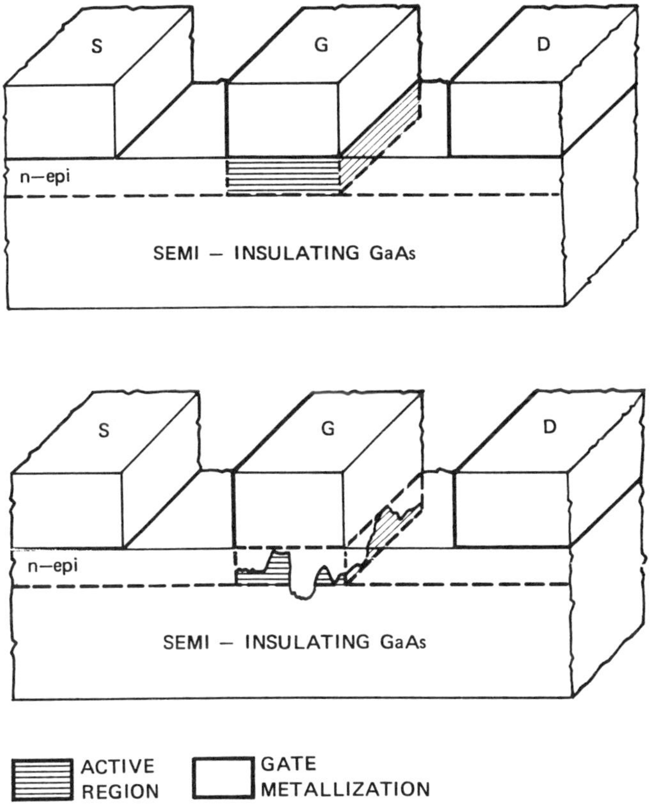

Figure 8.21 Schematic cross section of a MESFET showing the reduction in channel thickness caused by the sinking of the gate metalization into the active layer.

pads, and *vice versa,* have created quite a few problems, perhaps not yet fully resolved.

Figure 8.22 is an Au-Al phase diagram showing how at least five stoichiometric compounds are possible; and in particular, there is a eutectic point at a relatively low temperature (525°C). Hence, it is quite likely that phases such as Au_5Al_2 and Au_4Al will be formed even at normal operating temperatures, but, according to Philofsky [54], all five phases can be present.

The gold reserve consisting of the bond and the wire "recalls" the surrounding aluminum, with the consequent depletion of the gate metalization, and the formation of the so called *Kirkendall voids,* which can occur even at relatively low temperatures (< 300°C) [54] starting at hot points. The gate consists of extremely

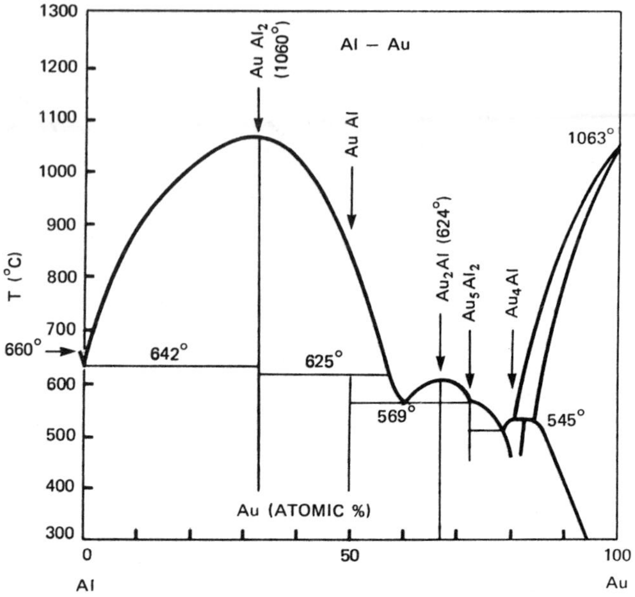

Figure 8.22 Au-Al phase diagram.

thin electrodes and can also lead very quickly to an open circuit, when the device is biased, due to the contribution of electromigration induced by the increased current densities from the reduction in the conductor's useful section. The phenomenon has catastrophic consequences for low-noise devices, whereas power devices, featuring many gate electrodes in parallel, exhibit a gradual reduction in efficiency.

Macpherson, Gleason, and Christou [55] have verified a gradual reduction in the output power in 1 W devices in an X-band tested under nominal operating conditions. In these devices, the Au-Al pair was present from the start, since the gate bonding pads were in aluminum and the interconnection wires made of gold. The analysis of the devices, which failed in times of the order of several tens of hours, showed holes in the gate metalization induced by the reaction between Au and Al catalyzed by the presence of Ga diffused through Al. White, Hewitt, and Turner [56], however, observed a degradation at longer intervals and obtained a MTF for a channel temperature of 70°C above 10^8 hours, but they agree on the need to avoid contact between Au and Al.

Since it is inadvisable to make the gate bonding pad with Al, one solution was to deposit a thick layer of Au on the top, interposing a suitable barrier between this and the underlying Al to prevent reaction between the two metals. The barrier was made from materials with good conductivity characteristics but limited reactivity compared to gold and aluminum. A satisfactory barrier, for example, may consist of Ti-Pt, since the reaction between Al and Ti prevents the formation of Al-Pt compounds, but the Pt separates the Au and Ti, which tend to interact, even though the Au tends to diffuse in the Pt [57].

However, the pressure exerted during the thermocompression bonding can lead to the mechanical failure of the barrier and, hence, can allow the two outer metals to come into contact. The same effect may also be caused by a slight misalignment of the masks during the fabrication of the multilayer structure. One way of avoiding this problem is to separate the gate metalization from the bonding pad and carry out the connection via a bridge of Ti-Pt-Au or, better still, by Ti-Pt or Cr as shown in Figure 8.23.

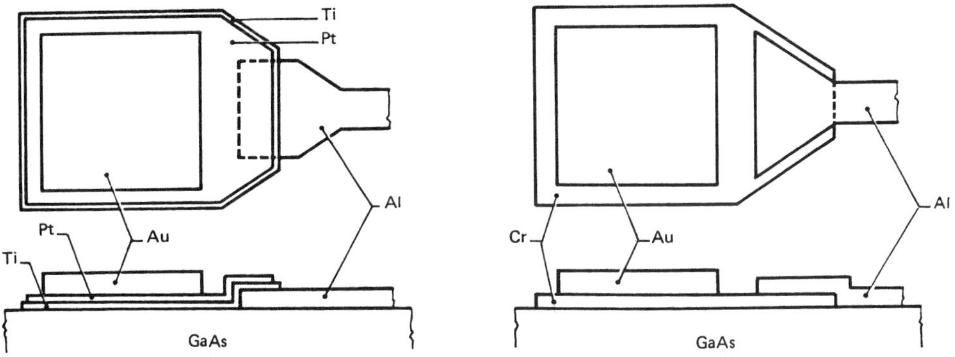

Figure 8.23 Examples of technological solutions adopted for the bonding pad of gate electrodes used to avoid the formation of intermetallic Au-Al compounds.

Irvin and Loya [58] have shown the risks involved in using an Al-Ti-Pt-Au system for the gate pad when the barrier becomes ineffective for the process defects described. The most common failure mechanism encountered in high temperature tests on low-noise MESFETs has been the formation of Au-Al phases, and considering the worst case, mean life values with applied bias have been obtained at 240 hours at 180°C and 94 hours at 250°C. In these cases, it is necessary to consider also the effects of electromigration, which probably intervenes at a second stage, as already mentioned. Irvin and Loya consider it correct to take account of both mechanisms in the calculation of the activation energy. The experimental data converge around a value of 1 eV. It is interesting to note that the MTF is from three to ten times higher for nonbiased devices than biased devices, which gives a fairly clear idea of the combined effect of the two phenomena.

In their search for a reliable metalization system, Huang et al. [59] observed failure times of the order of 20 hours and 40 hours respectively, for Au-Al and Pt-Al systems (with thicknesses 1000–7000 Å) heated in an inert ambient temperature of 240°C, proving once again the need for an effective integral barrier. They also tested the quality of Cr-Al, Mo-Al and Pd-Al systems. The structure proposed is Al-Ti-Cr-Pt-Au, where a time greater than 10^7 hours has been derived for the formation of intermetallic compounds at 100°C.

Lundgren and Ladd [41] have also observed a rapid degradation of the Al-Pt system (after 30 minutes at 300°C); furthermore, tests conducted on devices with the gate bonding pad having the structure shown in Figure 8.23b, with applied bias and at a temperature of 260°C, resulted in catastrophic failure conditions due to the formation of Au-Al compounds after 431 hours. The phenomenon would seem to result from the migration of the gold from the bonding pad through the Cr layer toward the Al layer; this would then be a secondary effect.

Christou, Cohen, and Macpherson [27, 60], having subjected devices with a Ti-Pt link between the Au gate bonding pad and the Al gate electrodes (the structure illustrated in Figure 8.23a) to high-temperature, accelerated tests, showed a net increase in the resistance associated with the gate. The Auger analysis showed an appreciable amount of oxygen (up to 20 atomic percent) in the layer of Ti, and arsenic (up to 20–25 atomic percent) in the Al layer. After a heat treatment at 200–210°C in a nitrogen atmosphere, the degraded devices recovered, at least partly, and the percentage of As in the Al was decidedly reduced. These factors have attributed the increase in resistance to the formation of an oxide or hydroxide of the Ti (TiO_2 $TiO(OH)_2$) that, during the high temperature annealing, is reduced by the As in the Al. The activation energy obtained for the oxidation of the Ti varies from 2 to 2.3eV [23] and, even if based on a rather limited number of samples, the mean time to failure at a channel temperature of 125°C has been estimated to be of the order of 10^4–10^5 hours. Hence, we could conclude that the probability of Au-Al intermetallic compound formation, keeping Al as the gate electrode material, is closely related to the quality of the device and process design. According to Irvin and Loya the application of appropriate screening procedures would take the failure rate down to a value below 50 FIT.

8.2.3.6 Effects of Humidity and Contamination

The history of Si device reliability is full of examples of faults caused by the concomitant presence of elements such as humidity, contaminants, and applied bias. The degradation effects of these factors are concentrated on the metalizations and manifest themselves catastrophically on both Al- and Au-based metalizations (see, for example, [61, 62]). Even the reliability of MESFETs, which incorporate metalization systems that are not too dissimilar to those used for Si devices and are always based on Au and Al, can be seriously damaged by phenomena such as electrolytic corrosion of the Al gate electrodes and Au migration.

MESFETs are generally encapsulated in a hermetically sealed package, individually, or as components of hybrid circuits, and mounted in chip form. The atmosphere inside the package may include water vapor with possible dissolved salts, caused either by a poor seal on the package or by a lack of control of the ambience in which the package was sealed. We should not forget that even the humidity absorbed by the package's internal surface can sometimes be sufficient. The contaminants (mainly halogens) may derive from inadequate cleansing of the process ambiences or from residues of photoresist still present on the device surface. The need for an effective protection to prevent any humidity coming into contact with the metalization is dealt with by Irvin and Loya [58]. A specially designed passivation layer should be deposited over the whole device surface and not just in the channel region; it should not exhibit defects, such as pin holes; and it should feature a small dielectric constant and low losses at microwave frequencies.

Tests carried out by Anderson, Christou, and Sleger [63, 64] have produced noteworthy results: the tested devices partly sealed in a hermetical package and partly not, with Au and Al base gates and with and without passivation, were subjected to temperature cycles (from $-10°C$ to $+100°C$ in times varying from 2–4 hours). The devices all were biased in direct current, with a choice of three different conditions, and the cycles performed in high relative humidity conditions (95 percent) in an ambience contaminated with Na^+ or Cl^- ions.

The electrical degradation encountered in Al gate devices has generally been an increase of the leakage current between gate and source, with a final short circuit condition ($I > 100$ μA) in devices biased after a number of cycles, falling between 2 and 22 for two types out of three. Essentially, three failure mechanisms were found:

1. Corrosion along the gate electrodes.
2. Electrolytic low-impedance paths created in the channel without the presence of corrosion.
3. Au-based metallic bridges between the gate and source electrodes with a consequent reaction between Au and Al.

The last phenomenon (see Figure 8.24a) is probably the final stage of the degradation induced by the second phenomenon, since the Auger microanalysis detected the presence of Na and Au in solution in the electrolytic path. The reaction between Au and Al has led to a progressive depletion of the gate metalization and the formation of a surface Au-Al layer extending from the gold source (the bridge) toward the bonding pad along the gate. The increase in current density in the short circuited zones seems to have caused a localized fusion of the gate due to the formation of hot spots.

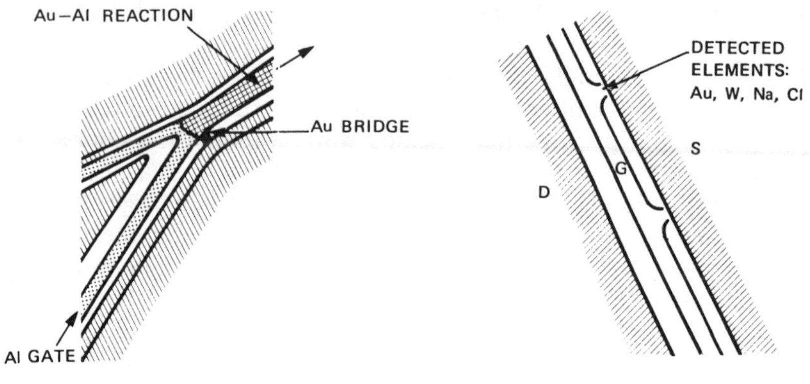

(a)

(b)

Figure 8.24 Schematization of failure mechanisms found on devices with Al gates and (a) Ti/W-Au gates, (b) after temperature cycles with applied bias.

Devices with Au/Ti-W-based gates have shown themselves to be less sensitive to the test conditions, while showing a gradual degradation in the case of applied bias, exhibiting an increase of I_{gs} and a decrease of I_{dss} and g_m. The leakage can be attributed to the formation of dendrites (see Figure 8.24b) between the gate and source electrodes, consisting essentially of Au grown in the presence of electrolytic paths created by Na^+ and Cl^- ions in solution with and deposition at the negative electrode (gate) of the products of the dissolution due to reduction on the basis of the reaction:

$$AuCl_2 + e^- \rightarrow Au + 2Cl^- \quad (8.3)$$

This phenomenon, which also is well known in Si technology, is described in detail in another chapter of this book.

An interesting conclusion of this work is that even devices covered by a SiO_2 passivation layer showed a significant tendency to fail in the same way as nonpassivated devices. This is an indication that silicon oxide passivation is not effective, whether due to defects such as cracks and pinholes through which the humidity can penetrate or problems in adhesion and reactions induced by the presence of contaminant ions that in relatively high humidity can increase the surface conductivity. The possibility of humidity penetrating through the oxide in different ways is indicated by the reduction in activation energy as relative humidity increases. We conclude that the presence of humidity and contaminant agents can at least be limited by proper process control; however the passivation problem remains, and this has major implications for the reliability of MESFETs even in relation to degradation mechanisms discussed elsewhere.

8.2.3.7 *Problems Related to Interface and Surface Conditions*

Frequency parameter drift has been observed on certain devices some time after they are switched on, in particular, a reduction in the gain associated with an increase in the amount of noise corresponding to a reduction in drain current (see, for example, [3]). This phenomenon did show signs of being reversible, however, since the drift parameters returned to their initial value some time after the device was turned off. This varied from lot to lot, in the sense that the drift and recovery times, virtually equal for the same device, varied from a few seconds to hours for devices taken from different lots. The cause of the drift has been explained in two different ways; first, due to the presence of deep traps at the interface between the active layer and the substrate, and second, due to the high density of surface states, with the consequent further modulation of the channel thickness.

To eliminate the effects attributable to defects or impurities at deep levels, next to the interface with the substrate, the solution has been to interpose a layer having high resistivity, grown epitaxially between the active layer and the substrate layer. This buffer layer, although it provides a qualitatively improved interface, is not always effective in eliminating the anomalies described, however. In effect, considering that the MESFET is only a thin film whose operation is determined by the flow of charges in a channel a few tens of μm thick, it is clear that even a minimal reduction in thickness, if not driven from the outside via the applied biases but induced by collateral effects, can signifcantly influence the efficiency of the device.

Itoh et al. [65, 66] introduced a channel thickness reduction $\gamma(x)$ such that:

$$d = d_0 [1 - \gamma(x)] \tag{8.4}$$

where, with reference to Figure 8.25a, d is the effective thickness of the active layer. They showed that a small percentage change in γ induces a more than appreciable reduction in the drain current and transconductance; the situation is illustrated in Figure 8.25b and 8.25c, where the Y axis gives the normalized value of I_{ds}—that is, $I_d = I_d(\gamma)/I_d(\gamma = 0)$—and of g_{m0}. This reduction is reflected significantly on the frequency characteristics (see Figures 8.25d and 8.25e) and has a much greater effect the higher is the operating frequency and the longer is the gate electrode.

The phenomenon can be accentuated by applying a bias to the substrate, as can be seen in the curve in Figure 8.26, where the two moments are clearly distinguished. These are the initial degradation of I_{ds}, which stabilizes at a value equal to about 90 percent of the starting value after approximately 10 minutes, and the variation induced by the application of a voltage equal to -5 V at the substrate 40 minutes after starting. The variation due to voltage is much higher than the first value, but it disappears once the bias is removed from the substrate.

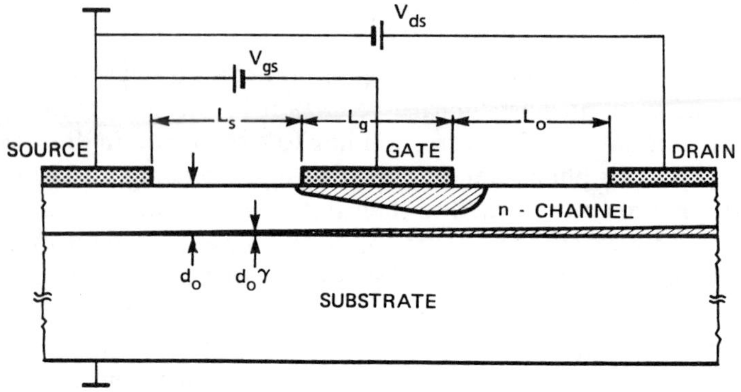

(a)

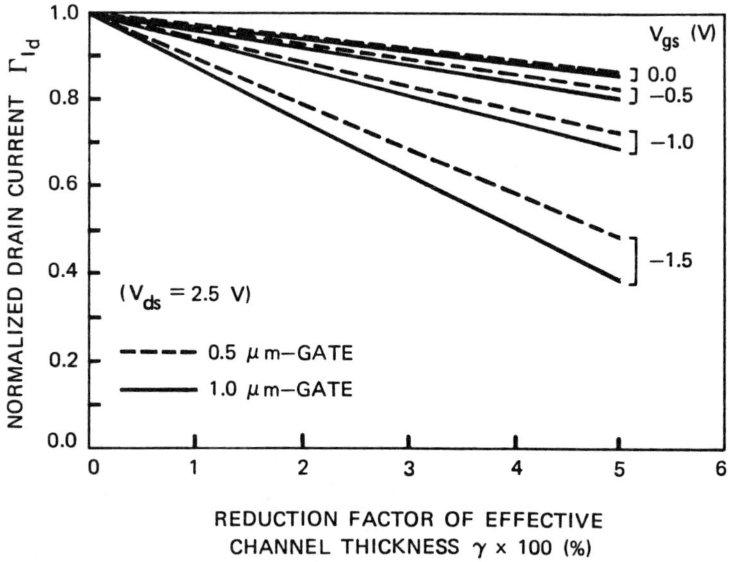

(b)

Figure 8.25 (a) Model used to analyze the interface effects on the electrical characteristics. The shaded areas show the extension of the depletion layer in the active layer. (b and c) Drain current and transconductance normalized *versus* the reduction factor of the effective channel thickness at different gate bias levels. (d and e) Dependence of S_{11} and S_{21} parameters on the channel thickness reduction factor in relation to frequency.

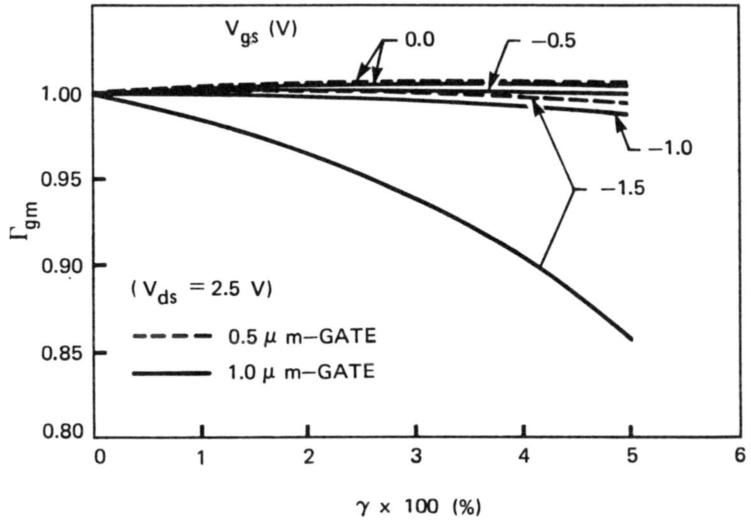

(c)

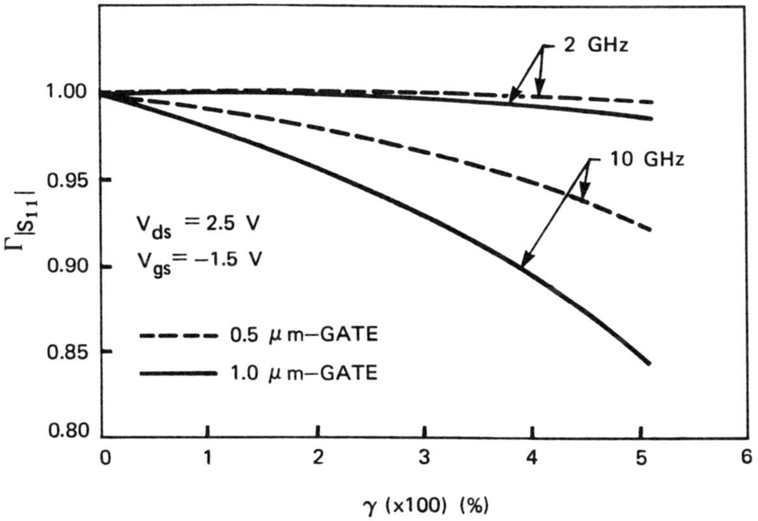

(d)

Figure 8.25 continued

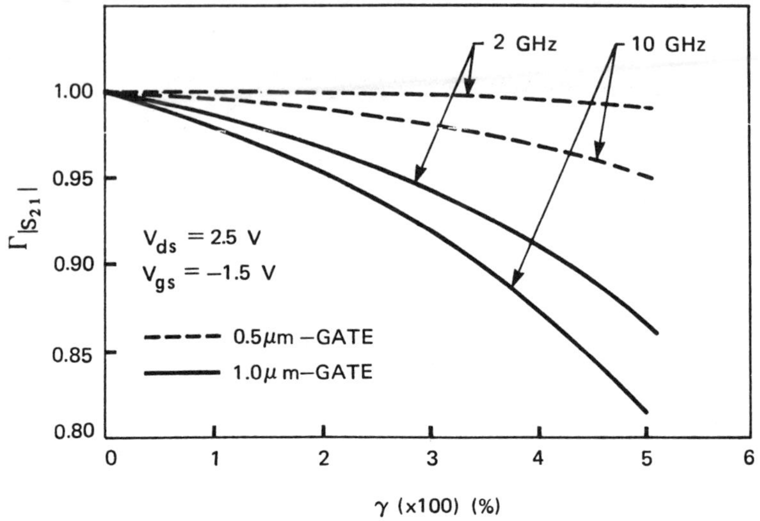

(e)

Figure 8.25 continued

This diagram verifies the dependence of the phenomenon on temperature and gives a constant activation energy, regardless of the drift value, equal to 0.82 eV.

Measurements using the *deep level transient spectroscopy* (DLTS) technique on devices with and without gates have attributed the reduction of drain current to the formation of negative spatial charge regions, due to the emission of holes by deep acceptors or by traps present at the interfaces, in the semiinsulating substrate or the buffer layer. The result is equivalent to the presence of a gate on the lower section of the active layer, and the phenomenon has been called *back-gating*. Since the bias of the substrate, or the buffer layer, may derive from the bias applied to the gate, Itoh and Yanai, [65] recommend the interposition of a dielectric layer between the gate pads and the substrate (or the buffer layer). Moreover, a long-term drift of I_d has been observed (with time constant at ten minutes at ambient temperature), which has been attributed to the presence of deep chromium levels at the interfaces with the substrate and the buffer layer. Finally, the nonuniform extension of the depletion region has been verified. It extends more along the interface toward the drain due to the electrical field generated by the bias (see Figure 8.25).

Omori and Wholey [67] have also explained the reduction of I_{dss} and the consequent degradation of other parameters, encountered during high temperature

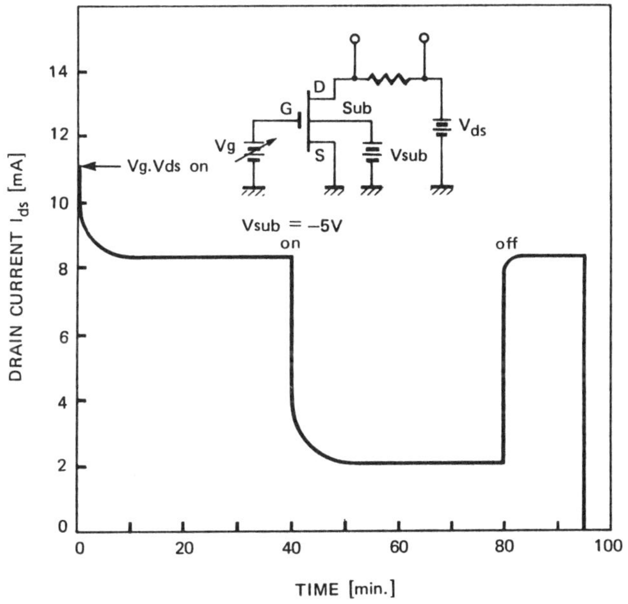

Figure 8.26 Example of drain current drift.

tests with applied bias, with the diffusion inside the channel of compensation impurities, such as Cr, coming from the substrate, which is enhanced by the high electrical field. The measurements revealed an appreciable variation in the concentration profile of the free carriers (see Figure 8.27), for which the quantity of Cr required has been calculated on the order of $4 \times 10^{11}/cm^3$. Chromium tends to diffuse rapidly and to accumulate in the regions incorporating defects or damage caused for example by implantation. However, they do not reject the hypothesis that the dopants diffuse from the channel into the substrate. This is also demonstrated by the fact that devices with the active layer grown by vapor phase epitaxy and doped with S exhibited, for the failure mechanism in question, a shorter mean life than devices implanted with Si. Sulphur, in fact, diffuses more rapidly than silicon in GaAs. The MTF calculated for this failure mechanism varies from 30 to 300 hours at 275°C, with associated activation energy of 1.5 eV.

As already mentioned, surface charges, in the form of mobile ions or electrons trapped in surface layers, can also cause a reduction in the drain current. In the absence of bias, these charges create a surface depletion zone between gate and drain and gate and source. The displacement of these charges toward the drain

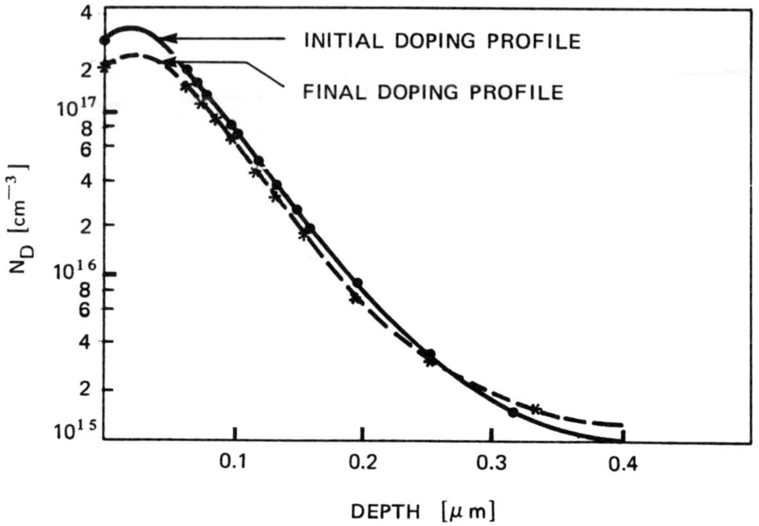

Figure 8.27 Initial and final channel doping profiles after the temperature tests with applied bias [59].

when the device is biased can aid the extension of the channel depletion zone toward the drain (see Figure 8.28) behaving like a gate side extension; hence, the name *side-gating.*

The presence of mobile ionic contamination has been observed by Dumas et al. [26] during aging tests performed on devices passivated with Si_3N_4 deposited via sputtering. In the case described, there was a rapid reduction in I_{dss}, which returned to the normal value after 200 hours storage at 150°C without applied bias and the electrodes short-circuited. The contamination can be eliminated with an appropriate surface treatment (cleansing with trichloroethylene or high temperature treatment in air) or passivating the device, for example, with polycrystalline GaAs, even if only in the active region.

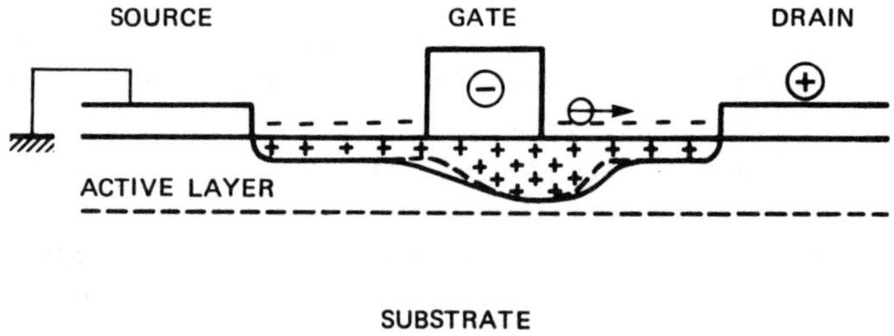

Figure 8.28 Schematization of side-gating phenomenon to explain the drift caused by the gate bias.

8.2.3.8 *Effects of Radiation*

The application range of GaAs MESFETs is expanding rapidly, as regards both discrete devices and integrated circuits. Integrated circuits in particular form a fast-emerging, promising technology that has assumed a strategic importance; this is because they seem to exhibit a satisfactory resistance to radiation, which makes them particularly suitable for use in space and in the military and industrial fields where they may be exposed to nuclear radiation. Radiation generally consists of flows of uncharged particles, or neutrons, charged particles, and high-energy photons (γ rays).

Exposure to high-speed neutrons tends to make the GaAs more intrinsic, due to the reduction in the concentration of the free carriers [68, 69]. Figure 8.29 compares the flow of neutrons required to achieve a transconductance degradation of 20 percent in GaAs and Si junction FETs.

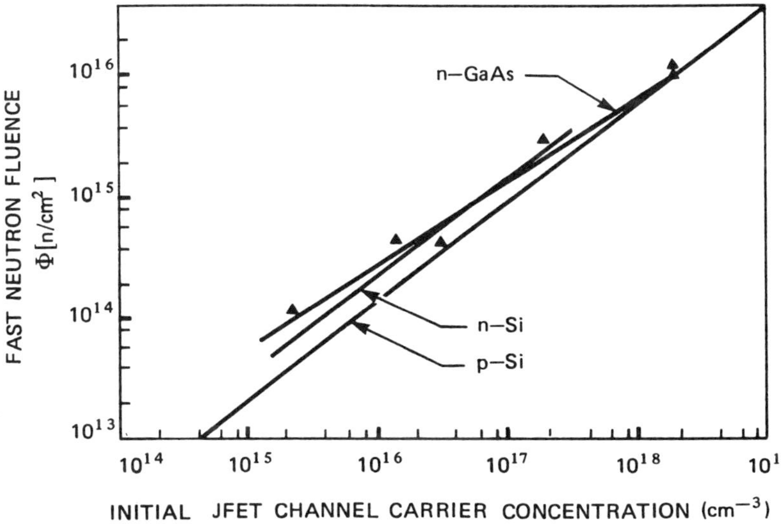

Figure 8.29 Fluence of high-speed neutrons required to reduce the maximum transconductance in junction FET from the initial carrier concentration.

The superiority of GaAs FETs is clear for low initial concentrations of carriers. It has however been demonstrated [68] that, for increasing flows of neutrons, the degradation of saturation transconductance is linear for enhancement-type junction devices having similar geometries and threshold voltages, but different doping profiles and junctions were obtained either through epitaxial growth or ion implantation, and it is virtually independent of the initial transconductance value.

However, the epitaxial transistors show a degradation rate 21 percent lower than for implanted devices. Data on the degradation of discrete MESFETs [70] show the same tendency with very small variations up to fluences of 10^{15}n/cm^2.

Considering that for concentrations of carriers on the order of 10^{16}/cm^3 the removal rate due to the effects of traps in the channel is on the order of 4 carriers/cm for each high-speed neutron/cm^2 [68] (see Figure 8.30) and that similar effects of current reduction in the channel can take place when the trapped charge density approaches the original doping density [69]; at these concentration levels, at least 2.5×10^{15}n/cm^2 would be necessary to disturb the channel conductivity.

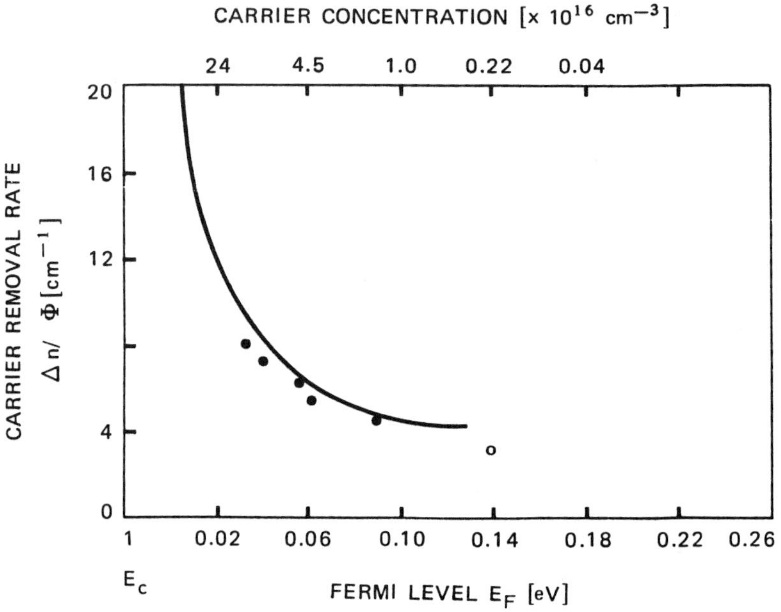

Figure 8.30 Carrier removal rate for n-type epitaxial GaAs *versus* carrier concentration: the hollow dot indicates data obtained by Stein; the filled dot, data obtained by Behle and Zuleeg [68]; the curve, data obtained by Aukerman et al. © IEEE, 1980.

Unlike Si MESFETs, the GaAs FETS with p-n or Schottky junction gates do not exhibit appreciable changes after prolonged exposure to ionizing radiation; up to total doses on the order of 10^7–10^8 rad (GaAs), the variations induced are minimal, whereas the changes in threshold voltage in the MOS due to build up of charge in the gate oxide become more sensitive at doses on the order of 10^6 rad (Si). Defining as rad the quantity of radiation that deposits an energy of 100 erg/g in the specific material, the conversion factor between rad (Si) and rad (GaAs) is 1.06.

In space applications, it is conceivable that cosmic radiation will have an effect mainly on integrated circuits rather than on discrete components. Specially radiative areas exist around the earth, known as *Van Allen belts,* in which charged particles are trapped by the earth's magnetic field. In the region closest to the earth, there is a net prevalence of protons, which represent 90 percent of the cosmic radiation, whereas further out, up to 20,000 km, belts consist basically of electrons, the maximum flow of which can be estimated on the order of 10^{10} electrons/cm^{2}s. The maximum proton flow is on the order of 10^4 protons/cm^{2}s. Other charged particles also exist in space, such as nuclei of heavy elements and α particles (nuclei of He with mass 4 and charge +2).

Takeuchi et al. [70] carried out irradiation tests with electrons on some 250 low-noise and -power MESFETs taken from different lots, with or without applied bias, and showed that variations of the dc parameters within ± 10 percent (I_{dss}, I_{ds}, g_m, V_p, V_{gf}) in the low-noise devices and within ± 14 percent (I_{ds}, V_p, V_{gf}) for power devices, with flows up to 10^{12} electrons/cm^2, or fluences of 10^{15} electrons/cm^2 corresponding to a total dose of 4×10^7 rad. The most significant technological parameters for the devices tested are shown in Table 8.2.

Table 8.2 Technological Characteristics of MESFETs Subjected to Irradiation Tests with Electrons [71]

	Low-Noise MESFET	*Power MESFET*
Substrate	Doped Cr, thickness = 150 μm	Doped Cr, thickness = 150 μm
Buffer layer	Thickness = 3 μm	Thickness = 6 μm
Active layer	Planar Thickness = 0.2 μm m = 2 × 10^{17}/cm^3	Recessed Thickness in recess = 0.3 μm external = 0.6 μm n = 10^{17}/cm^3
Gates	Al, L = 0.5 and 1 μm W = 300 μm	Al, L = 1 μm W = 1400, 2800, 5600, 11200 μm
Ohmic contacts	Au-Ge-Pt	Au-Ge-Pt
Passivation	1500 Å	6000 Å

The behavior of devices subjected to bombardment with heavy ions has been studied by Anderson et al. [72]. The irradiation of MESFETs, of different technological characteristics but all without passivation, with particles having an associated energy of 1.1 MeV for flows from 10^7 to 3×10^8 particles/cm^{2}s yielded a tolerance level estimate on the order of about 2×10^5 rad; this value, which is decidely low, according to Anderson et al., can give rise to permanent damage after about a year in devices situated on satellites orbiting in radiative belts where the flow of heavy ions, even induced by secondary reactions at package or device level, approximates or exceeds a value of about 2×10^{10} particles/cm^2. The main degradation observed has been a reduction in the drain saturation current with zero gate bias, due to the removal of carriers from the active region and consequent to the change in the crystalline structure; in some cases, burnout occurred.

Irradiation experiments with γ rays [70] have shown an increase in the noise figure of low-noise MESFETs, attributed to the introduction of levels responding only at microwave frequencies; variations have not been observed in the signal characteristics up to doses on the order of 10^8 rad (Si). The power MESFETs should have virtually zero sensitivity to γ radiation exposure.

8.3 MONOLITHIC MICROWAVE INTEGRATED CIRCUITS

Only in the last few years have *monolithic microwave intergrated circuits* (MMIC) emerged from the stage in which they were produced only by a few research laboratories, making an effort to demonstrate their feasibility to full-scale production [73]; indeed, though they are still far from being commonplace, they are now offered by the major manufacturers of GaAs components, almost always in parallel with a "foundry" service whereby the device can be designed and fabricated according to the customer's functional specifications. Although MMIC market forecasts for the next decade are often contradictory and their areas of application are far from well defined, it is widely believed that they will provide the same sort of impetus for microwave system development as the development of VLSI technology did for computers [74].

It is practically impossible to pinpoint the historical moment in which the idea was conceived of designing an integrated circuit explicitly for microwave operation. In fact, the years between 1952, when, according to Kilby [75], Dummer of the Royal Radar Establishment (G.B.) first had the idea of an integrated circuit based on semiconductor technology, to 1976, when Pengelly and Turner of Plessey developed the first fully monolithic GaAs X-band amplifier [76], were marked by a continuing series of small but important steps towards the MMIC. This is borne out by some of the most important trailing events in the development of GaAs technology.

The first high resistivity GaAs wafers obtained with a certain reproducibility via Cr doping were made in 1964 [78], whereas the first Schottky barrier MESFETs were produced by Mead at the California Institute of Technology, two years later [79]. The year 1967 saw the birth of the first microwave MESFET (Fairchild) [80] with 3 GHz maximum frequency, though its performance was still inferior to that of the bipolar Si devices of the time. The big leap forward came three years later, when a MESFET with 1 μm gate and 30 GHz maximum frequency was developed at the IBM research laboratories in Zurich [81]. A few years later, in 1973, the first significant results on GaAs power MESFETs were achieved and published [82, 83], thus firmly establishing this technology in all applications requiring very high speeds.

8.3.1 MMIC Technology

Though MMIC design methods and production technology have not yet reached a consistent level of maturity and there is thus no standard, universally known, manufacturing process, a few words concerning the technology are essential to an understanding of the problems which can affect these devices. From a conceptual point of view, an MMIC is no more than the natural evolution of a hybrid microwave alumina circuit where the substrate has been replaced by a semiconductor material (GaAs) whose characteristics allow production of the active and passive components that, in the hybrid approach, are subsequently mounted and wire bonded. The advantages in terms of weight, process tolerances, simple assembly, and uniform performance without the need for a tuning circuit are readily understandable, whereas the price for all this is the maintenance of a consistent, high standard of quality during the base material growth processes and the subsequent device production.

Figure 8.31 [84] is a schematic representation of a part of an MMIC, showing commonly used circuit elements. It will be noted that no specific processing stages were added to those already used in MESFET fabrication, as described at the beginning of this chapter. The only exception is that the resistors may be fabricated using a film of Cermet (a mixture of metal and dielectric) as well as by deposition of a metal-only film or by ion implantation into the GaAs. However, this type of resistor features considerable parasitic capacitance, which makes it unusable beyond 1 GHz. Metal resistors, which must employ materials with good adhesion to the substrate such as Cr, Ti, or Ni, generally exhibit lower resistivity than GaAs ones and greater insensitivity to temperature; furthermore, GaAs resistors are affected by current saturation with a consequent current-voltage characteristic nonlinearity. In some applications, however, it may be necessary to exploit these two properties to obtain resistance values that are much higher at microwave frequencies than the corresponding values in dc or for thermal compensation techniques.

Capacitors used in MMICs can be divided into two general categories: dc-block capacitors and tuning capacitors. In the dc-block capacitors, since the purpose is to allow only the RF signal to pass, the exact value is not particularly important and normally varies from 10 to 200 pF. These capacitance values are generally obtained with *metal insulator metal* (MIM) structures, where the dielectric consists of Si_3N_4, SiO_2, Ta_2O_5, or polyimide. As the dielectric is normally between .1 and .4 μm thick, it is important that the lower electrode on which it is grown be clean and smooth to minimize defects. In fact, most of the problems with this type of capacitor arise from the presence of pinholes in the dielectric or pointed irregularities on the electrode surface facing the dielectric.

Determining the exact capacitance is important for tuning or matching capacitors, where typical values range up to 1 pF. These units may be produced with

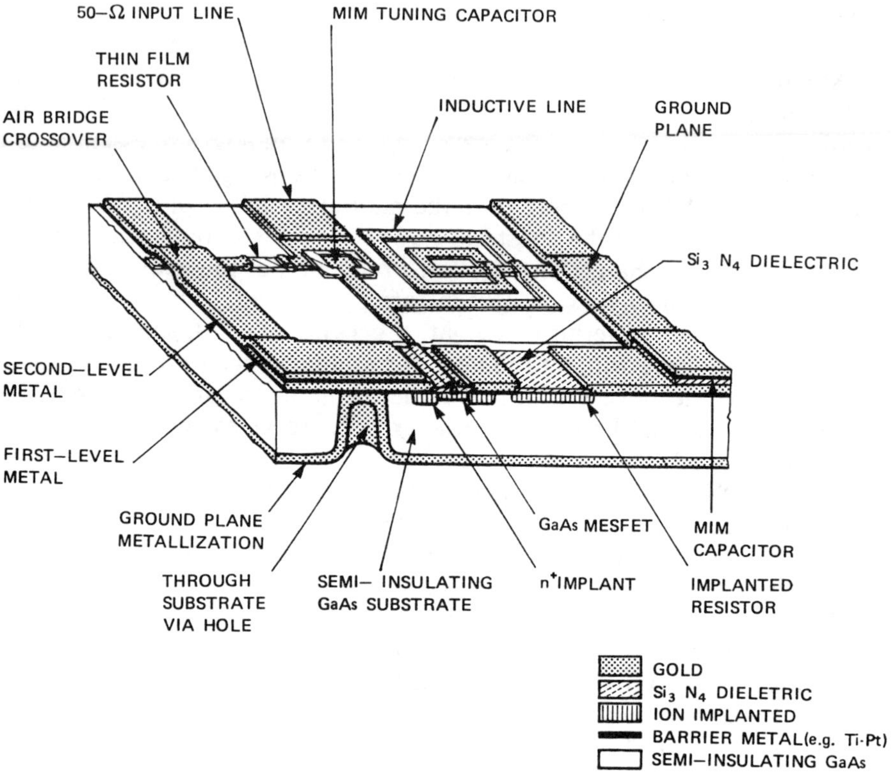

Figure 8.31 Schematic representation of an MMIC.

stubs parallel to the transmission line (usually microstrip) or with finger structure.

It must be borne in mind that both cases produce a distributed capacitance, which calls for accurate modeling if precise simulations are to be carried out. The use of accurate modeling techniques is equally essential in the case of inductors, which are generally produced with sections of line with a spiral or meander shape. The inductive effect comes from the mutual inductance caused by electromagnetic interaction across the various line sections. There is also a capacitive effect, due to interaction with the ground plane on the backside of the chip, which increases with frequency and must be taken into account in CAD programs.

8.3.2 Considerations Regarding MMIC Reliability

MMICs have provided extremely satisfactory performance, but a number of problems remain to be solved in connection with yield and reliability, particularly when the chip includes power devices. Integrating different components on a single chip offers a number of advantages in reliability over the hybrid approach; the number of wire bonds is reduced to the minimum required to connect chip and package, so that the reliability of the circuit as a whole depends on the reliability of the chip itself, which can be controlled through suitable processing techniques and screening.

Figure 8.32 shows a comparison between reliability data for discrete MESFETs and MMICs [85]. Very little reliability data has been published to date; however, if we also consider data published for digital integrated circuits, we can conclude that the same trends are operating as for Si devices, where the failure rate of a monolithic circuit is less than the sum of the failure rates of the discrete components manufactured using the same technology, which would be required to produce the same circuit [86].

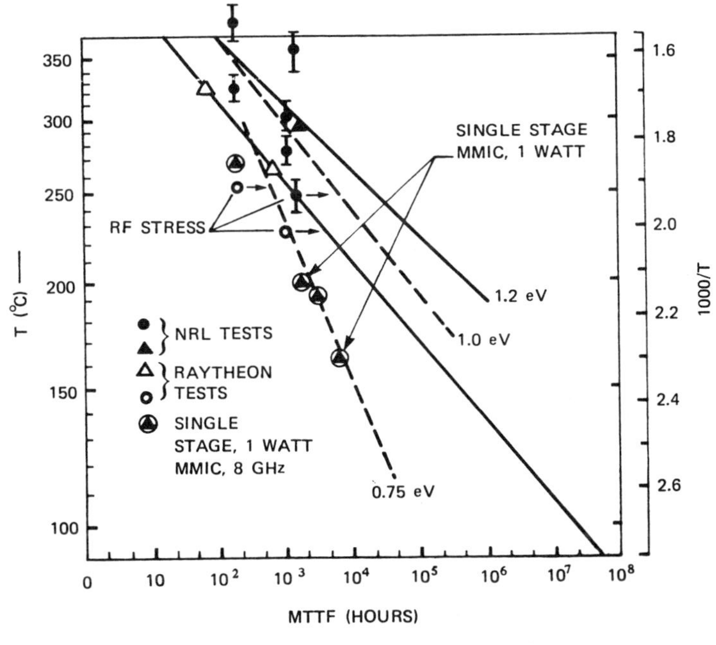

Figure 8.32 Comparative test data for discrete MESFETs and monolithic circuits.

When considering the factors that have the greatest influence on MMIC reliability, we often tend to forget design quality: a good amplifier, for example, must be designed to operate with supply voltages that may differ fairly considerably from the rated voltage without dangerous self-oscillation. Clearly, achieving such results depends on the precision with which the scattering parameters of the circuit elements can be measured and on the repeatability and uniformity of the fabrication process. The dominant factor in MMIC reliability continues to be the operating temperature of the various components, and mostly the active ones, as they dissipate more power and usually show higher failure rates than passive components and interconnetions.

By way of example, Katsukawa et al. [87] give an MTF of 3.5×10^7 hours at 130°C channel temperature for the MESFETs used in an S-band amplifier, and of more than 1.3×10^8 hours for the resistors of the same chip. Where power elements are present, thermal resistance can be reduced by thinning localized areas of the chip with a technique similar to the one used for making via holes. Though this makes processing the backside of the chip slightly more complex, it is doubtless worthwhile if we consider that reducing a MESFET's channel temperature by only 10°C means a more than threefold increase in mean time to failure (assuming an activation energy of 1.4 eV).

Another effective, low-cost measure involving no additional processing steps is to place electrostatic discharge protection diodes in parallel with the circuit inputs, which usually lead to the most delicate part of the MESFET, the gate [86]. Typical failure from electrostatic discharge is caused by damage to the channel between source and drain in MESFETs and rupture of the thin dielectric used for capacitors. Catastrophic effects may be promoted or worsened by uneveness or defects in the GaAs active layer or the oxide, respectively, which cause a local reduction in dielectric strength [88, 89].

Finally, it should be noted that the radiation resistance of MMICs makes them extremely attractive for aerospace and military applications. Data published to date [87, 90] are perfectly in line with the considerations made in section 8.2.3.8 and demonstrate MMIC's low sensitivity to neutron flow of up to 10^{15} n/cm^2 and to gamma radiation doses beyond 10^{17} rad (GaAs). The performance of MMICs is very similar to that of discrete MESFETs in this respect, and this is easily explained by their low level of integration.

8.4 HETEROSTRUCTURE FIELD EFFECT TRANSISTORS

In recent years, a new type of field effect transistor has excited increasing interest in competition with low-noise MESFETs. Now, this device has left the confines of the laboratory to become commercially available and features characteristics that make it extremely interesting for a number of applications, even though these presently are limited to areas where the chief requirements are high switching speed or operating frequency and low noise. Though structurally more complex than the MESFET, the heterostructure field effect transistor provides higher performance with essentially the same geometry. In fact, the extremely high mobility resulting from the formation of a two-dimensional electron gas leads to higher transconductance values (several hundred mS/mm) and, hence, to a low-noise level and high conversion gain.

By way of example, Figure 8.33 shows a comparison between two types of field effect transistor: a low-noise MESFET and a heterostructure FET produced by the same manufacturer [91]. The superiority of the latter component, despite its longer gate, is clear. The reliability of this new device, however, is still an open question.

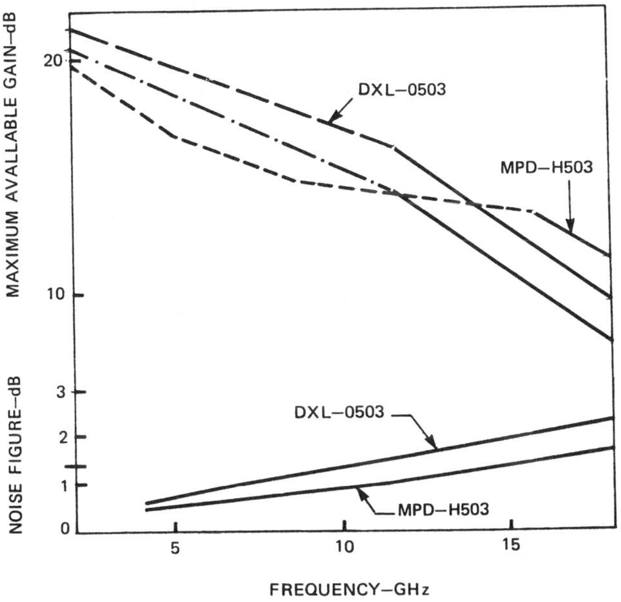

Figure 8.33 Comparison between noise figure and MAG of a low noise MESFET with .3 μm gate (DXL-0503) and an HEMT with .5 μm gate (MPD-H503) [91].

8.4.1 Technological Characteristics

The heterostructure field effect transistor has been given a variety of names, or rather definitions, in the several laboratories where it was developed more or less simultaneously. Each definition centers on one of the device's particular physical or structural properties:

High electron mobility transistor (HEMT), developed by Fujitsu.
Modulation-doped field effect transistor (MODFET), developed by Cornell University, University of Illinois, and Honeywell.
Two-dimensional electron gas field effect transistor (TEGFET), developed by Thomson-CSF.

Selectively doped heterostructure transistor (SDHT), developed by AT&T Bell Labs.

This device was made possible by the progressive improvement in epitaxial growth techniques and the increased purity of the materials used, chiefly substrates and sources. Control over the thickness and uniformity of the grown layers is of fundamental importance, because operation is based on the characteristics of an interface between a highly pure layer of, for example, undoped GaAs and a layer of heavily doped GaAlAs. The energy band diagram for a heterostructure of this type predicts that a triangular potential hole will be formed, given the abrupt discontinuity created by the different gap amplitudes of the two materials (see Figure 8.34).

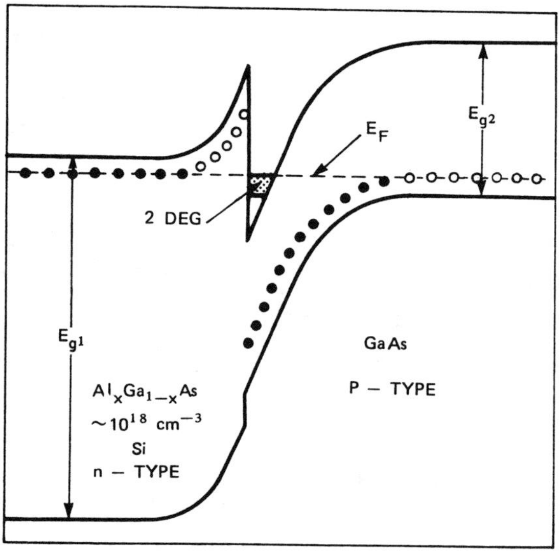

Figure 8.34 Band diagram of a GaAs-AlGaAs heterostructure showing neutral and ionized impurities and the two-dimensional electron gas (2 DEG).

As the Fermi level remains constant throughout the whole structure, at the discontinuity it lies inside the conduction band, which results in an accumulation of carriers (electrons in this case) in the states of this band that belong to the material with the highest electronic affinity; namely, GaAs. The electron gas so created has quasi-two-dimensional characteristics: it is confined within a thickness of less than 100 Å in the undoped semiconductor; and the carriers, as they are separate from the ionized impurities, feature extremely high mobility, particularly at low temperatures where scattering phenomena are much reduced.

At 300 K, carrier mobility is limited chiefly by interaction with optical phonons and can reach peak values on the order of 9000 cm^2/Vs; whereas at 4 K piezoelectric and with acoustic phonons, scattering mechanism dominates permitting values (shown in the literature, see [92]) of 5×10^6 cm^2/Vs to be reached. Even at more reasonable temperatures, the mobility values achieved are sufficient to guarantee, for example, extremely high carrier transit velocity between source and drain; that is, above 10^7 cm/s at 300 K and up to around 4×10^7 cm/s at 77 K for devices with gates less than 1 μm long. Transconductance is also very high: values above 400 mS/mm at 300 K have been achieved. This implies high gain with limited voltage swings and very short transit times under the gate.

The basic structure [93, 94] as shown in Figure 8.35 consists of four distinct layers, grown progressively on a semiinsulating GaAs substrate. The layers are as follows:

- Channel or buffer layer, consisting of undoped GaAs with thickness on the order of 1 μm; the two-dimensional electron gas forms within 100 Å from the surface of this layer.
- Setback or spacer layer consisting of undoped GaAlAs. The thickness of this layer varies from 300 to 600 Å and has a marked influence on carrier mobility, which rises as thickness increases. This rise in carrier mobility, however, is accompanied by a reduction in carrier concentration.
- Donor layer, consisting of GaAlAs with a high concentration of carriers ($> 10^{18}/cm^3$) achieved, for example, by Si doping during epitaxial growth. With the indicated doping level and a thickness in the order of 350 to 500 Å, a two-dimensional electron gas density of over $10^{12}/cm^2$ is produced.
- Cap layer, consisting of heavily doped GaAs ($> 10^{18}/cm^3$) with thickness ranging from zero to a few hundred Å. The function of this layer is to assist in the formation of an ohmic contact with very low associated resistance.

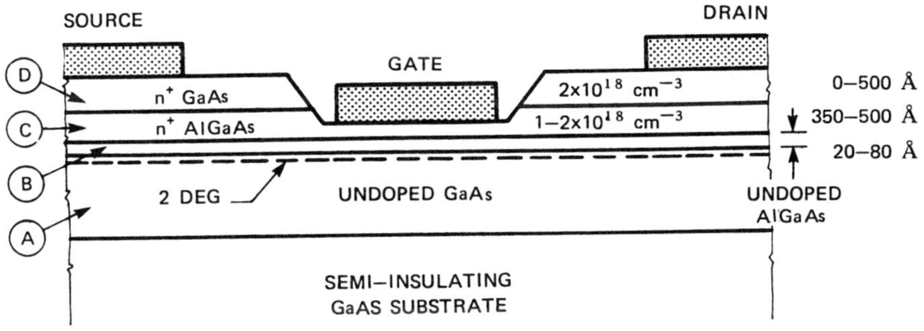

Figure 8.35 Cross section of a heterostructure field effect transistor showing typical layer thicknesses and doping levels.

The ohmic contacts and the gate electrodes are generally produced with metallic systems similar to those used for the MESFETs described earlier. The epitaxial growth techniques include *molecular beam epitaxy* (MBE), which provides the best results; *vapor phase epitaxy* (VPE); and *metal-organic vapor deposition* (MOCVD), which, though it has not yet entered into common use, has shown considerable potential. Liquid phase epitaxy is also used occasionally, though it provides less control over growth thickness than the other techniques. The systems capable of producing a two-dimensional electron gas that have received the most attention are shown in Table 8.3.

Table 8.3 Systems Capable of Producing Two-Dimensional Electron Gas [95]

High band gap	*Low band gap*	*Substrate*	*Growth techniques*
AlGaAs	GaAs	GaAs	MBE, MOCVD, LPE
AlGaAs	GaAs	Ge	MBE
AlGaAs	GaAs	Si	MBE
$In_{0.52}Al_{0.48}As$	$In_{0.53}GA_{0.47}As$	InP	MBE
$In_{0.52}Al_{0.48}As$	InP	InP	LPE
InP	$In_{0.53}Ga_{0.47}As$	InP	VPE, VLE, MOCVD
InP	$In_{0.75}Ga_{0.25}As_{0.50}P_{0.50}$	InP	MOCVD
GaAs	$In_{0.2}Ga_{0.8}As$	GaAs	MBE

These considerations apply to devices in which the charge carriers are electrons; that is, to the n channel transistor. In a GaAlAs/GaAs system (and in others such as Si/Ga_xSi_{1-x}, the two-dimensional gas may also consist of holes if the molar fraction of Al is brought to suitable values ($\geqslant 0.5$) during growth.

The mobility values achieved at 77 and 4 K are higher than 6,000 and 200,000 cm^2/V [95], respectively, but even higher values are expected. It is thus possible to produce complementary structures by combining p- and n-channel transistors to make circuits capable of operating at high speed and dissipating very low power under static conditions, as occurs with Si CMOS.

8.4.2 Reliability

Reliability problems of heterostructure field effect transistors can be divided into two categories. Basic problems include

Substrate defects,
Lateral diffusion of dopants,
Carrier freeze-out,
Persistent photoconductivity,
I-V characteristic collapse in the dark,
Ohmic contacts.

Process-related problems include

Ohmic contacts,
Schottky barrier formation,
Two-dimensional gas localization,
Channel passivation.

Surface defects are of major importance, particularly for materials grown by MBE. These "oval defects," whose dimensions range from 1 μm^2 to around 30 μm^2, are hillocks or faceted extrusions grown around the nucelation centers. Their origin is unknown, but (at least as far as the larger ones are concerned) they do not seem to be connected with substrate dislocations. As these defects stand out from the surface, they can cause problems during substrate lithography. If sufficiently large and located in the active area of a device, they can make it inoperative or cause severe degradation in characteristics.

When exposed to light at low temperatures (< 140 K), heterostructures generate a photo-current that persists even after the illumination is removed. The origins of this persistent photoconductivity would seem to lie in the photoionization of the silicon's deep donor levels (n-type doping species) in the $Al_xGa_{1-x}As$ layer if the molar fraction of the aluminum (x) is $\geq .2$ [96]. Recombination is not possible since at low temperatures and in the dark, the photoexcited electrons do not have sufficient energy to overcome the energy barrier created by lattice relaxation after ionization and the traps remain ionized. These traps are also responsible for donor layer carrier freezing at low temperatures, which causes a reduction in two-dimensional gas density and an increase in the threshold toward positive values [97].

Depletion devices have been observed to show a kind of instability that takes place at room temperature in the form of a hysteresis loop in the I-V curve (see Figure 8.36); this loop can be eliminated by exposing the FET to light. The phenomenon is explained by the traps, which cause a negative conductance regime and in turn cause the device to self-oscillate [98a].

I-V characteristic instability is detected at cryogenic temperatures and in the dark when the drain voltage is raised above a certain threshold, which originates a distorted curve as if a direct-biased diode was connected in series with the drain (see Figure 8.37). This problem again is likely to be caused by the presence of traps that, blocking the electrons, give rise to an electric field that reduces the current at low drain bias voltages. The phenomenon is linked both to the manufacturing process and to epitaxial growth conditions; in particular, if the drain n^+ region is sufficiently close to the corresponding side of the gate (< .4 μm), this problem can be eliminated [98].

Figure 8.38 shows the results of an accelerated test on discrete depletion transistors, for which MTBF was 2×10^3 hours with a 90 percent confidence level; 50 percent reduction in transconductance was adopted as the failure criterion [99]. It is interesting to note that the best results were obtained for the transistors with

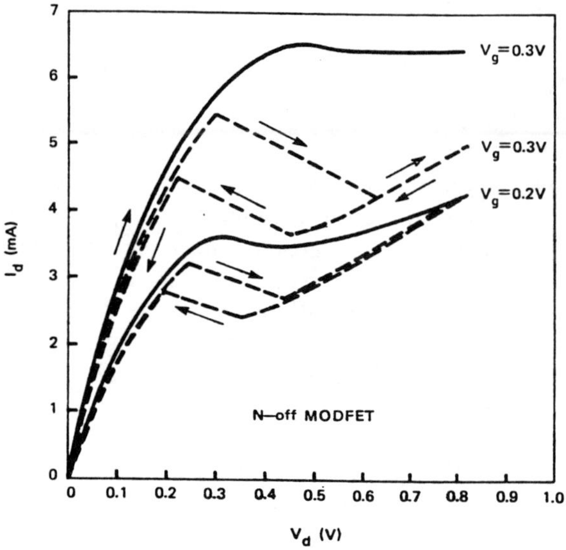

Figure 8.36 Hysteresis in enhancement MODFET I-V characteristics (dashed lines), eliminated by exposure to light (solid lines).

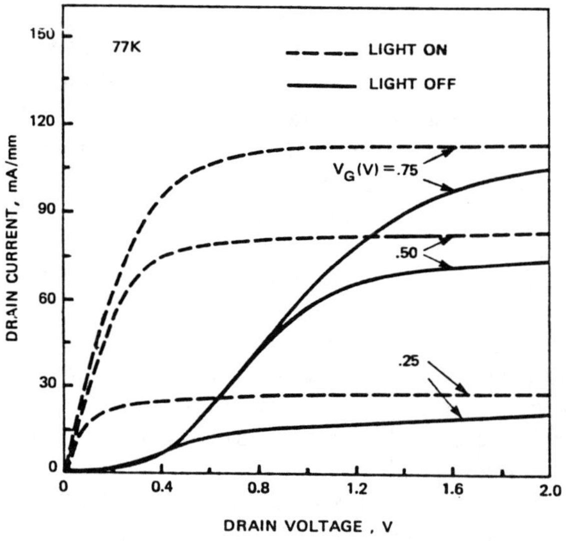

Figure 8.37 Distortion in I-V characteristic of an HEMT and restoration of normal conditions with light.

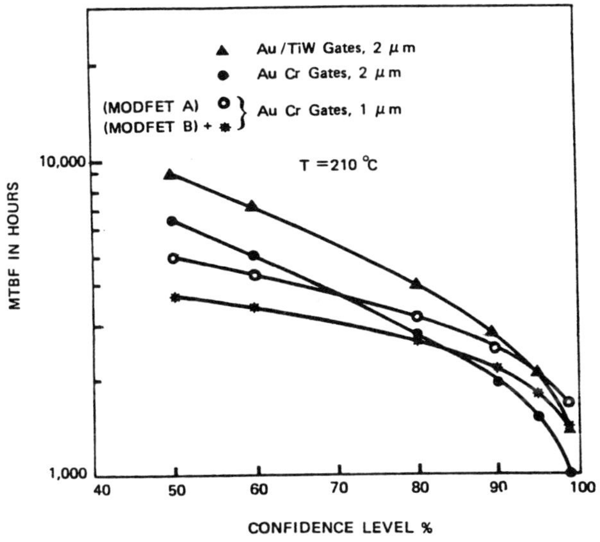

Figure 8.38 MTBF obtained during tests on discrete depletion HEMT devices.

TiW gate. Figure 8.39 shows a comparison between MTBFs for enhancement and depletion transistors tested with dc bias; enhancement FETs were shown to be distinctly less reliable. The main failure mechanism was Si migration from the donor layer at 210°C. This caused appreciable degradation in channel characteristics, leading to a heterojunction barrier reduction and two-dimensional gas deconfinement. Such migration is probably promoted by the electrical fields deriving from the bias and by composition gradients, as well as by temperature. Togashi, Takakuwa, and Kato [100] have observed Si diffusion from the donor layer to the undoped GaAs layer on MOCVD-grown devices, with mobility degradation and an increase in charge density above 600°C.

The sensitivity to burnout of heterostructure devices would seem to be greater than that of MESFETs [99, 101]. Favorable conditions for the phenomenon are inherent in the technology, given that the positive temperature coefficient region (the AlGaAs donor region) is very close to the metal contacts and the metal-semiconductor interdiffusion provoked by sintering causes the latter to move even closer to this region, which abounds in deep traps. Experiments performed with alpha particle irradiation on the structures shown in Figure 8.40 provoked burnout starting from a thermal runaway, originating in the donor layer by action of the hot electrons generated in the normal channel and in the parasitic channel, due to the presence of a surface layer of undoped GaAs on the donor layer. The thermal runaway is initiated when two metal filaments with different potentials arriving from the contact reach a critical distance in the AlGaAs layer.

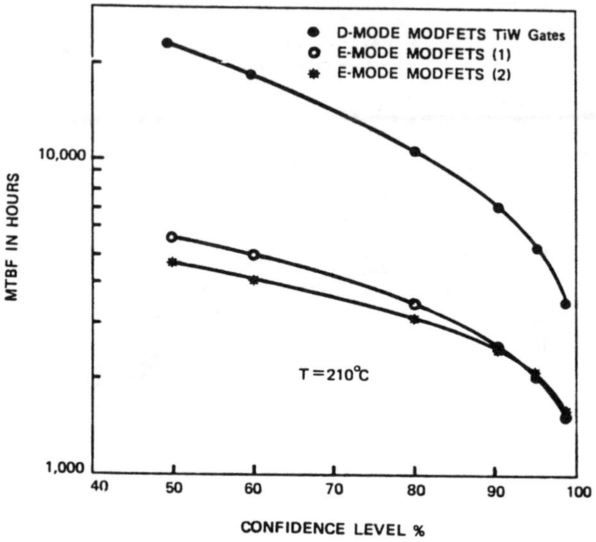

Figure 8.39 MTBF for depletion and enhancement HEMT devices on inverter circuits.

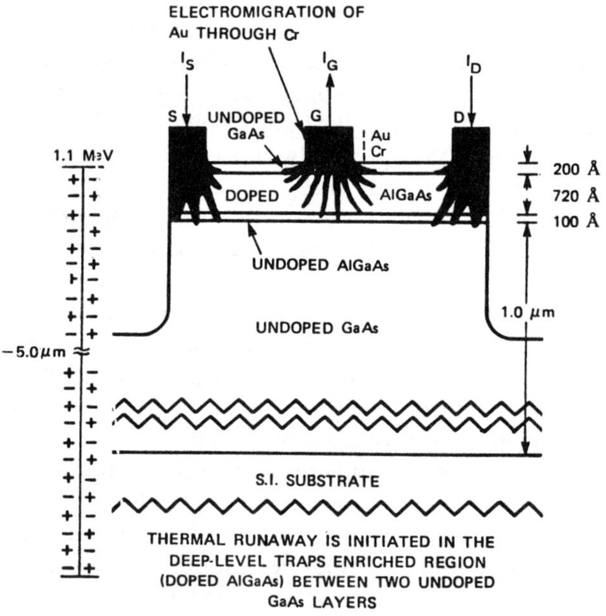

Figure 8.40 Burnout model for HEMT structures derived from alpha particle irradiation tests and simulation [10, 15].

A "double-recess" structure (that is, one in which the channel and gate are recessed from the semiconductor surface) would seem to be effective in reducing burnout sensitivity. Moreover, the use of metalization systems that are stable at high temperatures (chiefly including refractory metals such as Ti, W, Mo, and Ta either alone or alloyed) and silicides or amorphous interdiffusion barriers based on the same metals is advisable both for this purpose and for increasing resistance to *electrostatic discharges* (ESD).

8.5 CONCLUSIONS

Despite the significant improvements over the passing years, GaAs technology certainly has not yet reached the reliability levels typical of Si devices, and MESFETs are a case in point. Even if it seems that certain catastrophic failure mechanisms could be eliminated, using particular technological solutions or adequate screening procedures [102, 103] that nonetheless do not always seem to be so efficient [104], long-term degradation and, to a certain degree, instantaneous burnout are unresolved problems that require greater understanding [105]. This does not imply that the use of such devices for applications requiring a high quality standard is not advisable; low-noise MESFETs, in particular, have reached an acceptable settling level. The problem is still represented by the power devices that, apart from the exceptional performance achieved today, still require further study if the required objectives are to be met.

For the moment, the use of monolithic integrated circuits would seem to be restricted to fields where problems of cost, and hence of yield, rank second to the primary objectives of radiation resistance and better performance. Another primary objective, reliability, can be achieved through the most stringent possible production controls, at the moment only. In addition to military and aerospace applications, commercial applications are potentially numerous (receiving units of direct satellite television broadcasting and laboratory instruments are examples) but call for a reduction in costs that can be accomplished only when manufacturing techniques and processes have been refined sufficiently to increase output. In this case, the problem of reliability will become preeminent as these microcircuits are more widely used. As the manufacturing technologies are practically the same, what was said for MESFETs also applies to MMICs.

The history of heterostructure transistors is still too short for conclusions to be drawn. However, interest in the applications for this device is considerable in both the strategic and industrial fields and is directed beyond the discrete form, given that speed, low power dissipation, and the potential for high packaging density have caused efforts toward the development of digital integrated circuits to predominate from the start.

REFERENCES

1. F. Orito, Y. Seta, Y. Yamada, T. Ibuka, T. Okano, F. Hyuga and J. Osaka, "Large Size Dislocation-Free Gallium Arsenide Single Crystals for LSI Applications," *Technical Digest of the GaAs IC Symposium,* 1986.
2. R.J. Hamilton and N.K. Osbrink, "A GaAs FET Primer: Understanding these vital devices," *Microwave System News,* October 1982.
3. T. Irie, I. Nagasako, H. Kohzu, and K. Sekido, "Reliability Study of GaAs MESFET's," *IEEE Trans. Microwave Theory Tech.,* MTT-24, 1976.
4. H. Fukui, "Thermal Resistance of GaAs Field-Effect Transistors," *IEDM,* 1980.
5. S.H. Wemple and H. Huang, "Thermal Design of Power GaAs FETs," in *GaAs FET Principles and Technology,* J.V. DiLorenzo and D.D. Khandelwal, eds., Artech House, Dedham, MA, 1982.
6. G. Clerico Titinet and G. Manzone, unpublished manuscript.
7. J.C. Irvin, "The Reliability of GaAs FETS," eds., *GaAs FET Principles and Technology,*" in J.V. DiLorenzo and D.D. Khandelwal, Artech House, Dedham, MA, 1982.
8. S.P. Bellier, R.F. Haythornthwaite, J.L. May, and P.J. Woods, "Reliability of Microwave Gallium Arsenide Field Effect Transistors," *Proc. of the 13th Int. Reliability Physics Symp.,* 1975.
9. D.A. Abbott and J.A. Turner, "Some Aspects of GaAs MESFET Reliability," *IEEE Trans. Microwave Theory Tech.,* MTT-24, 1976.
10. S.H. Wemple et al., "Long-Term and Instantaneous Burnout in GaAs Power FET's: Mechanisms and Solutions," *IEEE Trans. Electron Devices,* ED-28, 1981.
11. F.A. Buot and K.J. Sleger, "Numerical Simulation of Hot-Electron Effects on Source-Drain Burnout Characteristics of GaAs Power FETs," *Solid-State Electron.,* 27, 1984.
12. W.R. Frensley, "Power-Limiting Breakdown Effects in GaAs MESFET's," *IEEE Trans. Electron Devices,* ED-28, 1981.
13. R. Yamamoto, A. Higashisaka, and F. Hasegawa, "Light Emission and Burnout Characteristics of GaAs Power MESFET's," *IEEE Trans. Electron Devices,* ED-25, 1978.
14. T. Furutsuka, T. Tsuji, and F. Hasegawa, "Improvement of the Drain Breakdown Voltage of GaAs Power MESFET's by a Simple Recess Structure," *IEEE Trans. Electron Devices,* ED-25, 1978.
15. S. Tiwari, L.F. Eastman, and L. Rathbun, "Physical and Materials Limitations on Burnout Voltage of GaAs Power MESFET's," *IEEE Trans. Electron Devices,* ED-27, 1980.
16. G.P. Schwartz, B. Schwartz, D. DiStefano, G.J. Gualtieri, and J.E. Griffiths, "Raman Scattering from Anodic Oxide-GaAs Interfaces," *Applied Physics Lett.,* 34, 1979.
17. J.M. Dumas, J. Paugam, C. Le Mouellic, and J. Y. Boulaire, "Long Term Degradation of GaAs Power MESFETs Induced by Surface Effects," *Proc. of the 21st Int. Reliability Physics Symp.,* 1983.
18. S. Igi et al., "The Effects of the Passivation Film on the Reliability of High Power GaAs MESFETS," *Proc. of ISTFA,* 1983.
19. M. Otsubo, Y. Mitsui, M. Nakatani and M. Wataze, "Degradation of GaAs Power MESFET's Due to Light Emission," *Technical Digest of the Intern,* Electron Device Meeting, 1980.
20. T. Kajimura, T. Kuroda, S. Yamashita, H. Todokoro, M. Nakamura, K. Mizuishi and J. Umeda, "Degradation of Ga(1-x) Al(x)As Visible Diode Lasers," *Applied Physics Letters,* 35, No. 12, 1979.
21. K. H. Kretschmer and H.L. Hartnagel, "XPS-Analysis of GaAs-Surface Quality Affecting Interelectrode Material Migration," *Proc. of the 23rd Int. Reliability Physics Symp.,* 1985.
22. H. Fukui et al., "Reliability of Power GaAs Field-Effect Transistors," *IEEE Trans. Electron Devices,* ED-29, 1982.

23. H. Fukui et al., "Reliability of Improved Power GaAs Field-Effect Transistors," *Proc. of the 18th Int. Reliability Physics Symp.*, 1980.
24. A.S. Jordan, J. C. Irvin, and W. O. Schlosser, "A Large Scale Reliability Study of Burnout Failure in GaAs Power FETs," *Proc. of the 18th Int. Reliability Physics Symp.*, 1980.
25. J. M. Dumas, D. Lecrosnier, J. Paugam and C. Vuchener, "Comparative Reliability Study of GaAs Power MESFETs: Mechanisms for Surface-Induced Degradation and a Reliable Solution," *Electron Lett.*, 21, 1985.
26. J. M. Dumas, D. Lecrosnier, and J. F. Bresse, "Investigation into GaAs Power MESFET Surface Degradation," *Proc. of the 23rd Int. Reliability Physics Symp.*, 1985.
27. A. Christou, E. Cohen and A.C. Macpherson, "Failure Modes in GaAs Power FETs: Ohmic Contact Electromigration and Formation of Refractory Oxides," *Proc. of the 19th Int. Reliability Physics Symp.*, 1981.
28. P.M. White, C.G. Rogers, and B.S. Hewitt: "Reliability of Ku-Band GaAs Power FETs Under Highly Stressed RF Operation," *Proc. of the 21st Int. Reliability Physics Symp.*, 1983.
29. J.C. Blair and C.R. Fuller, "Electromigration Induced Failures in Gold Films," *J. Applied Physics*, 43, 1972.
30. W.J. Slusark, G.L. Schnable, V. R. Monshaw, and M. Fukuta, "Reliability of Aluminum Gate Metallization in GaAs Power FETs," *Proc. of the 21st Int. Reliability Physics Symp.*, 1983.
31. N.M. Johnson, T.J. Magee, and J. Peng, *J. Vac. Sci. Tech.*, 13, 1976.
32. K. Katsukawa, Y. Kose, M. Kanamori, and S. Sando, "Reliability of Gate Metallization in Power GaAs MESFETs," *Proc. of the 22nd Int. Reliability Physics Symp.*, 1984.
33. E. Nagasawa and H. Okabayashi, "Electromigration and Ohmic Contact Properties of the Magnetron-Sputtered Al-2%Si Alloy Fillms," *NEC Research & Development*, No. 59, 1980.
34. I. Drukier and J.F. Silcox, "On the Reliability of Power GaAs FETs," *Proc. of the 17th Int. Reliability Physics Symp.*, 1979.
35. E.D. Cohen and A.C. Macpherson, "Reliability of Gold Metallized Commercially Available Power GaAs FETs," *Proc. of the 17th Int. Reliability Physics Symp.*, 1979.
36. P.A. Barnes and A.Y. Cho, "Nonalloyed Ohmic Contacts to n-GaAs by Molecular Beam Epitaxy," *Applied Physics Lett.*, 33, 1978.
37. J.V. DiLorenzo, W.C. Niehaus, and A.Y. Cho, "Nonalloyed and In Situ Ohmic Contacts to Highly Doped n-Type GaAs Layers Grown by Molecular Beam Epitaxy (MBE) for Field Effect Transistors," *J. Applied Physics*, 50, 1979.
38. D.V. Morgan, "Interdiffusion of Metal Films on Gallium Arsenide and Indium phosphide," in *Reliability and Degradation*, M.J. Howes and D.V. Morgan eds., John Wiley and Sons Inc., 1981.
39. K. Ohata and M. Ogawa, "Degradation of Gold-Germanium Ohmic Contact to n-GaAs," *Proc. of the 12th Int. Reliability Physics Symp.*, 1974.
40. A. Christou, "Solid Phase Formation in Au: Ge/Ni, Ag/In/Ge, In/Au:Ge GaAs Ohmic Contact Systems," *Solid-State Electron.*, 22, 1979, p. 141.
41a. J.E. Davey and A. Christou, "Reliability and Degradation of Active III-V Semiconductor Devices," in *Reliability and Degradation*, M.J. Howes and D.V. Morgan eds., John Wiley & Sons Inc., 1981.
41. R.E. Lundgren and G.O. Ladd, "Reliability Study of Microwave GaAs Field Effect Transistors," *Proc. of the 16th Int. Reliability Physics Symp.*, 1978.
42. G.S. Marlow, M.B. Das, and L. Tongson, "The Characteristics of Au-Ge Based Ohmic Contacts to n-GaAs Including the Effects of Aging," *Solid-State Electron.* 26, 1983.
43. K. Mizuishi, H. Kurono, H. Sato, and H. Kodera, "Degradation Mechanism of GaAs MESFET's," *IEEE Trans. Electron Devices*, ED-26, 1979.
44. K. Katsukawa, T. Takeuchi, K. Tokunaga, and I. Nagasako, "Failure Analysis and Reliability for X-Band Power GaAsFET," *NEC Res. and Develop.*, No. 71, 1983.

45. J.R. Waldrop, "Electrical Properties of Ideal Metal Contacts to GaAs: Schottky-Barrier Height," *J. Vac. Sci. Tech.*, B2, 1984.
46. A. Christou and H.M. Day, "Low-Temperature Interdiffusion between Aluminum Thin Films and GaAs," *J. Applied Physics,* 47, 1976.
47. K. Sleger and A. Christou, "Studies of Aluminum Schottky Barrier Gate Annealing On GaAs FET Structures," *Solid-State Electron.* 21, 1978.
48. J.D. Speight and K. Cooper, "Interlayer Diffusion Phenomena in Ti-Au Metallizations on n-Type GaAs at 250–450°C," *Thin Solid Films,* 25, 1975.
49. H.M. Day, A. Christou, and A.C. Macpherson, "Interdiffusion and Schottky Barrier Height variations in Au-W(Ti)/n-GaAs Contacts," *J. Vac. Sci. Tech.,* 14, 1977.
50. J.E. Baker, R. J. Blattner, S. Nadel, C.A. Evans, and R. S. Nowicki, "Thermal Annealing Study of Au/Ti-W Metallization on Silicon," *Thin Solid Films,* 69, 1980, p. 53.
51. C. Canali et al., "Power GaAs MESFETs: Reliability Aspects and Failure Mechanisms," *Microelectron. Reliability,* 24, 1984.
52. C. Canali, F. Castaldo, F. Fantini, D. Ogliari, L. Umena, and E. Zanoni, "Gate Metallization 'Sinking' into the Active Channel in Ti/W/Au Metallized Power MESFET's," *IEEE Electron Device Lett.,* EDL-7, 1986.
53. D. Ogbonnah and A. Fraser, "Reliability Investigation of 1 Micron Depletion Mode IC MESFETs," *Proc. of the 24th Int. Reliability Physics Symp., 1986.*
54. E. Philofsky, "Intermetallic Formation in Gold-Aluminum Systems," *Solid State Electronics,* 13, pp. 1391–1399, 1970.
55. A.C. Macpherson, K.R. Gleason, and A. Christou, "Voids in Aluminum Gate Power GaAs FETs under Microwave Testing," *Proc. of ATFA,* 1978.
56. P.M. White, B.L. Hewitt, and J.A. Turner, "Reliability Investigations of GaAs Power FETs with Aluminium Gate Metalization," *European Microwave Conf.,* 1978.
57. S.P. Murarka, H.J. Levinstein, I. Blech, T.T. Sheng, and M.H. Read, "Investigation of the Ti-Pt Diffusion Barrier for Gold Beam Leads on Aluminum," *J. Electrochem. Soc.,* 125, 1978.
58. J.C. Irvin and A. Loya, "Failure Mechanisms and Reliability of Low-Noise GaAs FETs," *Bell System Techn. J.,* 57, 1978.
59. C.L. Huang, F. Kwan, S.Y. Wang, P. Galle, and J.S. Barrera, "Reliability Aspects of 0.5 μm and 1.0 μm Gate Low Noise GaAs FETs," *Proc. of the 17th Int. Reliability Physics Symp.,* 1979.
60. E.D. Cohen, A.C. Macpherson, and A. Christou, "Reliability of Power GaAs FET's—Au Gates and Al-Au Linked Gates," *IEEE Trans. Microwave Theory Techn.,* MTT-29, 1981.
61. B. Reich and E.B. Hakim, "Environmental Factors Governing Field Reliability of Plastic Transistors and Integrated Circuits," *Proc. of the 10th Int. Reliability Physics Symp.,* 1972.
62. A. Shumka and R.R. Piety, "Migrated Gold Resistive Shorts in Microcircuits," *Proc. of the 13th Int. Reliability Physics Symp.,* 1975.
63. W.T. Anderson, A. Christou,.and K.J. Sleger, "Ionic Contamination-Humidity Effects on GaAs FETs," *Proc. of the 17th Int. Reliability Physics Symp.,* 1979.
64. W.T. Anderson and A. Christou, "GaAs FET Failure Mechanisms Due to High Humidity and Ionic Contamination," *IEEE Trans. Reliability,* R-29, 1980.
65. T. Itoh and H. Yanai, "Stability of Performance and Interfacial Problems in GaAs MESFET's" *IEEE Trans. Electron Devices,* ED-27, 1980.
66. H. Itoh, K. Ohata, and F. Hasegawa, "Influence of the Surface and the Episubstrate Interface on the Drain Current Drift of GaAs MESFET's," *IEEE Trans. Electron Devices,* ED-28, 1981.
67. M. Omori and J.N. Wholey, "Accelerated Active Life Test of GaAs FET and a New Failure Mode," *Proc. of the 18th Int. Reliability Physics Symp.,* 1980.
68. R. Zuleeg, "Radiation Effects in GaAs Integrated Circuits," *VLSI Microelectronics Microstructure Science,* 11, Academic Press, New York, 1985.
69. R.J. Chaffin, *Microwave Semiconductor Devices: Fundamental and Radiation Effects,* John Wiley and Sons, New York, 1973.

70. M. Takeuchi, H. Sasaki, N. Deguchi, Y. Nanko, and M. Shimodaira, "Electron Irradiation Tests on Small Signal and Power GaAs FETs," *Proc. of ISTFA,* 1982.
71. R.J. Gutman and J.M. Borrego, "Degradation of GaAs MESFETs in Radiation Environments," *IEEE Trans. Reliability,* R-29, 1980.
72. W.T. Anderson, A.B. Campbell, A.R. Knudson, A. Christou, and B.R. Wilkins, "Degradation in GaAs FETs Resulting from Alpha Particle Irradiation," *IEEE Trans. Nuclear Science,* NS-31, 1984.
73. D.N. McQuiddy, J.W. Wassel, J.B. Lagrange, and W.R. Wisseman, "Monolithic Microwave Intagrated Circuits: an Historical Perspective," *IEEE Trans. Microwave Theory Tech.,* MTT-32, September 1984.
74. C. Mahle and H.C. Huang, "MMICs in Communications," *IEEE Communications Magazine,* 23, September 1985.
75. J.S. Kilby, "Invention of the Integrated Circuit," *IEEE Trans. Electron Devices,* ED-23, July 1976.
76. R.S. Pengelly and J.A. Turner, "Monolithic Broadband GaAs FET Amplifier," *Electron. Lett.,* 12, May 1976.
77. M. Wittmer, T. Finstad, and M-A. Nicolet, "Investigation of the Au-Ge-Ni and Au-Ge-Pt system used for alloyed contacts to GaAs" *J. Vac. Sci. Technol.,* 14, No. 4, 1977.
78. G.R. Cronin and R.W. Haisty, "The Preparation of Semi-insulating GaAs by Cr Doping," *J. Electrochem. Soc.,* 111, July 1964.
79. C.A. Mead, "Schottky Barrier Gate Field Effect Transistor," *Proc. IEEE,* 54, February 1966.
80. W.W. Hooper and W.H. Lehrer, "An Epitaxial GaAs Field Effect Transistor," *Proc. IEEE,* 55, July 1967.
81. K.E. Drangeid, R. Sommerholder, and W. Walker, "High speed GaAs Schottky Barrier Field Effect Transistor," *Electron. Lett.,* 6, April 1970.
82. M. Fukuta, T. Mimura, I. Tujimura, and A. Furumoto, "Mesh Source Type Microwave Power FET," *1973 ISSCC Dig. Tech. Papers.*
83. L.S. Napoli, R.E. DeBrecht, J.J. Hughes, W.F. Reichert, A. Dreeben, and A. Triano, "High Power GaAs FET Amplifier—A Multigate Structure," *1973 ISSCC Dig. Tech. Papers.*
84. D.R. Decker, "GaAs Monolithic Microwave Integrated Circuits," in *VLSI Electronics Microstructure Science,* 11, Academic Press, New York, 1985.
85. A. Christou, "GaAs Reliability," Tutorial notes of the 24th Int. Reliability Physics Symposium, 1986.
86. P. Ho, T. Andrade, and E. Johnson, "GaAs MMIC Reliability Analysis and Its Impact on Microwave Systems," *MSN and CT,* August–September 1987.
87. K. Katsukawa, T. Kimura, K. Ueda, and T. Noguchi, "Reliability Investigation on S-Band GaAs MMIC," *IEEE Microwave and Millimeter Wave Monolithic Circuit Symp.,* June 1987.
88. Avantek Application Note AN-A003: "Electrostatic Discharge Damage in Transistors."
89. Avantek Application Note AN-A004: "Electrostatic Discharge Control."
90. M. Simons, "Radiation Hardness of GaAs ICs and Comparison with Si," *Military Microwaves Proc.,* 1984.
91. A. Swanson, J. Herb, and M. Yung, "First Commercial HEMT Challenges GaAs FETs," *Microwaves and RF,* November 1985.
92. J.H. English, A.C. Gossard, H.L. Stormer, and K.W. Baldwin, "GaAs Structures with Electron Mobility of 5×10^6 cm^2/Vs," *Applied Physics Lett.,* 50, 1987.
93. J.V. DiLorenzo et al., "Material and Device Considerations for Selectively Doped Heterojunction Transistors," *Techn. Dig. Int. Electron Device Meeting,* 1982.
94. R. Dingle, M.D. Feuer, and C.W. Tu, "The Selectively Doped Heterostructure Transistor: Materials, Devices and Circuits," *VLSI Electronics Microstructure Science,* 11, Academic Press, New York, 1985.

95. C.W. Tu, R.H. Hendel, and R. Dingle, "Molecular Beam Epitaxy and the Technology of Selectively Doped Heterostructure Transistors," *Gallium Arsenide Technology,* Howard W. Sams and Co., Indianapolis, 1985.
96. E.F. Schubert, A. Fischer, and K. Plogg, "Photoconductivity in Selectively n- and p-Doped Al (x) Ga ($1 - x$) As/GaAs Heterostructures," *Solid-State Electron.,* 29, 1986.
97. A.J. Valois, G.Y. Robinson, K. Lee, and M.S. Shur, "Temperature Dependence of the I-V Characteristics of Modulation-Doped FETS," *J. Vac. Sci. Tech.,* B 1, April–June 1983.
98. R. Fischer et al., "On the Collapse of Drain I-V Characteristics in Modulation-Doped FET's at Cryogenic Temperatures," *IEEE Trans. Electron Devices,* ED-31, August 1984.
98a. K.H. Duh, X.C. Zhu, A. Van Der Ziel and H. Morkoc, "Instabilities in MODFET's and MODFET Circuits," *IEEE Transactions on Electron Devices,* ED-31, No. 9, September 1984.
99. A. Christou et al., "Failure Mechanism Study of GaAs MODFET Devices and Integrated Circuits," *Proc. of the 23rd Int. Reliability Physics Symp.,* 1985.
100. K. Togashi, H. Takakuwa, and Y. Kato, "Reliability of Low Noise Microwave HEMTs Made by MOCVD," *Microwave J.,* April 1987.
101. F.A. Buot, W.T. Anderson, A. Christou, K.J. Sleger, and E.W. Chase, "Theoretical and Experimental Study of Subsurface Burnout and ESD in GaAs FETs and HEMTs," *Proc. of the 25th Int. Reliability Physics Symp.,* 1987.
102. T. Irie, H. Kohzu, and I. Nagasako, "Screening and Testing Techniques for GaAs Power MESFET's, *Proc. of ATFA,* 1979.
103. A.A. Immorlica and J.R. Michener, "High Voltage Screening of GaAs Power FET's: Effects on Burn-in Yield and Modes of Catastrophic Device Failure," *Proc. of the 23rd Int. Reliability Physics Symp.,* 1985.
104. K.J. Russell and J.K. Dhiman, "Power GaAs FET RF Life Test Using Temperature-Compensated Electrical Stressing," *Proc. of the 24th Int. Reliability Physics Symp.,* 1986.
105. J.M. Dumas, "Reliability Trends of GaAs MESFETs," *NATO Workshop,* Helsingor (DK), 1985.

Chapter 9
RELIABILITY OF OPTOELECTRONIC COMPONENTS

P. Montangero and A. Piccirillo

Emitter and photodetector reliability is becoming an issue of the utmost importance because traditional communication systems are being largely replaced by optical fiber networks. Their technology is not yet fully established, and the problems generated by these new devices are not comparable to the ones encountered with silicon semiconductors. We will indicate the basic physical operating principles, different technological solutions, and the main degradation mechanisms for each class of components, underscoring specific problems for life tests. Expected mean life ranges are also illustrated.

9.1 EMITTERS

The growth of optical fiber communication systems has led to the development of devices able to generate (LED, laser) and to detect (PIN, APD) an optical signal as well as physical carriers able to transmit such signals. Detector related issues are discussed in Section 9.2. The development of light sources clearly follows that of optical fibers, as these devices are basically required to have (1) a high optical power to be coupled in fiber and (2) the emitted wavelength in one region of minimum attenuation (Figure 9.1).

The earliest emitters were electroluminescent diodes with noncoherent emission, light-emitting diodes (LEDs). Their available optical power is on the order of magnitude of hundreds of microwatts and the emitted wavelength is in the 0.8–0.9 μm range. In the two decades following development of the LED, light emission has been dominated by coherent-emission lasers offering optical powers on the order of tens of milliwatts, which is why we shall be dealing more at length with lasers.

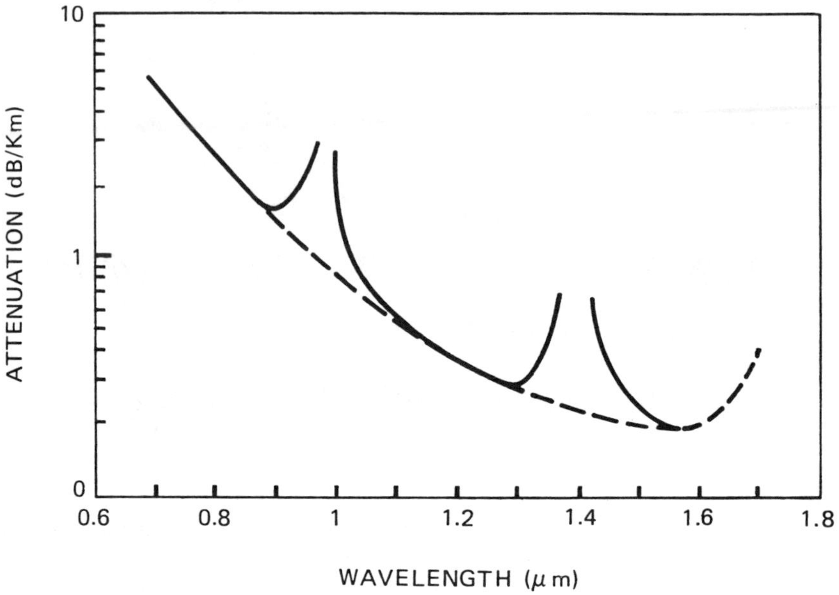

Figure 9.1 Silica fiber attenuation by wavelength.

Laser technology signifies the growth in number and sophistication of different structures. Some of them are manufactured by the same company, which makes an overall reliability study quite difficult.

In fact, degradation mechanisms may be specific either to the structure or to the manufacturing process of a given company. We therefore decided to focus on degradation mechanisms shared by all photoemitters irrespective of the details of their technological process. However, we did maintain one distinction: the 0.8–0.9 μm photoemitters that use GaAs as a basic material, and the 1.3–1.5 μm devices that use InP.

Finally, we divided reliability into its two main aspects: the mechanisms underlying degradation due to aging and quantifying reliability—whether and how we can currently estimate mean device life under given operating conditions.

9.1.1 Basic Physical Operating Principles

A light source, be it LED or laser, translates an electric signal into an optical one. The underlying physical principle is the hole—electron recombination, or an electron transition from one energy level to another, from the conduction to the valence band where the energy involved is released as a photon. This transition of levels may or may not be accompanied by a variation in the momentum. It is described as direct or indirect according to the presence or absence of such variation.

In Figure 9.2 we see the band structure as a function of the electron momentum $\overline{K}$ in both transitions: mean life is shorter in the case of direct transition, which is why it is favored.

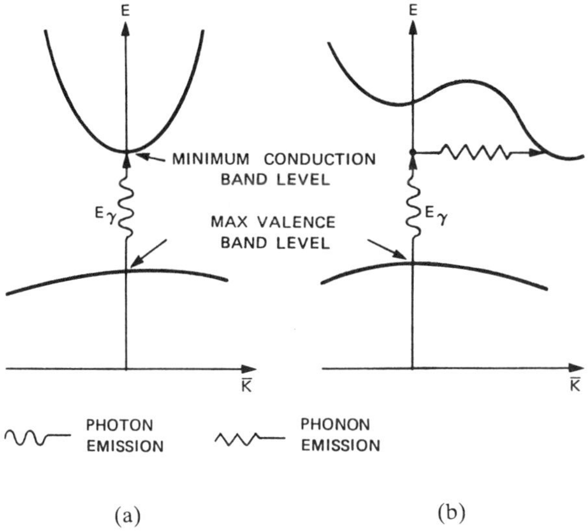

Figure 9.2 Types of transition: (a) direct, i.e., only photon emission; and (b) indirect, i.e., both photon and phonon emission.

Gallium arsenide, indium phosphide, and related compounds are used as materials for emitters, all of them being "direct gap." Other traditional materials such as germanium and silicon, which are used to manufacture semiconductor devices, cannot be used in this instance because of the "indirect gap."

Figures 9.3–9.4 and 9.5–9.6 illustrate an extremely simple but clear physical structure for both LEDs and lasers. Figure 9.7 sketches their emitting processes: spontaneous and incoherent in the case of LEDs, stimulated or coherent for lasers.

Radiative and nonradiative processes compete, and both must be considered when assessing device reliability, as photon emission variation often depends on the one prevailing over the other. A degradation mechanism in emitters may increase the percentage of nonradiative at the expense of radiative processes. Figures 9.8–9.9 give us a process outline for the most important radiative and nonradiative processes.

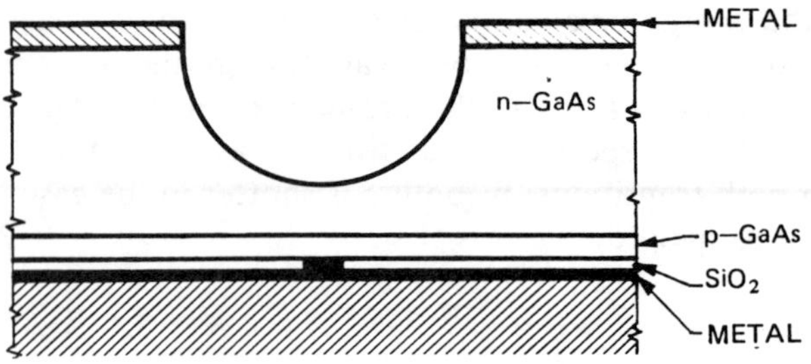

Figure 9.3 Cross section of a GaAs LED.

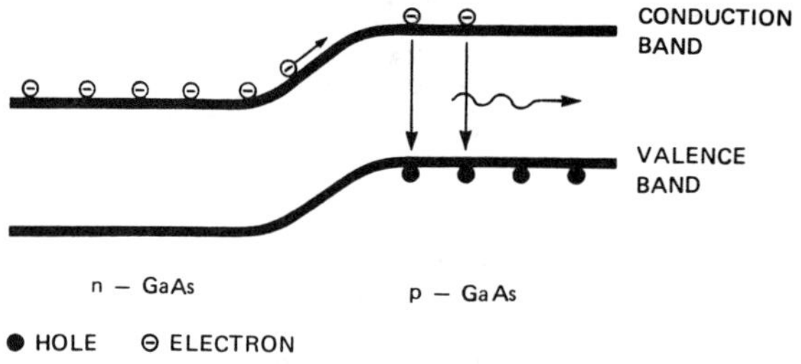

Figure 9.4 LED emitting process.

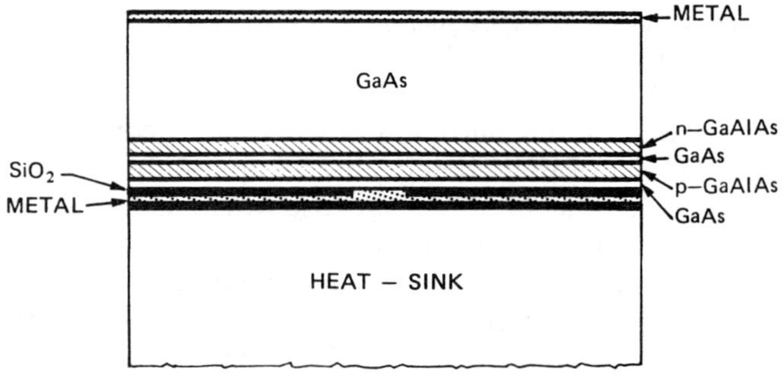

Figure 9.5 Cross section of an AlGaAs/GaAs laser.

9.1.2 Emitter Structure

As we said in the introduction, currently we believe lasers to be the best emitters for optical fiber transmission, especially in the case of long-distance transmission.

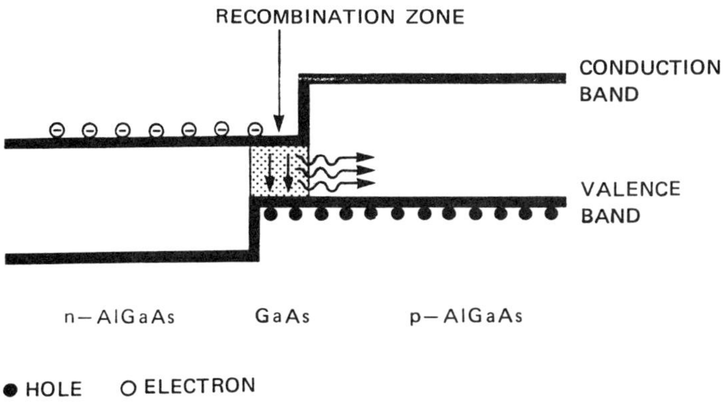

Figure 9.6 Laser emission process.

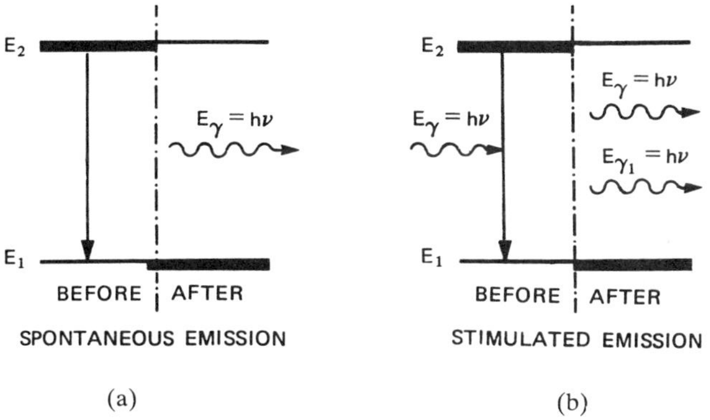

Figure 9.7 Emission processes: (a) spontaneous or incoherent; and (b) stimulated or coherent.

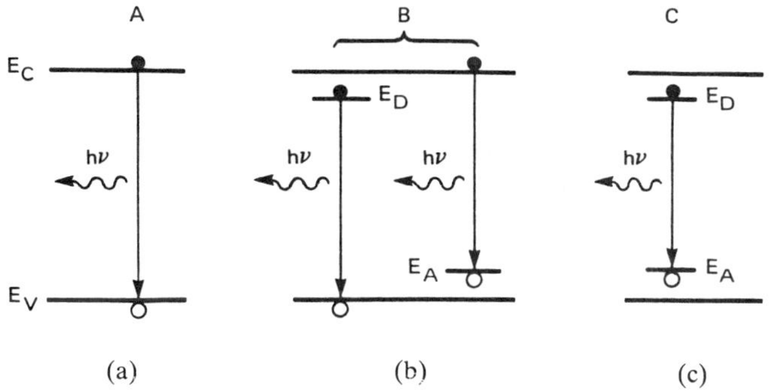

Figure 9.8 Radiative recombination processes: (a) two free charges; (b) one free and one bound charge; and (c) two bound charges.

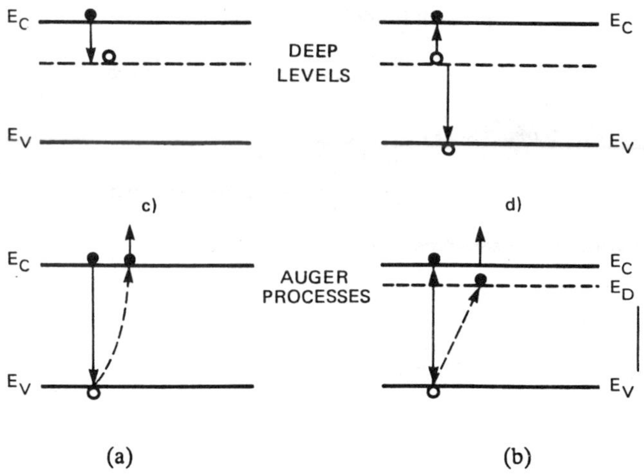

Figure 9.9 Processes of nonradiative recombination through deep levels: (a) electron capture; (b) electron reemission or recombination with a hole. Nonradiative recombination through Auger processes: (c) all carriers are free; and (d) one carrier is bound.

Current technology is still changing: no devices yet developed fully meet expectations, and optimization of existing performance is still being sought.

A laser is basically a Fabry-Perot cavity that resonates or reinforces some, and not other, wavelengths of the emitted radiation. Most of the reinforced light in the cavity is reflected back and forth between two parallel mirrors perpendicular to axis of the device (see Figure 9.10). Consider the light emitted on a plane parallel to the junction: there is a spectrum with more than one peak. These are associated to different wavelengths, and all surround a central maximum peak, the basic longitudinal mode (frequency) of the device (see Figure 9.11). We can also generate resonance transversally in the optical cavity, and in this case, for frequencies other than the basic one, too. On the whole, laser emissions will be both transverse and longitudinal multimode emissions, whereas the ideal emitter is single mode, longitudinal, and transversal, the single-frequency laser.

Transverse multimodality is eliminated by reducing the transversal size of the laser optical cavity and limiting the spatial distribution of carriers optically or electrically. Longitudinal multimodality is more complex, though *distributed feedback* (DFB) structures appear promising; a grating with a period sets the only frequency, which will be then reinforced in the laser optical cavity (Figure 9.12). The two facets at the end of the optical cavity no longer act as mirrors to reinforce the radiation.

The point we wish to make is that the device structure must be fully understood to understand why it does not function properly or when it is unreliable. A comprehensive knowledge will highlight the physical or technological causes.

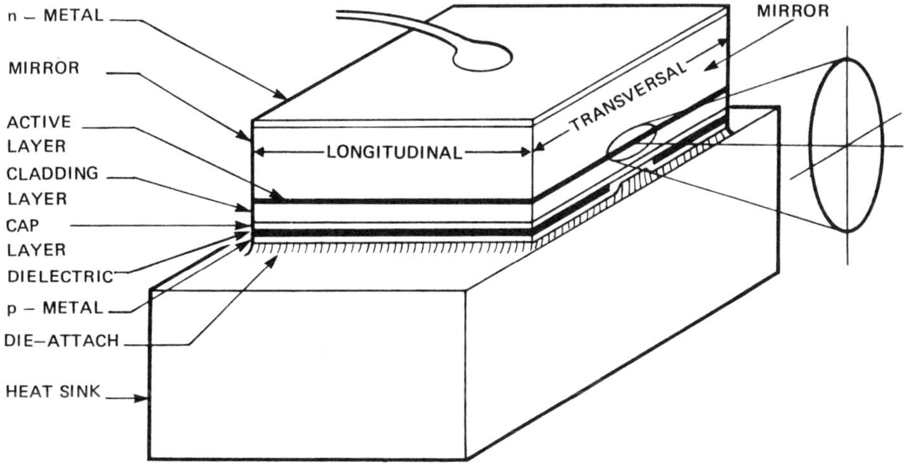

Figure 9.10 Basic structure of a laser diode.

Figures 9.5 and 9.13 show the structure of two GaAs lasers, the first multimodal and the second single mode, both emitting in the 0.8–0.9 μm range with a maximum optical output power of a few mW. The differences in the process are obvious: a heterostructure with a stripe confinement and a *transverse junction stripe* (TJS) laser. In the second case the active region is defined by the diffusion length of the carriers, which are forced to recombine within the active layer, as they encounter potential barriers along other paths.

Figure 9.14 shows a multimode InP laser lacking a contact layer, where confinement was obtained by deep diffusion in correspondence of the contact. Figures 9.15–9.16 show two InGaAsP/InP single-mode structures, called, respectively, *mushroom* and *ridge*.

Each structure has specific criticalities and will therefore have its own parameter drift and degradation processes, at times exclusive ones.

9.1.3 Emitter Electric Parameters

Degradation processes are studied because of their direct or indirect effects on the drift of electric parameters; this is when the user will notice device malfunctioning. Therefore, we discuss the two basic electric parameters for lasers and LEDs. However, especially where lasers are concerned, other optical characteristics, such as the emission spectrum and near and far field, should also be checked.

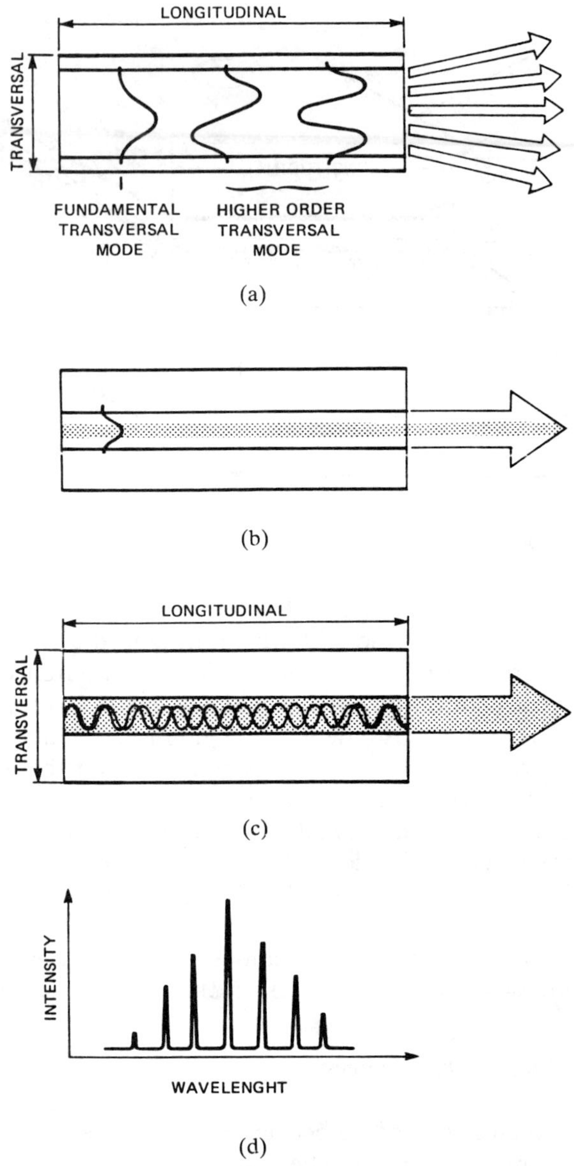

Figure 9.11 Laser multimodal emission: (a) transversal modes; (b) gaussian distribution of the beam; (c) longitudinal modes; and (d), the associated spectrum.

(a)

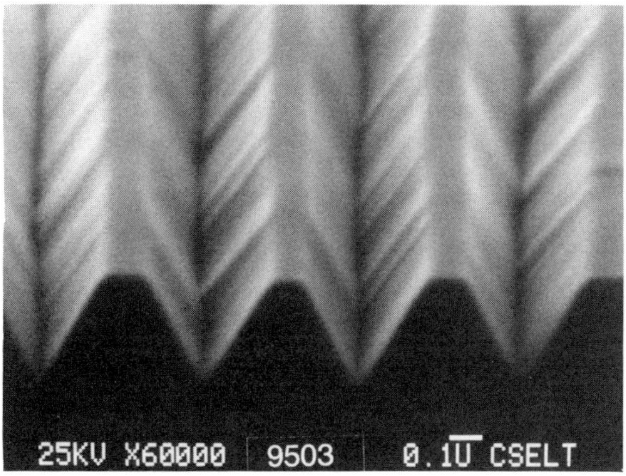

(b)

Figure 9.12 Ridge DFB laser: (a) chip; and (b) trapezoid grating.

9.1.3.1 Threshold Current in Lasers

A laser's threshold current is its most important parameter, and the following relation examines the effect of temperature and also other geometric-physical parameters of the device [1]:

$$I_{th} \sim g_{th} \frac{q \cdot d}{\tau} \exp[T_j/T_0] \tag{9.1}$$

where

q = electron charge;
d = laser active layer thickness;
τ = mean life of carriers;
T_0 = constant in K (temperature-material dependence);
g_{th} = gain, which can also be expressed as

$$g_{th} = \alpha + \frac{1}{L} \ln \frac{1}{R} \tag{9.2}$$

where

α = optical loss in cavity;
L = length of optical cavity;
R = mirror reflectivity;
T_j = junction temperature given by

$$T_j = R_\Theta \cdot P_d + T_{hs} \tag{9.3}$$

where

R_Θ = thermal resistance;
T_{hs} = heat sink temperature;
P_d = dissipated electrical power.

Technology experts aim at lasers with a low I_{th}, whereas reliability engineers are more interested in identifying what could cause these values to vary over time.

External efficiency is also important and can be expressed by

$$\eta_{ex} \sim \frac{1 - R}{L} \cdot \eta_i \exp\left[-\frac{T_j}{T_0}\right] \tag{9.4}$$

where

R = mirror reflectivity;

η_i = inner quantum efficiency.

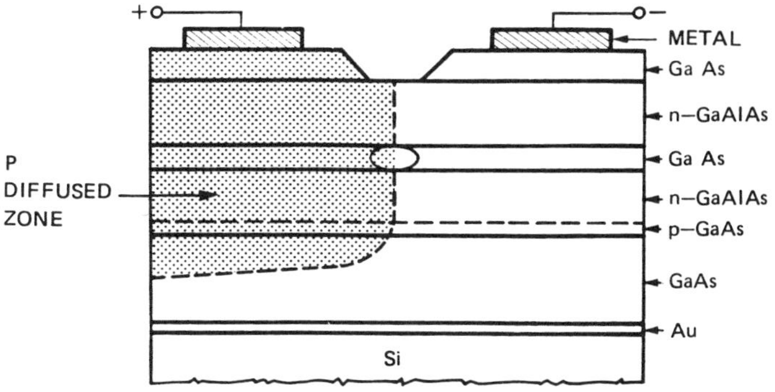

Figure 9.13 Section of a single-mode TJS laser chip.

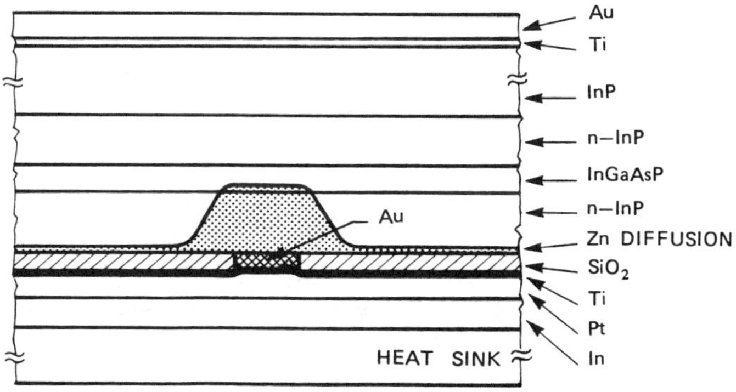

Figure 9.14 A multimodal laser chip with stripe confinement.

Figure 9.17 shows the I-V and L-I characteristics of a mushroom structure laser. Figure 9.18 shows L-I characteristics for the same laser at varying temperatures: when the injection current is the same, the higher the temperature the lower is the optical power.

LED emission is spontaneous, incoherent, and centered on a specific wavelength. The following parameter is taken as an indicator of its functioning properly; that is, its external radiative efficiency:

$$\eta_{ex} = L_{out}/P_{in} \tag{9.5}$$

where

L_{out} = output optical power;

P_{in} = injected electric power.

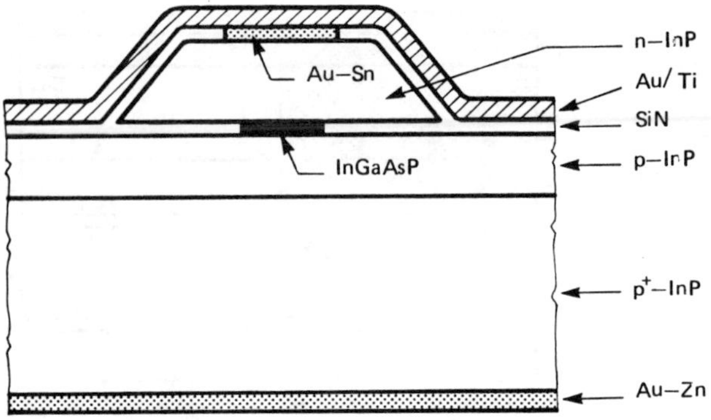

(a)

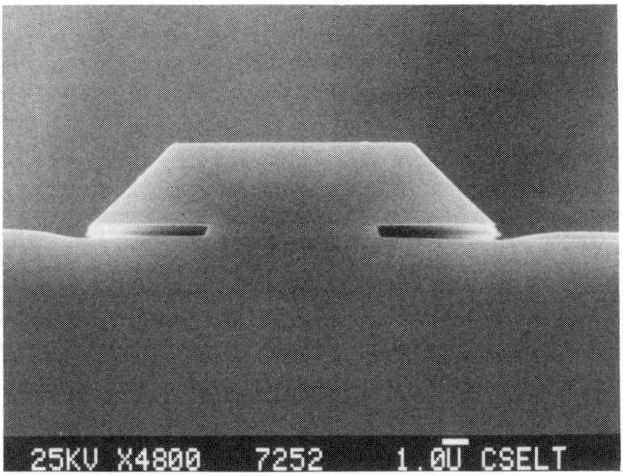

(b)

Figure 9.15 Single-mode mushroom laser: (a) structure; and (b) front view.

Figure 9.19 shows I-V and L-I LED characteristics.

Optoelectronic devices have only really been tested for reliability in the past two decades, a short time compared to electronic devices. This is why they have often not been treated comprehensively. As we described in the previous section, the reliability of a device basically concerns two aspects:

1. Identifying degradation mechanisms involving the whole device life and where possible, also understanding the dynamics-kinetics of the degradation process.
2. Evaluating mean life in relation to either a catastrophic failure or variation of the parameters that define the component beyond a given value set in

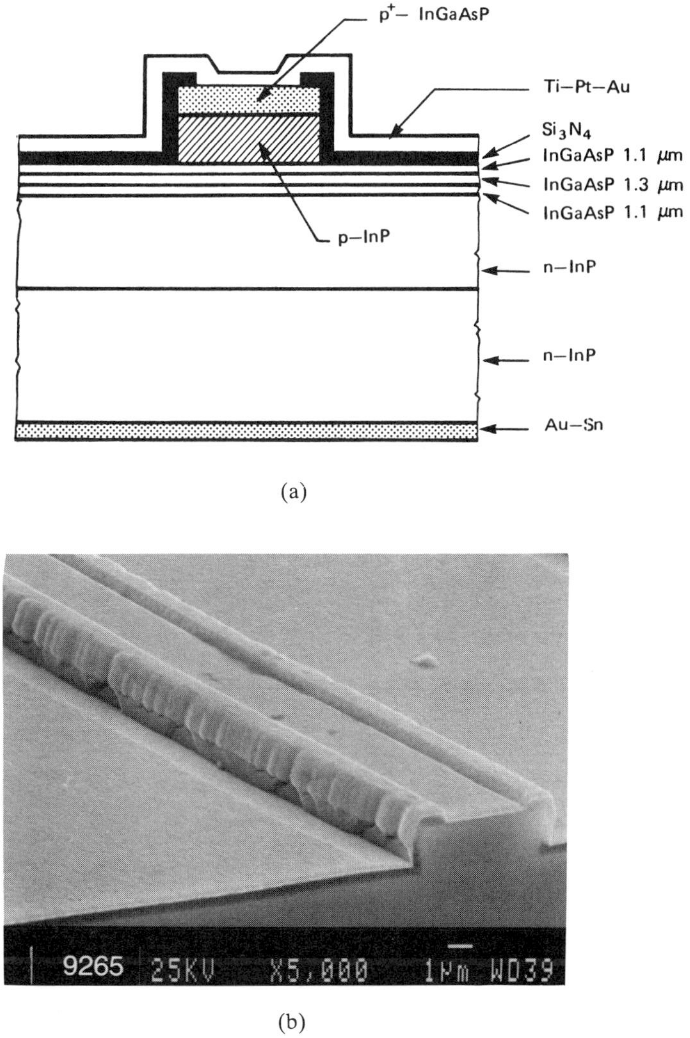

Figure 9.16 Single-mode ridge laser: (a) structure; and (b) perspective view.

relation to that specific system. As far as the latter is concerned, we should find the function relating mean life to degradation-accelerating factors, such as temperature or current, and then forecast mean life under actual conditions as accurately as possible.

When classifying degradation mechanisms, it is not easy to decide which are elated to the early phases of their life cycle and which to the wear-out period. In

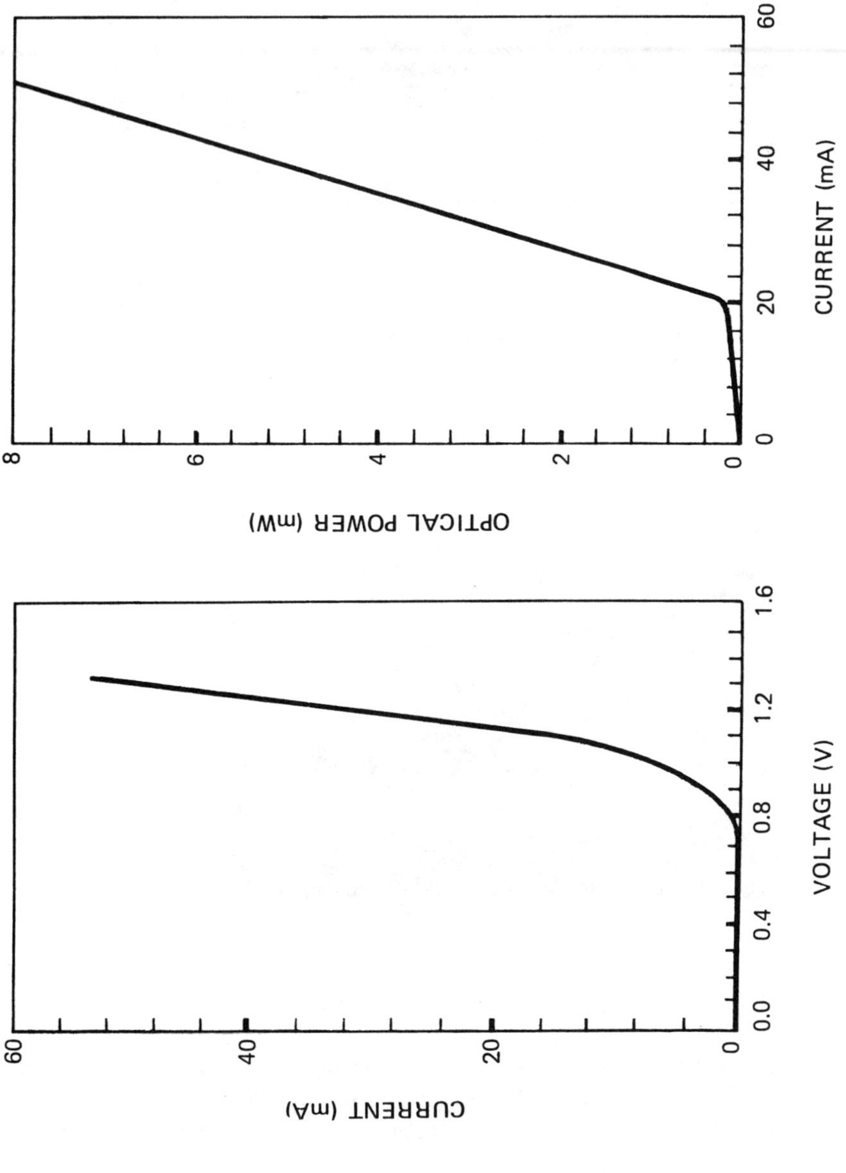

Figure 9.17 Example of laser I-V and L-I characteristics.

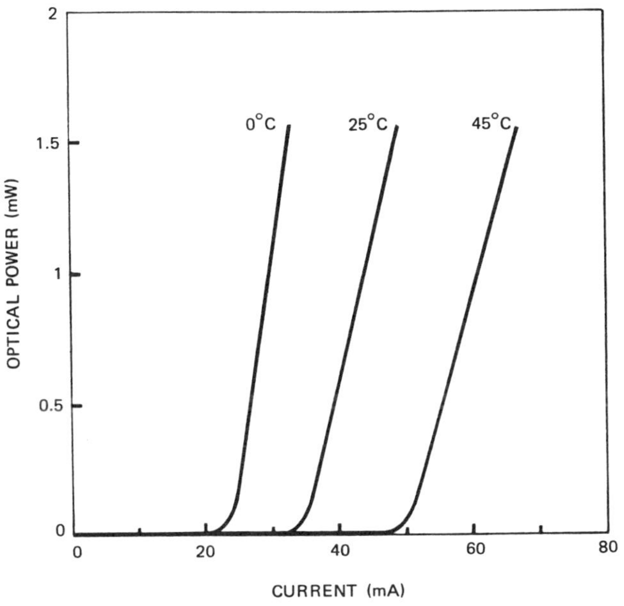

Figure 9.18 Variations of L-I characteristics with temperature.

fact in absolute terms the mean life of present devices is shorter than target values; that is, comparable to electronic devices. However, some of these degradation mechanisms may be irrelevant considering technological advances. These problems can thus be considered critical within the present technological framework though they may disappear in the near future.

In evaluating mean life, we consider that tests assessing this factor and its dependence on degradation-accelerating factors are usually performed in laboratories. Usually dependency, and the degradation processes it entails, is assessed by varying accelerating factors one by one. For LEDs we can design a set of tests where the various accelerating factors, usually temperature and current density, are kept distinct. With lasers, however, current density, temperature, and optical power all can accelerate degradation and cannot be separated; for instance, if we vary temperature, optical power and injection current cannot be kept constant. When carrying out tests maintaining optical power constant, the test will inevitably include acceleration as well, because of the various injection currents required for the various temperatures.

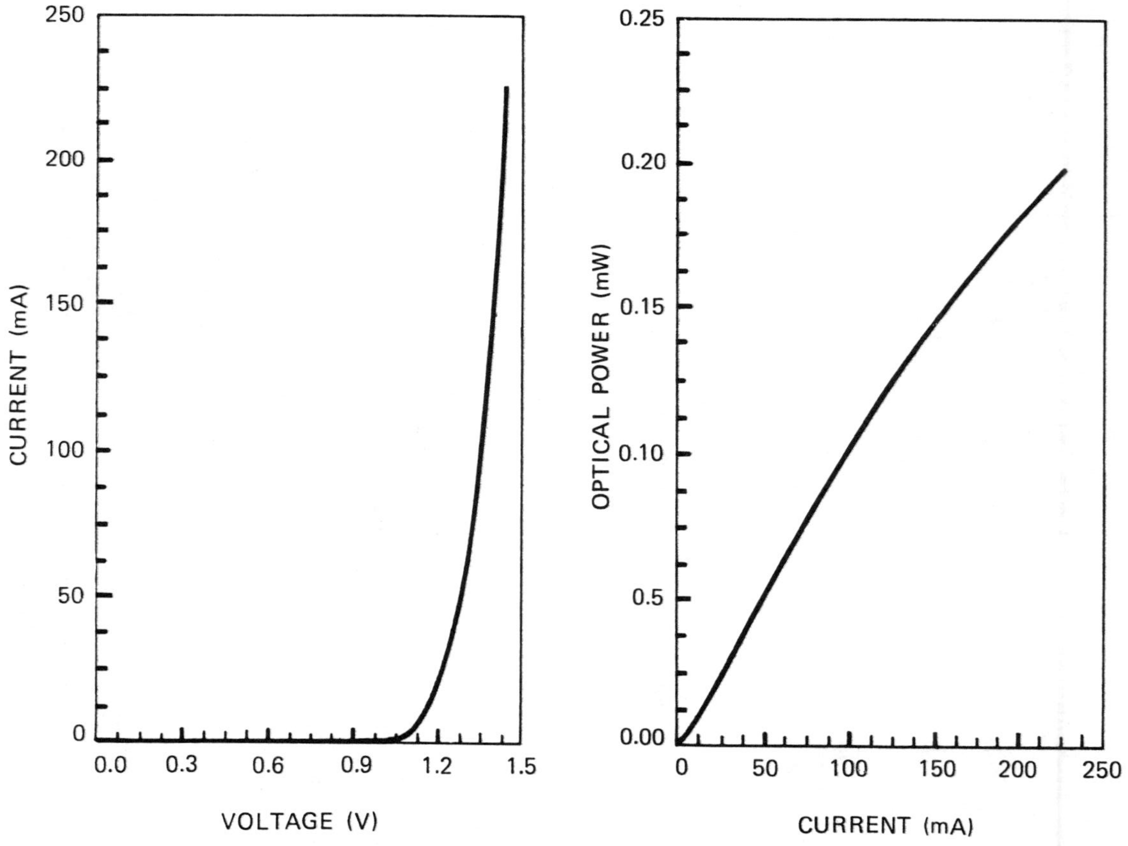

Figure 9.19 Example of LED I-V and L-I characteristics.

Constant optical power is advisable when the causes of degradation are unknown. However, we know some of the dominant degradation mechanisms and, therefore, also have a partial understanding of correlated accelerating factors. Compared to other factors, optical power appears less important when considering device degradation, and life tests no longer require strictly constant optical power. Table 9.1 illustrates correlations among the prevailing degradation mechanisms (see next subsection), the stresses that may be causing them, and device parameters.

Table 9.1 The Relationship between Degradation Mechanisms, Stresses, and Device Parameters

Degradation Mechanisms	*Parameters*	*Accelerating Factor*
Dark line defects	I_{th} (pulse)	Current density Temperature
Dark spot defects	I_{leak}	Optical power
Die-attach	R_{Θ}, $J_{th}(cw)$	Current density Temperature
Gold diffusion	R_s, R_{Θ} $J_{th}(cw)$	Temperature Current density

9.1.4 Degradation Mechanisms

As already pointed out in previous subsections, lasers and LEDs have different structures or are fabricated using different processes. Each of these may have peculiar critical points, though they all appear to share the following degradation mechanisms:

1. Formation and development of defective areas (laser and LED).
2. Mirror damage (laser).
3. Die-attach deterioration (laser).
4. Gold diffusion within semiconductors (LED and laser).

AlGaAs/GaAs and InGaAsP/InP emitter devices have different criticalities and therefore will be listed in two classes.

9.1.4.1 Formation and Development of Defective Areas

The presence of defects in the active area is undoubtedly a main cause of parameter drift or emitter malfunctioning. These defects may already be present in the substrate or form during the epitaxial growth of the structure or the manufacturing of the device. Finally, they may also develop following thermo-electric stimulation in the course of the operative life of the device [2].

Although not exhaustive, Table 9.2 summarizes the main defects in the two classes of compcnents, giving their origin and the impact on the electrooptical parameters of the device.

Table 9.2 Main Defects in AlGaAs/GaAs and InGaAsP/InP Components

Defect	*Origin*	*AlGaAs/GaAs*	*InGaAsP/InP*
Threading dislocations	Climbing Substrate dislocation	η reduced	No effect
Dislocation networks	Climbing Process stress Operative life stress	η reduced	No effect
Misfit dislocations	LPE-mismatch Epi/substrate Edge growth	No	η reduced J_{th} increase
Dislocation clusters	LEC-single crystal	No	η reduced J_{th} increase
In-inclusions	LPE-melt carryover	No	η reduced Shunt, unstable I-V, P-I

Note: LPE = liquid phase epitaxy; LEC = liquid encapsulation Czochralski.

Some of these defects have been virtually eliminated thanks to improved bulk and heterostructure growth techniques. However, because of the unavoidable lag, they do not yet enter into updated laser reliability expertise which is why malfunctioning may often be due to these causes.

Figures 9.20 and 9.21 [3], for instance, are *secondary electron microscopy* (SEM) cathodoluminescence images highlighting defects in two substrates purchased from different suppliers. Both surfaces appear to have a high defect density, and in Figure 9.20, we observe dislocation clusters, which will be discussed later. Figures 9.22a, 9.22b, and 9.22c show another cluster of dislocations as well as a hillock formation. By comparing the three images referred to the one heterostructural zone, we can realize how a defect in the substrate may remain in a quaternary layer and give rise to a hillock formation reaching up into the surface. Please see the references for a more detailed study of dislocations, which here are treated only briefly.

Because of LPE heterostructure growth, the substrate is carried over from one well to another in the crucible. When the separating septum is scratched, indium residues may be left on the surface and then segregated within layers that will have subsequently grown. This phenomenon, known as *melt carryover*, may

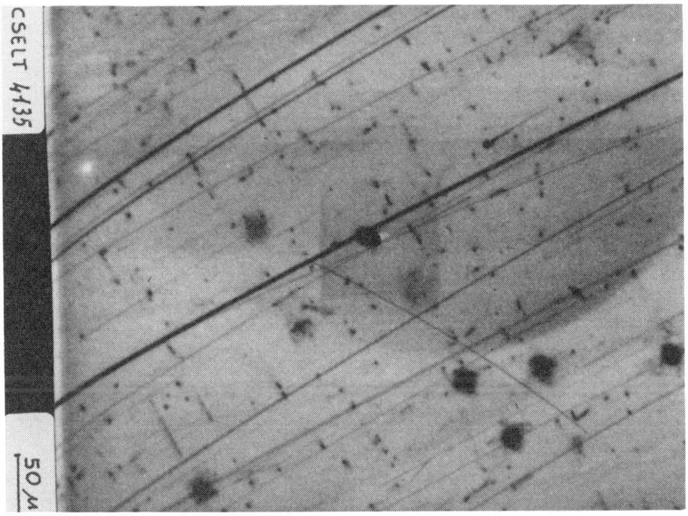

Figure 9.20 InP substrate with residual mechanical defects, clusters, and dislocations. ECL-Si 25 KeV.

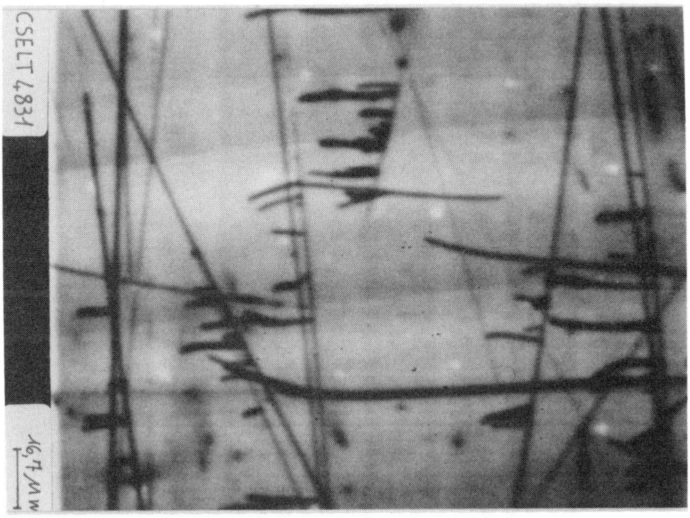

Figure 9.21 InP substrate with residual mechanical defects, native and induced dislocations, and growth striae. TCL-Si 25 KeV.

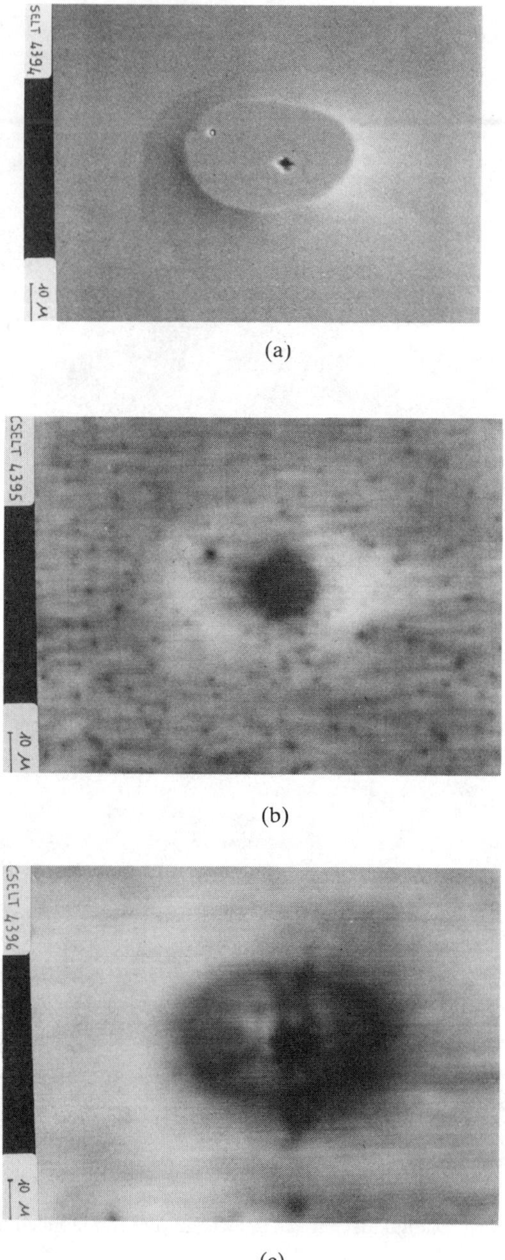

Figure 9.22 Hillock in a single LPE InGaAsP/InP heterostructure: (a) morphology; (b) defects in the quaternary layer TCL-Ge 25 KeV; and (c) defects in the substrate TCL-Si 35 KeV.

be enhanced by in situ cleaning of growth surface with an unsaturated indium melt [4, 5]. Indium inclusions may act as shunt paths for the current.

The bulk GaAs and InP material is universally grown using the LEC technique. In the case of InP because of the high vapor pressure of phosphorus, the melt, including the required doping, is encapsulated in the boron oxide, which has a high water content unless it undergoes a thermal treatment. Cockayne, Brown, and MacEwan [6] suggest the following cluster-forming model:

1. InP melt and water from the boron oxide react as follows:

$$2InP + H_2O(v) \rightarrow In_2O(v) + P_2(v) + H_2$$

2. Trapping the gaseous contents inside the grown crystal.
3. Releasing the stress generated by the high pressure of these gases by creating dislocation loops along the <110> directions. These "clusters" may be up to 50–100 μm in size. *Transmission electron microscopy* (TEM), SEM observations, and *x-ray topography* (XRT) prove that cluster centers were inclusion and precipitate free, whereas dislocation loops could be observed in the said directions [7].

When the heterostructure is grown on substrates with one or more clusters, dislocations may cross several layers and hillocks may be found up to the surface (see Figure 9.22). Section 9.1.5.4 will illustrate an example of gold diffusion of the metalization within the semiconductor, the diffusion itself being favored by a dislocations cluster.

The growth of heterostructures becomes possible when the lattice parameters of two bordering layers fit at growth temperature [8]. This is why we must have a given composition of quaternary and ternary layers. Small lattice mismatches are tolerated because the phenomenon is partly unavoidable and caused by different lattice thermal expansion parameters when passing from growth temperature ($\cong$ 600°C) to room temperature (InP: 4.56 10^{-6}°C^{-1}; InGaAsP: 5.42 10^{-6}°C^{-1}).

Thanks to the thickness of the active layer (0.15–0.2 μm), a quaternary-InP mismatch is usually not critical in standard lasers with LPE grown material. However, if "edge growth" [9] is present, local significant composition variations are possible, which mean higher interface mismatching with subsequent formation of edge dislocations along the <110> crystallographic directions. These dislocations appear as parallel lines often at equal distances from each other, in a formation known as *misfit dislocation*. Figures 9.23, 9.24, and 9.25 illustrate a misfit dislocation formation in a InGaAsP/InP laser. Figure 9.23 is a front view of the device, where we can easily detect that it is from a wafer edge. Figure 9.24 shows two images (SEI and EBIC) of the surface, where the just described pattern of parallel lines can be observed. Finally, Figure 9.25 shows the device surface before and after its cladding was partially removed: the pattern has disappeared, which suggests it was confined to the interface.

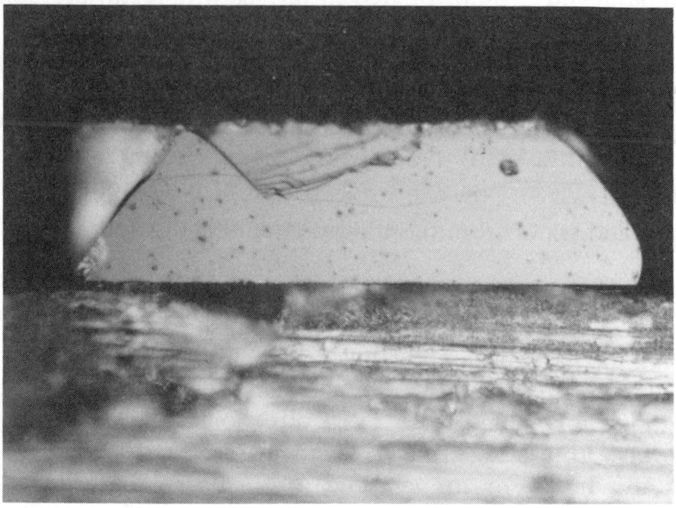

Figure 9.23 Front view of a laser from a wafer edge.

In terms of device performance, the presence of misfit dislocations means increasing threshold current in pulsed conditions and a greater sensitivity to temperature, as shown by low values of T_0 (43K). Abrahams [10] and Yamaguchi [11] proved the electric asymmetry of <110> and <1$\bar{1}$0> dislocations and a parallel increase of leakage current. The <1$\bar{1}$0> dislocations responsible for this behavior correspond to a half-plane, where the edge atoms are phosphorous giving rise to acceptor levels in the InGaAsP/InP interface.

GaAs devices are virtually dislocation misfit–free because of the smaller variability in lattice parameters of the AlGaAs ternary with composition and because the lattice thermal expansion parameter is only slightly different from GaAs (GaAs: 6.64 10^{-6}°C^{-1}; AlGaAs: 6.05 10^{-6}°C^{-1}).

Most III-V compounds crystallize in the blende structure consisting of two interpenetrating *Fcc* sublattices. The GaAs structure is obtained by placing Ga and As atoms alternatively in their sub-lattices. Each atom is surrounded by four neighboring ones at the apex of a tetrahedron.

Dislocation motion in a crystal may be conservative (a slip or glide movement) or nonconservative (with transport of material to or from dislocation). The latter is known as *dislocation climbing,* and thanks to it, passing through several layers of the device heterostructure becomes possible. It also means the area of dislocation becomes bigger and spreads. Furthermore, because of the dual nature of the lattice, climbing requires not one but two atoms to be emitted or absorbed.

If climbing decreases the extra half-plane, it is then termed *positive climbing* and in the opposite case *negative.* Climbing dynamics are illustrated in Figure 9.26

(a)

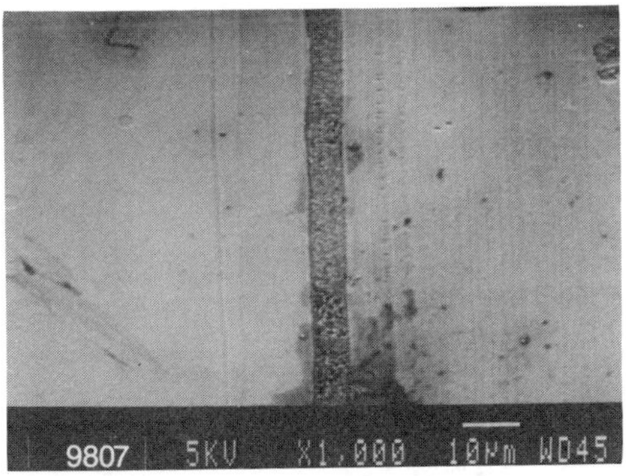

(b)

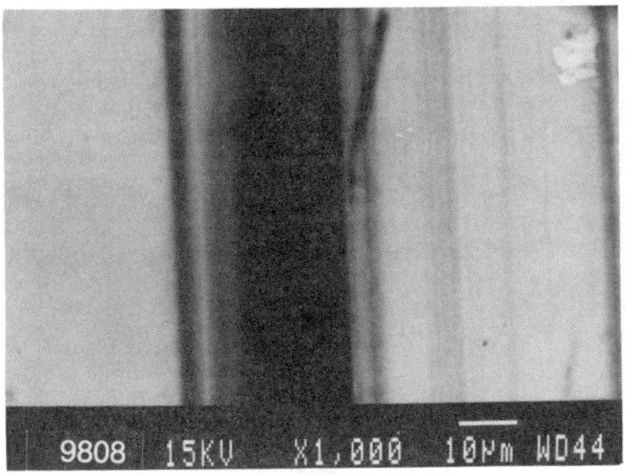

(c)

Figure 9.24 P side view of the laser in Figure 9.23: (a) light-microscope; (b) secondary electron image (SEI); and (c) electron beam induced current (EBIC) image.

(a) (b)

Figure 9.25 Light microscopy of the device in preceding two figures (a) before, and (b) after, a partial and selective etching of the cladding layer.

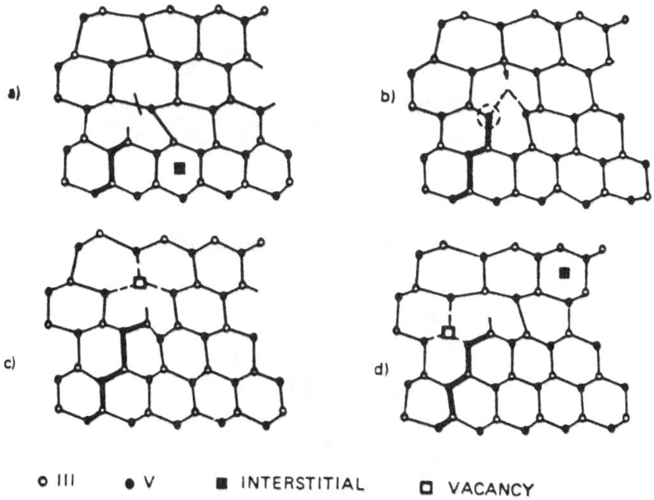

Figure 9.26 Climbing of a dislocation: (a) 60° dislocation; (b) the absorption of the interstitial atom and the subtraction of an atom from the other lattice; (c) correct configuration; and (d) final configuration.

on the usual projection plane <110> of the III-V structure [12, 13]. Figure 9.26a shows a 60° dislocation with a $\bar{b} = 1/2\bar{a}$ <110> Burgers vector and interstitial impurities near the core of the dislocation. If the interstitial impurity is absorbed and an atom is taken away from the other lattice leaving a vacancy (Figure 9.26b), the correct pattern is once again obtained (see Figure 9.26c). Figure 9.26d shows one last configuration, where the vacancy has moved to a position favoring climbing.

The series of conditions needed for climbing drastically reduce the probability of it actually taking place: we can reasonably expect few dislocations to spread by climbing. Climbing is not a conservative movement and therefore requires dislocations, vacancies or interstitial defects, as well as energy or a migration force to perform their movement.

Petroff and Kimberling [13] suggest dislocation climbing is possible thanks to the energy released by the nonradiative recombination events and hence cannot be eliminated. Woolhouse [14] focuses on the influence of outside stresses on this type of movement and suggests that dislocation movement is caused by electromigration, because of the high density of operating current of emitters, and by thermomigration, because of the strong thermal gradient in the device.

One interpretation of how defects and dislocations in particular affect and decrease radiative efficiency can be obtained by considering the distribution of energy levels within the material. Figure 9.27 shows a 60° dislocation highlighting the dangling bonds. The atoms along the dislocation line lack the first neighbors to saturate their bonds. However, they can interact and give rise to a local distribution of energy levels superimposed on the perfect crystal one. More specifically, at the dislocation, a localized excess of energy levels is generated in the valence band and a decrease in the conduction one. Furthermore, some discrete energy levels may be present inside the gap. The decrease in radiative efficiency may be due to these discrete energy levels inside the gap: carriers, such as electrons, injected into the device may be trapped by these energy levels because of the potential barrier generated at dislocation (Figure 9.28). Carriers are thus prevented from radiative recombination.

Figures 9.29 and 9.30 give two examples of dislocations formed during the operative life of emitters. Figure 9.29 is a SEM image, obtained using cathodoluminescence, of a GaAs LED emitting surface, whose structure is illustrated in Figure 9.4 and which failed during the accelerated life test. Chemical etching of the material exposed the zone near the junction where the dislocation pattern along <110> lines becomes visible [15]. Figure 9.30 shows SEM images, obtained using SEI and EBIC techniques, of the optically active area of an AlGaAs/GaAs laser mounted p-side up. Note the dark lines crossing the active area decreasing its efficiency.

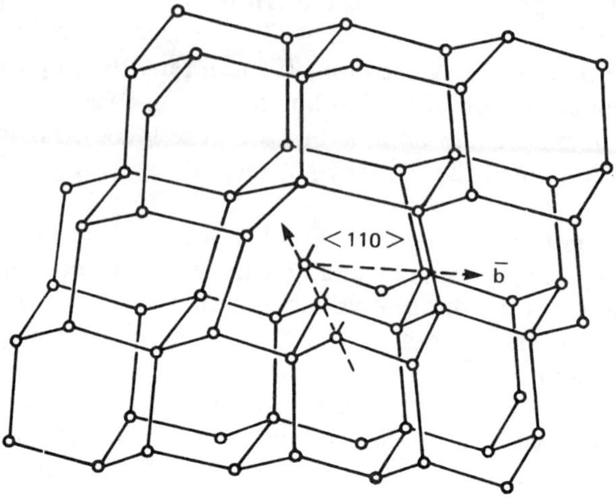

Figure 9.27 Example of 60° dislocation.

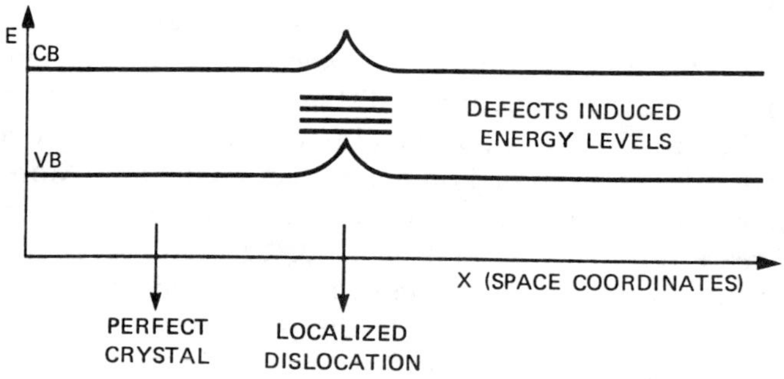

Figure 9.28 Energy level diagram.

We can virtually rule out a decrease in radiative efficiency following the development of climbing dislocations for InP lasers, at least with the present structures. Two causes appear to underlie climbing inhibition in InGaAsP layers:

1. In interstitial impurities that have migrated to the dislocation give rise to localized solid solutions, such as InGaAs. Kirby [16] has estimated their configuration energy as being roughly 0.31 eV. For 0.8–0.9 μm laser, Al impurities behave likewise, though in this case Kirby has estimated their energy at about 0.00 eV. In the first case, the As atom needed to climb is not available because of the configuration energy.
2. The absence of deep levels inside the gap at the dislocation.

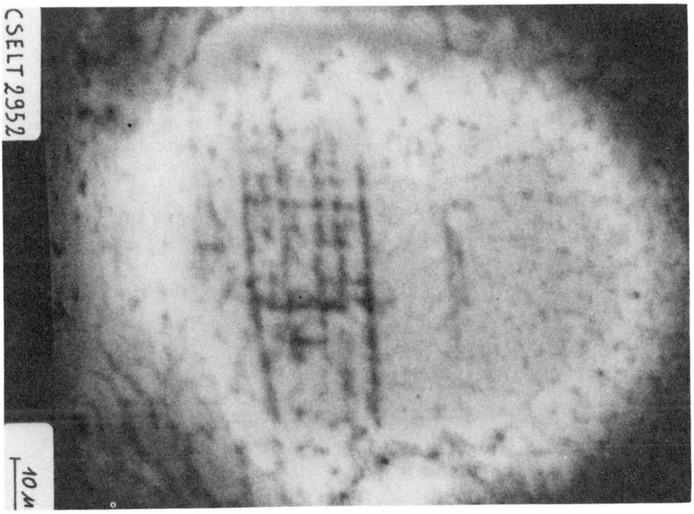

Figure 9.29 Dislocation patterns developed at the p-n junction of a LED diode. ECL image after 65′ chemical etching.

In the case of InP, level density due to "dangling bonds" increases nearing and within valence and conduction bands. This means there is a very low density of deep levels in the gap, making nonradiative recombination events unlikely and hence decreasing energy availability for defect motion.

9.1.4.2 Mirror Damage

Equations 9.1 and 9.2 express a relation of the dependence of laser diode parameters on mirror reflectivity. More specifically (9.1) and (9.2) require maximum reflectivity to obtain a minimum value for threshold current, whereas (9.4) requires a minimum reflectivity to have a maximum external efficiency (other conditions standing) to determine usable optical power.

Reflectivity plays an important role as far as reliability is concerned; that is to say, for a stable performance in the course of time. Hence, we must preserve the physical integrity of the mirrors to ensure the permanance of this parameter. The problem arose most of all in the early AlGaAs/GaAs lasers, where no mirror coating had initially been used. The "thermal runaway" mechanism is interesting in this respect, as it leads to the catastrophic failure of the laser illustrated in Figure 9.31.

Fabry-Perot laser mirrors are usually obtained by simply cleaving the crystal along specific crystallographic planes. A cleaved semiconductor surface is a defect

(a)

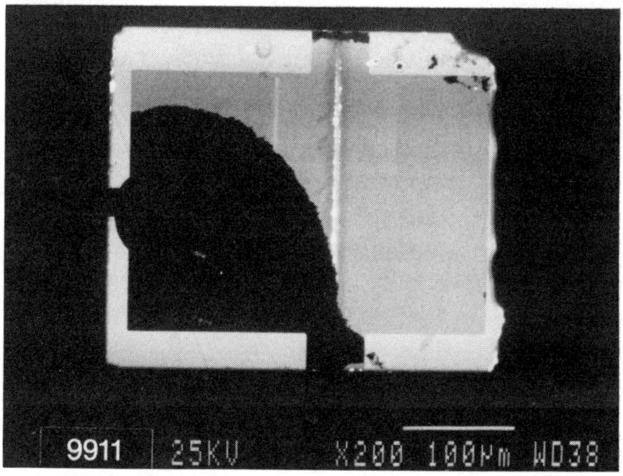

(b)

Figure 9.30 P side view of a degradated AlGaAs/GaAs laser: (a) SEI image after partial metal removal; and (b) EBIC image highlighting dislocations in the active zone.

from a crystallographic point of view: it is a strong alteration of the lattice resulting in a series of dangling bonds. These may associate with impurities and generate a high density of surface states. Henry et al. [17] associate the presence of these states to the triggering of the process, which in turn leads to the catastrophic

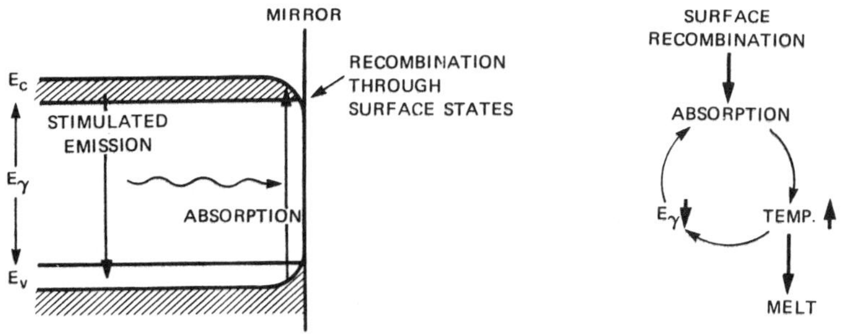

Figure 9.31 Model for mirrors catastrophic damage.

thermal damage to the mirrors. The dynamics leading to thermal runaway and subsequent GaAs melting can be summarized as follows:

1. Nonradiative recombination through fast surface states at the mirrors.
2. No population inversion next to mirrors and increase of stimulated radiation absorption coefficient.
3. Heating and, consequently, an increase in localized temperature.
4. A decrease in gap energy caused by the temperature ($E\gamma(T) = E\gamma(O) - \alpha T^2/(T + \beta)$ where T is temperature, and α and β two constants specific to each material) and consequent increase of absorption coefficient.

As this mechanism (nonradiative recombination → increase of the absorption coefficient → heating and temperature increase) sets in and the temperature enhances itself, soon GaAs may reach its melting temperature (~1500 K), thus clearly damaging mirrors permanently and causing the laser to fail. Figure 9.32 shows an AlGaAs/GaAs laser: on the mirror are signs of material melting in the active zone with its extrusion from the device bulk.

Currently mirrors are coated with oxides to overcome the problem: Al_2O_3 or SiO_2 are both used although the former is preferred because its thermal dilatation coefficient is nearer that of the GaAs (Al_2O_3: 7.58 10^{-6}°C^{-1}; SiO_2: 0.5 10^{-6}°C^{-1}; GaAs 5.5 10^{-6}°C^{-1} at 300 K). The thus obtained improvement is due to the saturation of dangling bonds of surface atoms with a dramatic drop of surface state density.

This degradation process appears not to touch InGaAsP/InP lasers, which consequently have uncoated mirrors, because of the greater thermal conductivity of this material (InP: 0.66 W cm^{-1} K^{-1} at 300 K; GaAs: 0.44 W cm^{-1} K^{-1} at 300 K) and to the lower rate of surface recombination (InP: $\leq 10^3$ cm s^{-1}; GaAs: ~ 10^7 cm s^{-1}).

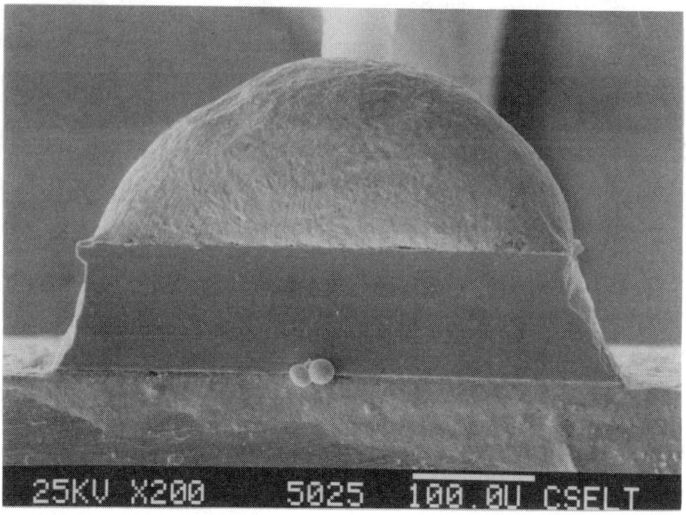

Figure 9.32 Front view of an AlGaAs/GaAs laser: catastrophic failure due to thermal runaway.

9.1.4.3 Die-Attach Deterioration

The threshold current density is an exponential function of junction temperature, which must consequently remain stable throughout device functioning. More specifically, it should not increase to avoid starting a closed process: increase of threshold current → increase of junction temperature → increase of threshold current, and so on, until the device fails.

Junction temperature, given by equation (9.3), depends on the laser thermal resistance; that is, on its ability to dissipate the heat produced within the diode (Joule effect plus nonradiative recombinations: $Pd = (1 - \eta_{ext})\, IV + I^2 R_s$). An R_θ increase causes a T_j increase with the consequences just described.

The total thermal resistance of the laser is given by the sum of the chip laser and the die-attach thermal resistance contributions:

$$R_\theta = R_{\theta \text{die}} + R_{\theta \text{die-attach}}$$

By measuring the temperature increase at the diode junction when it is fed and assessing it as a function of time, we end up with a dynamic assessment of thermal resistance. The method consists in measuring voltage variation at the junction, and it identifies the two components separately. When activated—usually in the first microseconds—the chip thermal resistance is the only one to be felt, as the thermal gradient only involves the active and contact zones of the device.

In the following microseconds, the thermal gradient reaches the heat sink and the system moves toward a stationary state, where the die-attach thermal resistance is felt, too. Various die-attach or heat sinks used with the one laser chip produce different thermal resistances, which causes a varying degree of operating stress.

Figure 9.33 shows the results of the dynamic thermal resistance measurements for two identical lasers: one die with a correct and the other with an incorrect soldering, the latter having voids. The temperature increase in the first microseconds is comparable in both, whereas at steady state the correctly soldered die will perform better.

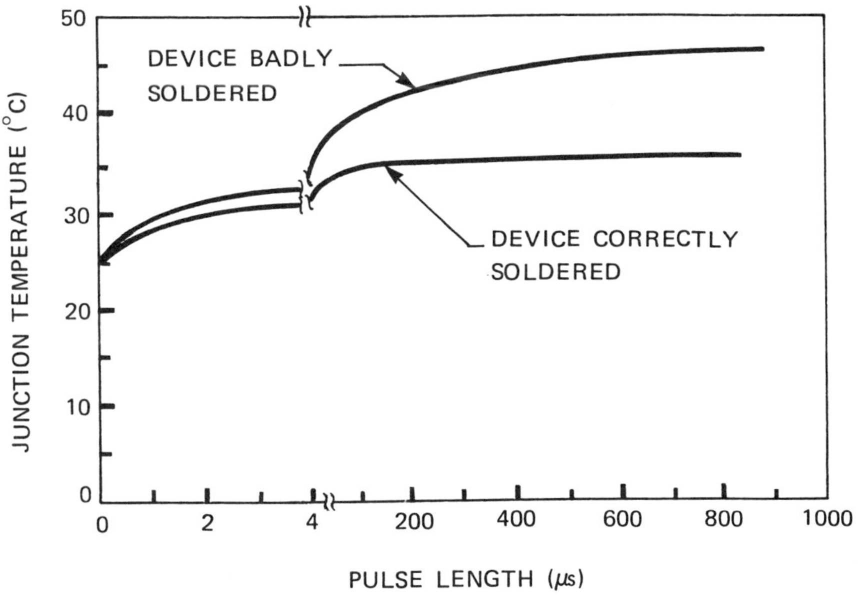

Figure 9.33 Dynamic thermal resistance of a device soldered correctly and badly.

Figure 9.34 compares two devices with the same chip laser and two different mounting technologies, one indium soldered on a copper "cube" heat sink, and the other gold tin soldered on a diamond heat sink. Once again the initial temperature rise is comparable ($R_{\theta \mathrm{die}}$ contributes the same), but the diamond heat sink reaches a stationary condition in a shorter time (< 4 μs) and the overall temperature increase is smaller than for the copper mounted one. Indium is often used as a die attach because of its low melting temperature (~156°C), which means the chip will not suffer any thermal stress. Currently, the choice is questioned, especially as indium tends to form intermetallic compounds with gold, employed

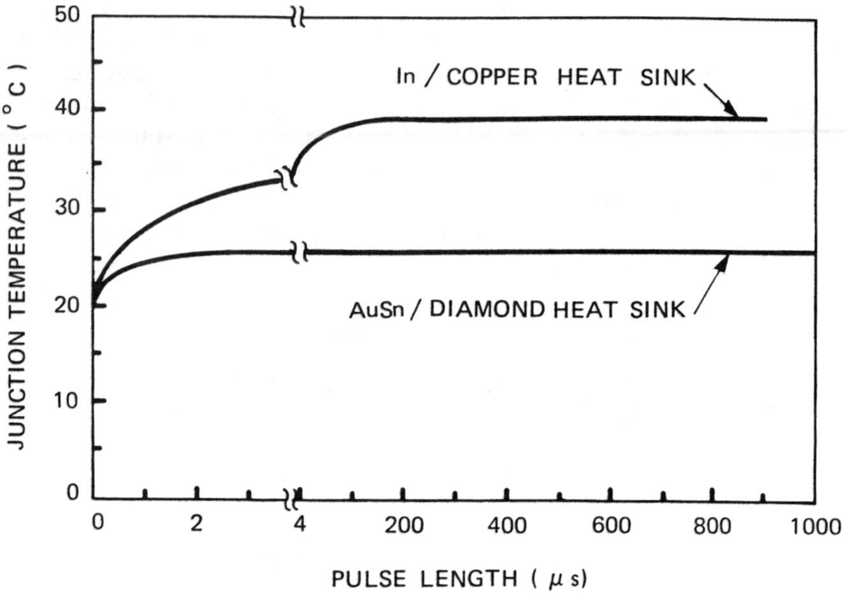

Figure 9.34 Comparison between two mounting technologies.

both as a coating for the copper heat sink and in the metalization. The In-Au interdiffusion with formation of voids and In-Au intermetallic phases is held to be responsible for the increased thermal resistance and has also been observed in devices that also operate at room temperature. Indium is also responsible for another failure mechanism: it causes sudden irreparable faults [18] due to indium whisker growth. This growth is caused by indium electromigration on the surfaces of the laser mirrors (see Figure 9.35). In the standard p-side down assembly, the indium whiskers can reach the active layer and short circuit the p-n junction.

9.1.4.4 *Gold Diffusion from p-Contact*

Metalization, and especially those on p-side, is a critical point in emitters and important in terms of technology. The geometry of the contact requires it to be small (the size of the contact stripe from 2 to 10μm) to confine current in a satisfactory way while offering low contact resistivity (ρ_c: 10^{-5}–10^{-6} Ω cm^2) to minimize heat dissipation due to the Joule effect. This is why, so far, metalizations have always been gold-based; that is, CrAu, AuZnAu, TiAu, and so on. However, although gold means a good contact from an electric point of view, it has been identified as a potential cause of emitter degradation due to the high diffusion coefficient gold has in III-V materials ($D = 10^{-16}$ cm^{2}s^{-1} for GaAs). This means

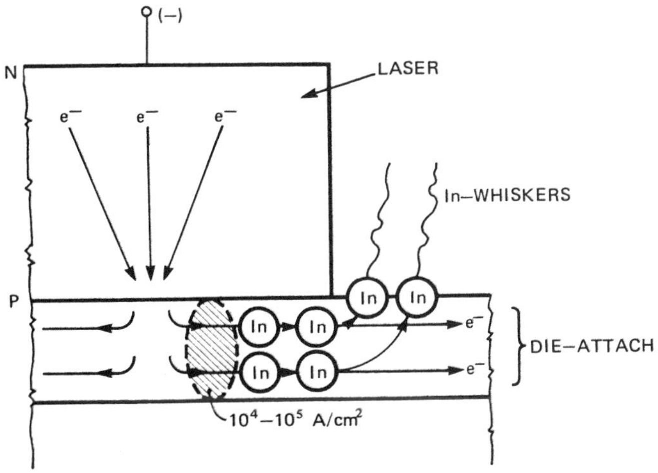

Figure 9.35 In die-attach degradation in a laser caused by electro-migration.

that, under working conditions when we have a current density on the order of $kAcm^{-2}$, it may migrate to the active layer, located only a few microns from the contact, by thermo-electromigration, producing localized metallic precipitates. These will, first of all, reduce radiative efficiency. Figure 9.36 shows the gold Auger profiles for the p-contact stripe of the two lasers illustrated in Figure 9.14. The first one refers to a device that has never been used, the second one to a device that operated for 6,000 hours [19]. Note how the gold has diffused within the InP in the latter. The presence of gold at 0.25 μm in the reference device is the result of the alloying to decrease contact resistance and increase metalization adhesion. Another example is illustrated in Figure 9.37: a "p" laser surface with a structure like the previously described ones at the end of the life test. Here, we note a lateral surface diffusion of gold. Current flow and contact geometry may explain the pattern.

Diffusion is not a uniform process, and its direction is affected by defects in the crystal, such as dislocations. In these cases, the crystallographic defect acts as a preferential channel where gold may diffuse more rapidly. Accumulation and precipitation (decoration of dislocations) may also take place. For instance, Figure 9.38 shows a couple of SEI and EBIC images from the p side of a degraded laser. The dark spot on the stripe shown by the EBIC picture is caused by a cluster of dislocations, like the ones in Figure 9.20. This defect is also found in the substrate and is duplicated during the epitaxial growth of the various layers, which means it must have been there from the beginning of the life of the device. As can be seen in Figure 9.39, by taking a 1 μm layer off the surface, gold is evidenced in

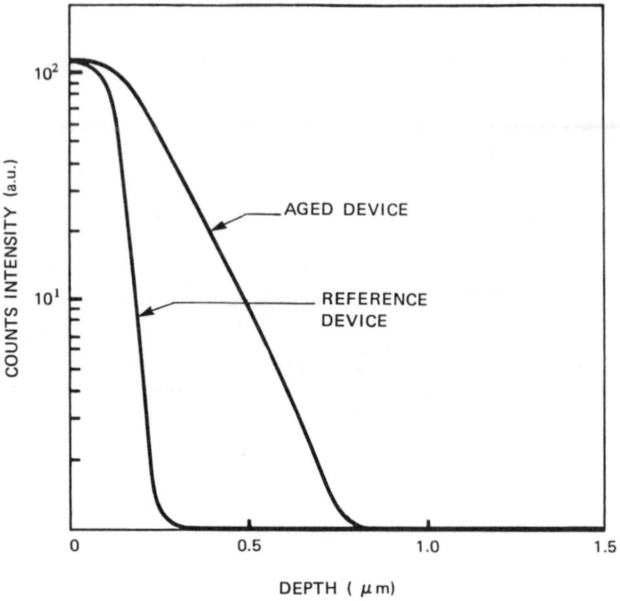

Figure 9.36 Gold Auger profiles of two laser stripes: (a) reference device; and (b) device after a 6,000 hour life test.

Figure 9.37 The p-side of a degraded laser. Note the lateral diffusion of gold from the stripe and the many thermal etches characteristic of the LPE growth of an InP layer.

(a)

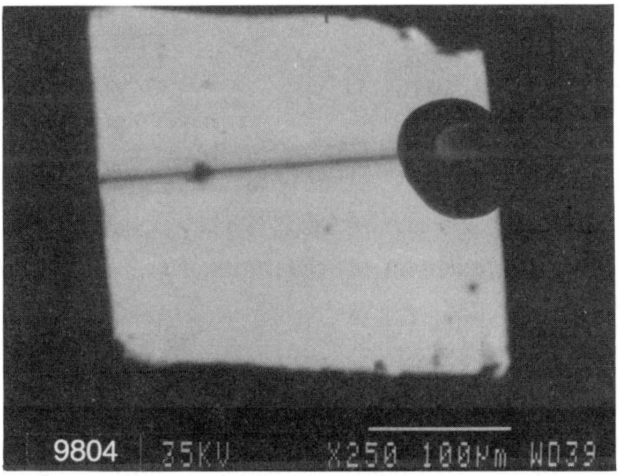

(b)

Figure 9.38 The p-side of a degraded laser: (a) SEI image; and (b) EBIC image.

the area corresponding to the dark spot visible in the EBIC. Gold diffuses more rapidly along cluster dislocations and accumulated in this area, causing a progressive increase of the threshold current in cw and of R_Θ and, hence, producing a permanent degradation of the device.

Gold diffusion takes place regardless of whether the emitter is LED or laser, which is why the problem has to be tackled and overcome rapidly. Current trends favor multilayer metalizations, such as TiPtAu, where Pt acts as blocking layer

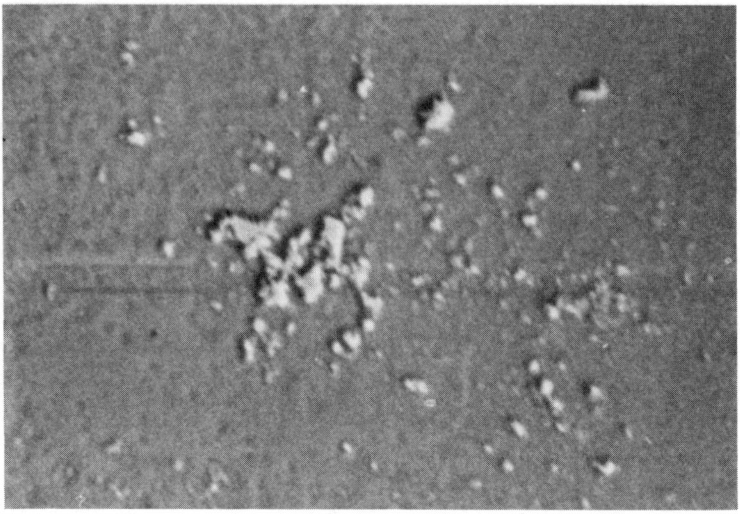

Figure 9.39 A SEI image of the dark spot in Figure 9.38 after ≈ 1 μm of surface material was removed. Observe the gold particles (1,000 X).

thus preventing gold diffusion. In this instance, there are Schottky contacts and a Zn diffusion in the top layer in contact with the metalization.

9.1.5 Quantitative Measurement of Reliability: Parameter Drift and Mean Life Evaluation

It is always critical to predict device mean life, or failure rate, which is what users often wish to know. This is especially true for photoemitters, where technology is still in a phase of rapid evolution and field data are lacking. Accelerated laboratory life tests are often the only source of information, though not altogether satisfactory: component overstressing can trigger degradation processes absent in normal operating conditions or *vice versa;* and other aspects present under these conditions, such as humidity or mechanical stress, might not be considerd. In any case, laboratory tests are an effective tool when testing component reliability: in fact, a device that fails in the course of accelerated life tests is unlikely to have a long life in the field. We shall not consider the incident of accidental catastrophic failures, but deal mainly with evaluation in the case of parameter drift. When evaluating, we should identify the following:

- One or more significant parameters.
- A failure criterion; that is, the maximum admissable variation of one or more parameters within the system performance.

- The availability of a population of devices that may be considered equivalent when carrying out a statistical analysis of test results.

These conditions are easily satisfied for LED emitters, thanks to their simple structure and a sufficient stabilization of the technology. An LED is amply described by its radiative efficiency, and there is only one failure criterion, its decrease to 50 percent.

For instance, Figure 9.40 shows the distribution of the failure observed in laboratory tests [20]. The log normal function proved to be the most adequate when describing the incidence of failures over time. Device mean life is determined by the "mode" of the function; that is, by the time required for 50 percent of failures to happen. In this case we were also able to define the relationship between mean life and the two factors accelerating device degradation, current density and temperature:

$$t_{50\%} = \frac{A}{J_F^x} \exp[E_a/KT_j]$$

where A and x are two numeric coefficients that depend on the device structure.

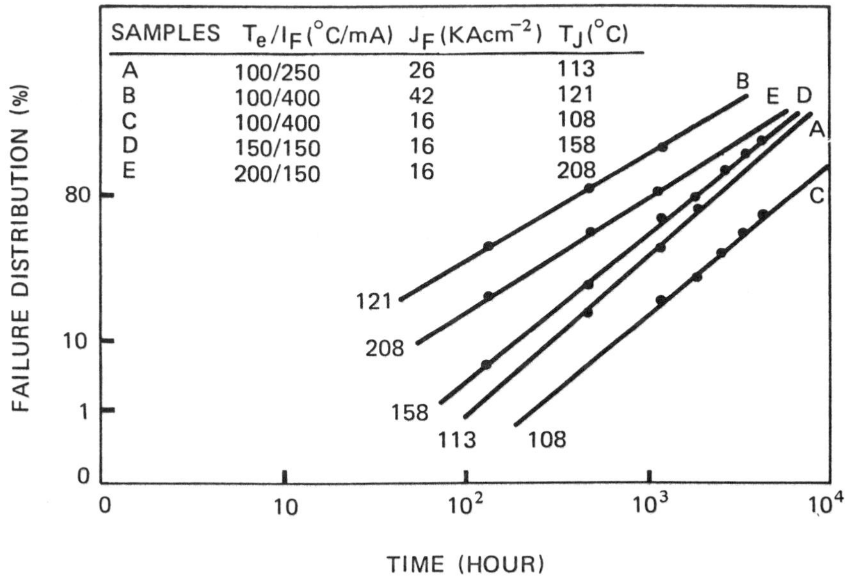

Figure 9.40 GaAs LED failure distribution by time in life tests, for different operating conditions.

A general malfunctioning criterion for lasers, defined in relation to system performances, is as follows: the "inability" of the laser to supply the amount of stimulated power or the quality of the optical signal required to maintain an error rate below preestablished values, usually 10^{-9} bit/s. Such a definition involves several parameters, the variation of which is associated to degradation in device performance. The parameters include an increase in threshold current, a decrease in differential quantum efficiency, an increase in leakage current, and triggering of higher order modes or variations in the front-back power ratio controlling the signal stability. We must select parameters so that they do not require an excessive number of tests yet are able to describe the "good behavior" of the device.

First, we have to define a failure criterion. Usually, as we shall see, threshold current is the first parameter to be checked, and hence the first constraint we introduce is a limit to the amount by which it can be increased. However, its priority must not be taken for granted as L-I linearity in the working zone is just as important. In this case, a quantitative criterion is not easily introduced; for instance, a qualitative criterion might be the absence of kinks or nonexcitation of higher-order modes.

With reference to the uniformity of devices, we would like to add that manufacturing is not well enough established to consider devices equivalent. Screening will help us reach a better population uniformity. When reliability requirements are very strict, we tend to assess each device separately, including its scattered behavior pattern.

Figure 9.41 shows activation energy values of roughly the same number of devices, showing quite a dispersion, from 0.78 eV to 1.28 eV [21]. Therefore, were we to take 1 eV as an average value this might incorrectly predict mean life values: a mean life interval might prove more significant and be likely to include the expected value.

Figure 9.42 shows AlGaAs/GaAs laser mean life improvements. We see that, in the early days, lasers only lasted a few hours, whereas currently we can reasonably expect a mean life of 10^6 hours. We lack assessments of InGaAsP/InP laser devices based on long-term field observations or accelerated life tests. However, values of 10^4–10^7 have been given; their extreme variability is due to both the marked diversification in structure and the fact that often the only available data come from the manufacturers.

A DFB laser can be realized with different structures. Several structures are already available commercially but few reports on their reliability. Generally speaking, a DFB laser will present the same weaknesses of other lasers with the same structure plus some characteristic problems due to the grating. The fundamental electrooptical parameter is still the threshold current, but both the single frequency wavelength and the side-mode suppression ratio are important in evaluating the performances of DFB over time: the single frequency wavelength can shift with time and the side-mode suppression ratio can decrease due to less effective action by the grating.

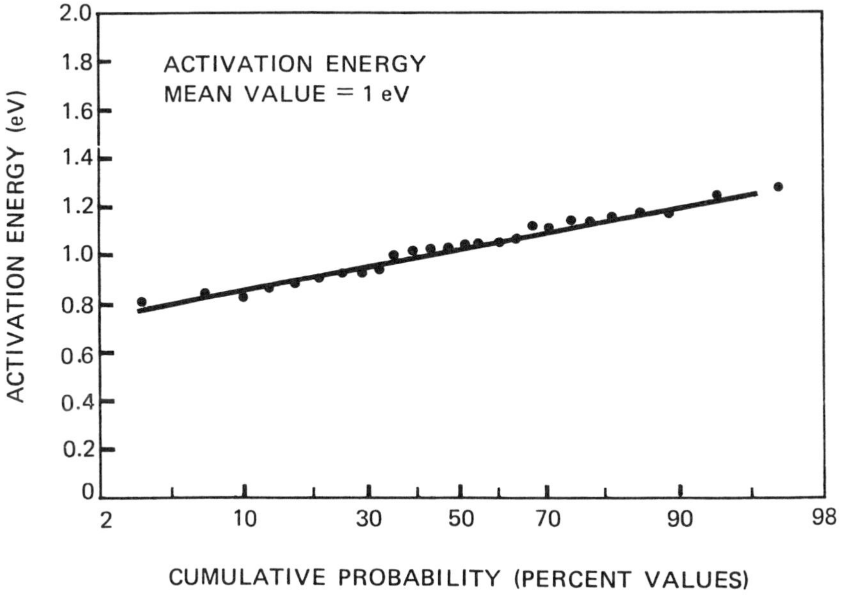

Figure 9.41 Distribution of E_{att} values obtained in temperature step-stress tests.

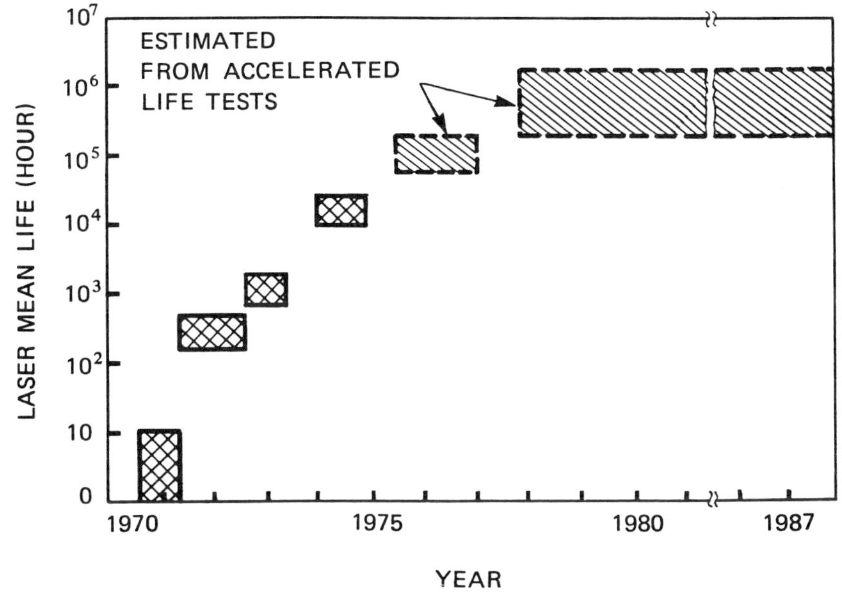

Figure 9.42 Laser mean life trends over time.

9.2 PHOTODETECTORS

In the field of photodetectors for optical fiber transmission, research and development is aimed at several needs, especially:

- Development of detectors capable of operating with present light sources; that is, at operating radiation frequencies. The choice of material for fabricating the device is mainly connected to this point.
- Devices capable of identifying even extremely weak signals, to limit the number of repeaters along the optical transmission line.
- Fast detectors, capable of detecting extremely short signals, fit for extremely high-capacity transmission systems.

The choice of the device structure depends primarily on requirements stated in the latter two.

PINs and *avalanche photodiodes* (APDs) are two classes of photodetector components. They differ in that APDs have an area of primary carrier multiplication characterized by a high electric field. In this section, we discuss their operating principles and describe a typical structure for each class of detectors. Following this, the discussion turns to photodetector degradation and measures of reliability.

The first light sources came in the 0.8–0.9 μm range, and silicon proved the most convenient material for photodetectors at that wavelength. The development of 1.3–1.5 μm emitting light sources initially required germanium, though later, thanks to improved performance, new materials were used, especially III-V compounds. Today, InGaAs detectors are used increasingly. Initially, there were only PIN devices, but the research and design of particular structures—such as *separated absorption multiplication* (SAM), and *separated absorption grading multiplication* (SAGM)—made possible avalanche devices with this material, too.

As mentioned earlier, detectors must be able to detect very weak signals (from hundreds of nW to tenths of μW), which requires as little leakage current (dark current) as possible. Figure 9.43 plots dark current as a function of the $E\gamma$ of the various materials. Dark current is inversely proportional to the energy gap [22], which is why germanium is a less attractive material than the III-V compounds for long wavelength detectors. Another reason is the high effective ionization ratio, 0.8 as opposed to 0.02 for Si and 0.25–0.5 for the InGaAs, responsible for device noise.

The electric analysis of APD and PIN photodetectors evidenced that most malfunctioning was due to an increase in dark current. This increase could be caused by the presence of contaminants in passivating oxides or by microplasmas at the junction.

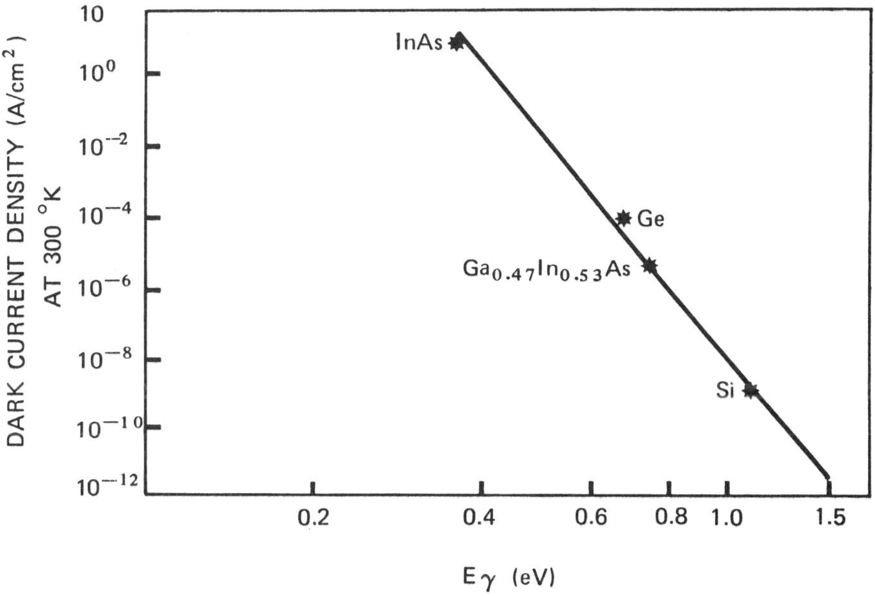

Figure 9.43 I_d as a function of $E\gamma$ for different materials.

9.2.1 Operating Principles of Typical PIN and APD Structures

The basic function of a detector is to translate an optical signal into an electric one. The physical principle underlying this is the generation of electron-hole couples, which happens when the energy of the incident photons is equal to or greater than the semiconductor energy gap ($h\nu \geqslant E\gamma$). Figure 9.44 illustrates this process for an APD photodetector. The interacting photon generates a primary electron-hole couple in the absorption region, which is separated by the electric field. The electron is then accelerated toward the multiplication zone, where new electron-hole couples are generated by impact, the avalanche effect.

Both PIN and APD devices can usually be made with "mesa" or planar structures whatever their material. They will also differ in terms of critical points and reliability. Figures 9.45 and 9.46 show two examples of structure: mesa and diffused planar junction PIN photodetector. When photosensitive areas are equivalent, the mesa structure offers a larger depleted surface area than the planar structure. Consequently, all the problems due to surface condition and dielectric coating will influence the former more than the latter.

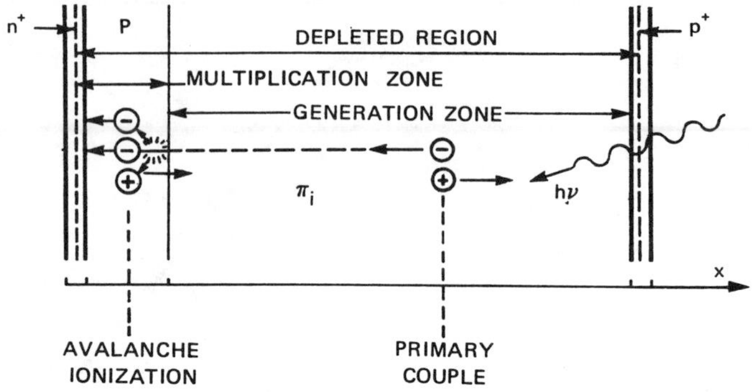

Figure 9.44 Functioning principles of an avalanche detector.

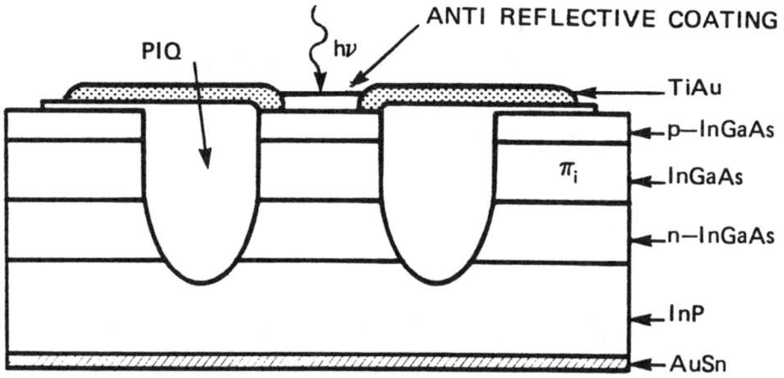

Figure 9.45 Cross section of an InGaAs PIN with a mesa structure.

Figure 9.47 shows an APD planar structure where generation and multiplication regions are not separated. Typically, these structures have a guard ring closing the junction to avoid premature breakdown at the junction edge. There is also a channel around the junction, called the *channel stopper* because it limits the surface extension of the depleted region.

9.2.2 Electric Parameters

Basic electric parameters, dark (I_d) and noise (I_N) currents, can be used to indicate that the device is working properly both in PINs and APDs; M, the multiplication or gain factor, only applies to APDs. Dark current is more important in both classes.

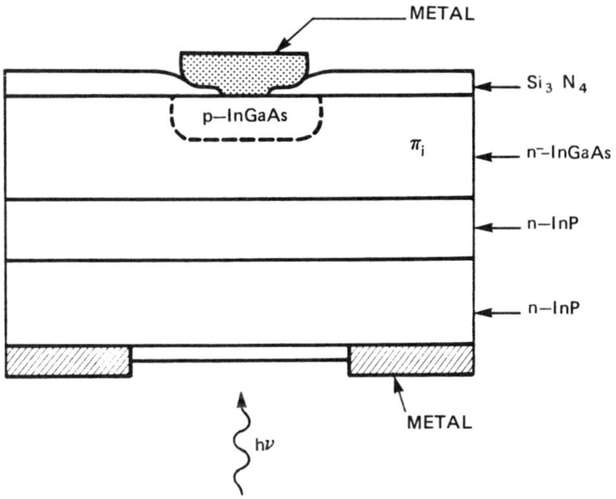

Figure 9.46 Cross section of an InGaAs PIN with a planar structure.

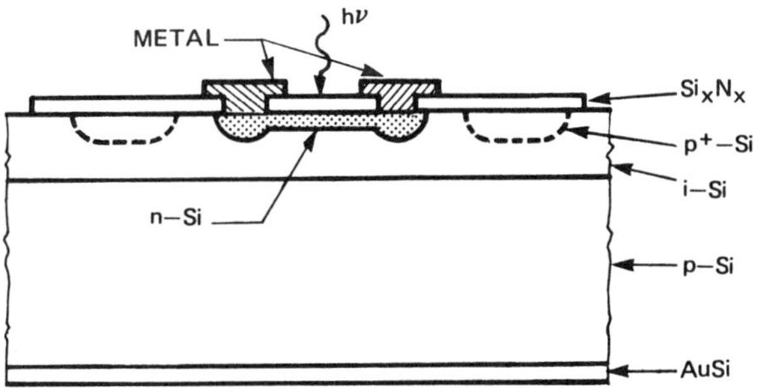

Figure 9.47 Cross section of a Si APD device with a planar structure.

9.2.2.1 Leakage or Dark Current

Dark current is made up of various components, each correlated, in one way or another, to one or more aspects of the material, the technological process, or other electric parameters. Figure 9.48 shows the dark current as a function of the reverse voltage for a InGaAs homojunction PIN with a mesa structure. The single components of the dark current are evidenced. The literature offers more detailed

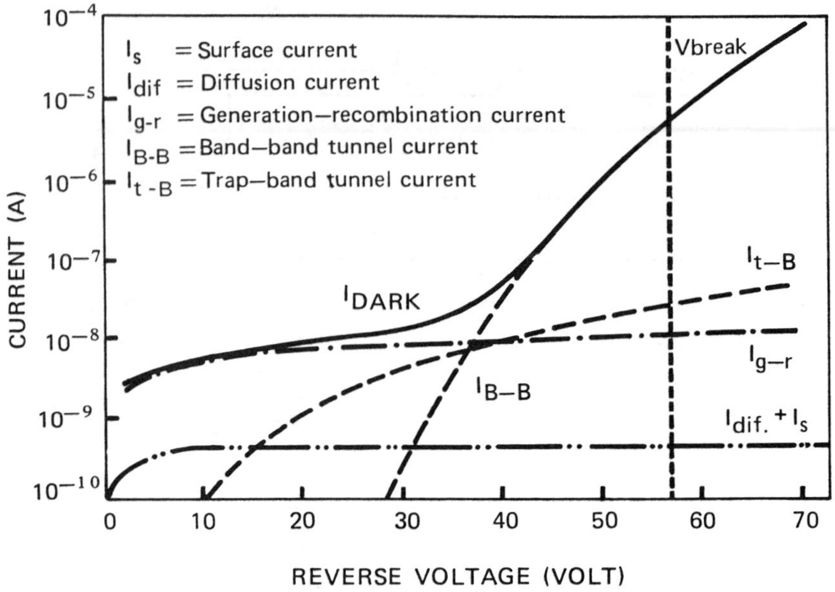

Figure 9.48 Dark current and its components in an InGaAs/InP PIN with a mesa structure.

information on the subject, and the following summarizes specific contributions [23]:

- Surface current is proportional to the exposed surface area of the depleted zone.
- Bulk current is proportional to the volume of the depleted zone.
- Current component due to anomalous multiplication from the presence of microplasmas.

According to various authors, surface current is often the prevailing component in a degradation process. Futhermore, in some cases, it may be intermittent, which clearly would generate problems in identifying device malfunctioning. This is why we present mathematical expression for surface current in a mesa structure:

$$I_S = \frac{1}{2} q n_i s_0 L W_j$$

where q is the electron charge; n_i, the intrinsic carrier concentration; s_0, the surface recombination rate; L, the length of the junction perimeter; and W_j, the width of the depleted zone.

The surface recombination rate is expressed as

$$s_0 = \sigma v_{th} N_{st}$$

where σ is the capture section; v_{th}, the thermal speed; and N_{st}, the density of surface states.

The intrinsic carrier concentration is expressed as

$$n_i = 4.9 \cdot 10^{15} \left(\frac{m_e^* m_h^*}{m_0^2}\right) T^{3/2} \exp(-E_\gamma/2KT)$$

where m_h^*, m_e^*, and m_0 are the effective hole mass, the effective electron mass, and the electron free mass. T stands for temperature, E_γ is the energy gap, and K is Boltzmann's constant. The previous relation can therefore be rewritten as

$$I_s \sim \sigma v_{th} N_{st} \exp(-E_\gamma/2KT)$$

evidencing the dependence of surface current on surface state density, which is important as it may vary with time and consequently be responsible for device degradation. This formula also illustrates the exponential dependence of this component on $1/T$ by means of $E_{\text{att}} = E_\gamma/2$ activation energy.

The components of bulk current are diffusion current, generation-recombination current, and tunnel (band-to-band, trap-to-band) currents. When we want to know which prevails or is responsible for device degradation over time because of some modification, then we should analyze the dark current–temperature dependency. In fact, if we write the various mathematical relations out, the result is

$$I \sim \exp\left(\frac{-E_\gamma}{nKT}\right)$$

where

$$n = 1 \text{ for } I_{\text{diffusion}}$$

$$n = 2 \text{ for } I_{g-r}, I_s$$

There is a very slight correlation between the tunnel component and temperature (a variation through E_γ) and a stronger one with the squared polarization voltage.

Generally speaking, we can see instances where more than one component plays an important role in determining dark current, in which case n value will be between 1 and 2. Figure 9.49 gives an example: a PIN InGaAs device where dark current activation energy is established at a time t = 0 and then after 2,000 operating hours. Note how the dark current increased after 2,000 hours though its

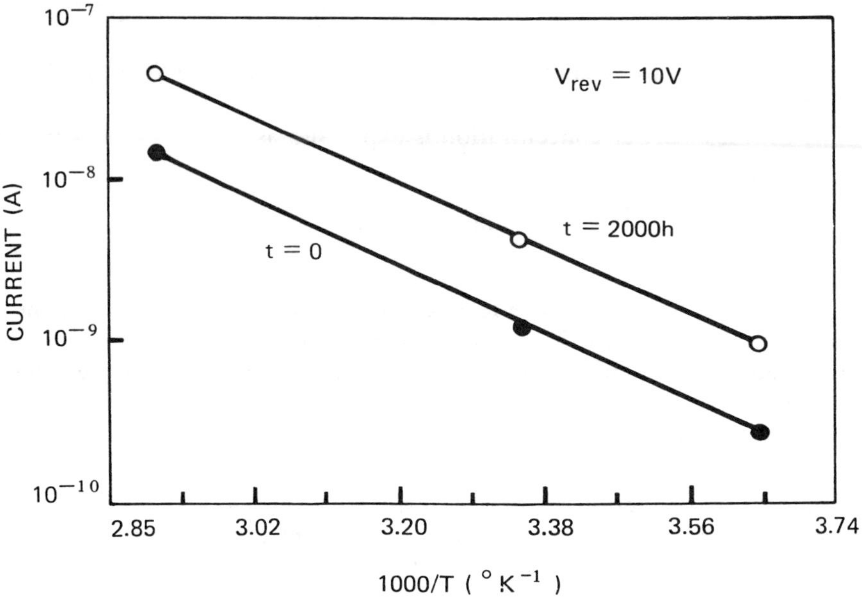

Figure 9.49 Dark current variation as a function of temperature for a PIN device at $t = 0$ and after 2,000 operating hours.

dependence on temperature remained unvaried. Clearly the mere dependence analysis of dark current and temperature is not enough to determine the weight of single components, though it can help when selecting possible interpretations of photodetector behavior.

When dealing with APDs, we should also consider multiplication (M) factor variation. Factor M is defined as the ratio between the number of outgoing charges leaving the multiplication zone and the number of incoming ones generated by the incident optical flow. M depends on the carrier ionization rate, and if ionization is only due to electrons, their relation can be described thus:

$$M = e^{\delta}$$

where

$$\delta = \int_{x_1}^{x_2} \alpha dx \text{ for } \beta \ll \alpha$$

and where x_1 and x_2 are the two multiplication zone limits. α and β refer to electrons and holes and stand for their probability to undergo an ionization process per length unit. Usually we refer to their ratio as the *effective ionization ratio*:

$$K_{\text{eff}} = \beta/\alpha$$

In terms of current signal, M is equal to the I/I_0 ratio where I is the current with and I_0 without multiplication.

9.2.3 Photodetector Degradation from Parameter Drift and Failure Mechanisms

Unlike lasers, where malfunctioning may be attributed to the degradation of several optoelectronic characteristics, overall dark currents are accepted as the sole critical parameter for photodetectors, which are made up of several components. We expect the component that increases the most to be different with different structures. (This is summarized in Table 9.3.)

Table 9.3 Current Component and Failure Mechanism for Planar and Mesa Structures

Structure	*Current Component*	*Failure*
Planar diffused APD/PIN	*I* bulk	Junction degradation Localized breakdown
Mesa	*I* surface	Oxide contamination

Microplasmas appear to be the main reason for the increase of bulk. From the electric point of view, a junction with microplasma maintains a localized breakdown voltage below the average value. The presence of microplasmas can be identified when the characteristic I-V has a linear slope, which may at times be negative. This is interpreted as the triggering resistive paths for carriers, which may saturate [22]. The presence of microplasmas is also confirmed by light emission.

The APD multiplication factor is enhanced by the presence of microplasmas. In fact, it has an exponential relationship with the carrier ionization coefficient; this in turn depends on the electric field, which in correspondence to microplasmas is higher than in the remaining junction depletion zone. Schockley et al. [24] showed the dependence of multiplication factor M in the proximity of V_{bd} is as follows:

$$1/M = [(V_{bd} - V_{op})/V_{bd}] \cdot n$$

where n is an empirical constant. In the presence of a significantly lower V_{bd} area (i.e., of a microplasma), the relationship no longer holds.

Chin, Chen, and Ermanis studied the malfunctioning of PIN devices, whose structure is described in Figure 9.46, and found microplasmas to be responsible for their behavior [25]. After several thousand hours of operation, the devices showed a marked increase in dark current. The failure analysis of these degraded devices has proved that such an increase is due to individual leakage paths located on the junction perimeter. Both biased and unbiased EBIC analysis has shown that leakage paths are associated to microplasmas (Figure 9.50). In fact, when biased, current anomalous gain sites correspond to microplasmas.

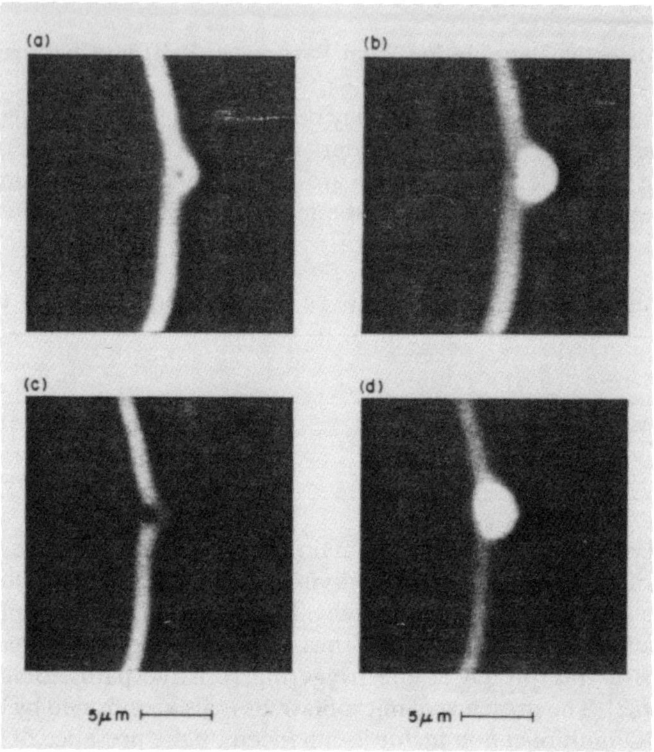

Figure 9.50 EBIC image of microplasmas at the junction: (a) before aging, without bias; (b) before aging, −20V bias; (c) after aging, without bias; and (d) after aging, −10V bias.

These PIN devices have a p-n planar diffused junction. Zinc diffusion takes place using a first layer of silicon nitride as a mask. If pinholes are present in this layer, there will be small diffused regions with a small radius of curvature (Figure 9.51). When these regions are less than half a micron from the actual junction edge and the electric field is very high, they will modify the junction, becoming microplasmas. In other cases, the development of microplasmas has been attributed to the enhanced zinc diffusion along threading dislocations. This may cause localized premature breakdown due to the anomalous electric field in correspondence to Zn spikes. As mentioned earlier, microplasmas represent resistive paths for carriers and will consequently generate an increase in current leakage.

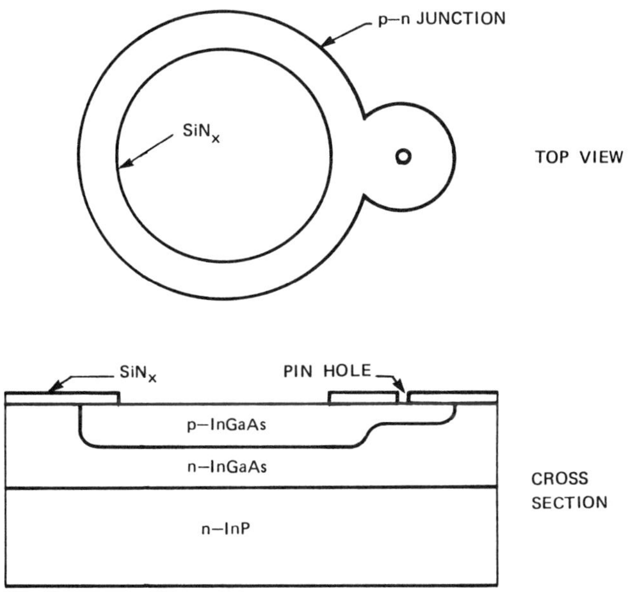

Figure 9.51 The pinhole effect in the SiN_x mask.

Mobile impurities trapped in passivating oxides and the degradation of the surface due to external agents (0_2, H_20) or to stress on the interface are held responsible for the increase of the surface component of dark current.

In the first case positive ions, such as Na^+ or K^+, may move toward the semiconductor-oxide interface. They may attract electrons from the semiconductor that in turn will cause an increase in the number of occupied interface states near the conduction band. As a result, generation-recombination processes will increase, at least in terms of the leakage current surface component [26].

Figure 9.52 illustrates the variations in dark currents as a function of time during APD step-stress tests [27]. The dark current suddenly increases in three devices. When measuring dark current as a function of multiplication factor of a degraded device (see Figure 9.53), we can see how the dark current increase is due to a nonmultiplied component (I_d independent from M), which means it is a surface one. The suggested failure mechanism attributes the dark current increase to the presence of ionic impurities in the passivating oxide that migrate to the oxide-semiconductor interface under the effect of the electric field.

Figure 9.54 shows a decrease in dark current as a function of the square root of time after removing bias. The linear variation in the square root of time is characteristic of a diffusive process. Futhermore, it confirms the presence of impurities that diffuse back after the bias voltage is removed and dark currents resume

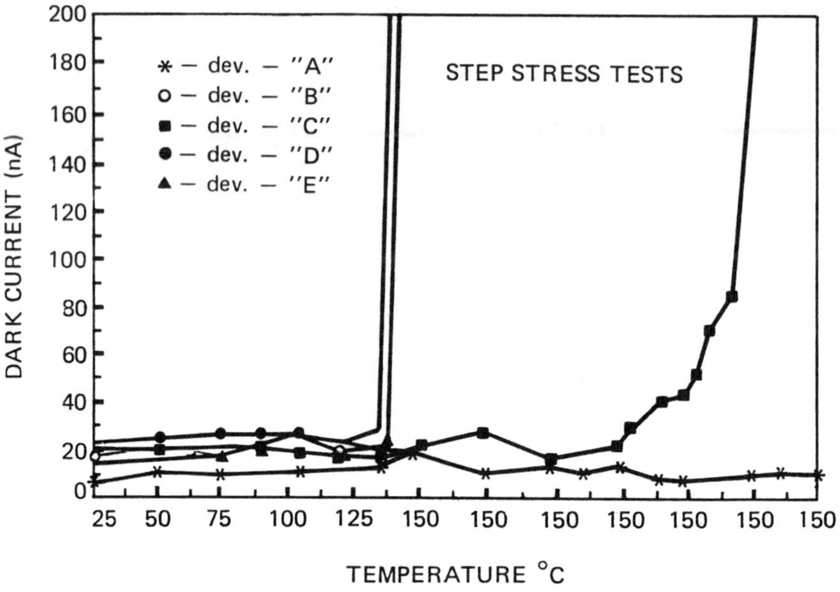

Figure 9.52 Dark current variations during temperature step-stress tests.

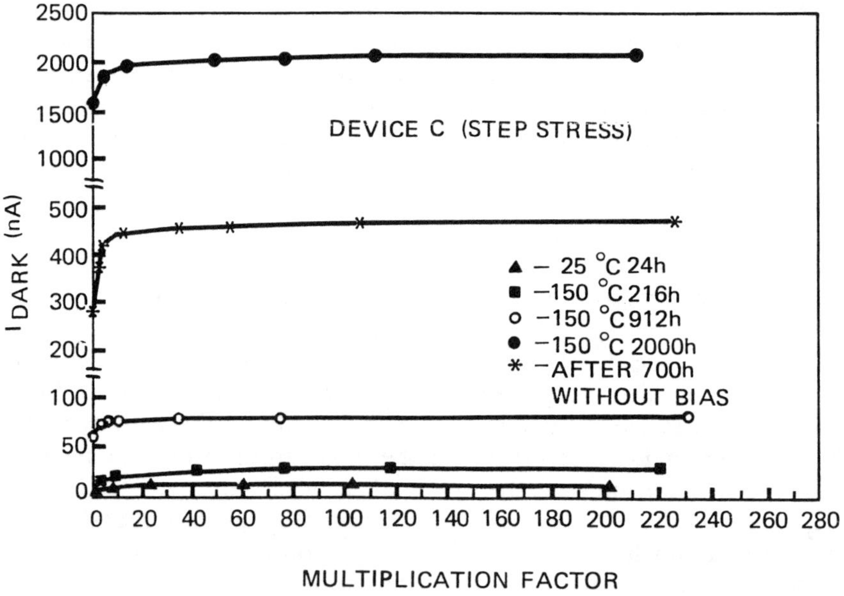

Figure 9.53 Dark current variations by multiplication factor during temperature step stress.

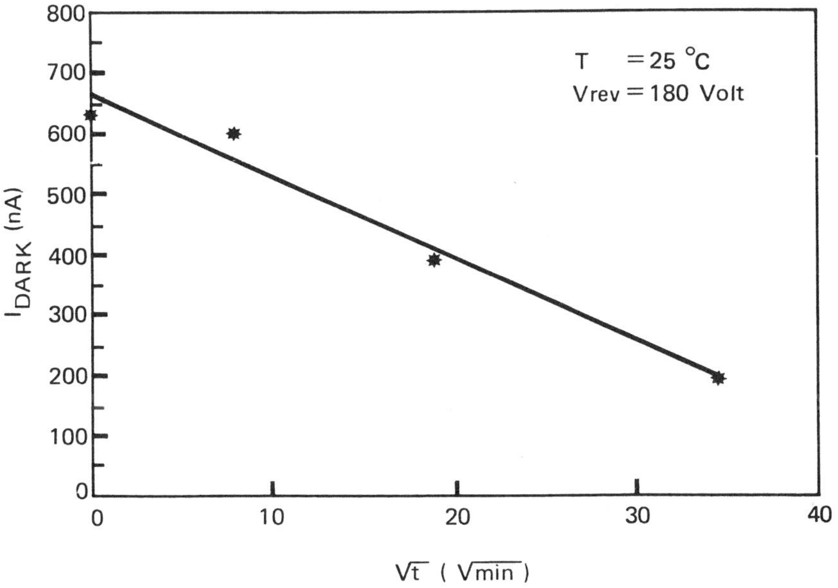

Figure 9.54 Dark current variations over time after the removal of the bias voltage.

initial values. To avoid this kind of degradation, we have to use extremely clean oxides and carefully cleanse the surface before deposition.

9.2.4 Quantitative Measurement of Reliability

Photodetectors are simpler than lasers, and dark current is the first and most important parameter of correct functioning. Therefore, laboratory life tests are planned for a forced increase, possibly activating impurities in oxides or stressing the weak points of the junction. This is why bias voltage and temperature are used jointly as the degradation-acceleration factors. An increase of leakage current is taken as a failure criterion. Its entity depends on the predicted component use and can vary tenfold from its initial value to 1 μA. The mean life of Si and Ge detectors falls somewhere between 10^6 and 10^8 hrs, whereas III-V material detectors have a 10^5–10^6 hr mean life.

REFERENCES

1. J.I. Pankove, *Optical Processes in Semiconductors,* Dover Publications, Mineola, N.Y., 1975.
2. A.K. Chin, *J. Crystal Growth,* 70, 1984, pp. 582–596.
3. M. Cocito, C. Papuzza, F. Taiariol, *Proc. of 3rd Conf. on Microscopy of Semiconductor Materials,* Oxford (USA) University Press, New York, 1983; and *CSELT Rap. Tech.*, 11, 1983, pp. 293–297.

4. S. Mahajan et al., *Applied Physics Lett.*, 41, 1982, pp. 266–269.
5. G. Fornuto and C. Papuzza, *Materials Chemistry and Physics,* 16, 1986, pp. 45–52.
6. B. Cocayne, G.T. Brown, and W.R. MacEwan, *J. Crystal Growth,* 51, 1981, pp. 461–469.
7. P. Franzosi, G. Salviati, M. Cocito, F. Taiariol, and C. Ghezzi, *J. Crystal Growth,* 69, 1984, pp. 388–398.
8. K. Nakajima, S. Yamazaki, S. Komiya, and K. Akita, *Jap. Applied Physics,* 52, 1981, pp. 4575–4582.
9. S. Yamazaki, Y. Kishi, K. Nakajima, A. Yamagouchi, and K. Akita, *Jap. Applied Physics,* 53, 1982, pp. 4761–4766.
10. M.S. Abrahams, J. Blanc, and C.J. Buiocchi, *Applied Physics Lett.*, 21, 1972, pp. 185–186.
11. A. Yamagouchi, S. Komiya, P. Ueda, K. Nakajima, I. Umebu, and K. Akita, *Proc. of Int. Symp. GaAs and Related Compounds Japan,* 1981, pp. 161–166.
12. F.R. Nabarro, ed., *Dislocations in Solid,* North Holland Press, Amsterdam, 1979.
13. P.M. Petroff and L.C. Kimerling, *Applied Physics Lett.*, 1976.
14. G.R. Woolhouse, *IEEE J. Quantum Electron.*, QE-11, 1975, pp. 556–561.
15. A. Antolini and A. Piccirillo, *CSELT Rapp. Tecnici,* 10, 1982, pp. 297–305.
16. P.A. Kirby, *IEEE J. Quantum Electron.*, QE-11, 1975, pp. 562–568.
17. C.H. Henry, P.M. Petroff, R.A. Logan, and F.R. Merrit, *J. Applied Physics,* 50, 1979, pp. 3721–3732.
18. K. Mizuishi, *J. Applied Physics,* 55, 1984, pp. 289–295.
19. R. De Franceschi, M. Liberatore, P. Montangero, and A. Piccirillo, *Proc. of 17th ESSDERC,* Bologna, Italy, 1987.
20. A. Antolini, C. Guasco, and A. Piccirillo, *Proc. of the 3rd Coll. Inter. Fiabilitie and Mainten.*, Toulouse, France, 1982.
21. B.W. Hakki, P.E. Fraley, and T.F. Eltringham, *AT&T Techn. J.*, 64, 1985, pp. 771–807.
22. T.P. Pearsall, *Proc. of 4th ECOC,* 1978.
23. S.R. Forrest, *IEEE J. Quantum Electron.*, QE-17, 1981, pp. 217–226.
24. R.H. Hait, A. Goetzberger, R.M. Scarlett, and W. Shockley, *J. Applied Physics,* 34, 1963, pp. 1581–1590.
25. A.K. Chin, F.S. Chen, and F. Ermanis, *J. Appl. Phys.*, 55, 1984, pp. 1596–1606.
26. J.F. Verwey, *Proc. of Conf. on Insulating Films on Semiconductors,* Durham, NC, 1979, pp 62–74.
27. F. Cucchietti, A. Gallesio, M. Liberatore, and A. Piccirillo, *IEEE J. Selected Areas in Comm.*, SAC-4, 1986, pp. 1121–1130.

Chapter 10

POWER DEVICE RELIABILITY

C. Cognetti, G. D'Espinosa, G. Fioravanti, E. Ghio, and L. Lonzi

10.1 INTRODUCTION

Power level in silicon devices has grown remarkably in recent years; the main driving forces are: (1) the growth in circuit complexity and, as a consequence, in integration level (LSI and VLSI); (2) the demand of power integrated devices with higher and higher current values.

The type of package affects power device reliability more significantly than is the case for signal devices. In fact, power device packages not only must protect the silicon chip against environmental pollution but must also sustain high heat flows. Power packages can be divided into two main families: metal cans and plastic packages. Due to their lower cost, plastic packages have almost completely replaced metal cans; for this reason, unless otherwise specified this chapter will deal essentially with plastic packages. From the point of view of power dissipation, plastic packages can be divided into three groups:

1. Low power packages (less than 1 W), which do not require an external heat sink ("dual in line" packages or similar ones); they are used for devices with very low power dissipation (signal devices).
2. Medium power packages (1–2 W), similar to the previous ones but able to substain higher heat flows. They often require an external heat sink, connected to the frame by the leads. They are used in signal devices with associated moderate power levels.
3. High power packages (3–20 W), in which a silicon chip is attached to a metal plate (integrated heat sink or "slug") that can be used by itself as a heat sink or connected to an external heat sink. Because of the integrated heat sink,

they are often referred to as *asymmetric packages.* Types 1 and 2 are also referred to as *symmetric packages.*

For better illustration of the main characteristics of power devices, an exploded view of a typical five-lead package ("pentawatt") is shown in Figure 10.1. Other packages for power ICs are similar to this one, only the number of leads changes up to a maximum of 23. The materials usually employed are

Integrated heat sink (slug): copper.
Bonding wires: aluminum or gold.
Leads: copper (usually tin-lead plated).
Molding compound: usually epoxy.
Silicon chip.
Die attach: metal alloys with a low melting temperature (200–300°C).

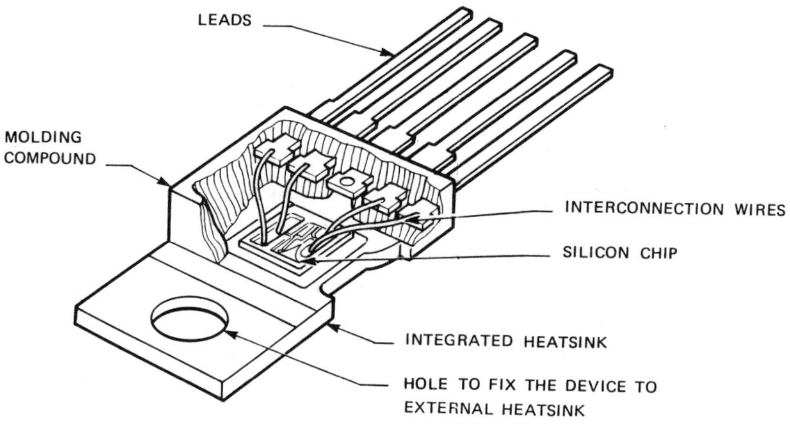

Figure 10.1 Internal view of a power integrated device (pentawatt package) showing its main constituents.

10.1.1 Main Sources of Stress for Power Devices

The main types of stress encountered by a power device during its operating life are

High current.
High voltage.
High temperature.
Mechanical stress (plastic packages).
Humidity.
Electrical overload.

Failure mechanisms related to the combination of the first three parameters will be discussed in this chapter, and the effects of each parameter is discussed in other chapters as well (e.g., the chapter on electromigration).

Plastic packages can be affected by mechanical and thermal stresses. When a power device is secured to an external heat sink and the two surfaces in contact (external and integrated heat sink) are not perfectly flat or if excessive torque strength is applied, severe mechanical stresses are transferred to the silicon chip. Stresses are also generated during thermal cycles due to different thermal expansion coefficents of the resin, the integrated heat sink, and the silicon chip. This may cause chip or resin cracks. Due to high power dissipation, power devices are usually subjected to wide thermal cycles during their operating life. These problems are typical of power devices and will be discussed later in the chapter.

The sensitivity to humidity of power devices [1], can be easily explained, considering that power packages have large surfaces of contact between the encapsulating resin and the integrated heat sink. These interfaces provide an easy path for humidity to reach the silicon chip. This sensitivity is even higher when the power device is stored in an unbiased state. Anyway, the phenomenon is not peculiar to power devices. The long-term effect is corrosion of bonding pads or metal stripes.

As regards electrical overload, we must consider that power devices are usually the devices at the lower impedance of an apparatus, and more exposed to electrical overload. In fact, efficient control of high external power levels requires keeping the internal impedance as low as possible. Any external overload (e.g., low-output impedance) is converted into high current, with potentially catastrophic effects on the device itself. Other remarkable phenomena will be analyzed later, especially the sensitivity of power devices to electrostatic discharge, the "second breakdown" phenomenon, and the presence of parasitic MOSTs in integrated power devices.

10.1.2 The Design of Power Devices

Design of power devices is optimized to control high electrical power levels in a certain time frame. To obtain the characteristics already discussed, the designer must optimize the following critical features: active areas (i.e., p-n junctions that control external power); the metal stripes that must carry high current values without any degradation; and the passivation, which must protect active zones against impurities activated by high temperature and the electrical field.

10.2 MECHANICAL AND THERMOMECHANICAL STRESS IN PLASTIC-PACKAGED POWER ICs

Plastic encapsulated ICs are affected by severe thermomechanical stress, mainly caused by the mismatched thermal expansion coefficients of the different packaging materials (Table 10.1). This results in drifts in the electrical parameters of the circuits (due to the piezoresistance effect of silicon) and metalization and passivation cracks during thermal cycling. These problems are encountered on both symmetrical packages, fully enclosing the silicon chip and die attach area (dual-in-line, micropackage), and asymmetrical packages, where the copper heat spreader to which the chip is soldered extends outside the package (Figure 10.1).

Table 10.1 Thermal Expansion Coefficients of Various Packaging Materials (ppm/°C)

Silicon	2, 3
Copper	16–17
"Kovar"/"Alloy 42" (Fe/Ni alloys)	7–9
Epoxy compound [$T < T_g$]	20–25
Epoxy compound [$T > T_g$]	65–80

Figure 10.2 shows the difference between these two types of packages. Moreover, in asymmetrical power packages, purely mechanical stress may originate when the device is secured to the massive external heat sink. In this phase, high intensity stress may arise if the external heat spreader is not flat; in the worst case, the silicon is damaged permanently. Therefore, experimentation is needed to improve the thermomechanical properties of the molding compound, minimize the effects related to piezoresistance; and improve the package structure to prevent external stress from entering the die pad area. So far, studies have led to the development of integrated strain gauges enabling measurement of stress intensity and CAD simulation, both valid tools for improving the reliability of plastic packaging.

10.2.1 Thermomechanical Stress from Packaging Materials

Table 10.1 shows the thermal expansion coefficients of silicon and more commonly used packaging materials. The definition of T_g (glass transition temperature) together with additional molding compound data are given in Appendix A of this chapter. The widely differing thermal expansion coefficients of the materials in Table 10.1 explains the thermomechanical stress generated on the circuit during the assembly process and for any temperature change during the working life of the device. Stress mainly emanates from the die attach to the metal package frame or the molding compound shrinkage.

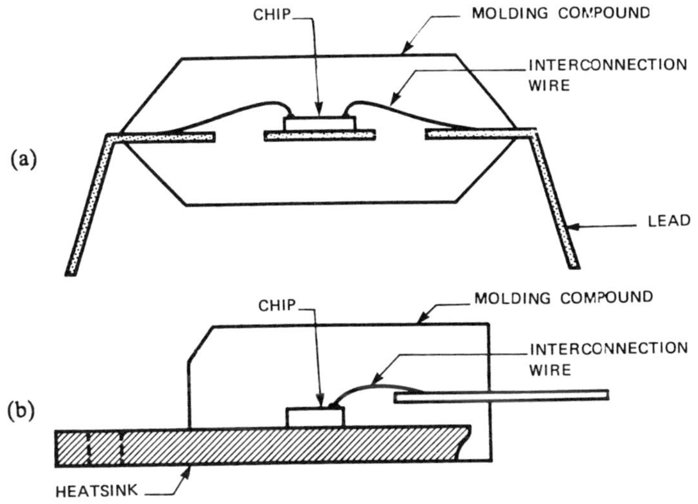

Figure 10.2 Low-power packages : (a) symmetrical; and (b) asymmetrical.

Stress From Die Attach

The die attach process consists in soldering the silicon chip to the package frame. This produces stress mainly due to the die-attach material, the frame material, the die-attach temperature, and the die size. Table 10.2 shows the stress data obtained using silicon strain gauges. On symmetrical signal packages, the frame material is a metal alloy having a low thermal expansion coefficient and low conductivity, and the die attach uses epoxy glue at a curing temperature ranging from 150 to 200°C. The stress associated with this type of die attach is practically negligible on account of the thermal expansion coefficient of the metal and the low die attach temperature involved. On symmetrical and asymmetrical power packages, where effective thermal dissipation depends on stable thermal resistance of the die attach, high-melting alloys on a copper frame are used. The alloys most commonly used are PbSn 95/5, PbAgSn, PbAgIn, and SnAgSb, which melt between 250 and 300°C. The high temperature and widely differing thermal expansion coefficients of copper and silicon originate stress of 250–300 kg/cm^2 on 6.25 mm^2 chips. AuSi alloys with a melting point of 370°C are now rarely employed for plastic IC packages on account of their high cost. When used, however, they were responsible for stress of 400–450 kg/cm^2.

Table 10.2 Stress Associated with Different Die-Attach Materials (Strain gauge area: 6.25 mm^2)

Material	*Stress* (kg/cm^2)
Epoxy glue (glue die-attach) T = 150–200°C	0–100
Tin-lead alloys (soft die-attach) T = 280–300°C	250–300
Gold-silicon alloy (hard die-attach) T = 370°C	400–450

Stress From Molding Compounds

Figure 10.3 shows the stress from molding compounds. At molding temperture, with the compound in liquid form, stress on the chip (excluding that originated by the die attach) is practically nil (zero stress point) but gradually increases as the compound sets and, after curing, presents a final value m proportional to the shrinkage factor (0.15–0.3 percent) (also defined in Appendix A). Stress increases further during annealing at 170–180°C, which completes the curing process. These are all chemical forms of stress. As the compound cools from curing to ambient temperature, thermomechanical stress S reflects the expansion coefficients (Figure 10.4): a sharp increase in $T > T_g$ and a more gradual increase in $T < T_g$. Thermomechanical stress may be expressed by the following equation:

$$S = \alpha_1 E_1 \Delta T_1 + \alpha_2 E_2 \Delta T_2$$

where $\Delta T_1 = T_g - T_{amb}$; $\Delta T_2 = T_{curing} - T_g$ and α_1, E_1, α_2, E_2 are the thermal expansion coefficient and moduli of elasticity, respectively, for $T < T_g$ and $T > T_g$.

Clearly, therefore, the realization of molding compounds putting minimum stress on the device (low-stress compounds) involves working on the shrinkage factor (0.4 percent is the minimum compatible with the workability of the compound) and particularly on the thermal expansion coefficient, modulus of elasticity, and glass transition temperature.

Figure 10.5 shows characteristic stress maps obtained using a 6.25 mm^2 test device consisting of a matrix of piezoresistive elements (see Appendix). The maps show the x and y components of stress S on the chip surface. Table 10.3 shows the maximum stress values, at ambient temperature, of different types of compounds, leading us to the following observations:

1. Stress, including that of low-stress compounds, is compressive, ranging from 1000 to 1500 kg/cm^2. This is extremely high, and it originates by shearing stress (not shown) parallel to the die surface.
2. Low-stress properties, as in compound 2 in Table 10.3 (formulated 5–6 years ago) were achieved by partially curing the compound. Heat treatment at 150°C for 20 hr (well below the working life of the device at that temperature) was sufficient to increase compressive stress by 10–15 percent.
3. Improvements aimed at reducing curing time presently cause deterioration in the thermomechanical properties of the compound: compound 4 in Table 10.3 presents stress levels characteristic of compound 1 of 8–10 years of age.
4. Low-stress properties are often accompanied by impaired reliability during damp tests. This may be caused by impaired grip between the compound and frame or contamination of the additives used to improve the thermomechanical properties of the compound. Figure 10.6 shows pertinent data taken from reference material.
5. As of 1985, manufacturers found that combined epoxy-silicon compounds reduced stress drastically. In some cases, compressive stress had been reduced 50 percent, and the linear thermal expansion coefficient was reduced to 18–19 ppm/°C. If productivity and reliability tests prove successful, this will have marked a major step forward in controlling the effect of thermomechanical stress on plastic packages.

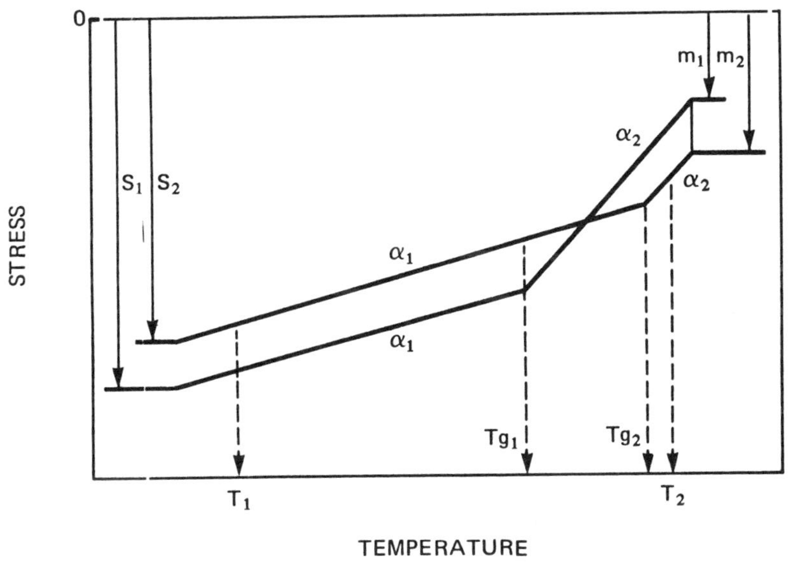

Figure 10.3 Internal stress in plastic packages for two molding compounds.

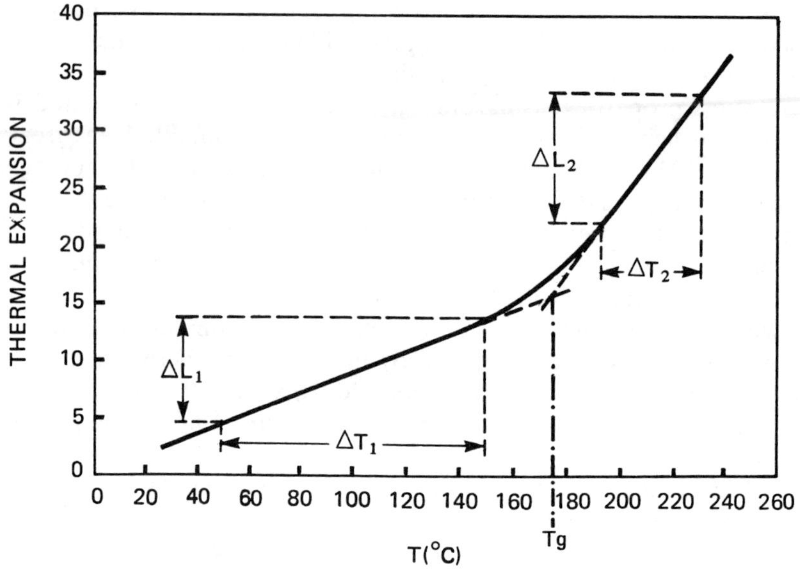

Figure 10.4 Thermal expansion of epoxy molding compound.

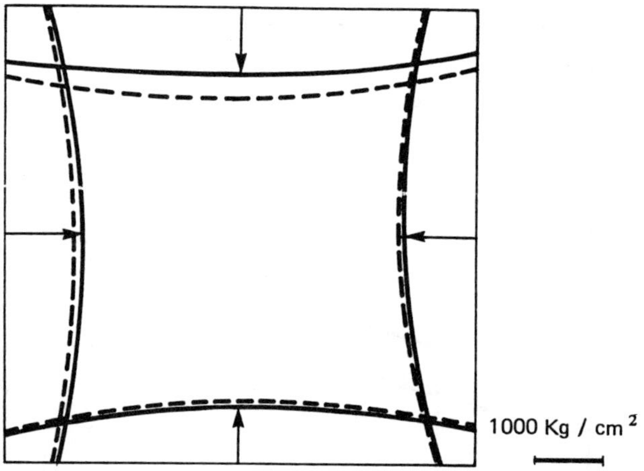

Figure 10.5 Typical stress maps for symmetrical and asymmetrical packages: unbroken line is dual in line; dashed line is double T0220 (multiwatt package).

Table 10.3 Maximum Stress of Different Molding Compounds, Measured by Integrated Strain Gauges

	ppm/°C	*Modulus of elasticity* (kg/cm²)	*Shrinkage* (%)	*Stress* kg/cm²	*After heat treating* 20 hr at 150°C (kg/cm²)
Compound 1	25	2500	0.6	1500	1500
Compound 2 (low stress)	22	1600	0.5	1250	1450
Compound 3 (low stress)	20	1400	0.45	1150	1200
Compound 4 (fast curing)	25	1400	0.5	1400	1500

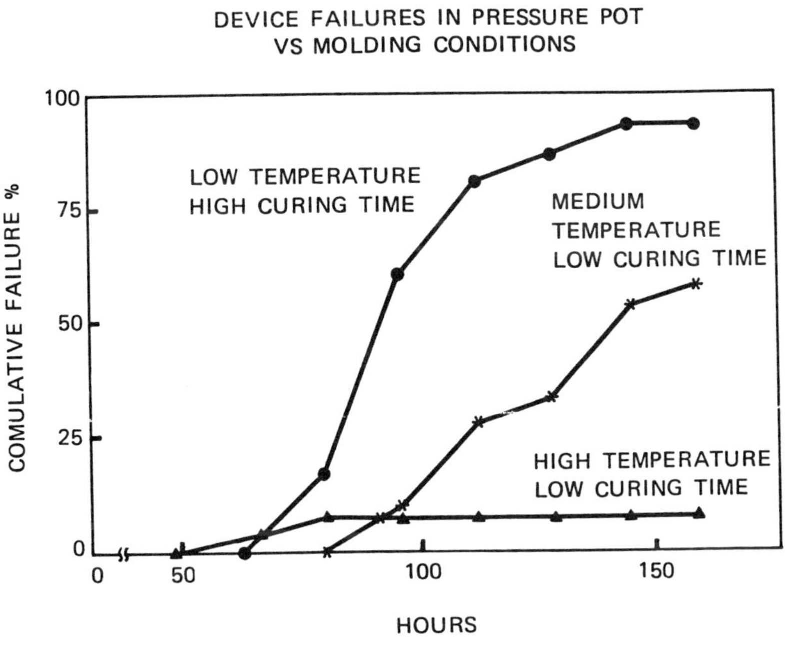

Figure 10.6 Correlation between stress intensity and failure rate in moisture resistance test (pressure cooker). Stress is lower for lower curing temperature.

10.2.2 Effects on Electrical Circuits

Silicon is characterized by high piezoresistance coefficients: carrier mobility, in fact, is seriously affected by external stress on the lattice, which therefore affects the absolute value of both p and n silicon resistors subjected to compressive or tensile stress. Variation $\Delta R/R$ equals piezoresistance coefficient matrix P_{ij} times stress tensor S_{ij}, and is a function of piezoresistance coefficients times stress components. In terms of quality, we may therefore assume:

$$\Delta R/R = F\{P_{ij} + S_{ij}\}$$

The effects of stress therefore may be minimized by reducing the stress or reducing the piezoresistance coefficient of silicon; that is, by using resistors oriented along directions less susceptible to stress. As available compounds are all subject to more or less the same compressive stress, invariably in excess of 1000 kg/cm^2, any response to circuit resistance control can only be achieved by means of the second solution. The silicon employed for the substrate of bipolar signal and power ICs is almost invariably obtained by cutting single crystals in two planes (111) and (100). Both offer advantages and disadvantages in terms of epitaxial lattice defects, surface state density, dopant diffusion speed, and oxide growth. A major advantage of silicon (100) is undoubtedly the relationship between the piezoresistance coefficient and doping agents p (e.g., boron) as shown in Figure 10.7. As shown in the figure, the stress sensibility of boron-doped resistors in plane (100) is high along axis <110> (coefficient 60) but falls to one-tenth (coefficient 6) along axis <100>. The same does not apply, however, to plane (111) because the piezoresistance coefficients are independent of the crystal axis and equal approximately 40. A high degree of stability can therefore be expected, subsequent to plastic packaging from boron-diffused resistors in plane (100) and along axis <100>.

The piezoresistance coefficient also depends on the temperature and, inversely, on dope concentration. Table 10.4 shows the drift of diffused bipolar IC resistors subsequent to die attach and molding. Important points to note are the dependence of the drift on orientation and dopant concentration, and especially, that the maximum variation in resistance is limited to a few percent.

IC design rarely calls for high-precision resistors in terms of absolute value, because the resistance dispersion of the diffused layers is 20–25 percent between wafers and 5 percent on the wafer itself (on resistors doped via ion implantation, these values drop to 5 percent and 1 percent, respectively). High-precision resistive voltage dividers are frequently demanded, however; and their performance may be optimized by exploiting appropriate design parameters (width, spacing, contact area, and particularly orientation). Figure 10.8 shows test layout consisting of four equal rating resistors, which may be measured singly, by means of Kelvin contacts,

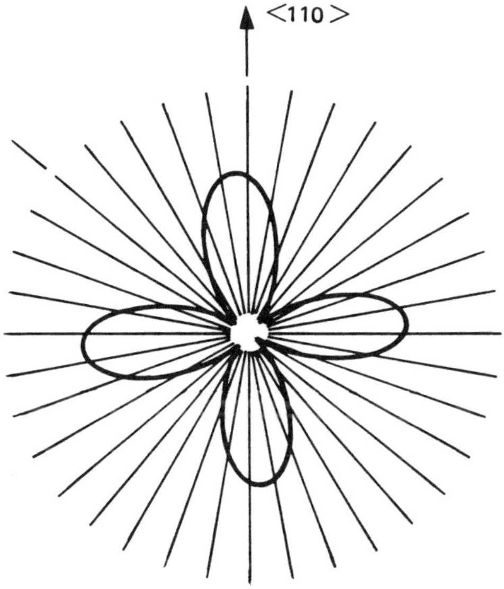

Figure 10.7 Piezoresistance coefficient in (100) p-silicon.

Table 10.4 Shift of Integrated Resistors, Diffused in the (100) Plane, After Various Packaging Processes

	Die-attach PbSn	*Die-attach AuSi*	*Molding and curing*
P 110	−3%	−4%	−2%
P 100	−0.5%	−0.5%	−0.3%
N^+	+1.1%	+2%	+1.5%

or arranged in the form of a divider. In the latter case, standard offset distribution is a good indication of the precision standard achieved by technology.

Figure 10.9 shows the test results of diffused resistors on plane (100); yielding a standard deviation on the wafer of approximately 0.025 percent. By comparison with dividers formed via ion implantation, which provides for greater uniformity, 0.05 percent may be attributed to photolithographic "spread," and 0.01 percent to varying sheet resistance on the wafer. This is an excellent starting point from which to work.

Subsequent to plastic packaging and thermal cycling, which affect the package structure (particularly grip between the compound and frame), sharp deviation in standard offset is observed only in the case of dividers on axis <110>. For those

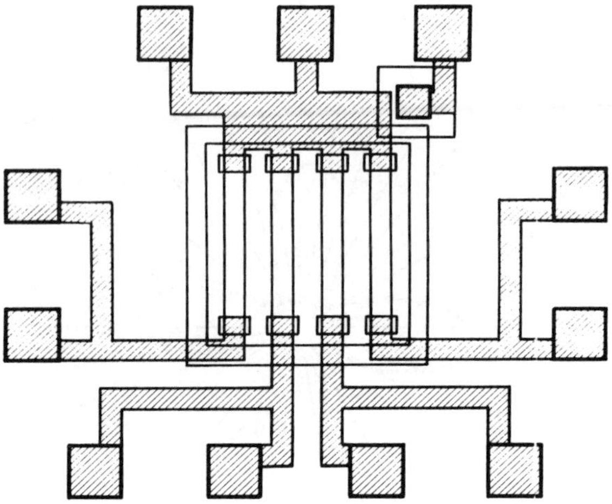

Figure 10.8 Test structure to evaluate the accuracy of integrated voltage dividers.

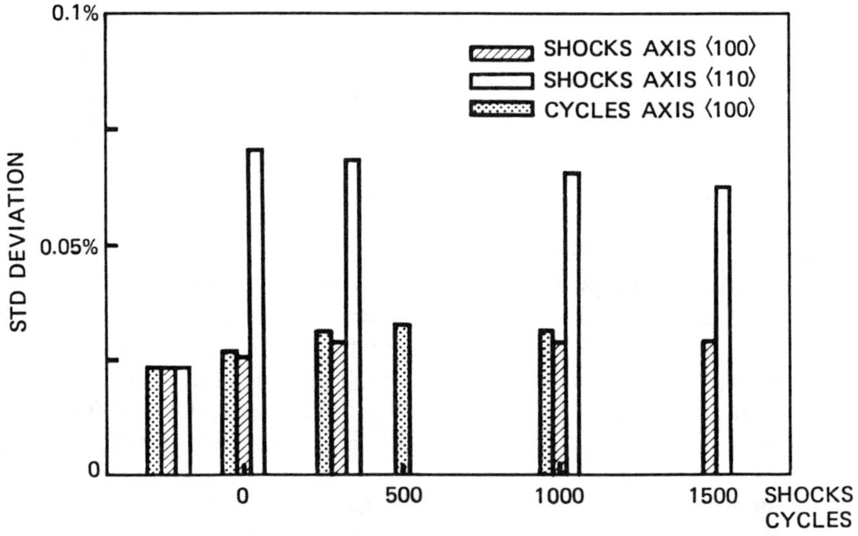

Figure 10.9 Offset standard deviation of voltage dividers diffused in the (100) plane, measured on wafer, after package molding and thermal cycles.

with <100> orientation, the variation is much lower, as per Figure 10.7. These provide for 0.1 percent precision, which is extremely useful, for example, to avoid trimming in the case of 9–10-bit A/D-D/A converters.

10.2.3 Effect of Thermomechanical Stress on Reliability

The effect of stress generated by molding compounds is best revealed during thermal cycling (thermal shocks and cycles) and thermal fatigue testing (see the accelerated tests chapter). During accelerated testing, the package structure is stressed thermomechanically, due to differing expansion and contraction of the various package materials. The chip surface and die attach are subjected to shearing stress, which results in damage to the aluminum tracks and passivation cracks; in the worst case, the silicon itself is also damaged. Figure 10.10 shows the shearing stress originated on the chip surface during thermal cycling.

Stress intensity depends on the thermal expansion coefficient of the compound; the distance from the barycenter of the plastic block (silicon die size); and the temperature range (particularly downward). The effect of stress also depends on the type, grain, and size of the inorganic compound filler employed.

As shown in Figure 10.10a, at low temperatures, the molding compound contracts toward the barycenter, thus exerting friction on the raised parts (tracks and passivation) of the die surface, to which the compound grips in much the same way as a plaster cast. The outward-facing portions of the tracks are thus crushed and displaced by an amount *d*, which increases alongside the number of cycles and the distance between the track and the barycenter of the block. This is an important factor in the case of large devices (Figure 10.10b).

Displacement may be limited by molding compounds with a low thermal expansion coefficient, as well as organic protective coatings between the chip and compound. When selecting the coating type and process, it is important to bear in mind its effect on the contact wires. During thermal cycling, the wires are also subjected to stress, which increases alongside the chip area, and they fracture after a number of cycles, depending on the package and wire location. Figure 10.11 shows the gradual distortion of a peripheral track during thermal cycling; and Figure 10.12, the typical effect of passivation cracks caused by the molding compound.

10.2.4 Mechanical Stress from Power Package Structure and Application

As already stated, high-power packages are asymmetrical, the die is soldered onto a 1.2–1.5 mm thick, integrated copper heat spreader, extending outside the package for connection to an external heat spreader. Contact is often made by torquing to 6–10 kg $\times$ cm. Rarely is the external heat spreader perfectly flat, due to both

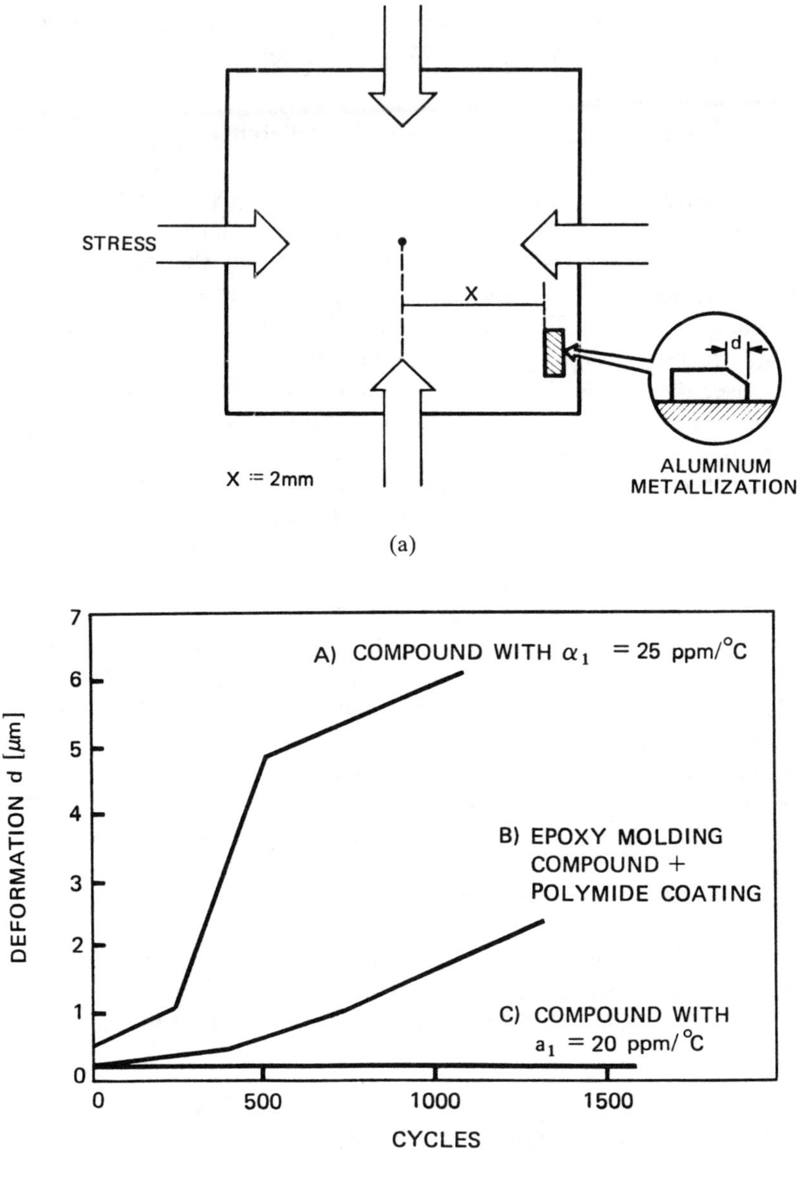

Figure 10.10 Shift of aluminum metalization during thermal cycles.

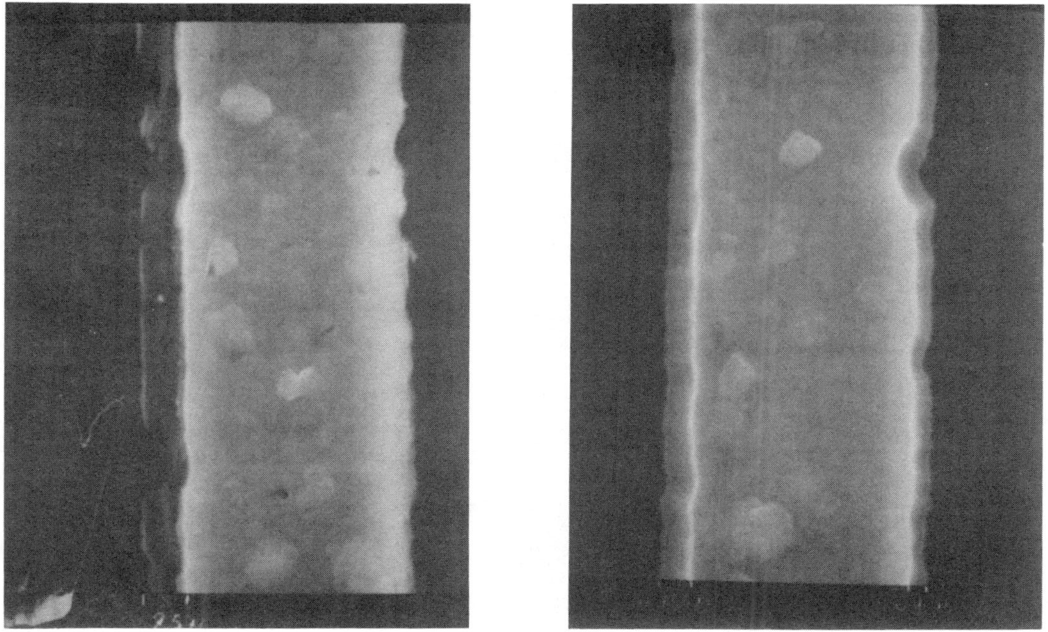

(a) (b)

Figure 10.11 Shift of aluminum metalization: (a) after 200 thermal shocks; and (b) after 500 thermal shocks.

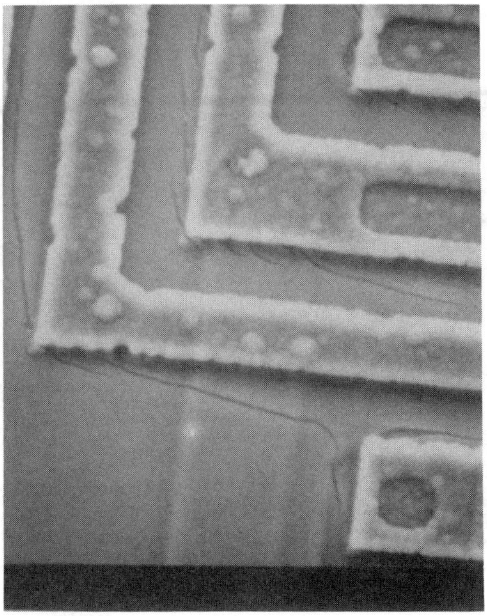

Figure 10.12 Cracks in passivation, due to thermomechanical stress of encapsulation, after 1300 thermal shocks.

current manufacturing practice and additional grinding costs. During torquing, tensile stress may be transmitted to the die, as the integrated heat spreader adapts to the imperfectly flat external surface. Stress intensity varies according to the dip on the heat spreader and the applied torque, and it may even exceed the ultimate tensile stress of the silicon (2500 kg/cm^2), thus fracturing the single crystal and possibly the active circuit portion. In the first case, damage is catastrophic, whereas the second usually results in damage parallel with the die surface, which increases thermal resistance and may be detected by timing intervention of the internal thermal protection circuit (which usually falls from 25 to 15--10 s). The average reject percentage due to this type of failure ranges from 1000 to 5000 ppm, depending on the quality of the external heat spreader.

One way to control it is by adjusting the design of the package structure by adopting materials with more suitable mechanical properties (e.g., low stress epoxy compounds), or by adjusting the package structure to prevent transmission of stress between the screw region and the die pad. By simulating operating conditions with a distorted external heat spreader, the mechanical properties of power packages are studied. This, known as a *torque test* is illustrated in Figure 11.25 of the next chapter. The test apparatus consists of a steel block with a groove ranging in depth from 20 to 100 μm. The test pieces are screwed to a constant 15 kg × cm torque on blocks with increasingly deep grooves. They are operated at an electrical performance determined to obtain failure statistics indicative of the mechanical properties of the package. The stress originated during the torque test, using the usual integrated strain gauges, is also measured. To sum up, therefore, the mechanical properties of power packages may be determined by

1. Mechanical torque testing.
2. Statistically determining the performance of integrated power circuits subjected to increasingly severe test conditions; performance data should also include thermal protection circuit intervention time indicating cleavage parallel with the die surface.
3. Measuring stress on the die by means of strain gauges.

10.3 ESD PROTECTION IN BIPOLAR ICs

The problem of IC protection against *electrostatic discharge* (ESD) was formerly limited to *metal oxide semiconductor* (MOS) circuits, due to the intrinsic weakness of gate oxide. The recent trend toward greater integration and compactness, however, has resulted in bipolar ICs also affected: small area transistors may be damaged by ESD on the base-emitter junctions.

The problem of ESD, whose seriousness from manufacturing standpoint can be expressed in terms of roughly 2 percent less efficiency for more critical processes, presents two widely differing aspects: one relating to the design of a straightforward integrated structure for increasing the intrinsic strength of the circuit; and the other to ESD handling precautions during manufacture, assembly, shipping, and actual use.

In terms of ESD resistance, all types of circuits present a *damage voltage* (V_{dam}) beyond which discharge results in failure due to voltage (MOS) or current (bipolar ICs). An idea of the V_{dam} values involved is given in the next chapter; this section deals with a few straightforward protection structures for bipolar ICs. Clearly, all-round ESD protection can be achieved only by bearing in mind both aspects of the problem: during handling, for example, discharge may be as high as 30 kV, at which level circuit protection alone is not enough. In this respect, antistatic materials (shipping tubes, bracelets, and floor and workbench materials for discharge grounding) are indispensable for preventing the formation of high ESD levels. Antistatic precautions alone, on the other hand, are insufficient to ensure low-voltage protection. The only solution to the problem, therefore, is in combining both protection structures and handling precautions.

10.3.1 Standard Models and Targets

For solving the problem from the circuit standpoint, models are employed for simulating specific types of discharge. Three general models have been suggested:

1. Human body model (HBM).
2. Charged device model (CDM).
3. Field induced model (FIM).

Simple movements on the part of an individual may produce a substantial charge. When a charged individual touches a device, as during manual assembly, part of the energy stored in the body may be transferred to the device in the form of discharge. This situation is simulated by the HBM model. The CDM model simulates the charge on the frame or package generated, for example, by the device sliding into shipping tubes. The FIM model simulates an uncharged device in proximity to an electric field. The three different situations are simulated by means of RC or RLC circuits, an example of which is shown in Figure 10.13.

The ESD model used most frequently is an HBM RC circuit having resistance and capacitance values ranging from 1000 to 2000 Ω and 100 to 250 pF (Figure 10.14). Demand is increasing among IC users for protected devices, most of which are covered by MIL STD 883C, which defines an RC circuit simulating discharge between two pins on the device, the others being disconnected. The test apparatus provides for adjusting pulse fall time by means of interchangeable cells for different R_2 and C_1 values. The characteristic of each device determines which of two discharge sensitivity classes it belongs to and the pins in need of protection. Devices with base inputs or small area transistor emitters are generally more sensitive to discharge, and in some processes present damage voltages of around 400 V. Circuit protection must therefore be provided for increasing damage voltage over and above standard targets.

10.3.2 Effect of ESD on Transistors

The principal ESD failure mechanism in bipolar ICs is localized overheating in the depletion layer of a PN junction, which melts the silicon. Testing PNP and NPN transitors with different connections and areas (22.6 mils^2 for PNP; 15 and 51.14 mils^2 for NPN) showed greater sensitivity of the junction during low-voltage processes. In this case, damage voltage is 400 V using a 1500 Ω, 100 pF RC cell (Table 10.5).

When less than total, failure produces an increase in base current I_B and therefore a fall in gain on the component. Degradation of electrical characteristics is shown in Figure 10.15, which shows the inverse characteristic of a minimum area NPN transistor before discharge and after discharge of 500 V and 2500 V. Degradation of the junction is shown clearly in Figure 10.16.

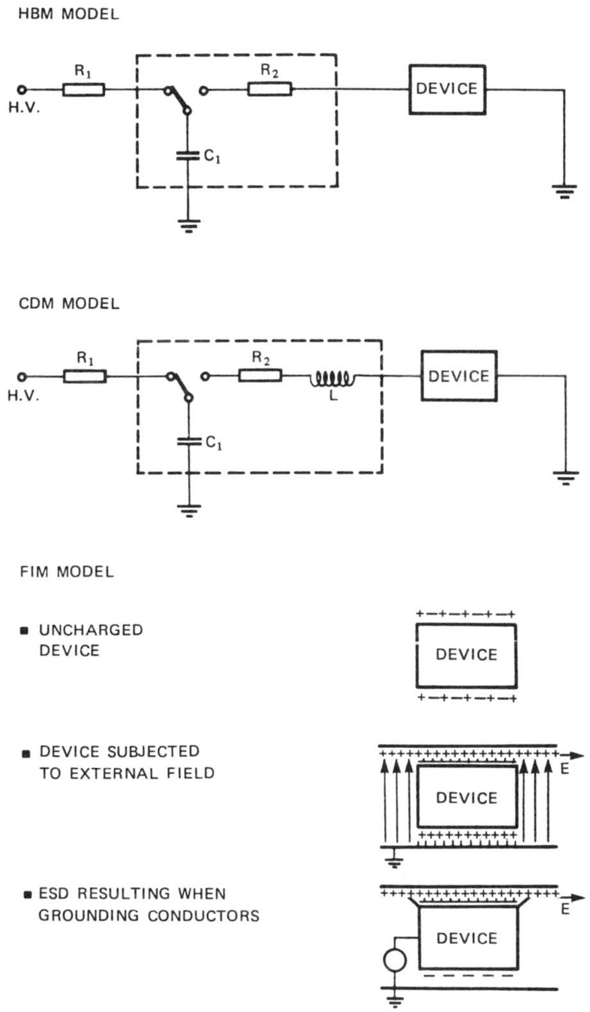

Figure 10.13 Models for the simulation of electrostatic discharge.

10.3.3 Basic Requirements of Protection Structures

Protection structures short ESD pulse voltage to ground or to the supply voltage to limit the current entering the critical junction. Effective protection poses many, often conflicting, requirements, and the normal practice is to strike the best possible compromise among them. For example, small areas (think of the number of protection structures required at one time on a device) conflict with protection strength,

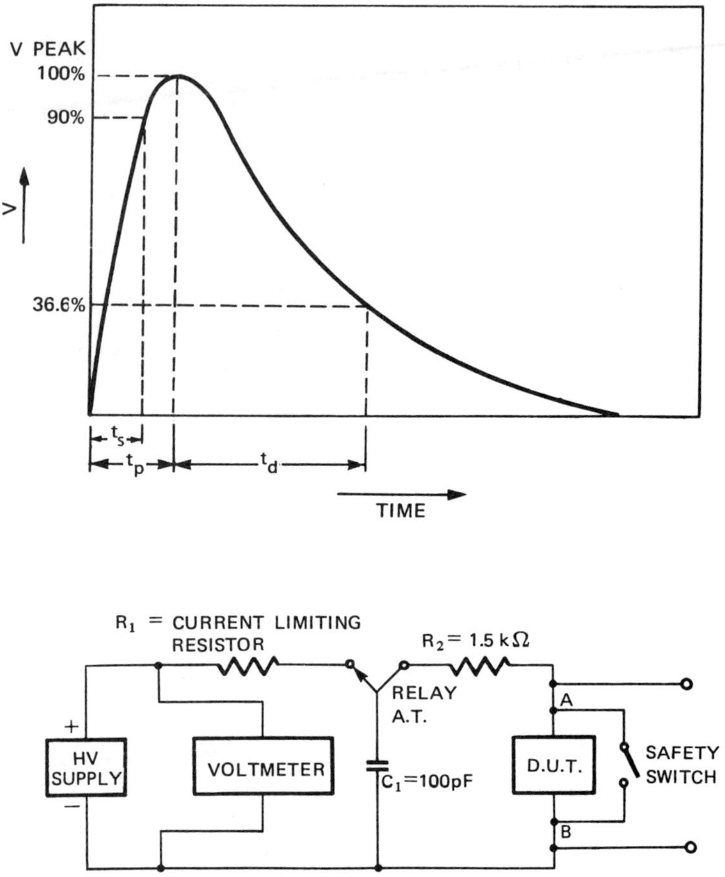

Figure 10.14 Pulse and test circuit for the evaluation of ESD protection: t_s = rise time; t_p = peak time; and t_d = decay time.

which is directly proportional to area: small areas generally imply particularly ESD-sensitive junctions. Intervention speed versus structure speed is another important factor: the low parasitic capacitance and ground resistance of small areas are essential for achieving high-speed intervention. Still another important point to bear in mind is the "transparency" of the protection, which, under normal operating conditions, must in no way affect circuit performance. Furthermore, any solution must be valid regardless of the bipolar process involved.

Table 10.5 Damage Voltage for Bipolar Transistors

		DAMAGE VOLTAGE		
		NPN	*NPN*	*PNP*
(+) *Terminal*	(−) *Terminal*	*Area* = 15 mils^2	*Area* = 51.14 mils^2	*Area* = 22.6 mils^2
B	C	5000 V	25000 V	1250 V
C	B	1500 V	2250 V	4750 V
B	B	2250 V	25000 V	400 V
E	B	400 V	2000 V	2500 V
C	E	3750 V	25000 V	750 V
E	C	2250 V	25000 V	750 V
B-E	C	4000 V	25000 V	1230 V
C	B-E	1750 V	4000 V	25000 V
E	B-C	400 V	2750 V	2000 V

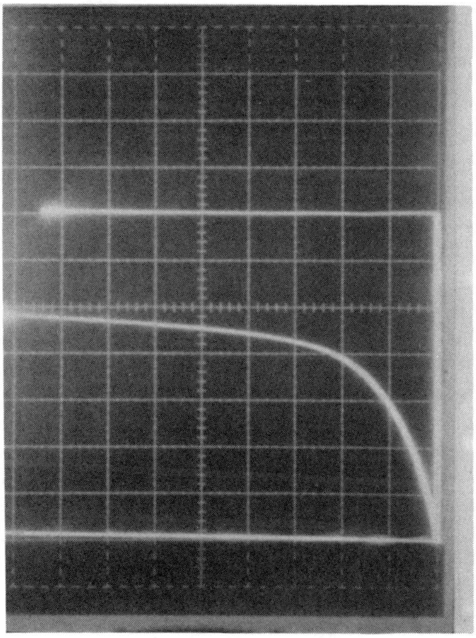

Figure 10.15 Base emitter reverse I-V characteristics, before and after ESD test.

10.3.4 Results

On the basis of these requirements, various circuit configurations were analyzed and compared using a particularly ESD-sensitive structure with V_{dam} = 400 V for negative spikes and a minimum area transistor base input. The structure, a typical application of which is sync separation in the TV field, was used to compare various configurations commencing from the same criticality level. As shown in Figure 10.17, the structures analyzed were parallel types, such as simple diodes with different base-emitter or base-collector connections, Schottky diodes, BV_{cer} transistors, and phantom emitter transistor and resistor. Figure 10.17 shows the area and damage voltage of the structures as measured by a curve tracer, for determining no change in the electrical characteristics of the device (in particular, no shorting or softening of the direct and inverse characteristic) and by bench testing the device after discharge. Both tests are indispensable in that failure of one does not necessarily imply failure of the other. V_{dam} therefore is established as the voltage at which at least one test fails.

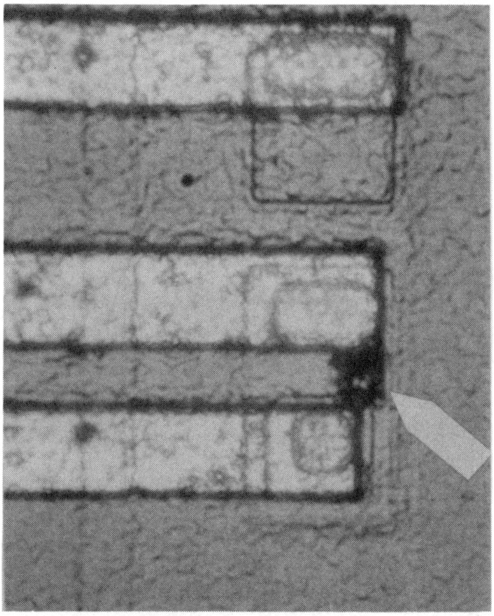

Figure 10.16 Base emitter failure after ESD test.

The most interesting find is undoubtedly that the V_{dam} of the protected device ranges from 3000 V to 9000 V, depending on protection, as compared with 400 V on the unprotected device. Also, a sufficient number of solutions are available to enable appropriate selection for any type of device.

ESD PROTECTION STRUCTURE	DAMAGE VOLTAGE	SIZE
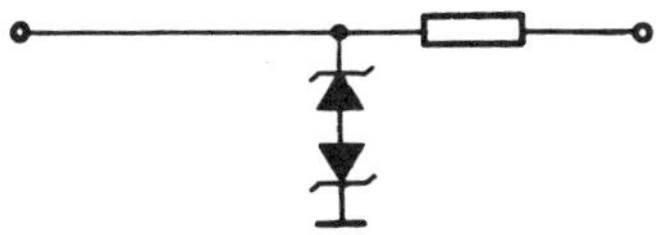	3000V	80 $mils^2$
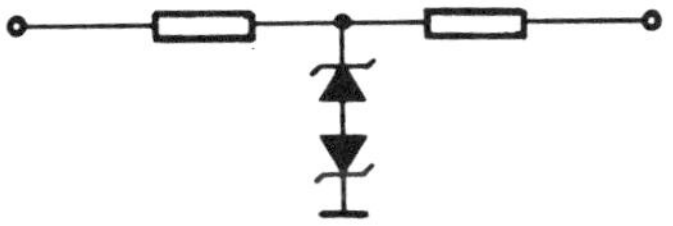	3000V	107 $mils^2$
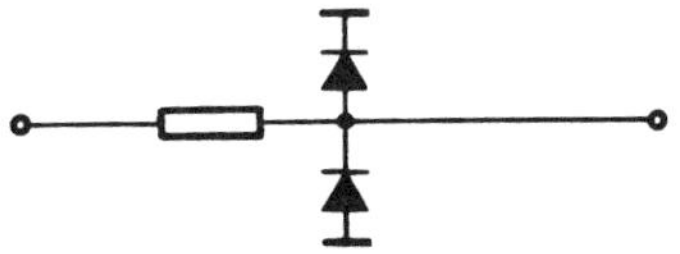	3000V	57 $mils^2$
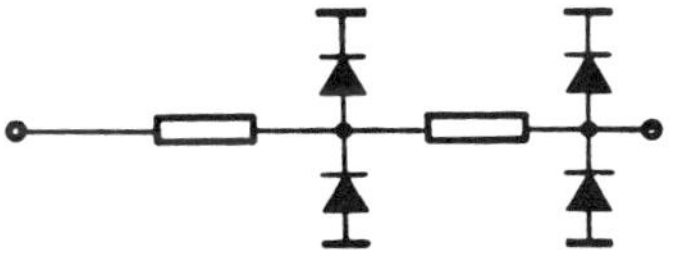	5000V	100 $mils^2$
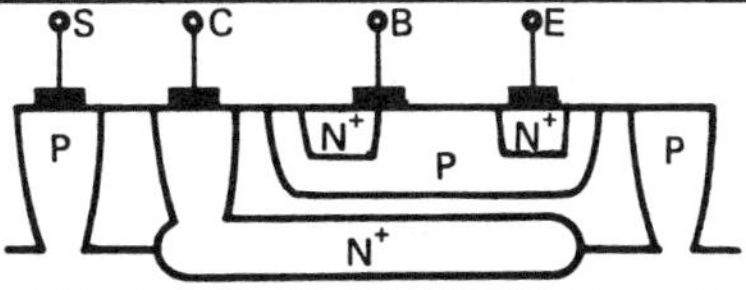	9000V	19.5 $mils^2$
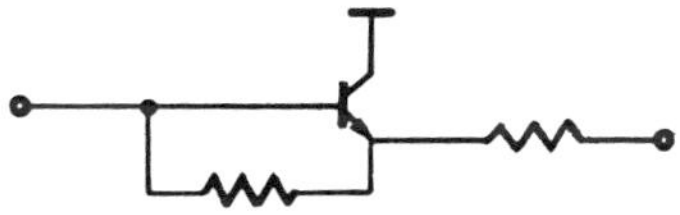	3000V	60 $mils^2$
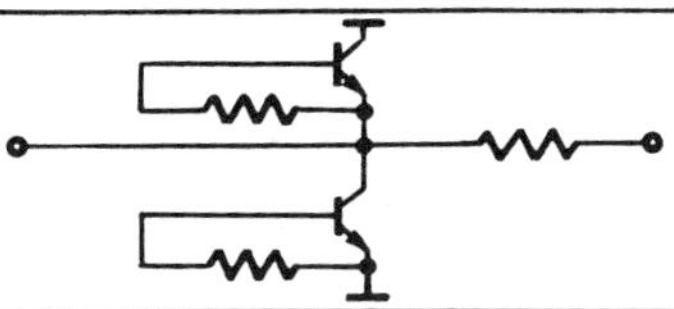	5000V	65 $mils^2$

Figure 10.17 Damage voltage for different protection structures.

10.4 SECOND BREAKDOWN

Two different failure modes, typical of power transistors, are both called *second breakdown*. The first one is called *forward second breakdown* because the base current flows in the normal direction (for instance, enters into the base of an NPN transistor): this breakdown can happen in the normal amplifier operation. The second one is called *reverse second breakdown* because the base current flows in the opposite direction (for instance, exit from the base of an NPN transistor): this breakdown can happen in switched operation on an inductive load.

10.4.1 Forward Second Breakdown (I S/B)

Figure 10.18 shows the operational limits on a power transistor caused by the duration of applied pulses. The area under each line, called the *safe operating area* (SOA), is limited by the following parameters:

Maximum collector current.
Maximum applied bias between collector and emitter.
Maximum power dissipation (segments at 45°).
Forward secondary breakdown (segments with slope greater than 45°).

The forward secondary breakdown causes a so-called thermal runaway in a point of the transistors: current density and temperature increased jointly due to a positive feedback effect until local melting. The most critical areas are those where, even under normal operating conditions, current density and temperature are higher than average. This is due to the following:

- The forward voltage drops in the base, due to base current, making the base emitter junction voltage different from point to point all along the junction; VBE is greater, particularly in the emitter peripheral zones, therefore the current gets higher values (current crowding).
- The voltage drops in the metalizations are not negligible and cause similar effects.
- Eventual voids in the joint die-frame create areas with higher thermal resistence.
- The central area of the die is warmer than peripheral ones because it is in less favored conditions for the dissipation of created heat.

These affects cannot be eliminated, despite design objectives for current and temperature uniformity; for example, interdigitated geometry.

A power transistor may be considered an aggregate of many elementary transistors that operate under different conditions. When a power pulse, $P = V_{CE} \cdot I_C$ is applied to the transistor, the following phenomena occur in each elementary transistor:

1. The junction temperature increases. Since the electrical conductivity of the silicon increases jointly with the temperature, the temperature increase causes a current (and therefore power) increase:

$$dP_1 = V_{CE} \cdot dI_C$$

2. The temperature increase causes an increase of power dissipation:

$$dP_2 = dT/R_{th}$$

If $dP_1 < dP_2$, a thermal equilibrium is obtained. If, on the contrary, $dP_1 > dP_2$, the temperature and the current increases enhance each other and thermal runaway occurs, causing transistor destruction. The ratio

$$S = \frac{dP_1}{dP_2} = R_{th} \cdot V_{CE} \cdot \frac{dI_c}{dT}$$

is the thermal stability index. R_{th} is the thermal resistance of the device (see Appendix B of this chapter). For $S < 1$, the operation is steady. For $S > 1$, the thermal runaway occurs.

The following actions are used to avoid thermal runaway:

1. Introduction of regulation integrated resistors ("ballast") in series with each elementary transistor emitter. These resistors are part of a properly designed emitter mask, and so they do not cost extra. Thus, the elementary transistors, whose trend is higher current density, are submitted to a negative feedback, since the large voltage drop in the ballast resistor decreases V_{BE} and therefore the current.
2. To optimize the die attach by elimination of voids that locally increase the thermal resistance.
3. Reduced emitter doping producing a more resistive emitter but with decreased gain.

10.4.2 Reverse Second Breakdown (E S/B)

Figure 10.19 illustrates schematically a switching type operation of an NPN transistor on inductive load. Switching the base circuit to the off position, the stored energy in the inductance $L = (1/2\ LI^2)$ begins to temporarily dissipate in the transistor. The collector-emitter voltage gets the sustaining value V_{CEx}.

The current of holes generated through multiplication in the base-collector depletion regions is divided. The first part leaves the base and, due to voltage

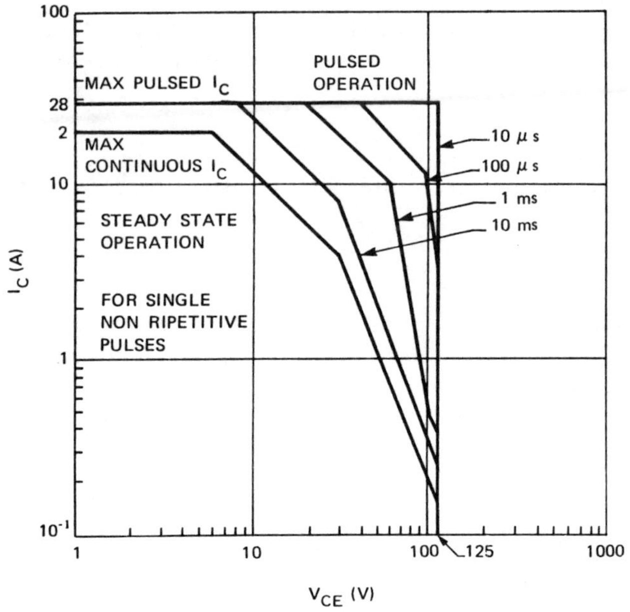

Figure 10.18 Limits on power operation caused by applied pulses duration.

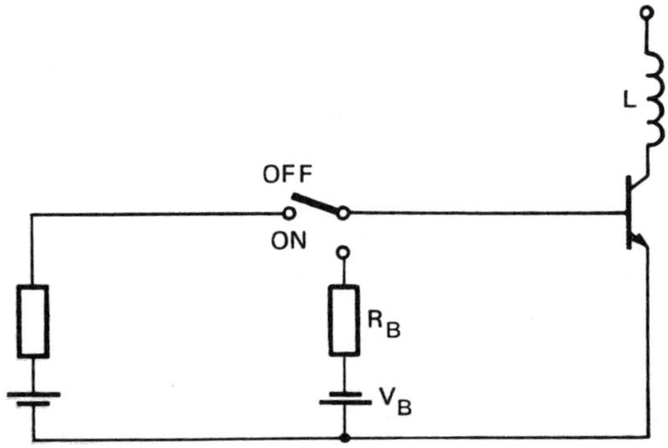

Figure 10.19 Circuit of switching operation of a power transistor on inductive load.

drops in the base and in the resistor R_B, holds the base central region more positive than the relevant emitter region: in this central region, emitter current crowds, and the region is in inverse proportion to base outgoing current I_B. The second part, current generated through multiplication in the base-collector depletion regions recombine in the base, holding the transistor in gain operation. The extremely high current density in the central region changes the trend of electrical field in the base-collector junction.

The derivative of the electrical field versus the distance from the base is given by the Poisson formula:

$$\frac{d_{Ec}}{d_x} = \frac{9}{\epsilon}\left(N_d - \frac{J_c}{qv}\right)$$

where N_d is the dopant concentration in the collector, J_c is current density; q is electron charge; and v is electrons speed through the depletion region.

When J_c increases, the field slope decreases until zero (the area between field plot and abscissa axis does not change because V_{CB} is constant, (see Figure 10.20). With a further increase in J_c, the field slope becomes positive and increases more and more until the field E_c in the central region reaches the critical value of avalanche multiplication, causing device destruction.

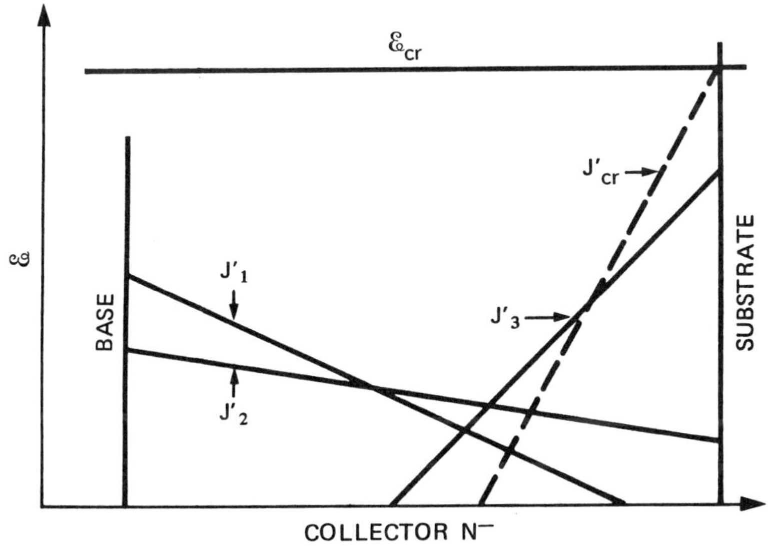

Figure 10.20 Trend of electrical field in the base-collector junction under reverse second breakdown conditions.

The energy that can be tolerated by the transistor decreases with increasing I_B because, as mentioned, the central region where emitter current crowds is in inverse proportion to I_B (smaller region involves higher current density). Therefore, such energy decreases by increasing V_B and decreasing R_B. Since the reverse second breakdown is an internal phenomenon the transistor package remains cold.

In discrete power transistors, one way to increase the resistance to reverse second breakdown is the introduction of one or more collector layers with a dopant concentration that is intermediate between the region N^- and the substrate N^+, but this solution cannot be applied to the integrated power transistor, since it would involve a strong reduction of epi layer-insulation breakdown. A widely used method is a Zener diode connected in parallel between transistor collector and emitter: this diode absorbs most of the energy stored in the inductor, so protecting the transistor.

10.5 PARASITIC MOS

Bipolar integrated circuits may present critical structures very sensitive to ionic contamination in which heavy degradations can be observed. These degradations are enhanced by applied voltage and high temperature and are mainly observed in plastic encapsulated circuits. Ionic contaminants could be contained in the molding compound itself, or they could be released during the various phases of assembly process, for instance, by handling or inadequate rinsing. Typical structures very sensitive to ionic contamination are lateral PNP transistors (Figure 10.21). The effects of foreign ions on such structures will be discussed, as well as the methods used to minimize device degradation.

10.5.1 Structures and Electrical Characterization of Lateral PNP Transistors

From failure analysis conducted on plastic encapsulated power integrated circuits that formerly were subjected to a high-temperature bias life test, V_{CEO}-I_{CEO} characteristics of some lateral PNP transistors were heavily degraded (Figure 10.22a). It was also noted that 6–8 hours at 150°C produced a complete recovery of such characteristics (Figure 10.22b). Various configurations of lateral PNP transistors were tested to explain the failure mechanism. Devices were passivated with silicon glass (PSG) and assembled both in plastic packages and in metal cans.

Among the various structures were those having a base channel covered by a metal layer electrically connected with the emitter region (Figure 10.21b) and those with uncovered base channel (Figure 10.21a).

The bias life tests conducted at high temperature on devices assembled in metal cans never degraded, whereas tests on plastic encapsulated devices led to significant improvement in the reliability of bipolar power devices.

Plots of I_{CEO} as function of stress time, for the two structures, is reported in Figure 10.23. By looking at this figure, we can extract the following points:

1. The increase in I_{CEO} is very significant in structures with an uncovered base channel, whereas structures with a covered channel do not show any further appreciable degradation after a slight initial increase.

2. For structures with a covered channel, the initial increase in I_{CEO} (by a factor 2 every 10°C) is expected from theoretical considerations.
3. For structures with uncovered channel, we observe three phases: an early increase in I_{CEO}, a partial recovery, and a subsequent further degradation.

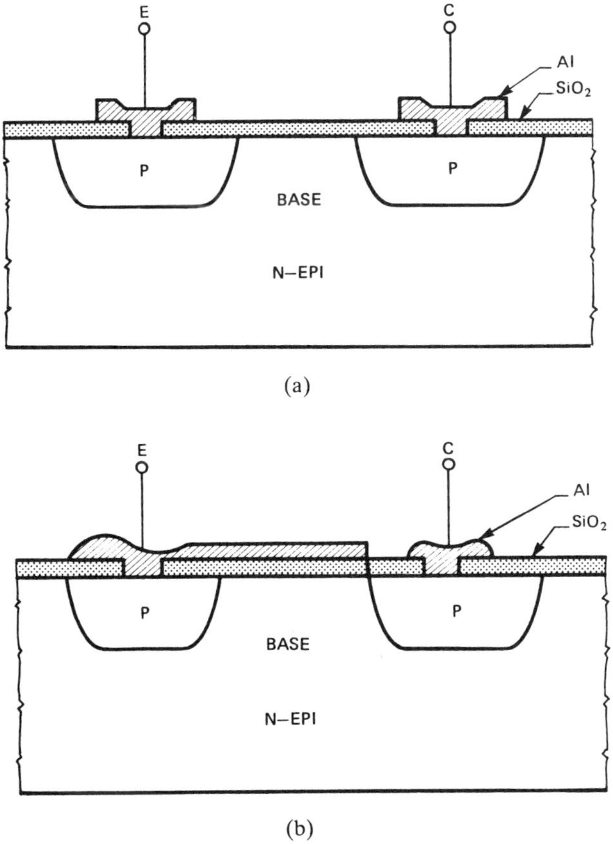

Figure 10.21 Cross section of lateral PNP transistors: (a) uncovered channel; and (b) covered channel.

The degradation observed in structures with uncovered base channel come from surface ionic contamination. In fact, epoxy compounds normally used for device encapsulation and die bonding, actually contain ions like Na^+, K^+, Cl^- as contaminants. Resins are in direct contact with the silicon chip. In the presence of high temperatures and applied voltage, the highly mobile ions tend to accumulate at the resin-oxide interface, thus promoting inversion effects at the silicon surface. By looking at the lateral PNP configuration in Figure 10.21 we see that the ions causing the observed inversion are negative ions coming from molding compounds.

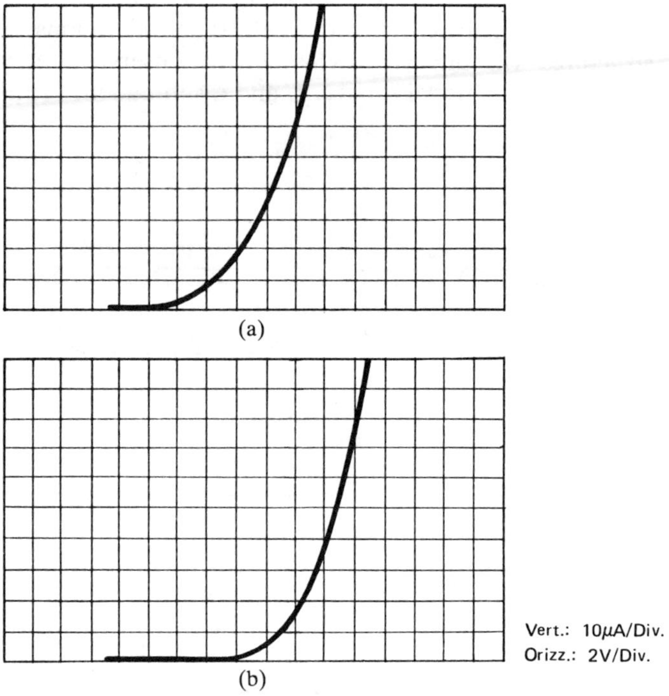

Figure 10.22 V_{CEO}-I_{CEO} characteristics of a lateral PNP transistor in a plastic package: (a) anomalous curve; and (b) correct curve obtained after 8 hours at 150°C.

The degradation can be explained by the build up of a surface conduction on top of silicon oxide, which extends the negative collector potential along the base channel. This causes the formation of a "phantom gate," which is responsible for the inversion at the silicon surface. By covering the base channel with a metal (Al) plate connected to the emitter contact (which is always the electrode at higher positive potential of a PNP transistor), this inversion can be prevented. Anyway, inversion effects can be observed on the lateral PNP transistor (or on similar structures) whenever a negatively biased metal stripe runs over a base channel.

In order to minimize device degradation due to inversion, we recommend the following actions:

- Well-studied design rules for the various layers.
- Use of resins with low ionic content.
- Elimination of the main sources of contamination in the various phases of assembly process.

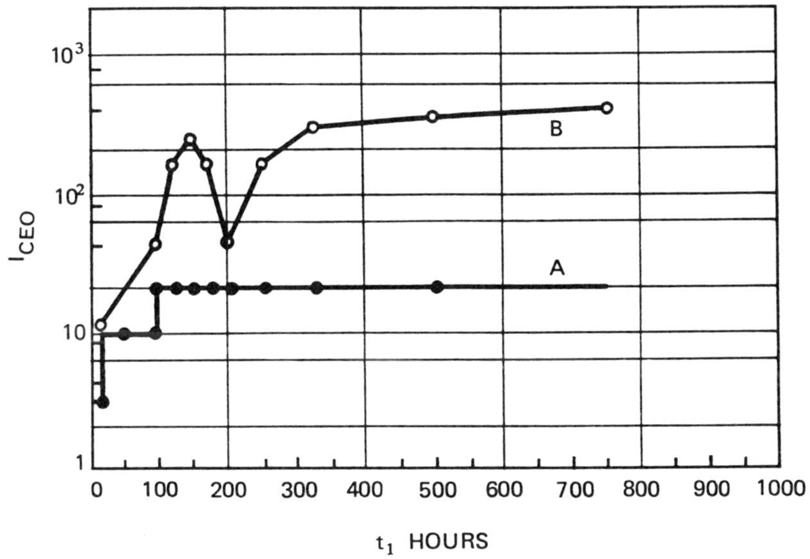

Figure 10.23 Typical I_{CEO} behavior over stress time for the two types of "lateral" PNP devices: *A* is a covered channel; and *B* is an uncovered channel.

10.6 APPENDIX A: RELIABILITY OF MOLDING COMPOUNDS

10.6.1 Introduction

In recent years, the evaluation and development of molding compounds has become increasingly sophisticated. A complete examination of these methods is beyond the limits of this book. Therefore, we shall discuss only what directly concerns the device reliability.

Epoxy molding compounds are now widely used in the fabrication of silicon devices. In the past, the use of siliconic molding compounds was essentially due to their purity; today, they are used only in a few special applications for a poor adhesion to the metal surface of the frame, due to water formation in the curing phase. So, humidity and foreign ions can easily penetrate into the device from the external environment.

Many efforts were made by resin suppliers in order to improve the main characteristics of molding compounds: purity, adhesion to the frame, internal stress, and curing time. Modern epoxy resins can be worked easily and offer a high level of purity, good mechanical resistance and thermal conductivity, a low thermal expansion coefficient up to 150°C, low molding shrinkage (0.4–0.6 percent), low stress, and limited cost.

A complete evaluation of a molding compound requires both a detailed chemical and physical characterization and a complete set of reliability tests on dedicated devices. The main chemical and physical evaluations are

1. Evaluation of polymerization kinetics and study of the influence of the "degree of cure" on resin physical properties. This evaluation is of high importance, because a low degree of cure make devices less reliable.
2. Determination of thermal decomposition kinetics, both in inert atmosphere and in air, and evaluation of thermal stability.
3. Determination of ionic content on water extract to simulate the release of contaminants during humidity tests.
4. Evaluation of electrical and thermal properties (for high-power applications).
5. Evaluation of workability and productivity in an assembly pilot line environment.

Reliability tests are usually conducted on proper test vehicles, in order to compare the compound under evaluation to the standard compounds.

10.6.2 Methods of Characterization

Polymerization Kinetics

Epoxy compounds used to encapsulate silicon devices are fabricated from thermosetting resins, usually in powder form at room temperature. Polymerization is obtained via transfer molding: cross linking among polymeric chains is thermally activated and promoted by pressure. Polymerization reaction is highly exothermic and produces a complex polymer with good mechanical properties and thermal stability at normal device operating temperatures. The percentage of links actually formed, with respect to the total admitted, is called *degree of cure.*

Usually, after transfer molding, resin is only partially cured; so a post-molding cure of many hours at a proper temperature is necessary to complete the polymerization. Typical process conditions are

- Transfer molding: temperature 170–180°C, pressure 50–70 kg/cm^2, curing time 90–150 s.
- Post-molding cure: temperature 170–180°C, curing time 8–16 hours.

A good degree of cure improves all the compound characteristics that may directly influence device reliability; for example, dielectric properties, mechanical strength, adhesion to the frame, resistance to humidity.

The polymerization kinetics and the degree of cure can be determined by a *differential scanning calorimeter* (DSC), which operates as follows: A specimen of the material to be analyzed is put in a small furnace and a blank is placed in an identical reference furnace. The two furnaces are heated simultaneously by temperature ramps at a given heating rate and any change in thermal energy (whether endothermic or exothermic) of the sample, with respect to the reference, is continuously monitored.

In the case of an epoxy resin, a typical DSC trace of an uncured sample is a broad exothermic peak, having its maximum at about 170–180°C (Figure 10.24, uncured sample); the heat of reaction is given by the peak area. Figure 10.24 also shows typical DSC traces of samples analyzed after the various thermal treatments indicated at the side of each trace. The peak area for each trace represents the degree of cure after the related treatment.

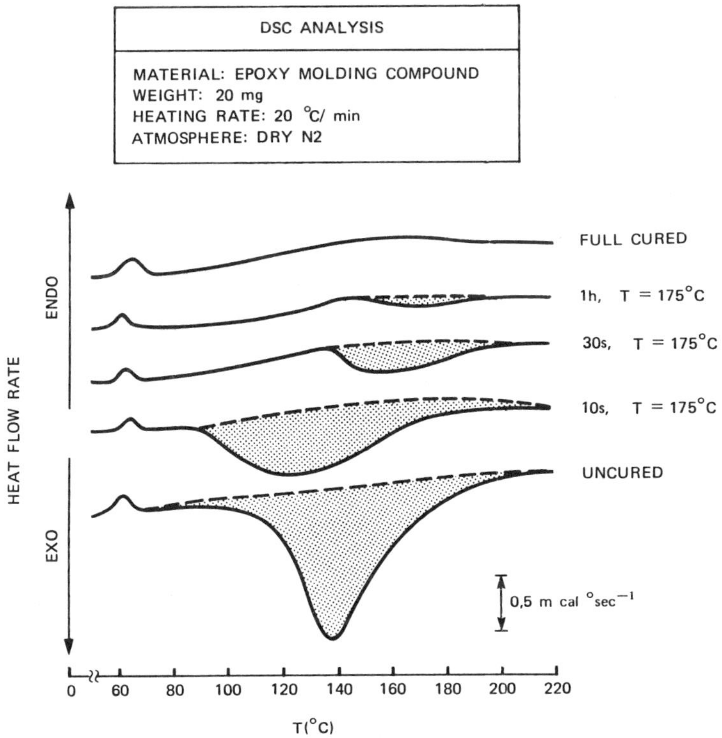

Figure 10.24 Curing kinetics of molding compounds by DSC plots.

The width and profile of the curve for the uncured sample depend not only on the base resin, but also on the various additives used (catalysts, mold release agents, fluidizers, *et cetera*). Since peak repeatability for a certain compound is excellent, DSC analysis is a good method of identifying and checking epoxy molding compounds.

Thermal Stability

Thermal stability has become a very important property of molding compounds. In fact, they are usually stable at temperature values higher than device operating range (200–220°C), but users very often subject plastic encapsulated devices to accelerated reliability tests or high temperature soldering cycles.

Thermogravimetrical analysis is the best method to evaluate the thermal stability of encapsulating compounds. The weight loss is recorded as a function of temperature, at a given heating rate and furnace atmosphere. If an oxidizing atmosphere is used, it is possible to identify the type of compound under test by its inorganic residue. In fact, the inorganic residue is typically 65–75 percent for epoxy compounds, and 90–95 percent for siliconic compounds. Typical decomposition profiles of two different epoxy compounds are shown in Figure 10.25. Furnace atmosphere is first nitrogen (up to 800°C) and then air. Figure 10.25 reveals that the decomposition process consists of three phases: (1) decomposition in nitrogen of base resin (at about 400°C); (2) breakdown of molecular chains; and (3) organical residues burning (in air) at about 800°C. The final portion (horizontal straight line) of the curve represents the amount of inorganic material.

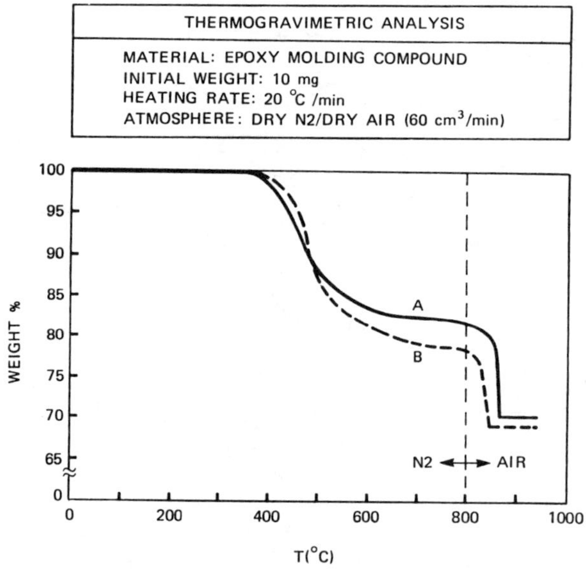

Figure 10.25 Weight loss curves of epoxy molding compounds.

Thermal Expansion Characteristics

Thermal expansion properties of molding compounds are directly related to mechanical stresses on the die under severe thermal conditions (thermal shocks and cycles, thermal fatigue). For this reason, *thermal expansion coefficient* (TEC) is one of the most important physical parameters of encapsulating resins (see also section 10.2 of this chapter).

Thermal expansion curves of compounds are determined by using a dilatometer, which operates from 20°C to 250°C. A typical thermal expansion plot was already shown in Figure 10.4. Note the increase of TEC above the glass transition temperature (Tg) in that figure. Tg is usually defined as the intersection point between the two almost linear portions of the thermal expansion curve. The Tg value, for a certain compound, is a function of the degree of cure, and plots of Tg *versus* cure time are widely used to optimize curing conditions.

Still referring to Figure 10.4, thermal expansion coefficients below and above the glass transition temperature are defined as

$$\partial_1 = \frac{\Delta L_1}{\Delta T_1 \cdot L_0}$$

$$\partial_2 = \frac{\Delta L_2}{\Delta T_2 \cdot L_0}$$

where ∂_1 = TEC of molding compound in temperature range below Tg; ΔT_1 = a temperature range in the almost linear portion of the thermal expansion curve below Tg; ΔL_1 = change in length of specimen due to temperature variation ΔT_1; L_0 = length of specimen at room temperature; ∂_2 = TEC in temperature range above Tg; ΔT_2 = a temperature range in the almost linear portion of the curve above Tg; ΔL_2 = change in length of specimen due to temperature variation ΔT_2. Typical thermal expansion coefficients and glass transition temperatures of epoxy molding compounds are

$$Tg = 150\text{–}170°\mathrm{C}$$

$$\partial_1 = 20\text{–}23*10^{-6}\mathrm{C}^{-1}$$

$$\partial_2 = 60\text{–}80*10^{-6}\mathrm{C}^{-1}$$

For state of the art low-stress molding compounds, ∂_1 is lower: $16\text{–}18*10^{-6}\mathrm{C}^{-1}$.

Determination of Ionic Content

Chemical analysis of molding compound is usually conducted on water extract in order to quantify the content of hydrolizable ions. The presence of high mobility ions such as Na^+, K^+, and Ca^{++} is potentially dangerous for the stability of the electrical characteristics of devices in high-temperature life tests. The presence of hydrolizable ions, such as Cl^-, can promote metal corrosion in accelerated temperature-humidity tests.

Electrical Characteristics

Electrical parameters of molding compounds are measured in view of the application field of devices (high-voltage, high-power, high-frequency, *et cetera*). The main electrical parameters usually reported in compound data sheets are dielectric constant, dielectric strength, dissipation factor, arc resistance, volume resistivity, and surface resistivity.

10.7 APPENDIX B: THERMAL RESISTANCE AND THERMAL IMPEDANCE

10.7.1 Brief Description of Heat Transmission

In order to better realize how heat propagates from the inside toward the outside of an IC, we briefly examine some concepts about heat transmission. As is well known, heat propagates according to three physical mechanisms: conduction, convection, irradiation. Let's analyze the laws these phenomena obey.

Conduction

The main relationship of conduction heat transmission is the Fourier law. In the monodimensional case, it can be expressed: $q = \mathrm{KA}\ \mathrm{d}T/\mathrm{d}x$. The thermal power (quantity of heat transmitted in the time unit between two opposite isothermal surfaces of area A) is directly proportional to the thermal gradient between the two surfaces with a constant which is the thermal conductivity K of the material, measured in

$$\frac{\mathrm{K\ cal}}{\mathrm{h \cdot m \cdot {}^\circ C}} \quad \text{or} \quad \frac{\mathrm{W}}{\mathrm{m \cdot {}^\circ C}}, \quad \frac{\mathrm{W}}{\mathrm{cm \cdot {}^\circ C}}$$

A minus sign means the heat flows in direction opposite the thermal gradient: according to the second principle of thermodynamics, heat flows from higher temperature points toward lower temperature points. The thermal conductivity is very high in metals and very low in air. Table 10.6 shows some reference values. When heat flows at steady state conditions between two flat, parallel surfaces, distant Δx, the thermal gradient and thermal power do not change relative to time, therefore:

$$\frac{q}{\Delta T} = \frac{A}{\Delta x} \mathrm{K}$$

and the conduction thermal resistance will be

$$R_{th} = \frac{\Delta T}{q} = \frac{\Delta x}{\mathrm{K} A}$$

Table 10.6 Thermal Conductivity of Some Materials

Materials	K-W/cm°C
Copper	3.96
Air	1.5×10^{-4}
Epoxy resin	$5 \div 14 \times 10^{-3}$
Pb/Sn 95/5	0.35
Al	2.38
Au	3.15

Convection

The following relationship describes this phenomenon:

$$q_c = h_c A \Delta T$$

The thermal power transmitted from a surface of area A at a temperature T_s to a fluid at a temperature T_F, in a point distant from the surface, is proportional to the temperature difference with a parameter called the *mean convection thermal conductance* or *convection heat transfer coefficient,* and is expressed in W/m^2°C. This parameter changes considerably from point to point; therefore, the mean value is considered.

The evaluation of h_c is rather complicated and depends on many parameters: surface geometry, speed and physical properties of the fluid, temperature difference between fluid and surface, *et cetera*. A convection thermal resistance can be defined in a way similar to that for conduction:

$$R_c = \frac{1}{h_c A}$$

Some h_c reference values are listed in Table 10.7.

Table 10.7 Convection Heat Transfer Coefficient

	W/m²°C
Free air	6–30
Forced air	30–300
Forced convection oil	60–1700
Forced water	300–12000
Boiling water	3000–6000

Irradiation

The following law describes this phenomenon and it refers to a full radiator or black body:

$$q = \sigma A_1 T_1^4$$

The heat quantity q radiated from a full radiator of area A_1 is proportional to the fourth power of the absolute temperature (T^4) in kelvins, the proportionality constant σ is the Stefan-Boltzmann constant. If the black body is held in a black cavity at temperature T_2, the heat exchange that takes place is

$$q = \sigma A_1 (T_1^4 - T_2^4)$$

Bodies are not full black (gray), hence:

$$q = \mathrm{E}\sigma A_1 (T_1^4 - T_2^4)$$

where E is the emissivity of the gray surface: ratio between the energy emitted by the gray body and the black body.

Taking into account the emissivity of bodies under examination and their geometrical relationships, we have

$$q = \sigma A_1 F_{1-2} (T_1^4 - T_2^4)$$

where F_{1-2} is an experimental coefficient that takes into account just the shape, size, and mutual positions of bodies.

An irradiation thermal resistance would be defined:

$$q = Ki\,(T_1 - T_2')$$

Ki is expressed in W/°C or K cal/h°C, where Ki is an irradiation thermal conductivity and T_2' a reference temperature that can coincide with T_2.

We can conclude that the thermal resistance is

$$Ri = \frac{T_1 - T_2'}{\sigma A_1 F_{1-2}(T_1^4 - T_2^4)}$$

measured in °C/W or h°C/K cal. Table 10.8 lists some emissivity values.

Table 10.8 Emissivity of Some Materials

Material	*Emissivity at 38°C*
Bright Al	0.04
Bright Copper	0.04
Oxidized Copper	0.87
Bright steel	0.17
Oxidized steel	0.85

10.7.2 Definitions of Thermal Resistance and Capacity

Very simply, thermal resistance for an IC is defined as the ratio between the junction temperature increase when a power Pd is applied and the value of the power itself ($R_{th} = \Delta T_j/Pd$). This definition relates to the measurement of the parameter and includes all the heat transmission aspects: conduction, convection, and irradiation.

For conduction, heat transmits inside the package; for convection and irradiation, mainly outside. It is necessary, then, to define two different concepts of thermal resistance: thermal resistance junction to case and thermal resistance junction to air. The first parameter is used in packages that need a heat sink; the second one is used in all other cases.

Basically, for devices with a power higher than a few W (2 : 3W), which need a heat sink inside the package, the thermal resistance is expressed as follows:

$$R_{thj-a} = R_{thj-c} + R_{th-d}$$

where

R_{thj-a} = thermal resistance junction to air

R_{thj-c} = thermal resistance junction to case

R_{th-d} = thermal resistance of the heat sink

For devices up to 2 W, where the only way to dissipate heat is through the structural elements of the package itself (frame, resin), only a resistance junction to air (R_{thj-a}) is defined. Some R_{thj-c} and R_{thj-a} values, referred to most used packages, are listed in Table 10.9.

Table 10.9 Thermal Resistance of Some Usual Packages

	R_{thj-c}	R_{thj-a}
Plastic packages:		
Multiwatt	3	45
Pentawatt	3.5	60
FIN-DIP	10	70
POWER-DIP	3–10	50–80
DIP Cu		60–75
DIP Alloy		120–210
DIP Alloy Multileads		75–135
Mini-DIP Cu		100
μ-PACKAGE-SO	90–140	140–200
(μW-SO)	35–50	90
Ceramic packages:		
8 leads		160
14–16 leads	110–140	
KYOTO Multileads		70

Thermal Capacity

A usual definition of heat capacity, like that for thermal resistance, does not exist. This perhaps is due to difficulties in experimental evaluations or to the absence of a tradition considering such a parameter. As a matter of fact, even the resin suppliers fail to give data on specific heat.

From thermodynamics, the thermal capacity C_K of a thermally insulated body at initial temperature T_0, is defined as ration between heat quantity ΔQ supplied by a source (for a time t), and the temperature increase ΔT which the body undergoes:

$$C_K = \frac{\Delta Q}{\Delta T}$$

Heat capacity is measured in J/°C and can be expressed as $C = Cs \cdot V \cdot y$; that is, multiplying the specific heat C_s by the weight $V \cdot y$ of a body. The specific heat is the heat quantity necessary to rise by 1°C an unitary weight body.

Before describing the equivalent electrical models, Table 10.10 specifies the equivalences between thermal and electrical quantities. This table demonstrates the comparative values between electrical and thermal models, allowing easy movement from one to the other.

Table 10.10 Equivalence between Electrical and Thermal Quantities

Electrical quantities			*Thermal quantities*		
Quantity	Symbol	Unit	Quantity	Symbol	Unit
Electric charge	Q	Coulomb	Heat quantity	Qt	J
Electric potential	V	V	Temperature	T	°C
Current	$I = \Delta Q/\Delta t$	A	Power	$P = \Delta Qt/\Delta t$	W
Ohmic resistance	$R = V/I$	Ω	Thermal resistance	$R_{th} = \Delta T/\Delta P$	°C/W
Electrical capacitance	$C = \Delta Q/\Delta V$	Farad	Thermal capacity	$C_K \Delta Qt/\Delta T$	J/°C
Time Constant	$T = RC$	s	Time constant	$T = R_{th} \cdot CK$	s

Equivalent Electrical Models

Two electrical models, which are referred to in Figure 10.26, are equivalent to the thermal model. In the first of these models (Figure 10.26a), Si is silicon; *D* is the die attach; *I.H.* and *E.H.* the internal and external heat sinks; and *H.A.* is the Heat sink–air interface. The model was obtained using the Fourier equation with suitable approximations, starting from a distributed parameters network. Every cell is referred to external ambient and represents the share of thermal resistance that heat meets cell by cell to flow from silicon toward external ambient. The last element, that represents the interface between external heat sink and external ambient, takes into account convection and irradiation and is the largest share of the chain.

Although all the elements together (silicon, die attach, heat sinks) make at most 1°C/W, *H.A.* alone can be worth 4 : 5 times the value of all the others; obviously, this depends on the size, shape, and material of the external heat sink.

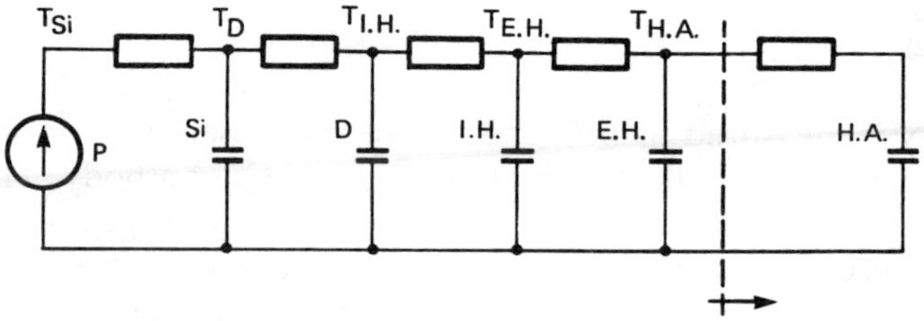

(a)

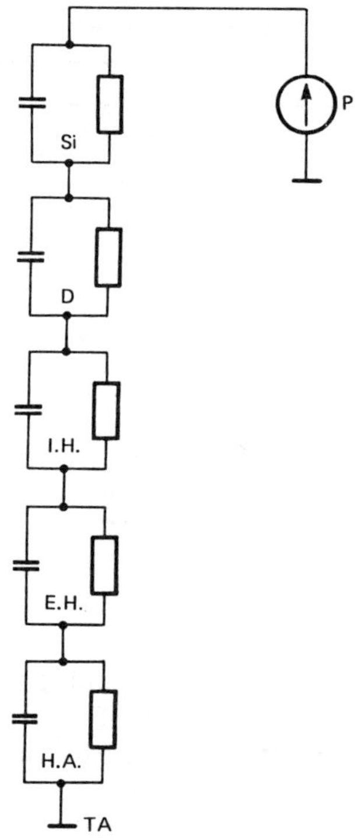

(b)

Figure 10.26 Electrical models equivalent to the thermal model: (a) cell equivalent model; and (b) cells in series.

The second model (Figure 10.26b) is formed by cells placed in series. This model, which was developed after the previous one, is suitable to represent the behavior of our system, as all the elements of our structure are identified cell by cell, and their thermal aspects are very clearly explained. Heat flow from silicon to external ambience is represented by current flows into the different cascade cells. Obviously, the values of single elements are different and, therefore, so are the time constants; nevertheless:

$$\Sigma R_{Si} = \Sigma R_{Pi}$$

$$\Sigma C_{pi} = \frac{1}{\Sigma 1/C_{si}}$$

In practical applications, the scheme of Figure 10.26b can be simplified, as in Figure 10.27, where we have only four elementary cells: silicon, internal heat sink, external heat sink, external ambience. Some reference values of R_{th} and the time constant are given beside each element. Among values shown, some were evaluated theoretically (silicon), others experimentally and are meant to be an example to understand the magnitude order of times in the different cells.

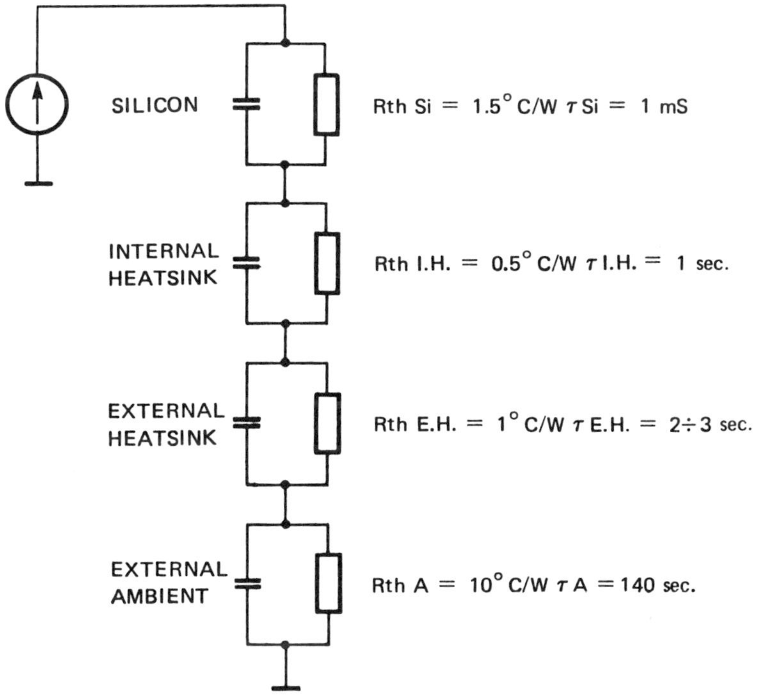

Figure 10.27 Simplified cell model for an asymmetric structure.

Package Pulse Thermal Response

We now consider the behavior of temperature in intervals (t_1) when a pulse is applied (temperature increasing phase: $nT \leqslant t \leqslant nT + t_1$):

$$
\begin{aligned}
TJ = {} & Po\, R_{\mathrm{Si}}(1 - \mathrm{e}^{-t_1/\tau\mathrm{Si}}) \cdot \frac{1 - e^{-nT/\tau\mathrm{Si}}}{1 - e^{T/\tau\mathrm{Si}}} \\
& \cdot e^{-t}/\tau\mathrm{Si} + Po \cdot R_{\mathrm{Si}}\,(1 - e^{-t}/\tau\mathrm{Si}) \\
& + PoR_{\mathrm{Cu}}(1 - e^{-t_1/\tau\mathrm{Cu}}) \qquad (10.1) \\
& \cdot \frac{1 - e - nT/t\mathrm{Cu}}{1 - e^{T/\tau\mathrm{Cu}}} \cdot e^{-t}/\tau Cu \\
& + Po \cdot R_{\mathrm{Cu}}(1 - e^{-t}/\tau\mathrm{Cu} + Po \cdot R_{A(1-e^{-t})}/\tau A) \\
& \cdot \frac{1 - e^{-nT/\tau A}}{1 - e^{T/\tau A}} \cdot e^{-t/\tau A} + Po \cdot R_A(1 - e^{-t/\tau A})
\end{aligned}
$$

For practical reasons, from now on, calculations refer only to an elementary cell. On peak point of the pulse ($t = t^1$), we have

$$
\begin{aligned}
T_{\mathrm{jmax}}\,(t = t_1) = Po\, R_{\mathrm{Si}}(1 - e^{-t}1/\tau\mathrm{Si})\, & \frac{1 - e^{-}(n + 1)\; T/\tau\mathrm{Si}}{1 - e\; T/\tau\mathrm{Si}} \\
& + \ldots + se\;(n + 1)\;\mathrm{T} \gg \tau \qquad (10.2)
\end{aligned}
$$

$$
T_{\mathrm{Jmax}} = Po\, R_{\mathrm{Si}} \left(\frac{1 - e^{-t}1/\tau\mathrm{Si}}{1 - e^{-T/\tau\mathrm{Si}}} \right) + \ldots \qquad (10.3)
$$

This occurs, for instance, after a suitable number of cycles with $T = t$. Similarly, we can consider the behavior of temperature of intervals t_2, between one pulse and the next (temperature decreasing phase: $t_1 < t < t_2$):

$$
T_{\mathrm{J}} = Po\, R(1 - w^{-t}1/\tau) \cdot \frac{1 - e^{-(n+1)t/\tau}}{1 - e^{-T/\tau}} \cdot e^{-}\, \frac{t - t_1}{\tau} + \ldots \qquad (10.4)
$$

At the point ($t = T$):

$$
T_{\mathrm{J}}(T) = Po\, R(1 - e^{-t}1/\tau) \cdot \frac{1 - e^{-(n+1)t/\tau}}{1 - e^{-T/\tau}} \cdot e^{-}\, \frac{t - t_1}{\tau} + \ldots \qquad (10.5)
$$

Also when $(n + 1)T \gg t$, we have

$$T_J(T) = Po\,R(1 - e^{-t}1/\tau) \cdot \frac{1 - e^{-(n+1)t/\tau}}{1 - e^{-T/\tau}} \cdot e^{-}\frac{t - t_1}{\tau} + \ldots \tag{10.6}$$

Now if we consider $T \ll t$, we can develp the Taylor series: Equation (10.3) becomes: $T_{Jmax} = t_1/T\,R\,Po + \ldots$ Equation (10.6) becomes: $T_{Jmin} = t_1/T\,R\,Po + \ldots$ This means that, when the thermal time constant is much longer than the electrical constant, due to electrical signal, the thermal "ripple" ($T_{Jmax} - T_{Jmin}$) tends to zero. It is very important not to have thermal signals in ICs, for reasons of stability. By maintaining the structural aspects of a package within the proper dimensions it is then possible, within certain limits, to avoid several problems connected with thermal aspects.

10.7.3 Measurement Methods

Several methods were conceived to measure a semiconductor device's thermal resistance: the optical (chip temperature scanned by an infrared system) the chemical (a thermosensitive coating on the chip), and the physical (thermocouples nearby the chip); but only the electrical methods allow R_{th} measurements precise and practical enough to perform easily on a sufficient sample size. These include measurement through the substrate diode or through the Kit part. We describe these two methods, as they are the ones most used by both IC suppliers and customers.

The ideal R_{th} measurement for power ICs would be the one that simulated the behavior under field conditions and permitted measuring the junction temperature at different dissipating power levels through a temperature sensitive element. This contrivance has not yet been designed by anyone dealing in IC field.

"Substrate Diode" Method

This method utilizes the substrate diode or diodes present in every IC, both as a dissipating element and a thermosensitive element. First, a V_{be} calibration of the diode is performed, in a temperature range from 25°C to 110°C at a current level of 1 mA. This operation generates a graphic (straight line) that illustrates the V_{be} level at each temperature. The basic operating principle of this method is shown in Figure 10.28. Briefly, the component is heated for a time T_{on} of about 70 ms up to the required Pd level (1 W at most) and for a time $T_{meas} = 275\ \mu s$ the component V_{be} is measured.

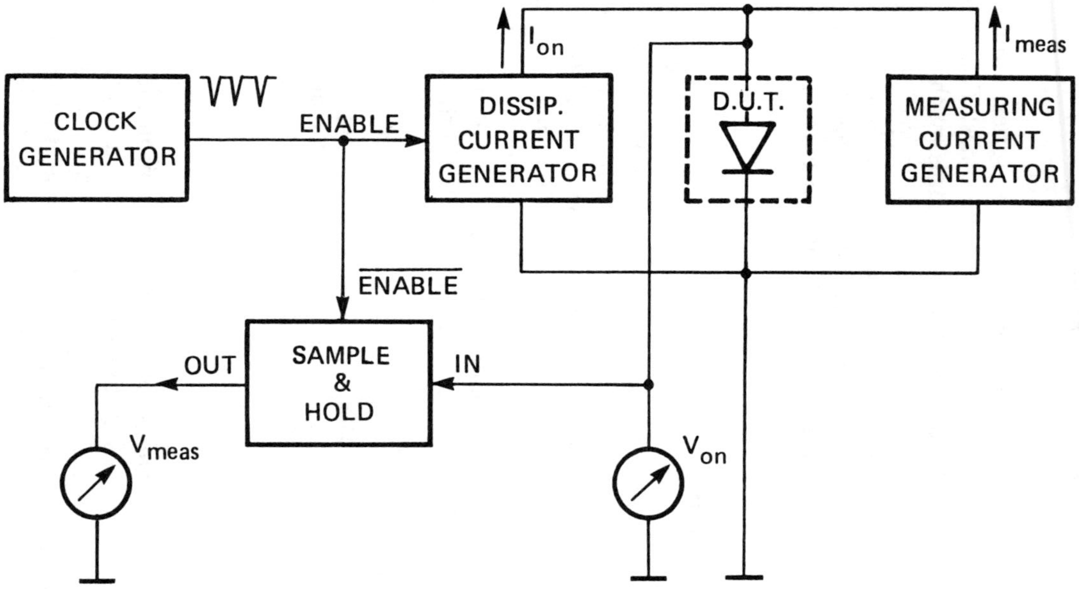

(a)

Figure 10.28 Substrate diode method of measuring thermal resistance: (a) test circuit; and (b) current waveform.

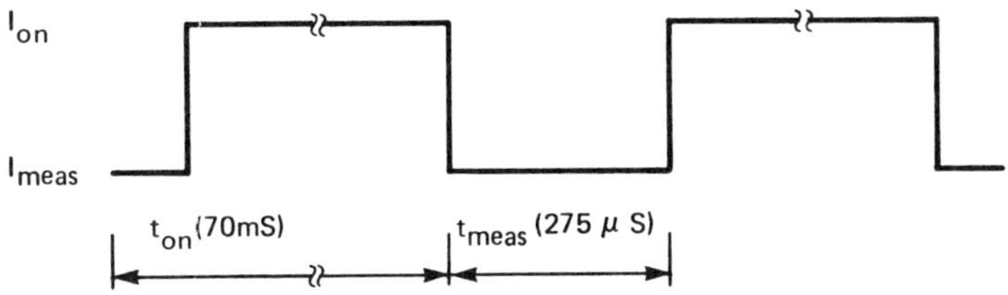

(b)

Figure 10.28 continued

The V_{be} value, taken in dissipation and calibration state, leads to the corresponding ΔT through the correlation constant $K(-2\ \text{mV/°C})$. We can then write:

$$R_{th} = \frac{\Delta V_{be}/K}{Pd}$$

The main advantages of this method include: its amenability to measurement, as every IC has a substrate diode; satisfactory accuracy; worldwide use, allowing comparisons; and its codification by MIL-STD 883C, method 1012.

Its main disadvantages include the need to use a low dissipating power, only 3 W at most insufficient in power ICs; R_{th} values a little lower than in actual use, as it utilizes silicon areas larger than those involved in real application and at lower temperature; and general lack of correlation with real conditions.

Kit Part Method

This method follows the previous one, trying to improve the critical aspects. The measurement concept is shown in the block diagram of Figure 10.29, which is basically identical to that of Figure 10.28, with the substrate diode replaced by a transistor (kit part). The temperature sensitive element, used then as thermometer, is the base-emitter diode, which must be precalibrated.

This method approximates the real dissipating situation better than the previous one, by allowing dissipating powers up to 20 W and representing an improvement in R_{th} characterization. In fact, the kit part used is almost always a power transistor of the interested IC. Obviously, this method has the drawback of requiring a transistor structure, not always available; therefore, making comparisons difficult.

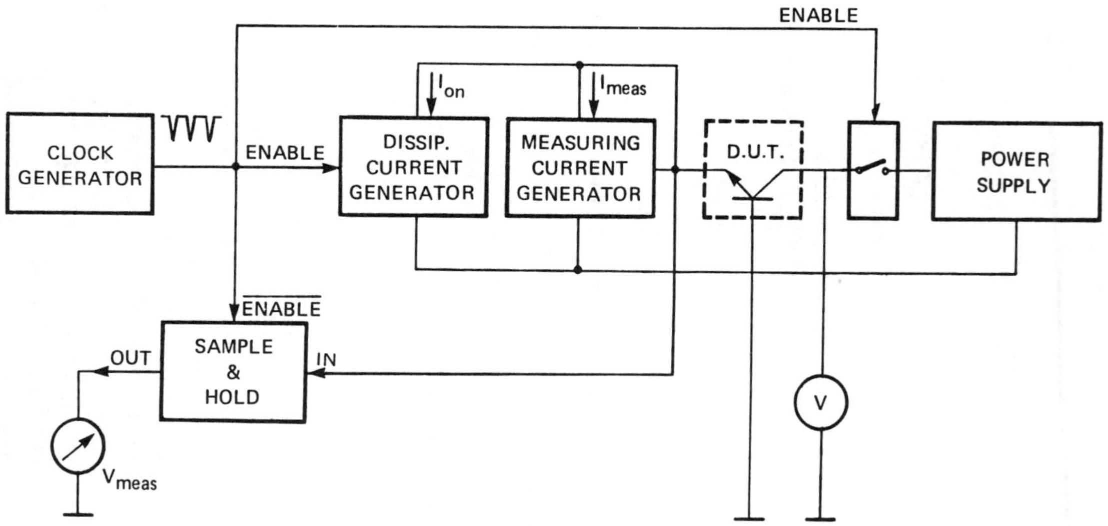

(a)

Figure 10.29 Kit part method of measuring thermal resistance: (a) test circuit; and (b) current waveform.

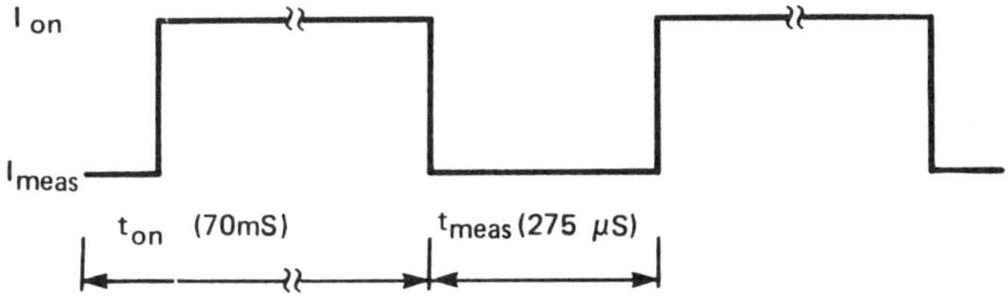

(b)

Figure 10.29 continued

10.7.4 Final Considerations

Surely, thermal resistance is one of the most interesting parameters for anyone dealing with a power IC. Until recently, data books very often reported values of this parameter laconically without any reference, an unwarranted amount of incompetence and superficiality on an element of such importance. Due to a growing interest for package thermal problems, characterizations of thermal resistance are now in greater and greater demand. Successful techniques are even being developed to measure thermal impedance.

In conclusion, tests of thermal resistance ought to resemble real use conditions as much as possible. That is, measurement must be taken at the right power levels, with correct mechanical coupling with an external heat sink, and at an external ambient temperature corresponding to the highest foreseen. Only in this way can we evaluate correctly the junction temperature of the device and forecast its failure rate.

REFERENCES

1. P. Antoguetti, ed., *Power Integrated Circuits, Physics Design and Application,* Chapter 8, "Reliability," McGraw-Hill, New York, 1985.
2. Y. Kanda, "A Graphical Representation of the Piezoresistance Coefficient in Silicon," *IEEE Trans. Eltron. Devices,* ED 29, 1982, p. 64.
3. J.L. Spencer et al., "New Quantitative Measurements of IC Stress Introduced by Plastic Packages," *Int. Reliability Physic Symp. Proc.* 1981, p.74.
4. J.L. Spencer, "Calculating Stress and Mobility in Silicon Chips Using Strain Gauge Measurement," *TI Science Engineering J.* 1, 1980, p.34.
5. C.S. Smith, "Piezoresistance Effect in Silicon," *Physics Rev.,* 94, 1954.
6. K.M. Schleiser, "Piezoresistance Effects in Plastic Packaged IC's," *Proc. of the Int. Conf. Consumer Electronics,* 1982, p.28.

7. C. Cognetti et al., "Accuracy of Plastic Packages Diffused Voltage Dividers," *Hybrid Circuits,* 8, 1985, p.19.
8. C. Cognetti et al., "Power Packages for Bipolar IC's with Improved Mechanical Performance," *Proc. of the IEPS Conf.* 1985, p. 85.
9. F.F. Villa, "La rottura secondaria nei circuitti integrati di potenza," *Elettronica e Telecomunicazioni,* n. 3, 1984.
10. J.C. Harrison, "Control of the Encapsulation Material as an Aid to Long Term Reliability," *Microelectron. Reliability, 16, 1977.*
11. G. D'Espinosa and L. Lonzi, "L'Affidabilità nei Circuiti Integrati di Potenza," *Elettronica Oggi, 11, 1983.*
12. A. Peticlerc, *Electronique Physique des Semiconducteurs,* Editions du Tambourinaire, Paris, 1962.
13. F. Kreïth, *Principî di Trasmissione del Calore,* Ed. Liguori.

Chapter 11
ACCELERATED TESTS

G. Boccaletti, F.R. Borri, G. D'Espinosa, G. Fioravanti, and E. Ghio

11.1 DEFINITION AND SCOPE

Tests on semiconductors are carried out under accelerated conditions when the stresses are in excess of those found in normal use [1, 2]. Such tests are considered valid only if the failure mechanisms experienced during the accelerated stresses are the same as those experienced during normal use. The purpose of accelerated testing is to cause the "normal" types of failures to occur earlier and, thereby, decrease the total time required to obtain reliability information of the device under study [3].

The parameters changed to accelerate the life of the devices may be the environmental ones such as temperature and humidity, the electrical ones such as voltage and current, or even those of a mechanical nature such as vibrations and shocks.

11.2 TEST CARRIED OUT BY INCREASING THE TEMPERATURE

11.2.1 Arrhenius' Law

Temperature is one of the environmental parameters most often accelerated because the reaction rates of many physical and chemical failure mechanisms are greatly affected by temperature. Accelerated conditions are generally produced by raising the temperature.

The mathematical model frequently used in describing the mechanism reaction rate *versus* temperature is based on the law of Arrhenius, which was originally conceived for describing the kinetics of a chemical reaction:

$$R = R_0 \exp(-E_a/kT) \tag{11.1}$$

where

R = reaction rate;

R_0 = constant;

E_a = activation energy;

k = Boltzmann's constant = 8.63×10^{-5} eV/K;

T = absolute temperature.

In general, the effect of the temperature on a device to pass from a normal status to a failure status is evident from Figure 11.1. Assuming R_1 (reaction rate) corresponds to a temperature T_1 and R_2 is relative to T_2 and assuming, at the moment, a reaction linear with time, the parameter that determines the device passing from a "good" status to a "bad" status is expressed by R_1t_1 at the temperature T_1 and by R_2t_2 at the temperature T_2.

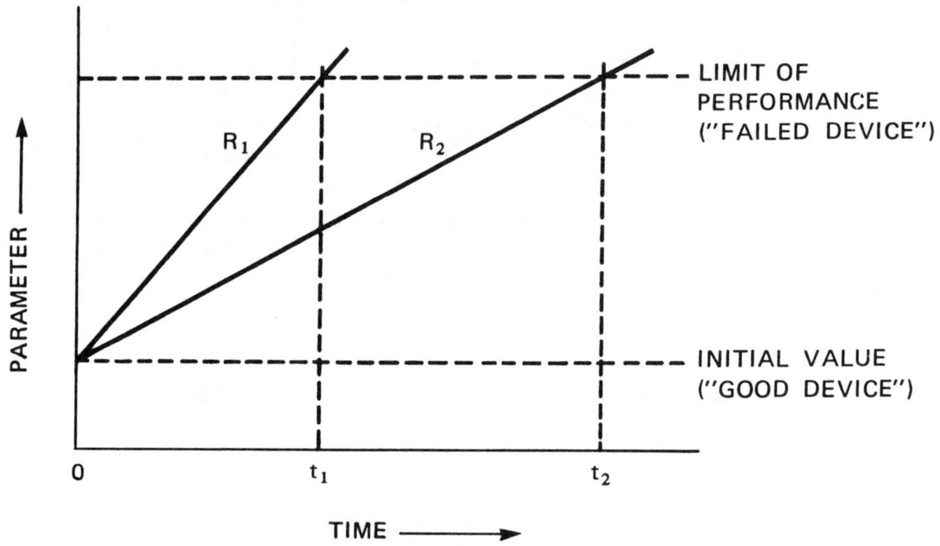

Figure 11.1 Switching a device from a "good" to a "bad" state with two different speeds.

Because $R_1t_1 = R_2t_2$ = constant, a generic time can be formulated from equation (11.1):

$$t_f + A \exp(E_a/kT) \tag{11.2}$$

Assuming a reaction proportional to t^n, instead of a linear one means that $R \cdot t^n$ = constant, from which:

$$t_f = A \exp(E_a/n\ kT) \tag{11.3}$$

Equation (11.2) for a generic time can also be expressed in the following form:

$$\ln t = \ln A + (E_a/K)\,(1/T) \tag{11.4}$$

The relationship (11.4) evidences that the logarithm of life time depends linearly on the inverse of the absolute temperature.

The activation energy assumes therefore the meaning of this linear slope: as E_a increases, the dependence of the ln t on the reverse of the temperature also increases.

If we plan accelerated tests on two lots of devices of the same type at two different temperatures T_1 and T_2, according to (11.4), the time to failure in the two cases can be expressed respectively by

$$t_1 = A \exp\,(E_a/kT_1)$$

$$t_2 = A \exp\,(E_a/kT_2)$$

where, assuming T_1 is higher than T_2, we will have t_1 lower than t_2.

The relationship

$$t_1/t_2 = \exp[E_a/K(1/T_1 - 1/T_2)] \tag{11.5}$$

represents how the life of the device shortens with increases of temperature and defines the acceleration coefficient in temperature of the failure mechanism characterized by an activation energy E_a.

Increases in E_a also increase the acceleration coefficient, which means, as already said, that it increases the dependence of this phenomenon on the temperature. We can determine E_a with the same procedure; that is, by knowing the times to failure at two different temperatures.

Analyzing (11.3) again, the experimental data of failure time provides no indication whether the phenomenon depends linearly on the time, as we previously obtained $E_a^1 = E_a/n$, where E_a is the apparent activation energy containing the dependence n on the time and is the effective activating energy of the physical chemical phenomenon determining the failure.

When more failure mechanisms coexist, the relationship that connects time to failure t_f and $1/T$ is no longer linear, with (11.4), but curvilinear. In this case, as it will be illustrated later, the data regarding the different failure mechanisms have to be treated, separately [26].

11.2.2 Constant Stress Tests

Constant stress tests are usually carried out on at least three lots of the same population of devices at three different temperature levels. To obtain E_a, as shown, it is theoretically sufficient to carry out tests at two temperature levels, provided there is the certainty about a sole failure mechanism. The third reference in the Arrhenius diagram of Figure 11.2 is set to confirm this hypothesis, by being in line with the other two.

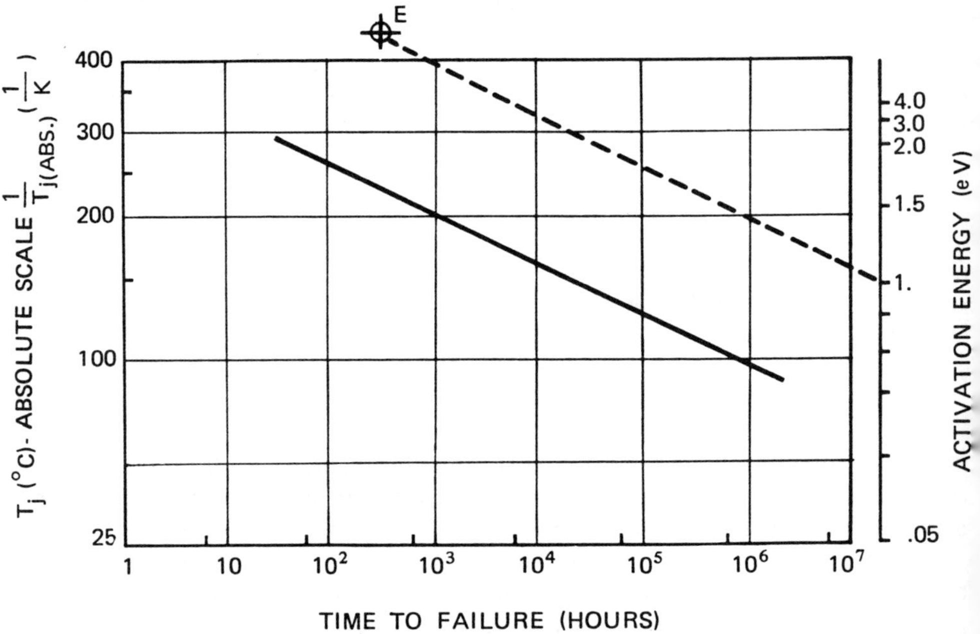

Figure 11.2 Arrhenius plot: In this example, $E_a = 1$ eV.

To do the intermediate measures, the devices are cooled to the ambient temperature (in case they are polarized) with applied voltage, in order not to modify the distribution of possible contaminating mobile charges accumulated during the test because of the sole temperature. Failures and the parameter variations found are hence recorded.

As experimentally verified [4], the failure distribution in the time follows a log normal law; as a consequence of this, the cumulative percentage of found failures, transferred on a normal scale (in the tests carried out at different temperature levels) is a linear function of the logarithm of the number of test hours (Figure 11.3). The lines of the plot are parallel because the basic hypothesis, confirmed by the practice on which accelerated tests are based, is that the distribution trend of the time to the stress only modifies the average life of the distribution. Keep in mind, however, that there often simultaneously is a main population and an anomalous one formed by the devices with intrinsic manufacturing defects, called the *freak population.*

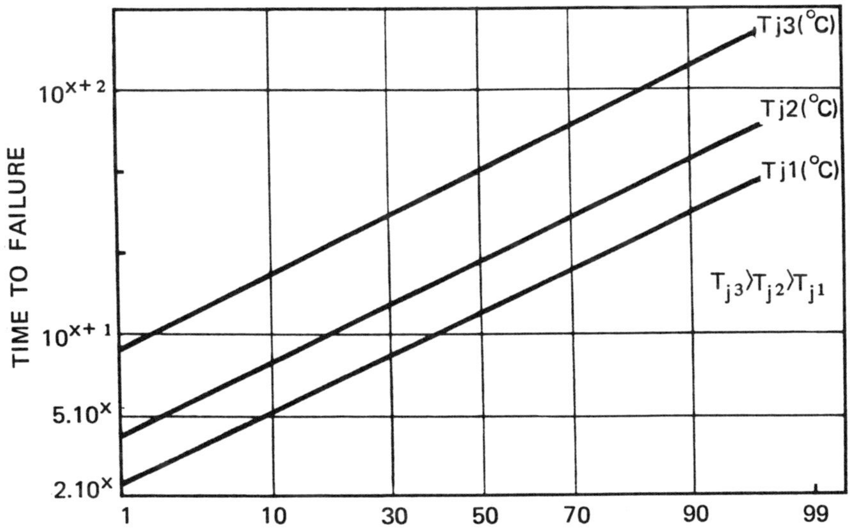

Figure 11.3 Plot of cumulative failure versus time to failure.

This population causes some failures in the first hours of life because they are weaker, consequently the time trend of these failures has to be analyzed separately from that of the main population, characterized by different failure mechanisms. The fact that the weak population can be easily eliminated by very short tests is important from an industrial point of view, as the overall reliability of the lots of devices submitted to these tests, named *screening tests,* increases. The overall distribution of failures in a bimodal population is shown in Figure 11.4.

The very different times to failure characterizing the two populations enables separate analysis of the relative data, to obtain the various activating energies E_a.

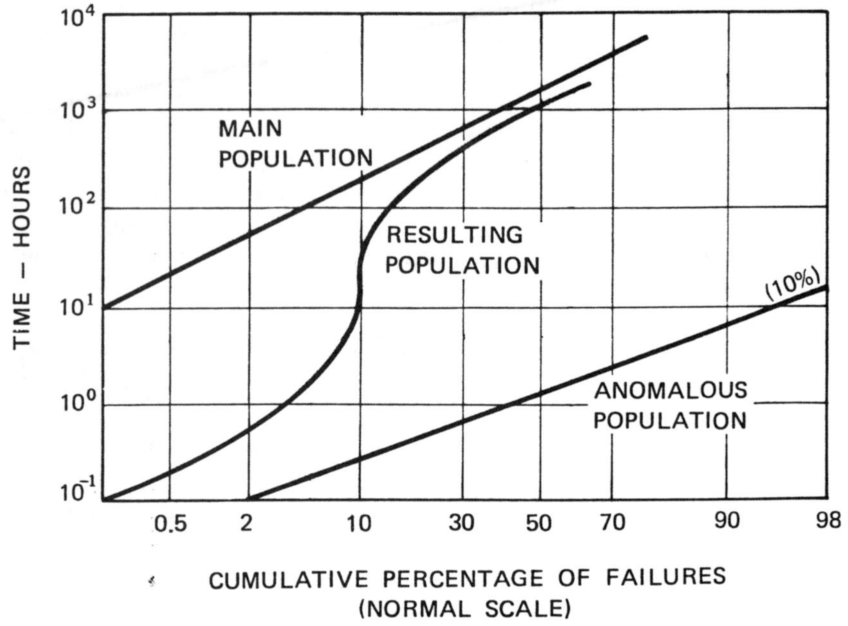

Figure 11.4 Plot of cumulative failure versus time to failure in a bimodal population.

The diagram of Figure 11.3 demonstrates how to obtain the Arrhenius diagram of Figure 11.2, with reference to the desired failure percentage. In fact, Figure 11.3 shows the correspondance between time to failure and cumulative failure percentage for each test temperature (T_j).

The slope of the straight line in the Arrhenius diagram is, as shown by (11.3), proportional to the activation energy E_a relative to the degradation (failure) mechanism prevalent in the specimen. E_a is determined by reading, in Figure 11.2, the value on the intercept point of the parallel out of E to the straight line relative to the considered failure percentage by the graded value on the right side of the diagram.

11.2.3 Step Stress Tests

In step stress tests, progressively increasing stresses are applied to the same sample for constant time intervals. The fundamental assumption in this kind of test is that the failure probability at each time interval is independent on previous steps, regardless of previous history. Figure 11.5 shows the way of procedure; stress S_1, which can be of any type but usually is temperature, is increased at each time interval x until all samples fail.

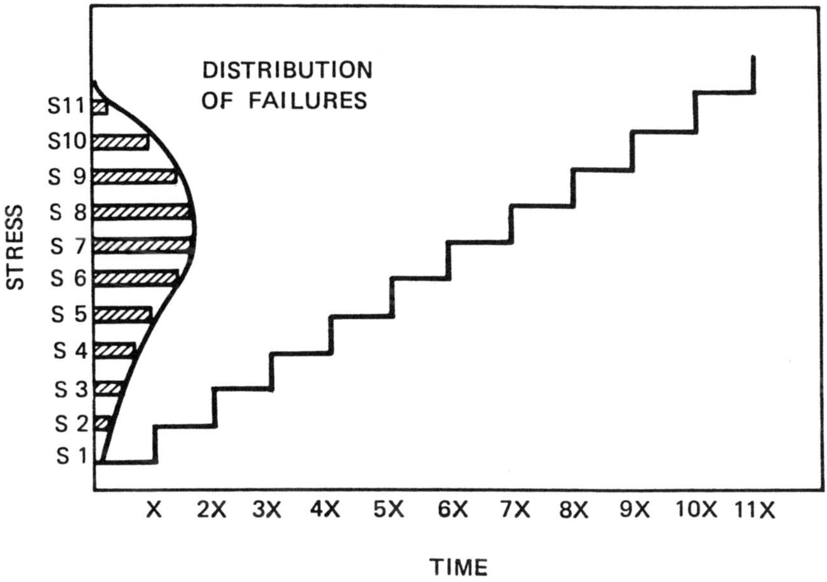

Figure 11.5 Step stress test.

This distribution of data is much more precise with larger samples and smaller amounts of stress increase at each step. Because the effects of the exposure at high temperatures are cumulative, a compromise is necessary, and the gap between the various temperatures has to be kept equal or higher than 25°C. The overall percentage of failures *versus* the stress level for a step stress test is shown in Figure 11.6.

Three different regions are evident: the first one corresponds to the failure at poor stress levels (inferior at S_1) of the freak population, the second one represents the main population and is kept as a baseline reference of the average response to the stress; the third, at last, corresponds to the new failure mechanisms arising from too much stress (greater than S_3), which causes quick failure of all specimens. The diagram in Figure 11.6 is ideal; in the practice, differences between the various regions can be less evident or some region may not be present.

By comparison with constant stress tests, step stress tests generate data on the product reliability more quickly; on the other hand, these data are less precise, owing to the simplified hypotheses taken into account (each step not influenced by previous ones). The step stress tests, therefore, are carried out as comparison tests, enabling a quick comparison of samples of different productions or as preliminary activity before more complex and expensive tests at constant stress.

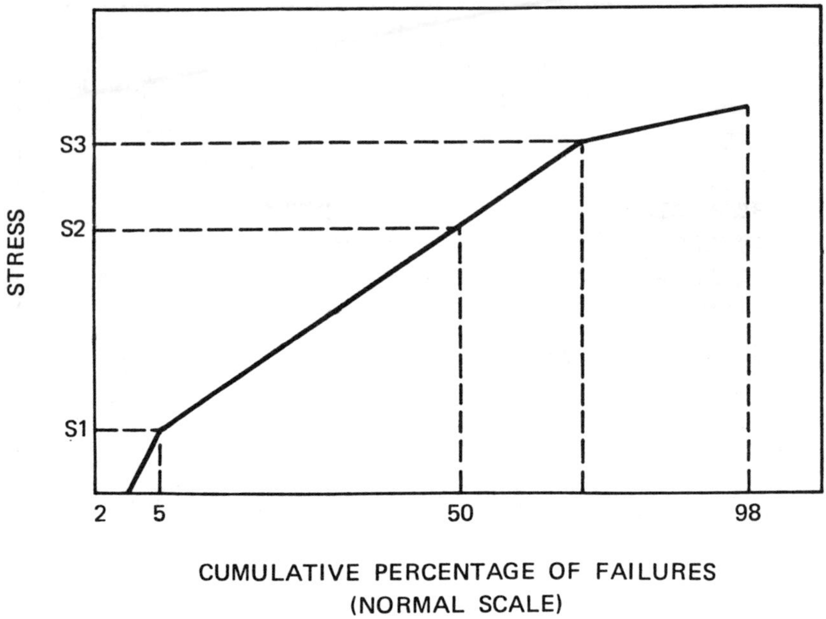

Figure 11.6 Plot of cumulative failure after step stress test.

In fact, when planning a scheme of accelerated tests, it is necessary to be sure before that the stress levels chosen are not so high as to induce failure mechanisms not typical of normal applications, in other words, with reference to Figure 11.6, stress levels should be verified as between S_1 and S_3.

The devices that failed the preliminary step stress tests represent a useful reference for the failure modes to be found during the constant stress accelerated tests. The step stress tests are pretty easy to accomplish and do not require a great many samples: in fact, 10–30 units may be representative.

11.2.4 Application of Accelerated Tests

The following example shows the application of these concepts. The text vehicle is a radio frequency NPN transistor having internal metalization layers of titanium, platinum, and gold (BFQ88A) assembled in a ceramic package (Microx). The accelerated tests here described have been performed in the SGS Reliability Laboratory to evaluate the main failure mechanism of this device and the associated activation energy [27]. These were the test conditions described in Figure 11.5. The junction temperatures T_j and collector currents I_c (at V_{ce} = 8V = cost) for the three tests are indicated in Table 11.1.

Table 11.1 Stress Parameters for Life Test

Test number	T_j (°C)	I_c (mA)
1	340	100
2	306	88
3	293	84

During the tests various parameters of the transistors have been monitored. In particular, the variation of gain has been checked. The maximum variation of h_{FE} allowed is 25 percent of initial value ($\Delta h_{FE}/h_{FEO} \times 100 = 25$), according to the SGS standard reliability program (SURE program). The results of the three tests are recorded in Tables 11.2, 11.3, and 11.4. The results of the three tests performed at different junction temperatures are plotted in Figure 11.7. From the same figure it is possible to determine the time for 50 percent of failures for the three different junction temperatures and to construct the Arrhenius plot (Figure 11.8). The three experimental points are well aligned; this confirms the unicity of failure mechanism (electromigration in this case). From the same Arrhenius plot it is possible to determine the value of activation energy ($E_a = 1.8$ eV) typical for this type of metallization (titanium, platinum, and gold).

Table 11.2 Life Test Results on RF NPN Transistors ($T_j = 340$°C, 21 devices under test)

(*progressive*) *Failed devices*	*Time to failure* (hours)
1	2.5
2	2.5
3	2.5
4	10.5
5	10.5
6	10.5
7	10.5
8	10.5
9	10.5
10	10.5
11	10.5
12	15.5
13	15.5
14	25.5
15	25.5
16	27.5
17	32.5
18	32.5
19	32.5
20	32.5
21	46.5

Table 11.3 Life Test Results on RF NPN Transistors (T_j = 306°C, 28 devices under test)

(*progressive*) *Failed devices*	*Time to failure* (hours)
1	15
2	30
3	32.5
4	52
5	70
6	70
7	70
8	120
9	120
10	120
11	120
12	120
13	173
14	173
15	173
16	173
17	173
18	173
19	173
20	179
21	179
22	203
23	250
24	250
25	250

Table 11.4 Life Test Results on RF NPN Transistors (T_j = 293°C. 21 devices under test

(*progressive*) *Failed devices*	*Time to failure* (hours)
1	121
2	121
3	121
4	121
5	179
6	199
7	247
8	247

Another application of these techniques evaluates the E_a of the CMOS integrated circuits, carried out in SGS [5]. Three types were taken as characteristic of the family (4007, 4017, and 4027). These were tested at temperature steps from

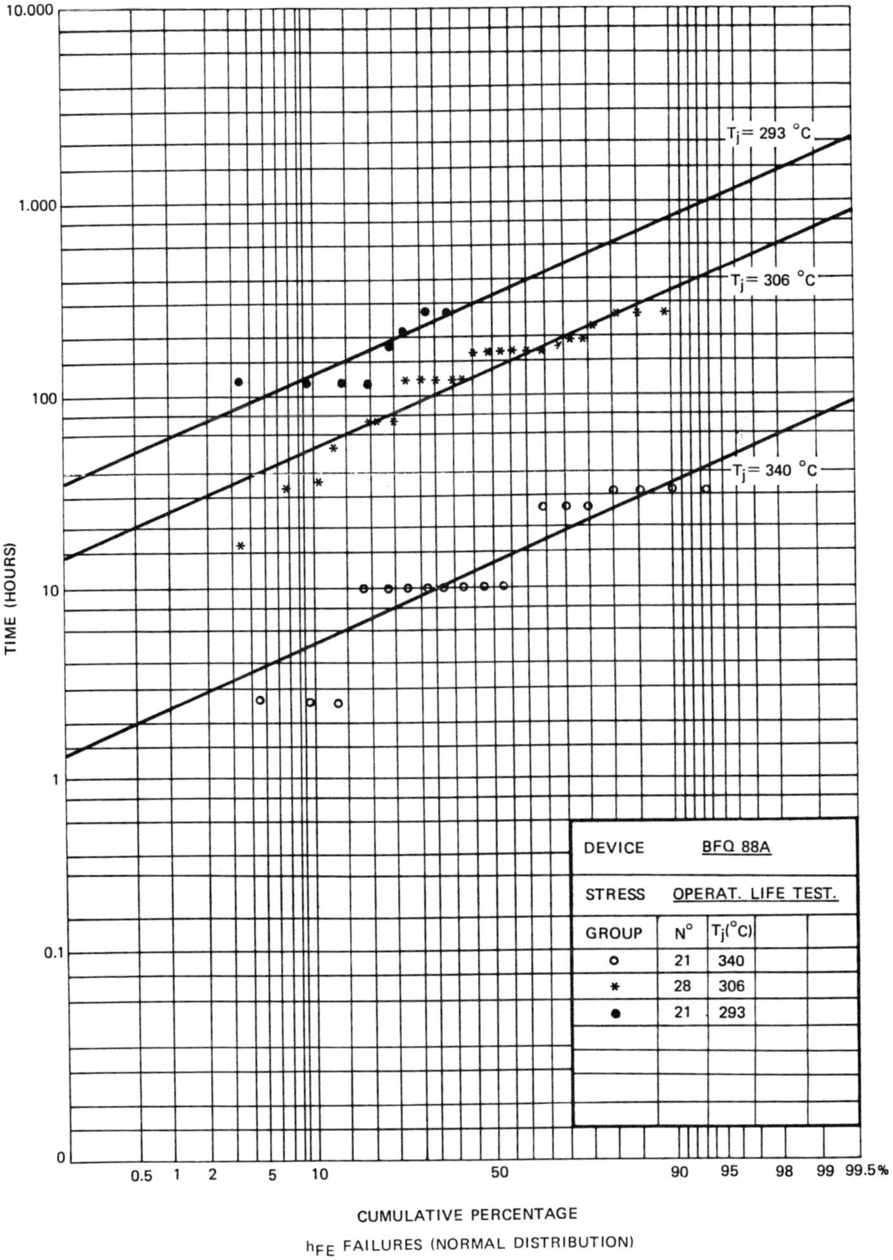

Figure 11.7 Plot of cumulative failure versus time.

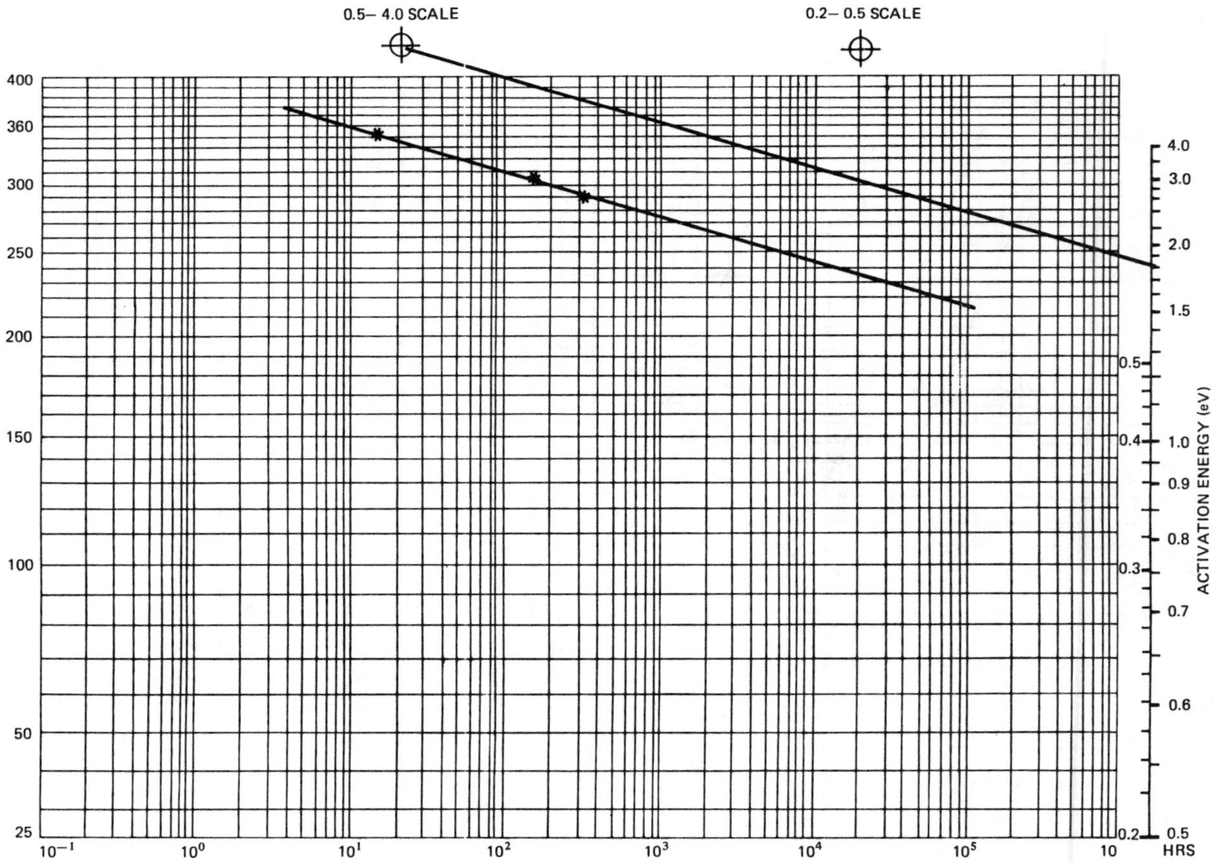

Figure 11.8 Arrhenius plot.

150 to 325°C, to locate the more adequate temperatures for the accelerated life tests and the more fit and representative parameters to be checked.

Each step lasted 16 hours, and after the electrical control, the specimens were again stressed for the same period at room temperature by 25°C higher than the one of the previous steps with the same supply voltage. At 275°C, the devices 4007 and 4027 showed some latch up and the device 4017 showed a very high leakage current at 325°C. On the basis of these results, we fixed the limits for the constant stress tests at the three temperatures: 175°C, 200°C, and 225°C for each of the three types under consideration. The Arrhenius diagram built up with these data gave a value of $E_a = 1.1$ eV.

11.2.5 Experimental Values of the Activation Energy

Each failure mechanism is characterized by its own activation energy and, therefore, is accelerated differently by temperature [4, 6–9]. Table 11.5 summarizes the main failure mechanisms for silicon devices. In this table, there is an indication about the elements contributing to the failure process, which parameters have to be increased to increase the stress (and the failure rate), as well as some activation energies of the phenomena.

Table 11.5 Main Failure Mechanisms of Silicon Devices

Sensitive element	*Failure mechanism*	*Relevant stresses*	*Activation energy*
Dielectric & interfaces Si-dielectric	Accumulation of surface charges	T, V	$E_a = 1.0$–1.05 eV
	Breakdown	T, V	$E_a = 0.3$–0.6 eV
	Injection of charges	T, V	$E_a = 1.3$ eV
Metalization	Electromigration	T, J	$E_a = 0.5$–1.2 eV (Al) $E_a = 1.8$ eV (Au) J^n; $n = 1$–4
	Corrosion (Al)	H, V, T	Many models for H $E_a = 0.3$–0.6 eV
	Contact degradation	T	$E_a = 1.0$ eV
Metal interfaces	Intermetal (Au, Al) compounds	T	$E_a = 1.0$ eV (Au-Al)
	Fatigue		Temperature cycles

Note: V = voltage; T = temperature; J = current density; and H = humidity.

These failure mechanisms are present in different percentages in different devices, depending on the technology, the manufacturer, usage conditions, and so on. The values of the activation energy quoted from the literature are also different, even for the same failure mechanism, because they are dependent on variables in the manufacturing process.

In the practice, sometimes the failure of some devices must be estimated by extrapolating them from life tests performed at a unique temperature (therefore, without any evaluation of E_a), or the trend of the early failures *versus* time must be estimated for different applications and different manufacturers (by averaging the different values of E_a). This is possible, at least in first approximation, using only two activation energies: the one for the phenomena connected to the early failure and the other for the useful life. For this reason the MIL STD 883 indicates a unique value of $E_a = 0.44$ V for early failure and 1 eV for the useful life.

11.3 ACCELERATED HUMIDITY TESTS

11.3.1 Introduction

The importance of humidity tests in the evolution of plastic package reliability has increased in the last decade. The tests evaluate the adherence of the resins to the frame material, their permeability, and their purity (ion contents), or better yet, they measure the efficiency of the package in reacting to humidity and show (with or without bias) various defects related to charge mobility increase with relevant accumulation or to generation of surface currents (leakage currents) with consequent electrolithical corrosion of the device's metal stripes (generally aluminium) until their complete interruption (open circuit).

Although the Arrhenius law (linking junction temperature and failure rate) is very well known and generally accepted even if with certain limitations, there is no similar law with the same level of confidence and verification linking failure rate and humidity stresses. Still, there is a certain standardization in humidity tests. In fact, almost all semiconductor manufacturers and users are familiar with tests, using a hot, humid environment with bias applied in the following conditions:

85°C and 85 percent of *relative humidity* (RH) for *temperature humidity bias* (THB):

121°C and 100 percent of RH at 2.08 Atm (pressure cooker).

In spite of this standardization, the calculation of the real accelerating factors versus the conditions of normal application of semiconductors in the field is no easy job: this is demonstrated by the various models for their evaluation.

An additional difficulty is the defining "normal" environmental conditions and therefore the real humidity level (difficult to establish); for instance, according to the US Weather Bureau, in the summer months and near some coasts of the

United States, the average temperature is 32°C with 65 percent of relative humidity; Peck and Zierdt [10] gave the same reference value in one of their publications on the humidity tests acceleration.

The test THB 85/85 is the best known and most often used. It is still used to study the failure mechanisms of the devices in plastic packages. The application of bias gives the additional stress for the electrolithical corrosion of the metalization.

In any case, the bias applied should be the nominal voltage in order to avoid anomalous acceleration, taking the additional care to adopt the minimum power dissipation condition. In this way, the worst case is met (a high power dissipation, in addition to reducing the relative humidity, opposes the entrance of humidity into the chip surface). As a reference, a group of experts at Matsushita E.C. recommends the test THB 85/85 with on/off cycles and duty cycle depending on the power dissipation involved according to the following table [11]:

$Pd \leqslant 100$ mW	on always
100 mW $\leqslant Pd \leqslant 1$ W	on/off $= \frac{1}{2}$
1 W $\leqslant Pd \leqslant 1.5$ W	on/off $= \frac{1}{4}$
$Pd \geqslant 2$ W	on/off $= \frac{1}{7}$

Due to the long duration of THB 85/85, generally performed for at least 1000 hours, other more accelerated humidity tests have been introduced. One test becoming more popular is the *high accelerated stress test* (HAST) that is practically performed through a pressure cooker with bias application and unsaturated vapor conditions, where both temperature and humidity can be varied [12]. Conditions such as 110°C/90 percent RH, 130°C/85 percent RH, and 140°C/85 percent RH can be obtained.

A very interesting application of these tests in the various hot and humid environments is the experimental evaluation of the various models linked to the humidity tests, or better the evaluation of the correlation between the reliability parameters (median life) and the test conditions (temperature and humidity).

Various models related to accelerated humidity tests are shown in the following pages [2]. Some of them are mentioned for reference only, because they correlate poorly to field results; some others are shown with graphic representation as well as an equation, to calculate the acceleration factors related to different ambient conditions.

11.3.2 Vapor Pressure Model

J.L. Flood [13] and F.N. Sinnadurai [14] demonstrated the following relationship between the median life ($t_{50\%}$) and the vapor pressure:

$$t_{50\%} = A \exp \frac{B}{V_p - K} \tag{11.6}$$

where A and B are constants; V_p is vapor pressure; and K is constant (131 mmHg).

Figure 11.9 shows results related to operational amplifiers by Motorola encapsulated in epoxy B resin, submitted to biased humidity tests considering corrosion as the failure mechanism. The fitness of the results is very good for the conditions 85°C/90 percent RH and 30°C/75 percent RH (24 mmHg), while no good relationship is present for the condition 121°C/100 percent RH and 2.08 Atm, typical for pressure cooker test. The results of this model are more conservative than the results coming from the Reich-Hakim and Eyring models, discussed later. As a reference, the plot of Figure 11.10 allows us to identify the vapor pressure as a function of temperature and relative humidity.

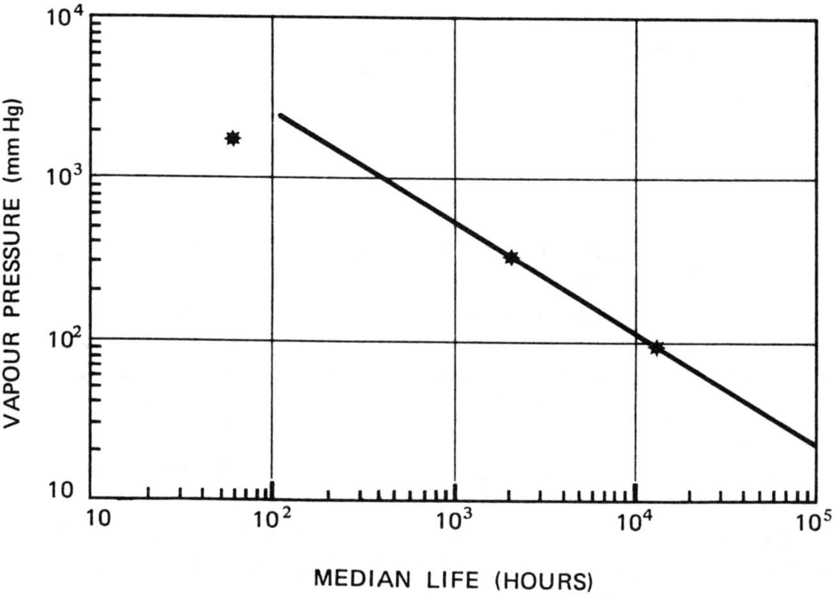

Figure 11.9 Vapor pressure model.

11.3.3 Reich Hakim Model

This model [15] shows the relationship between the median life and the sum of temperature (°C) and relative humidity (RH percent). Figure 11.11 plots the median life *versus* temperature-humidity bias conditions for operational amplifiers encapsulated in epoxy B resin according to the following exponential law:

$$t_{50\%} = \exp - [A + B(T°C + RH\%)] \tag{11.7}$$

The acceleration coefficient for different temperature-humidity conditions is

$$A(T/RH) = \exp\left[-\frac{E_a}{K}\left(\frac{1}{T_0 + RH_0)} - \frac{1}{T + RH}\right)\right] \tag{11.8}$$

where $E_a = 1.19$ eV and K = Boltzmann constant; T_0 and RH_0 are the reference accelerated conditions.

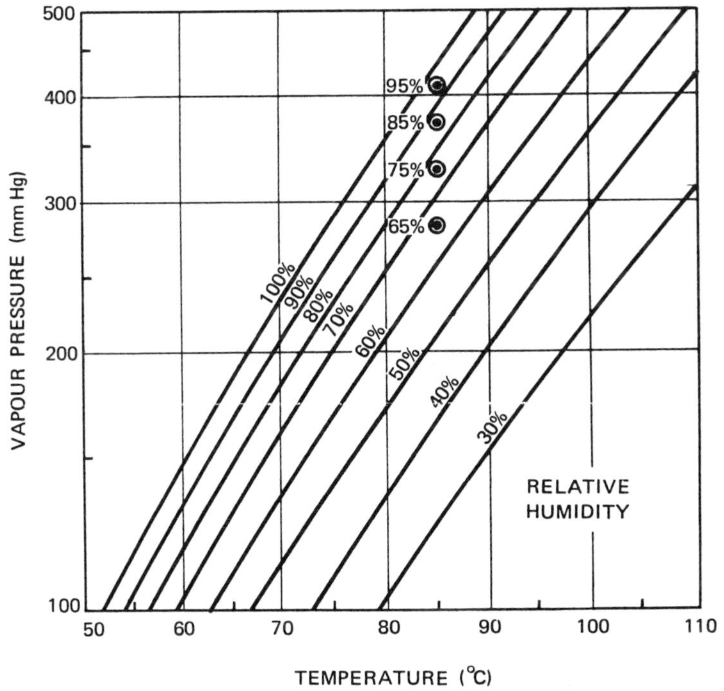

Figure 11.10 Relationship between vapor pressure and temperature for constant relative humidity values.

11.3.4 Lawson Model

In this model, developed by Lawson [16], the main hypothesis is that the accelerating factor due to the combined effects of temperature and humidity has a linear

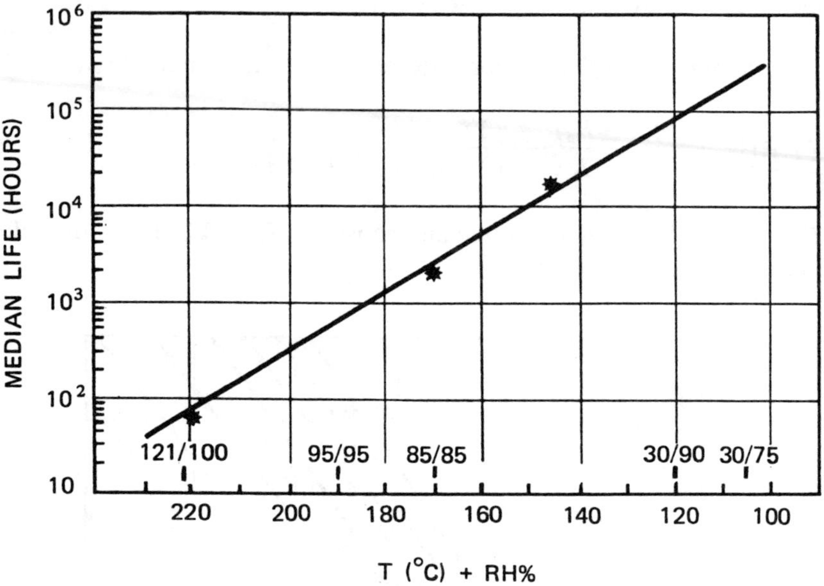

Figure 11.11 Reich-Hakim model.

relationship with temperature (as per Arrhenius law) and a nonlinear relationship with relative humidity.

If A is the total acceleration coefficient and A_1 and A_2, respectively, the acceleration for temperature and relative humidity, their relationship is the following:

$$A(T/\mathrm{RH}) = A_1(T) \cdot A_2(\mathrm{RH}) \tag{11.9}$$

where

$$A_1(T) = \exp\left[-\frac{E_a}{K}\left(\frac{1}{273 + T} - \frac{1}{273 + 85}\right)\right]$$

and

$$\ln A^2(\mathrm{RH}) = K'(85^2 - \mathrm{RH}^2)$$

where K' is a constant and K is Boltzmann's constant.

The fitness of this model is very good considering

$$E_a = 0.4 \text{ eV and } K' = 5.57 \cdot 10^{-4}$$

The failure mechanism in this case is also metal corrosion.

11.3.5 Eyring Model

This model [17] modifies the Arrhenius equation by taking into consideration parameters related to thermomechanical phenomena. Therefore, the Eyring model is suitable both for environmental and temperature stresses. The Eyring equation is as follows:

$$t_{50\%} = A \exp\left[\frac{E_a}{KT} \cdot F(V) \cdot F(\mathrm{RH})\right] \tag{11.10}$$

where the first part practically represents the Arrhenius law, whereas the remaining parts represent, respectively, the stress factor related to humidity and applied voltage.

The model can be better defined in this way:

$$t_{50\%} = A \exp\left(\frac{E_a}{KT}\right) \cdot \exp\left(\frac{B}{\mathrm{RH}} E_{ca}^{-1}\right)$$

where $B = 304$ and E_{ca} is the electrical field in the area of a possible corrosion (if, for instance, the distance between metal stripes is 0.4 mils, and applied voltage of 30 V creates an E_{ca} of $30/0.4 = 75$ V/mil). Figure 11.12 shows the plot related to tests performed on operational amplifiers adopting the Eyring model with median life as a function of

$$E_a/K \cdot 1/T(\mathrm{K}) + B/\mathrm{RH}\% \text{ with } E_a = 0.65 \text{ eV}$$

This model gives a median life of 60,000 hours for the condition 30°C/90 percent RH and 185,000 hrs for 30°C/75 percent RH, and shows a small deviation for the pressure cooker condition, where it is necessary to introduce a corrective factor of 2 to 2.7.

Because the model is based on the thermodynamics of chemical processes, it also can be applied to different failure mechanisms by changing only the E_a value.

This model was studied by IBM [12] and the accelerating factor is the following:

$$A(T/\mathrm{RH}) = \exp \cdot \left[\frac{-E_a}{K}\left(\frac{1}{T_0} - \frac{1}{T}\right) - B\left(\frac{1}{\mathrm{RH}_0} - \frac{1}{\mathrm{RH}}\right)\right] \tag{11.11}$$

where $E_a = 0.3$ eV;

$B = 296$;

K = Boltzmann's constant; and

T_0 and RH_0 are the reference conditions.

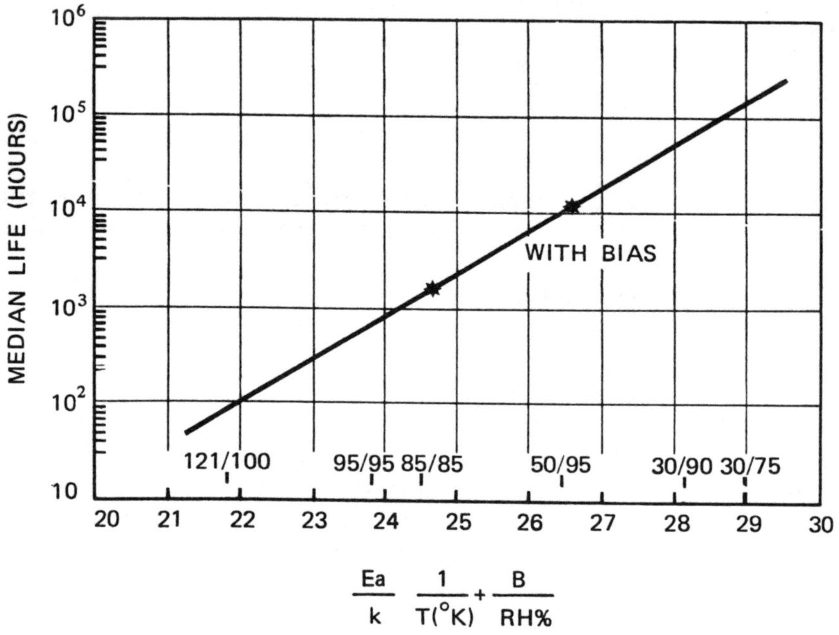

Figure 11.12 Eyring model.

11.3.6 Peck Model

This model [18] separately calculates the acceleration created by temperature and humidity. Figure 11.13 plots the acceleration factor (under constant relative humidity condition) as a function of temperature normalizing the factor at 85°C.

Figure 11.14 plots the acceleration factor (under a constant temperature) as a function of relative humidity, normalizing the factor at 85 percent RH. This model is based on the multiplication of the two factors to calculate the total accelerating factor. The model can also be represented by this equation:

$$A = \exp\left[-\frac{E_a}{K}\left(\frac{1}{T_0} - \frac{1}{T}\right) \cdot \left(\frac{RH_0}{RH}\right)^n\right] \qquad (11.12)$$

$E_a = 0.54$ eV, $n = 4.5$.

Naturally, all the described models have been obtained considering worst case conditions with minimum power dissipation and with continuous application of temperature, humidity, and supply voltage (no on/off conditions). Therefore, in case of a certain amount of power dissipation, some variation to the reference conditions must be applied; in fact, in this case, the real junction temperature and the reduced RH on the chip surface must be calculated (see Figure 11.15 for these values).

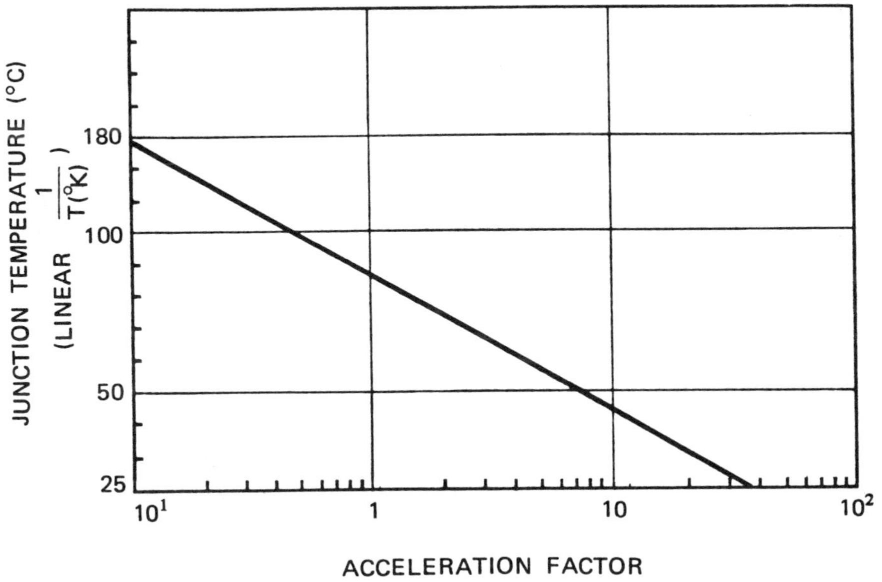

Figure 11.13 Peck model, acceleration factor due to temperature (RH constant).

As shown earlier, there are several empirical models for accelerated humidity tests (some models have not been even mentioned). This is because a well-defined estimation of reliability (coming from humidity stressing) is generally difficult on plastic packaged devices, and this is quite visible in Figure 11.16, where the relationship between humidity and life time is given for a big variety of models on this subject.

In addition, it is also very difficult for one model to be suitable for all the molding compounds, for the different packages (different frames, bonding processes, wires, and so on), for the various dice passivations, and for the different reliability testing conditions (temperature higher than 100°C with additional stress due to pressure). Unfortunately, often there is no field data to verify the laboratory

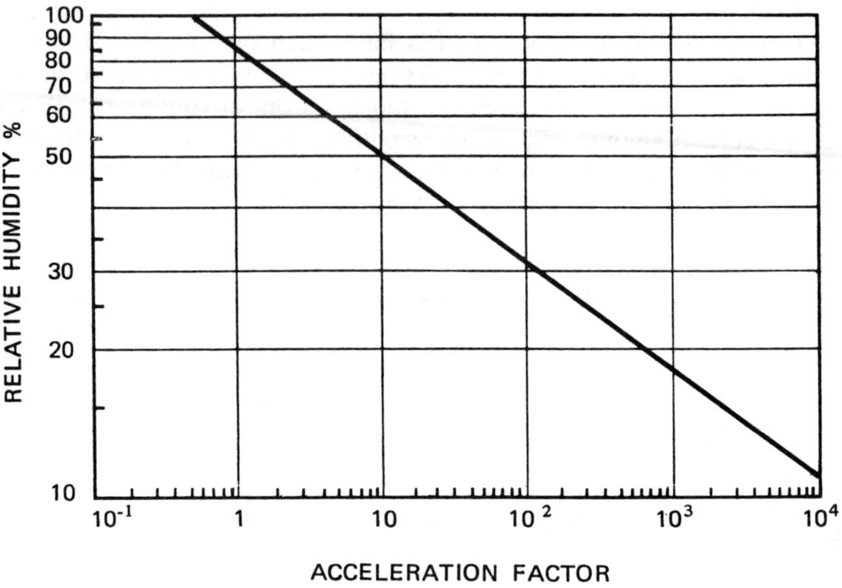

Figure 11.14 Peck model, acceleration factor related to RH (temperature constant).

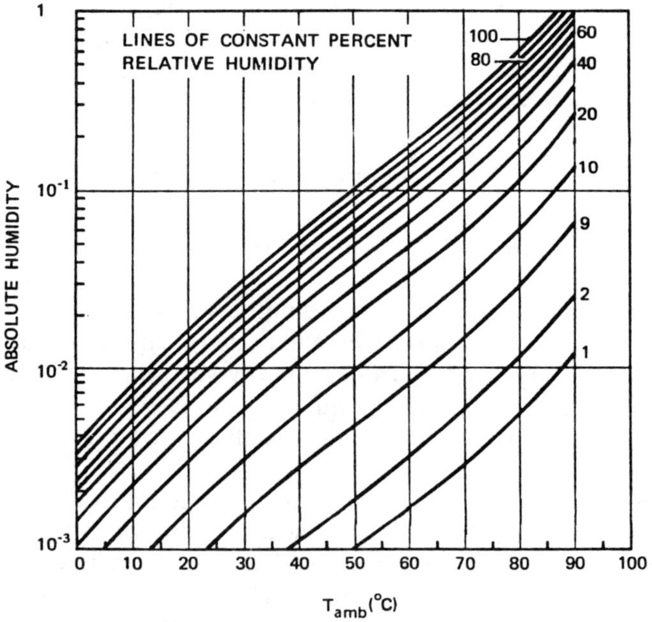

Figure 11.15 Psychrometer plot showing the relationship between absolute humidity and temperature for a constant relative humidity.

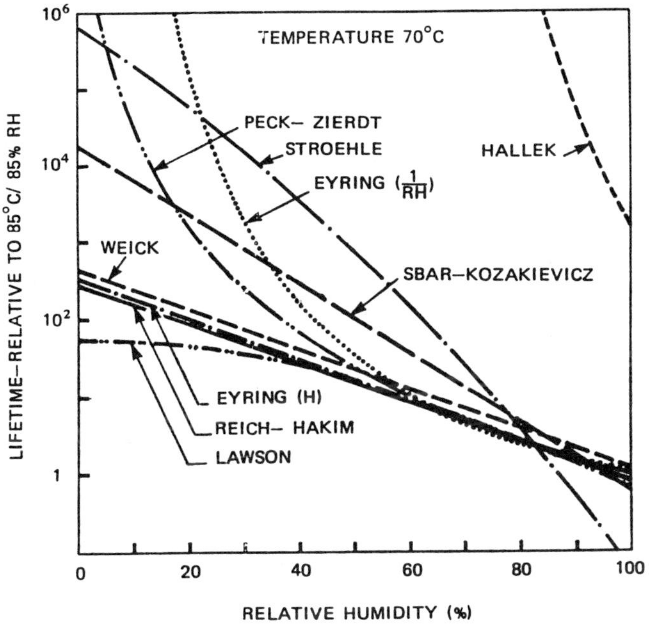

Figure 11.16 Relationship between humidity and life time predicted by models (temperature 70°C).

reliability test data, and when available, they offer very poor information on real applications.

11.3.7 Model Validity

In order to establish the validity of these models L.M. Ericsson performed an extended evaluation submitting 1000 samples of CMOS in plastic package to accelerated humidity tests in various conditions: 85°C/85 percent RH, 65°C/95 percent RH, 65°C/85 percent RH, and 54°C/95 percent RH. A satisfactory fitness has been demonstrated for the Reich-Hakim (Figure 11.17) and RH^2 models. The fitness of these two models was also coming from HAST.

This test at $T = 110°C$, $RH = 90$ percent (as per British Telecom application) shows a theoretical accelerating factor of 7.2 *versus* the reference condition 85/85; the experimental (laboratory) results show an accelerating factor ranging from 6.2 to 6.4, which is very close to the theoretical one (Figures 11.18–11.21). Table 11.6 lists some acceleration coefficients coming both from reliability laboratory tests and dedicated literature.

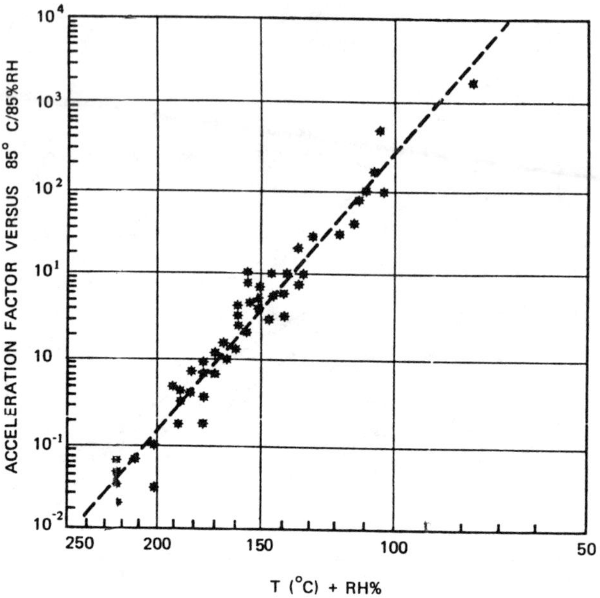

Figure 11.17 Fitness to Reich-Hakim model for CMOS devices submitted to humidity tests under various conditions.

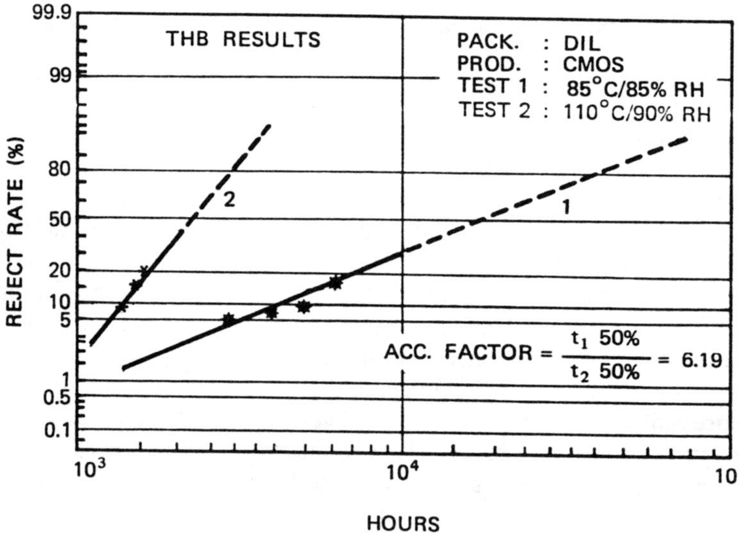

Figure 11.18 Experimental results on CMOS devices submitted to different humidity test conditions, THB (85°C/85 percent RH) and HAST (110°C/90 percent RH).

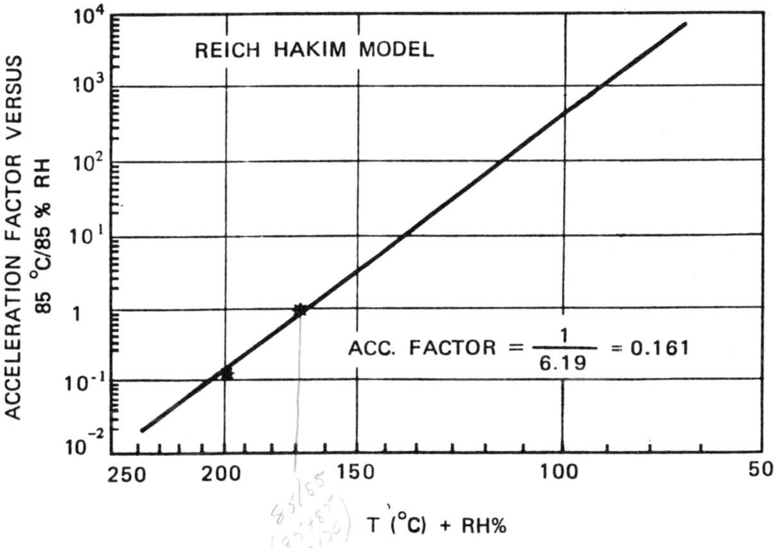

Figure 11.19 Acceleration factor of HAST (110°C/90 percent RH) *versus* THB (85°C/85 percent RH) according to Reich-Hakim model, for CMOS devices.

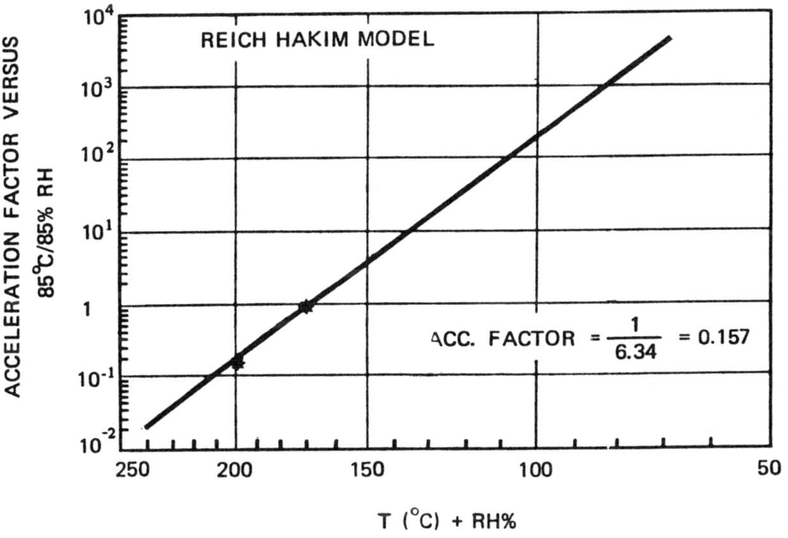

Figure 11.20 Experimental results on HSCMOS devices submitted to different humidity test conditions, THB (85°C/85 percent RH) and HAST (110°C/90 percent RH).

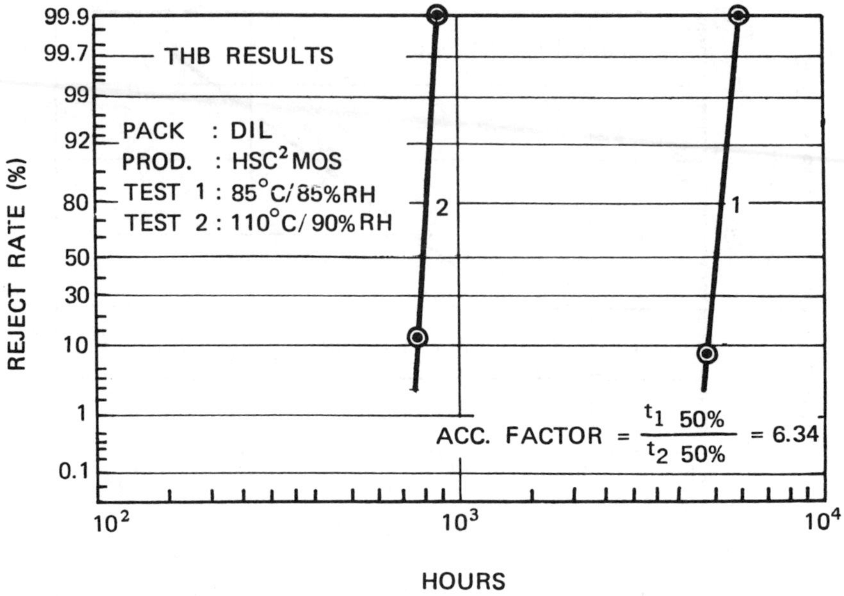

Figure 11.21 Accelerating factor of HAST (110°C/90 percent RH) *versus* THB (85°C/85 percent RH) according to Reich-Hakim model, for HSCMOS devices.

Table 11.6 Acceleration Factors for Humidity tests

	Theoretical (Reich-Hakim)	*Experimental SGS*	*Literature*
PTHB/PTH	14	1.4–6.3	3–10
PTHB/THB	20	8.8–11.6	8–35
PTHB/TH	280	> 40 > 140	45–200
PTH/THB	1.5	0.6–8.1 TIP.1.3–1.7	2–12
PTH/TH	20	> 21 > 42	15–20

Note: TH = Temperature humidity 85°C/85 percent RH; *THB* = Temperature humidity bias; *PTH* = Pressure temperature humidity; and *PTHB* = Pressure temperature humidity bias.

11.3.8 Application Example

As already mentioned, humidity can reach the silicon surface of the integrated circuits, creating possible metal stripe corrosion, increases in leakage current, and other typical wearout failures; for applications in telecommunication, it is very important that these failures do not appear in the first 40 years of field application, where the typical conditions are generally 30°C and 80 percent RH.

If we refer to the Reich-Hakim model, the acceleration coefficient of the test 85/85 with bias (see Figure 11.19) is 100; therefore to simulate the field conditions, one reliability test in condition 85/85 should have a duration of 40 years divided by 100, that means 3500 hours without failures (freak population excluded). Since the single component during its application is submitted to heating both by its own power dissipation and the surrounding heat of other components, the relative humidity on the chip, as a consequence, decreases.

For instance, starting at a temperature of 30°C a ΔT = 10°C reduces the RH from 80 to 43 percent (see Figure 11.15); therefore, the condition 30°C/80 percent RH becomes 40°C/43 percent RH and the accelerating coefficient of the test 85/85 increases to 1000. As a consequence, the duration of the test in condition 85/85 is reduced to 40 years/1000, or 350 hours. In addition, if the device, during the test 85/85, is functioning with a certain power dissipation, as already mentioned, the relative humidity decreases.

For example ΔT = 10°C brings the condition 85/85 to 95/85 and the acceleration coefficient *versus* 40/43 is 1000/3.48, or 287. The result is a test duration at 95/85 of 40 years/287, or 1200 hours. Similar calculations can be done starting from other environmental conditions. Table 11.7 lists the accelerating factors *versus* the reference condition 85/85 for different *T*/RH conditions.

Table 11.7 Acceleration Factor Variation, Using a 85/85 Nominal Condition and Power Dissipation of the Device

ΔT(°C)	0	1	2	3	4	5	6	7	8	9	10
T/RH	85/85	86/82	87/79	88/76	89/73	90/70	91/67	92/65	93/63	94/60	95/58
A	1	1.15	11.33	1.54	1.78	2.06	2.39	2.57	2.77	3.22	3.48

11.3.9 Life Test at Low Temperatures

The reliability tests at high temperatures with high humidity have to prove, in a short time, the failure mechanisms on devices in plastic packages. The static life test at low temperatures (0°C or less) performed on cavity packages can induce corrosion (in an accelerated way *versus* the ambient temperature) due to the environmental conditions very near to the dew point.

The following results have been obtained from tests performed during 1976 (according to internal SGS results) on CMOS, respectively, at 25°C and 2°C:

Test at 25°C: 4 failures out of 1000 samples after 2500 hours.
Test at 2°C: 9 failures out of 112 samples after 596 hours.

The estimated acceleration factor between the two conditions is about 100.

To minimize the problem, the humidity inside the ceramic package was reduced. Suitable techniques have been developed to measure and control the humidity content in ceramic packages, and now this accelerating factor can be considered obsolete for devices sealed according to modern manufacturing techniques and control criteria.

11.4 VOLTAGE ACCELERATED TESTS

The median life and the failure rate of semiconductors depend on the applied voltage; in fact, the induced electrical field concentrates, in the junction region, the mobile charges eventually present, originating various phenomena of electrical degradation. Generally, tests at cannot be performed at very high voltages without exceeding the breakdown voltage; therefore, the laboratory tests are performed at voltage values not much higher than those used in the normal application conditions. These results do not need strong extrapolations to compare with field conditions.

In addition, since the failure mechanisms involved are various and differently influenced by voltage, the voltage is poorly used as an accelerating factor and the literature on the matter is limited. The reference model mainly used comes from Eyring and also includes a thermal stress effect [19]:

$$\lambda = AT_j[\exp(-B/KT_j)][\exp(CV + DV/KT_j)]$$

where λ is the failure rate; T_j is the junction temperature; V is the voltage; and A, B, C, and D are constants to be calculated experimentally. The main difficulty with this model and the reason it is so rarely used is the need to calculate these constants experimentally. Another model studied for transistors is from Kemeny [20]:

$$\lambda = [\exp(C_0 - E_a/KT_j)][\exp C_1(V_{CB}/V_{CB\text{MAX}})] \qquad (11.13)$$

where C_0 and C_1 are constants; V_{CB} and $V_{CB\text{MAX}}$ are respectively the base-collector bias applied to the transistor during the test and the maximum applied voltage; E_a is the activation energy of the Arrhenius equation (in fact, if V_{CB} = constant, equation (11.13) equals the Arrhenius law).

For the Kemeny law, the C_1 constant has values ranging from 1.5 to 2.3 according to the $V_{CB\text{MAX}}$ of the transistor; therefore, the plot of the failure rate as a function of the voltage base-collector is shown in Figure 11.22.

For MOS devices, some experts have found that the failure rate depends on the exponential 4 of the applied voltage. Also, the results of various tests performed

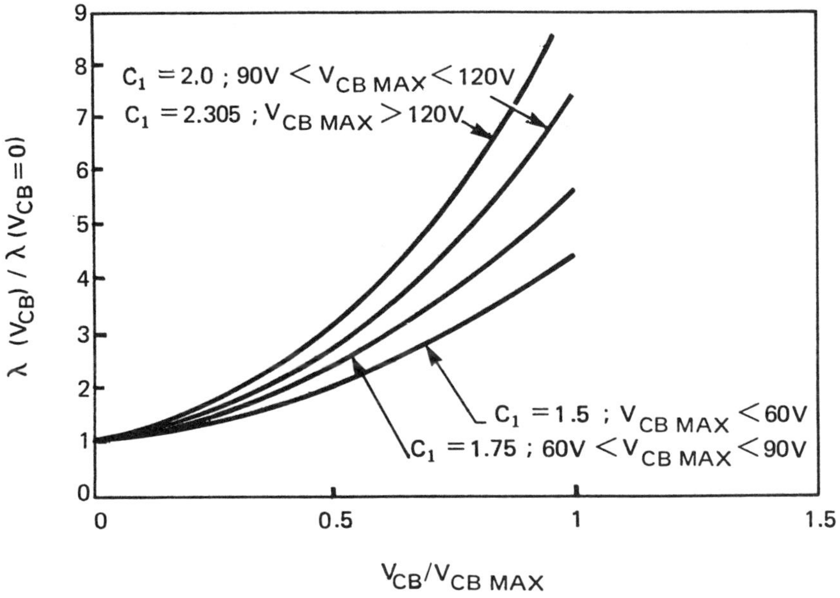

Figure 11.22 Transistor failure rate as voltage function (Kemeny equation).

in SGS on CMOS A series (tests at 5, 10, 15 V) and CMOS B series (tests at 10, 15, and 20 V) confirmed this law [21]:

$$\lambda = AV^4 + BV \tag{11.14}$$

This is represented in the plot of Figure 11.23 (λ_0 is the value of λ at 5 V for the *A* series and at 10 V for the B series).

Naturally neither equation (11.13) nor (11.14) can be applied to bipolar integrated circuit; even though (11.13) may seem more suitable than (11.14), it refers mainly to the breakdown of oxide, which is well related to MOS technologies but less important for bipolar ICs. The conclusion also coming from an evaluation of Figures 11.22 and 11.23 is that the variations of the failure rate *versus* voltage is generally a bit more than 1 decade, unless the process thresholds are not exceeded.

The Reliability Analysis Center (RAC) also has included a voltage factor π_V [22] in its model of reliability predictions. This factor is 1 for all technologies excluding CMOS when the applied voltage exceeds 12 V. In practice, the multiplying voltage factor used in the reliability prediction model (MIL HDBK 217) has the values listed in Table 11.8, where V_{DD} is the recommended maximum

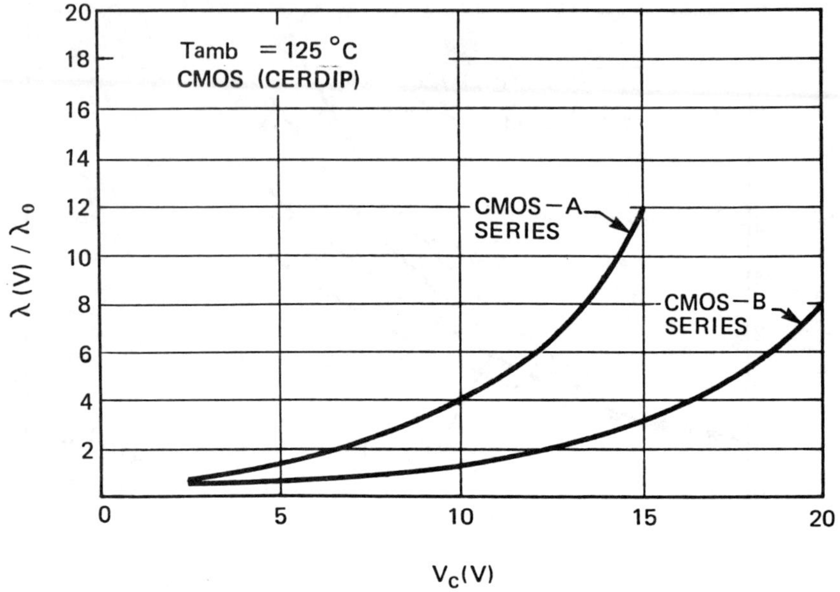

Figure 11.23 Influence of voltage on failure rate of CMOS devices.

Table 11.8 Multiplying Voltage Factors from MIL-HDBK-217E

Technology	π_V
CMOS, $V_{DD} = 5$ V	1
CMOS, 12 V $< V_{DD} <$ 15.5 V	Equation 1
CMOS, 18 V $< V_{DD} <$ 20 V	» 2
Other technology	1

Notes: Equation 1: $\pi v = 0.110 \exp(X)$, where $X = 0.168\ V_s\ (T_j + 273)/298$;
Equation 2: $\pi v = 0.068 \exp(X)$, where $X = 0.135\ V_s\ (T_j + 273)/298$;
V_s is applied bias and T_j is the junction temperature.

supply voltage. For CMOS according to RAC, the failure rate *versus* applied voltage is plotted in Figure 11.23.

The failure due to electrostatic discharges can be classified in two ways:

- Voltage failures, typical of MOS devices, that show as the breakdown of thin oxides.
- Power or current failures, typical of power bipolar devices, due to overheating in the regions of the p-n junction. The ranges of critical voltages to induce ESD failures for the different product families (independent from the typical failure model) are listed in Table 11.9.

Table 11.9 Critical Voltage Values to Induce ESD Failure in Different Electronic Devices

	Volts
VMOS	30–1800
MOSFET	100–200
GaAsFET	100–300
EPROM	100
JFET	140–7000
SAW	150–500
OP AMP	190–2500
CMOS	250–3000
Schottky Diodes	300–2500
Film Resistors (Thick, Thin)	300–3000
Bipolar Transistors	380–7000
ECL (PC Board Level)	500–1500
SCR	680–1000
Schottky TTL	1000–2500

In order to reduce the ESD failure rate, two main groups of precautions are used:

- Input protection networks directly designed in the integrated circuit structure.
- Limitation and control of the electrostatic discharges in the working stations and during the handling and shipping of the components.

To verify the effect of the input protection circuits, accelerated tests simulate the effect of electrostatic discharges under different conditions. A typical example is indicated in the MIL-STD-883 method 3015. For more details on this matter mainly with reference to the power and current failures, see Chapter 10 on power devices.

11.4.1 Resistance to Electrostatic Discharges

The possible damage to semiconductors (MOS and CMOS mainly) from *electrostatic discharges* (EDS) is well known. The human body can accumulate variable electrostatic charges; after walking on a carpet, values of 35,000 V have been measured (see Table 11.10). However, voltage is not the only parameter necessary to be defined; the value of the capacitance associated to the human body together with the voltage defines the energy available for each discharge:

$$E = \frac{1}{2}CV^2$$

Normally, $C = 100$ pF.

Table 11.10 Electrostatic Voltages Generated by a Human Body under Different Environmental Conditions

	Electrostatic voltages	
Means of static generation	10–20% *relative humidity*	65–90% *relative humidity*
Worker at bench	6,000	100
Vinyl envelopes for work instructions	7,000	600
Walking over vinyl floor	12,000	250
Work chair padded with polyurethane foam	18,000	1,500
Common poly bag picked up from bench	20,000	1,200
Walking across carpet	35,000	1,500

Another main parameter is the contact resistance, since the total energy dissipation in the discharge is shared between this resistance and the impedance of the device; the specialized literature refers to values ranging from hundreds to $10^5\ \Omega$. The simplest model used to simulate the discharge is a capacitance C charged at a voltage V through a resistor R in series (Figure 11.24).

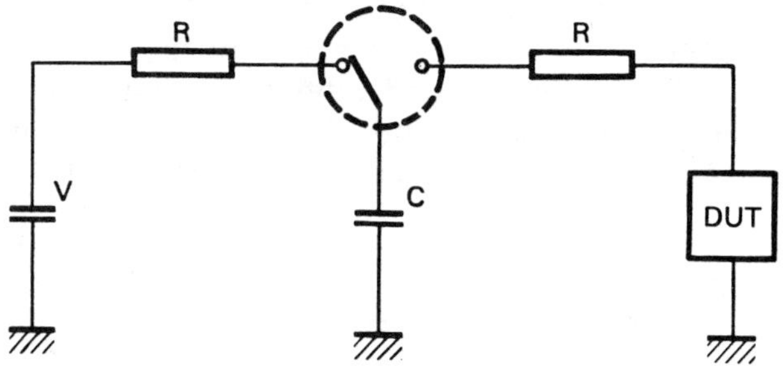

Figure 11.24 Human body model.

11.5 TESTS INCREASING MECHANICAL STRESSES

Table 11.11 lists tests of this type that are generally applied to the semiconductors. Depending on the ways of creating the mechanical stress, we can subdivide the tests into three groups: application of mechanical load (steady or variable); acceleration; or cyclic variations of temperature.

We must point out a substantial difference from the tests that accelerate temperature and humidity. In these last tests, the trend is to speed up phenomena that can occur over far longer times under application conditions. In the mechanical

Table 11.11 Mechanical Tests

Type of mechanical stress	*Screenings*	*Nondestructive checks*	*Destructive checks*
Mechanical load (steady and variable)		External leads pulls	External leads pulls
		Screw test	Internal wires pull (wire pull test)
			Lid torque
		External leads bending	Bending till breakage
		External leads torque	Torque till breakage
Acceleration	Centrifuge	Variable frequency vibration	
	Mechanical shock	Fatigue vibration	
	Particle detection by impact noise	Ultrasonic	
		Bumps	
Cyclic temperature changes	Limited number of thermal cycles (generally from 5 to 15)	Some hundreds of thermal cycles and thermal shocks	Thermal cycles and thermal shocks till wear out
		Thermal fatigue	Thermal fatigue till wear out

test, extrapolation of the results to the application conditions is difficult because the behavior of the material radically changes at some stress thresholds; for instance, the elasticity limit for steady state stresses or the fatigue breaking load for dynamic stresses. Therefore, the target of this test is merely a check of the components' resistance to predefined stresses (far larger than those forecasted in the application) in order to control the process stability and the incidence of worker defects assuming a priori that passing the tests guarantees the resistance to these typical stresses in the useful life of the components. This assumption is greatly justified by the experience collected over more than twenty years by manufacturers and users of semiconductors.

11.5.1 Detection of Defective Units ("Screening")

The applied stress levels are far lower than the resistance limit of the standard components, but they are sufficient to cause destructive effects on components affected by worker errors or material defects.

The MIL STD 883 includes in the screenings the centrifuge at 300.000 G (gravity acceleration). Most ICs in hermetically sealed packages (ceramic or metal can) connect the die and external leads with aluminum wires whose diameter is one mil or a bit more; therefore, the order of magnitude of stresses applied to the wires is some hundredth of gram; that is, 2 orders of magnitude lower than the stress levels needed to interrupt the connections in the pull test (see Table 11.10). The destructive effects of such screening are limited to detachment or breakage of bonds with nearly zero adhesion or excessive crushing.

Therefore, the centrifuge is useful to decrease early mortality. In the case of gold wire of one mil diameter or a bit larger, the test is more severe because the specific weight (and therefore the stress applied at 30,000 G) is at least seven times greater than for aluminum. Figure 11.25 reports the results of a step stress centrifuge test on devices in hermetic packages with 2 mil gold wires. The elimination of substandard units (freaks) started between 5,000 and 7,000 G, whereas the failures of standard units started at 130,000 G.

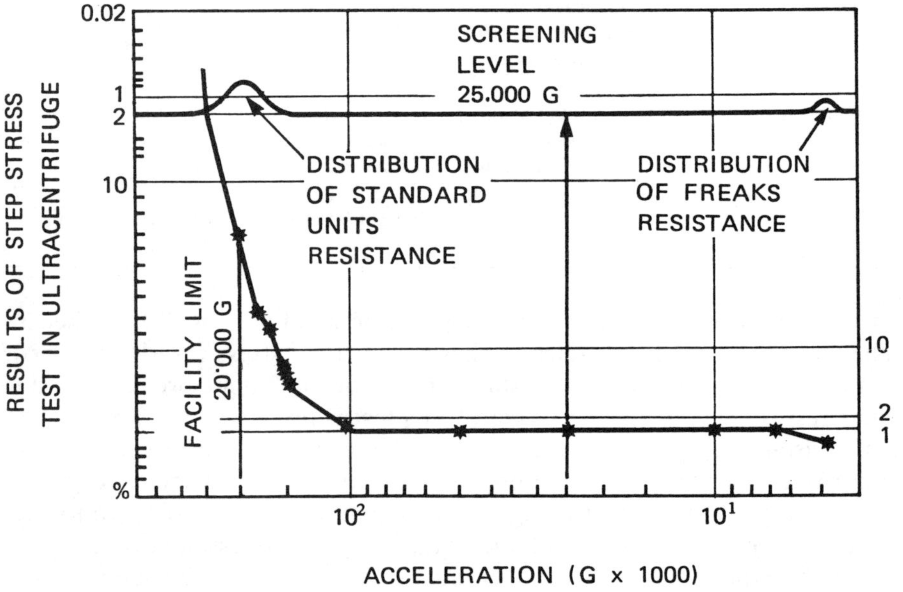

Figure 11.25 Step stress centrifuge test for hermetic devices.

On hermetically sealed power transistors that use aluminum wire of up to 10 mils diameter, the centrifuge stress at 30,000 G is some grams, but the limit of resistance to pull test is some hundred grams (MIL STD 883), therefore, we may apply the considerations already reported for ICs with 1 mil wire. Similar considerations may be done on the mechanical shocks: MIL STD 883 does not require peak values greater than 30,000 G and normal requests are 1,500 G or 3,000 G.

W. Fitch [23] studied the influence of different screening conditions on the subsequent behavior of ceramic and plastic dual-in-line transistors in the long thermal cycles test, finding an influence of screening conditions only on the hermeticity of glass-sealed ceramic components. Screening for a limited number of cycles mainly eliminates devices with weakened or scarcely adherent bonds (in the plastic packages particularly, such defects can cause intermittency) and devices with broken die.

11.5.2 Limit Checks

The stresses are applied close to the resistance limits of standard quality components without reaching them, in order to check whether eventual process shifts lowered these limits, or the stresses are increased until they exceed those limits.

The test of variable frequency vibration from 20 Hz to 2000 Hz, requested by MIL STD 883, can hardly cause wire resonance in the hermetic packages with few mils wires. According to Harman [24] the minimum frequency that could cause resonance is 3,000 Hz for gold wires and more than 10,000 Hz for aluminum wires. Centrifugal and vibration stresses in the applications seldom reached a value sufficient to damage bonds, and other, larger components of the equipment are subject to breakage before damage of standard bonds.

Also it is very unlikely that the fatigue vibration test requested by MIL STD 883 (96 hours at 60 Hz at peak acceleration 20 G) would destroy the wires. On the contrary, destructive effects have been repeatedly observed after ultrasonic washing; wires break close to the bonds because of fatigue after less than 1 minute, with applied frequencies in the range from 15.000 Hz to 40.000 Hz. This is why the ultrasonic tests never gained worldwide acceptance in screening.

The breaking load to fatigue for cycled or pulsed stresses generally decreases according to a linear trend *versus* the logarithm of cycles number (Wohler [25], plotted in Figure 11.26). However, under some given stresses threshold, the fatigue breakage occurs over too long a period to measure. In the vibration tests (excluding the ultrasonic washing), taking into account the limited values of peak acceleration, we may suppose that the stress levels stand far under such limits.

In order to check the fatigue of internal wires, generally thermal air-to-air cycles or thermal liquid-to-liquid shocks are used. The shocks require far less time than cycles, but they are much more expensive due to the high cost of the special liquid that must be used.

Tests up to 7000 cycles between −65°C and +150°C have been performed in SGS on different types of ICs assembled in different packages, detecting the following phenomena: under the same test conditions the transistors in plastic packages will wear out sooner than those in hermetic packages, due to the thermomechanical stresses of the resin on the wires. The stress applied by the resin to an internal wire due to a temperature change ΔT is

$$F = K \cdot \Delta T(\alpha_r - \alpha_s)$$

where α_r and α_s are the coefficients of thermal expansion of the resin and the wire, and K is a constant depending on characteristic and size of the wires.

The wires in the hermetic devices break during thermal cycles mainly due to the alternating flexion stresses at bonds, caused by wire loop changes from expansion and shrinking of the materials.

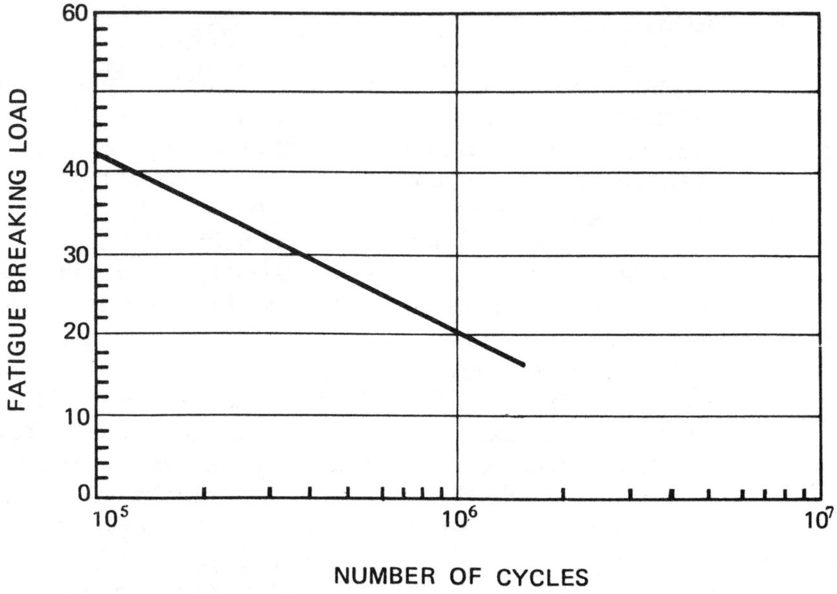

Figure 11.26 Mechanical fatigue breaking load.

Comparing the results obtained on plastic packages with wire lengths of 4.7 mm, 6.3 mm, and 7.1 mm showed that the resistance to thermal cycles was as great in shorter wires. The phenomenon also was observed on plastic packages, but it was less evident.

In order to check fatigue in the areas of soft solder die attach, another type of test was used. Soft solder die attach uses soldering alloys with low melting point, like Pb-Sn to bond the die to the frame. The number of cycles prior to fatigue and the temperature sweep (scanning) ΔT are linked by empiric relationship:

$$\log N = A + \frac{B}{\Delta T(\alpha_M - \alpha_S)}$$

where α_M and α_S are the thermal expansion coefficients of base metal and silicon, and A and B are constants depending on the layout and characteristics of the materials.

Evaluation of the resistance to fatigue in the die-attach area are particularly important for large chips that must dissipate high power. The cyclic temperature scanning is obtained by alternating high power periods with zero dissipation and forced cooling period, in a test called *thermal fatigue*.

Figure 11.27 is a typical pyramid diagram that links power dissipation, case temperature scanning, and number of cycles to fatigue breakage. Using power dissipation speeds up the test and enhances detection of devices with a defective die attach, which are submitted to more thermal scanning under the same conditions of dissipation.

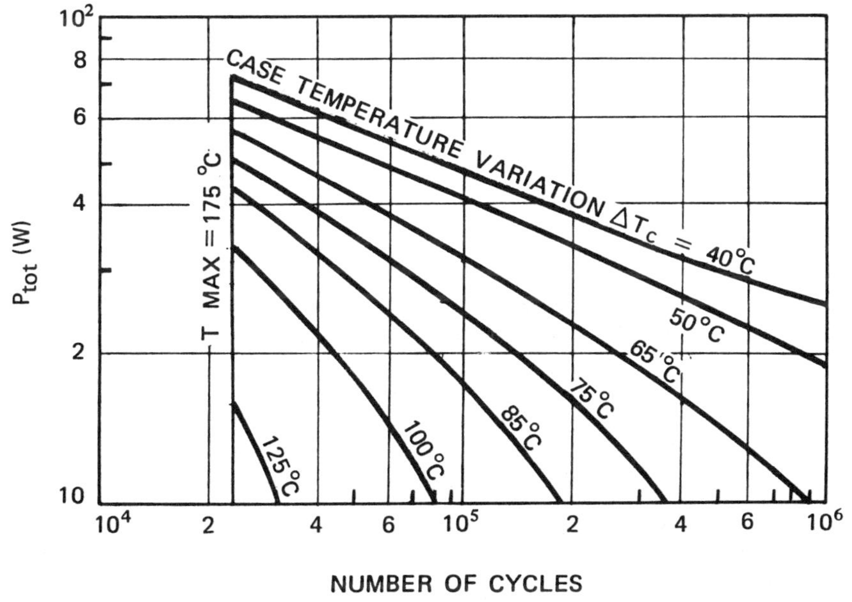

Figure 11.27 Failure due to thermal fatigue.

In addition to the nondestructive limit checks, we must mention the screwing tests on asymmetrical plastic power packages. These tests are made on grinded step blocks to simulate a worst case situation, where heat sinks that are not flat are used by the power devices users (Figure 11.28).

When screwing the device to the heat sink, this lack of flatness causes resin to arise between die and molding, tangent to mechanical stresses that are perpendicular to the die surface. Added to the compression stress the die is submitted to by the resin after molding and curing, the stress from asymmetry can cause cut stresses that can break the die. For a more complete matter dealing with mechanical tests, see Chapter 10 on power devices.

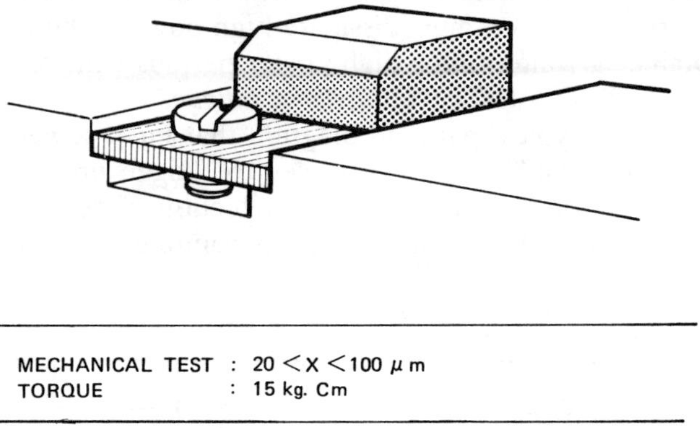

Figure 11.28 Screw test on a plastic package.

These destructive mechanical tests record the stress values that cause breakage:

- Wire pull test to check bonding operation.
- Die shear test to check die-attach operation.
- Lid torque on ceramic packages to check the lid-base sealing.

REFERENCES

1. I. Bazovsky *Principî e metodi dell'Affidabilità,* Etas Kompass, 1969. [Originally, *Reliability, Theory and Practice,* Prentice-Hall, Englewood Cliffs, NJ, 1961.]
2. G. Boccaletti, F. Borri, G. D'Espinosa, and G. Fioravanti, "Accelerated Reliability Testing," CNR, Scientific Technical Seminar, 3rd Course, Castromarina, September 1980, 1981.
3. D.S. Peck and O.D. Trapp. *Accelerated Testing Handbook,* Technology Associated, 1978, 1979, 1980, 1981.
4. D.S. Peck, "Semiconductor Reliability Predictions from Life Distribution Data in Semiconductor Reliability," Reinhold, New York, 1961.
5. G. Fioravanti, W. Beldi and C. Baldisserotto, "C-MOS Evaluation Report Under Estec Contract N°21222/73 AK," Internal Report SGS-Ates, 1975.
6. L.E. Terry and R.W. Wilson, "Metallization Systems for Silicon Integrated Circuits," *Proc. of the IEEE,* September 1969.
7. G.M. Johnson and M. Stitch, "Microcircuit Accelerated Testing," *Proc. of the IEEE,* 1977.
8. J.R. Edwards, E. Fuller, G. Allman, and M.L. Malwah, "VMOS Reliability," *Reliability Physics,* 1978.
9. T. Irie, I. Nagasako, H. Kohsu, and K. Sekido, "Reliability Study of MOSFETS," *IEEE Trans. on Microwave Theory-Tech.,* MTT24, 1978.
10. D.S. Peck and C.H. Zierdt, "Temperature, Humidity Acceleration of Metal-electrolysis Failure in Semiconductor Devices," *Reliability Physics,* 1973.
11. T. Ajiki, M. Sugimoto, H. Higuchi, and S. Kumada, "A new cyclic biased THB test power dissipating IC's," *Reliability Physics,* 1979.

12. J.E. Gunn, R.E. Camenga, and S.K. Malik, "Rapid Assessment of the Humidity Dependence of I.C. Failure Modes," *Proc. Reliability Physics,* 1983.
13. J.L. Flood, "Reliability Aspects of Plastic Encapsulated Circuits," *Reliability Physics,* 1972.
14. F.N. Sinnadurai, "The Accelerated Aging of Plastic Encapsulated Semiconductor Devices in Environments Containing a High Vapour Pressure of Water,"*Microelectronics and Reliability,* 1974.
15. B. Reich and E.B. Hakim, "Temperature-Humidity Acceleration Factors for Plastic Transistors and Integrated Circuits," *Solid-State Technology,* September 1972.
16. R.W. Lawson, "The Accelerated Testing of Plastic Encapsulated Semiconductor Components," *Reliability Physics,* 1974.
17. S. Glasstone, et al. *The Theory of Rate Processes,* McGraw-Hill, New York, 1941.
18. D.S. Peck and C.H. Zierdt, "The Reliability of Semiconductor Devices in the Bell System," *Proc. of the IEEE,* February 1974.
19. N.R. Mann, R.E. Shafer, and N.D. Singpurwalla, *Methods for Statistical Analysis of Reliability and Life Data,* John Wiley and Sons, New York 1974.
20. A.P. Kemeny, "Experiments Concerning the Life Testing of Transistors," *Microelectronics and Reliability,* 1971.
21. F. Borri and G. Termini, "Results of the Reliability Studies and of the Improvements Made by SGS-Ates to Its COS/MOS Family SGS-Ates," *Reliability Report,* 1978.
22. H.C. Rickers, "LSI/Microprocessor Reliability Prediction Model Development," RADC TR 79-97 Final Technical Report, March 1979.
23. W.T. Fitch, "Extended Temperature Cycling of Plastic and Ceramic IC's with Thermal Shock Preconditioning," *Reliability Physics,* 1976.
24. G.G. Harman, "Metallurgical Failure Modes of Wire Bonds,"*Reliability Physics,* 1974.
25. P.G. Forrest, *Fatigue of metals*, Addison-Wesley, Reading, MA, 1983.
26. W. Nelson, "Graphical Analysis of Accelerated Life Test Data with a Mix of Failure Modes," *IEEE Trans. on Rel.,* October 1975.
27. A. Spano, "MX Package, Microwave Devices," *SGS-Attes Reliability Report,* R.P. 502, 1980.

Chapter 12
MICROANALYSIS TECHNIQUES
M. Cocito

12.1 GENERAL COMMENTS ON MICROANALYSIS TECHNIQUES

The characterization of materials is a key function in any industry. Characterization is a general term that indicates a set of coordinated activities designed to achieve an exhaustive picture of the properties of a material. In microelectronics, the basic properties are structure (morphology, crystallographic structure, defects), composition (stoichiometry, chemical bond, impurities), and electrical properties (carrier concentration), transport properties, Schottky and p-n barriers, heterojunctions). The characterization of materials plays an important role in the research, development, and fabrication of a microelectronic device.

Requirements during the research stage are a study of the correlation between the material's basic properties and the electrical (optical, magnetic) parameters desired and the development of new methods for measuring the properties of the material.

Requirements during the development stage are application of the measurement methods to monitor changes in the material's properties during the technological processes, development of reliable measurement benchmarks for characterizing the properties of the material, and assessment of the quality of the materials purchased from different suppliers.

Requirements during the fabrication stage are management of the measurement benchmarks, processing the data, and analysis of defective or faulty devices.

The characterization of materials can provide averaged information on the whole volume considered or direct information on a selected microvolume. In the second case, called *microanalysis,* we say that the characterization exhibits spatial resolution. Spatial resolution can be lateral (X, Y) and vertical or depth (Z). The family of characterization techniques capable of providing spatial resolution is known as *microanalysis techniques*.

The increase in miniaturization and complexity of semiconductor technology poses severe yield and reliability problems. This situation necessitates general use of microanalysis techniques. In daily practice, the nature and complexity of the technological problems involve different types of information with varied spatial resolution and sensitivity requirements.

There are numerous microanalysis techniques and there is a host of acronyms to be found in the literature. However, they do share the same physical principles and instrumental solutions. The sample to be studied is irradiated with a beam of primary particles (electrons, ions, photons). The beam particles interact with the sample atoms. The beam-sample interaction produces the generation of secondary particles (electrons, ions, photons). The secondary particles emitted are detected (or analyzed) and their intensity (or their energy or mass spectrum) constitutes the analytic signal and provides information on the portion of sample analyzed. This description is illustrated in Figure 12.1.

The portion of sample analyzed is not where the secondary particles are generated (interaction volume), but the part from which the secondary particles are emitted. The emission volume is obviously less than or equal to the interaction volume, and the difference increases the smaller the escape depth of the signal. The emission volume determines the spatial resolution.

Because the analytical information is only obtained from a small sample, the two techniques of scanning (Figure 12.2) and ion erosion (Figure 12.3) are generally used to analyze more extensive volumes.

In scanning-based techniques, the beam of primary particles is finely focused, so that only a small circular area on the sample surface is irradiated at any one time. The beam of primary particles is scanned in synchrony with the luminous spot of a cathode ray tube. This creates a biunivocal sequential correspondence between the sample spots and the spots on the viewing screen. The luminous intensity of each of the screen spots is controlled by the intensity of the secondary particle signal. The result is a map of the signal intensity in relation to sample position.

In techniques based on ion erosion, the beam of primary particles is directed to the center of the area irradiated by an ion beam. The ion beam causes sputtering-induced erosion and produces a continual renewal of the surface struck by the primary particles. Simultaneous or sequential analysis allows the intensity of the secondary particle signal to be monitored in relation to sputtering time (or depth beneath the surface). This is known as the *sputtering profile.*

Tables 12.1 and 12.2 illustrate the principle of operation and the most common acronyms of the main microanalysis techniques.

When we speak of microanalysis techniques we tend to stress their spatial resolution capacity. We should not forget, however, that spatial resolution has a price. Lateral resolution cannot, in scanning-based techniques, be better than the

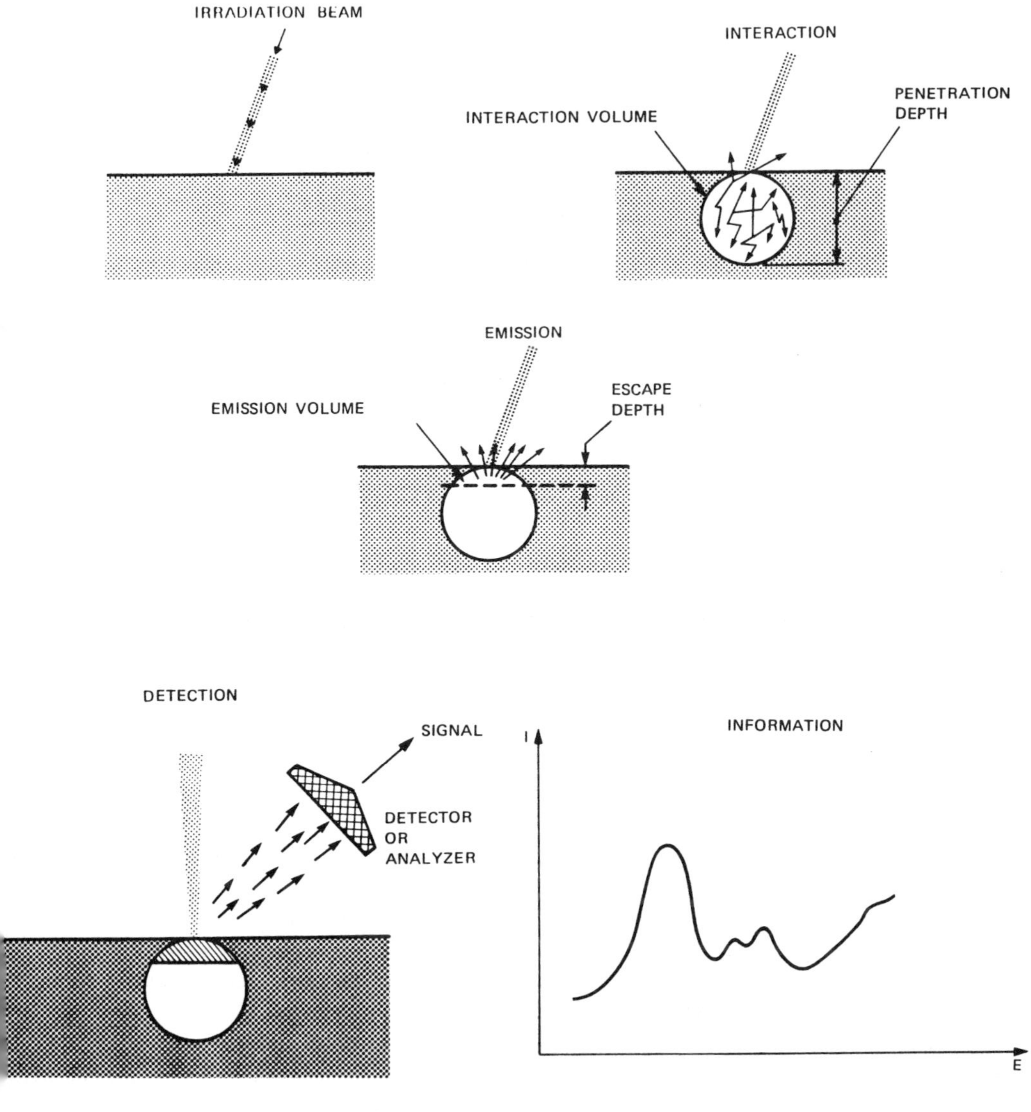

Figure 12.1 Schematic description of the operating principles of the microanalysis techniques

diameter of the primary particle beam; hence, to improve lateral resolution it is necessary to reduce the diameter. Figure 12.4 shows that a reduction in diameter necessitates a reduction in current. Therefore, lateral resolution is paid for in terms

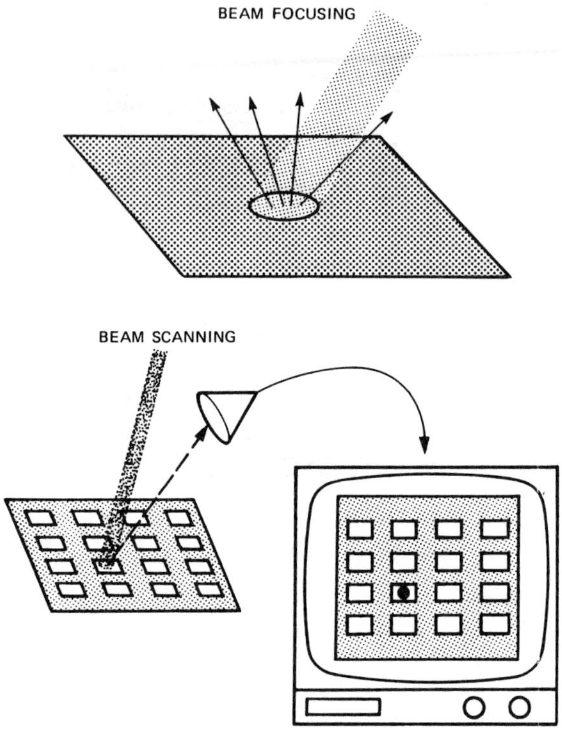

Figure 12.2 Principle of techniques using scanning

of sensitivity and accuracy for a given analysis time or in analysis time for a given sensitivity or accuracy.

The versatility of some of the innumerable microanalysis techniques has allowed their use in the most disparate technological sectors while other techniques have been confined to special fields. The first category includes *scanning electron microscopy* (SEM), *transmission electron microscopy* (TEM), *electron microprobe* (EMP), *scanning auger microprobe* (SAM), *photoelectron spectroscopy* (ESCA), *ion backscattering spectroscopy* (BS), and *secondary ion mass spectroscopy* (SIMS). These techniques have been available commercially for more than twenty years and are standard equipment in any laboratory involved in physical characterization with spatial resolution of materials, processes, and technologies for microelectronics.

This chapter will deal with the SEM, EMP, and SAM techniques, which are the most versatile for the study of electronic semiconductor components. See the summaries from [1–5] for initial information about the ESCA, BS, and SIMS techniques.

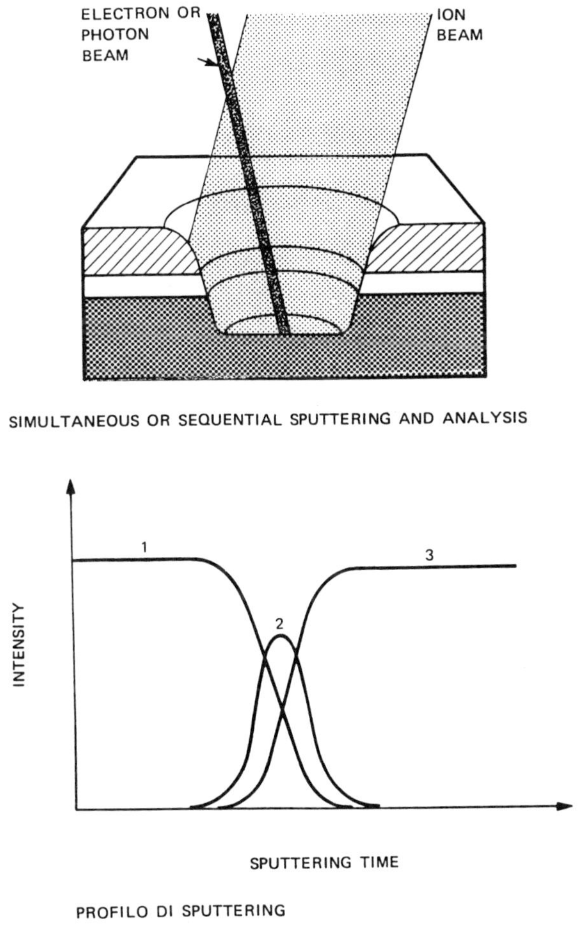

Figure 12.3 Principle of techniques using sputtering

12.2 SCANNING ELECTRON MICROSCOPY

12.2.1 Principle of the Technique

The SEM principle, in its commonest mode of operation, is illustrated in Figure 12.5. The electrons emitted by a tungsten filament are accelerated by a voltage normally falling between 0.2 and 50 kV and enter an electron optical column consisting of two or three magnetic lenses. These lenses produce a fine beam of electrons focused on the surface of the sample.

Table 12.1 Acronyms of the Main Microanalysis Techniques

Technique	*Acronym*
Electrons:	
Appearance potential spectroscopy	APS
Auger electron spectroscopy	AES
Electron microprobe	EMP
Electron microscopy/diffraction	TEM/D
Low energy electron diffraction	LEED
Scanning Auger microprobe	SAM
Scanning electron microscopy	SEM
SEM with Si solid state detector	SEM-Si (Li)
Scanning transmission electron microscopy	STEM
Ions:	
Field ion microscopy	FIM
Glow discharge mass spectroscopy	GDMS
Glow discharge optical spectroscopy	GDOS
Ion backscattering spectroscopy	BS
Ion induced X-ray fluorescence	IIXF
Ion luminescence spectroscopy	SCANIIR
Ion microprobe mass analysis	IMMA
Ion scattering spectroscopy	ISS
Particle induced nuclear reactions	PAA, NAA
Secondary ion mass spectroscopy	SIMS
Photons:	
Ellipsometry	ELL
Optical interferometry	OI
Optical microscopy	OM
Optical spectroscopy (IR, emission)	OS
Photoelectron spectroscopy (X-ray, UV)	ESCA, PES
X-ray diffraction	XRD
X-ray fluorescence	XRF
X-ray topography	XRT

Two pairs of deflection coils cause the beam to scan the sample surface in the form of a square frame. The currents flowing through these coils are also made to flow through the corresponding deflection coils of a CRT, so that a similar but larger frame is produced on the screen. The result, in sequential form, is a bi-univocal spot-by-spot correspondence between the fluorescent screen and an area of the sample surface.

The beam electrons striking the surface of the sample cause the emission of secondary electrons. These reach a collector and the resultant current is amplified and used to modulate the brightness of the CRT. This means that there is a match between the number of secondary electrons collected by a generic spot on the surface in question and the brightness of the corresponding spot on the screen.

Table 12.2 Operating Principles of the Main Microanalysis Techniques

Excitation *Emission*	*Electrons*	*Ions*	*Photons*
Electrons	AES TEM/D LEED SAM SEM STEM		ESCA PES
Ions		FIM GDMS BS IMMA ISS SIMS	
Photons	APS EMP SEM-Si (Li)	GDOS IIXF NAA SCANIIR	ELL OI OM OS XRD XRF XRT

The factors causing changes in the number of secondary electrons collected from various points of the object, and hence producing a contrast in the picture, will be discussed later, but they are essentially topographical. The magnification is determined by the ratio of the size of the scanning area on the CRT screen to the sample surface.

The good lateral resolution and significant depth of focus that can be obtained enable the surface topology of bulk samples to be observed at high magnification with surprising three-dimensionality. It would, however, be simplistic to view the SEM as just a superior optical microscope. It is of fundamental importance to stress that, because it is not necessary to focus the "radiation" emitted by the sample, any signal generated by the interaction between the beam and the sample can be used to give an image on the CRT. It is simply a question of the signal being collected with sufficient speed and converted, where necessary, into an electric current.

The most interesting types of signal in the field of electronic materials and devices are secondary electrons, Auger electrons, backscattered electrons, absorbed electrons, characteristic X rays, electrovoltaic effects, β conductivity, and cathodoluminescence. These signals determine various modes of SEM operation

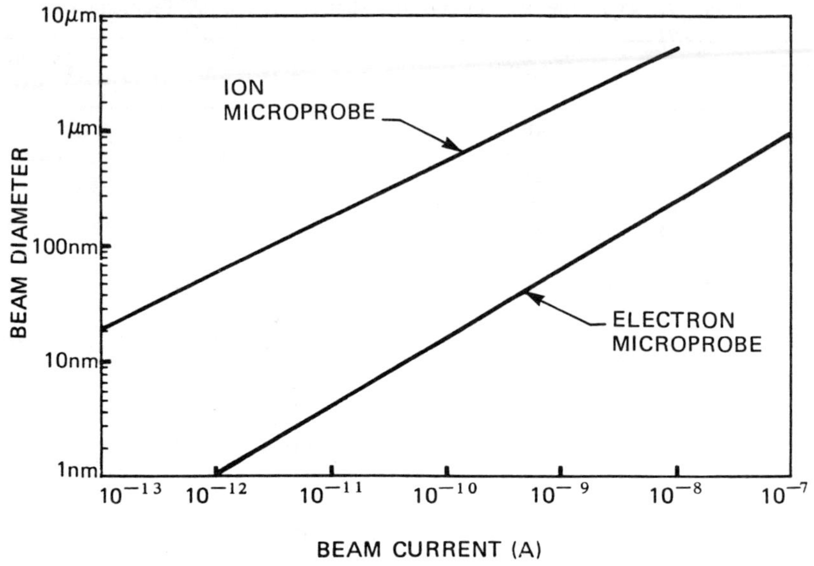

Figure 12.4 Diameter-current relation for electron and ion microprobes

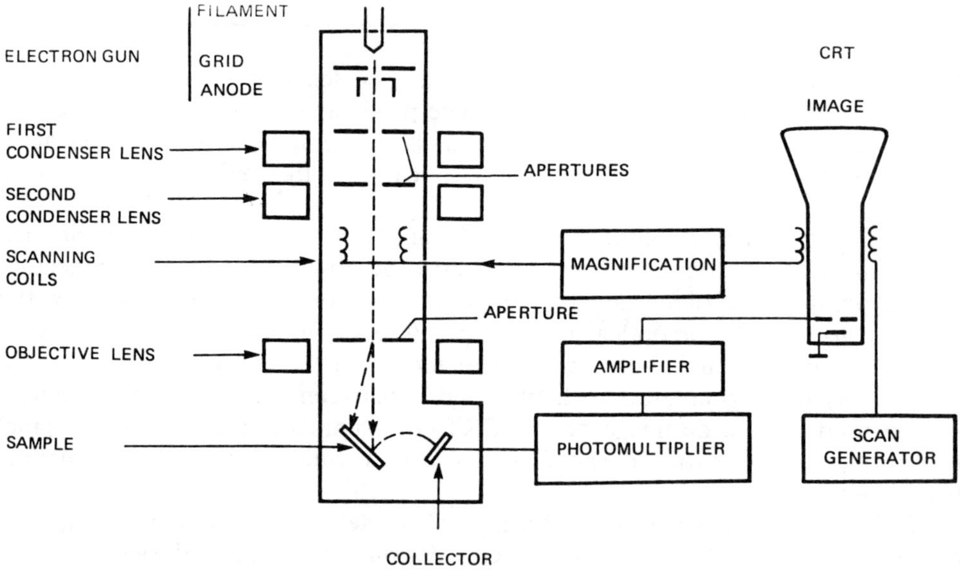

Figure 12.5 Block diagram of the scanning electron microscope

that provide information on the physical-chemical properties of the sample: morphology, chemical composition, distribution of the surface electrical potential, presence and properties of spatial charge, inversion and depletion regions, current multiplication phenomena, minority carrier diffusion length, variations in resistivity, properties of emission due to luminescence, presence of recombination centers, and so on. The SEM hence represents an extremely versatile electron beam analytic system.

12.2.2 Modes of Operation

Figure 12.6 illustrates the various types of signal that determine the different SEM operating modes.

All the electrons emitted by the sample with energies ranging from 0 to about 400 eV are collected for secondary electron images. This range includes real secondary electrons, Auger electrons, and low-energy backscattered electrons, with a net prevalence of real secondary electrons. These electrons originate at a maximum depth of about 5–10 nm beneath the surface. They are the images most widely used and yield information on the topography of the sample surface and the distribution of the electrical potential on the surface.

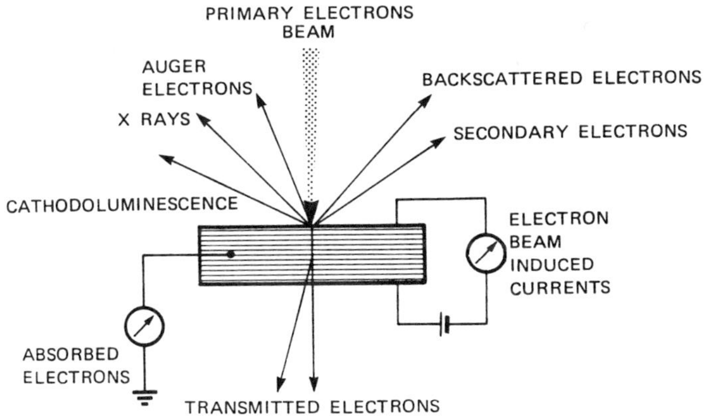

Figure 12.6 Various types of signals emitted following the electron-material interaction

The Auger electrons emitted by the sample are energy analyzed for the Auger electron images. These electrons have energies normally ranging from 50 to 3000 eV and come from an average depth of about 2 nm beneath the surface. This mode of operation provides information on the chemical composition (concentration and spatial distribution of the elements) of the outermost sample layer and, associated with the sputter etching, information on the chemical composition of the underlying layers.

The backscattered electron images are obtained by collecting the beam electrons that penetrate the sample and then leave following a wide-angle deflection or several small-angle deflections. These electrons have quite a wide energy distribution, but the majority of them lose only small amounts of energy; they come from depths of several microns. These images provide information on the topography of the surface and the chemical composition (mean atomic number) of the deeper surface layers.

The current flowing between the sample and ground to keep the sample electrically neutral is used as the signal for absorbed electron images. Normally, the intensity of this current is about half that of the primary electron current. These images are, in cases of practical interest, complementary to backscattered and secondary electron images, and provide information on the chemical composition (mean atomic number) of the sample.

The signal used for characteristic X-ray images is represented by X rays emitted by the sample, analyzed on the basis of their energy or wavelength. The X rays of analytical interest have energies ranging from 0 to 40 eV and come from a depth of a few microns. This mode of operation enables the identification of the chemical elements present in the sample and provides information on their concentration and spatial distribution.

In semiconductor materials, the electron beam generates a certain number of electron-hole pairs, which under given conditions can be collected at the contacts with two points of the sample and produce a current in a suitable external circuit.

It is necessary to distinguish between the charge collection current in the presence of an external bias and that found without this bias. In the first case, the external bias provides the driving force required to separate and collect the charge carriers, and we speak of β conductivity. In the second case, there is a charge collection current only if the electron bombardment produces an internal driving force in the sample; these internal driving forces are produced by electrovoltaic effects. There are two distinct electrovoltaic effects: barrier and bulk. In both cases, an electrical field inside the sample separates the carriers generated by the beam; this produces a difference in the Fermi level for the electrons at the two contacts providing the driving force. The barrier electrovoltaic effect is produced by the electrical fields at the p-n and Schottky barriers; the bulk electrovoltaic effect is created by the fields due to the disuniformities of distribution of the impurities inside the semiconductor material.

The barrier electrovoltaic effect provides information on barrier position and properties, current multiplication phenomena, crystallographic defects, and the diffusion length of the minority carriers. The bulk electrovoltaic effect and β conductivity essentially provide information on the variations in resistivity.

In one class of semiconductor materials the electron-hole pairs generated by the beam can recombine radiatively, producing photons usually in the visible and near infrared spectrums; this phenomenon is known as *cathodoluminescence.* Cathodoluminescent images can be obtained using suitable detectors from all materials emitting in the range 0.3–3 μm with efficiencies greater than 10–4 to 10–5. This mode of operation can provide information on the presence of crystallographic defects and composition or doping disuniformities, the mean life of minority carriers, and the distribution of temperature within the device.

12.2.3 Lateral Resolution Limitations

The best resolution that can be achieved in any particular operating mode depends on a certain number of factors; diameter of incident electron beam, effects of elastic diffusion of incident electrons inside the sample, and signal-noise ratio.

The resolution cannot be less than the diameter of the electron beam striking the surface of the sample. Any finer detail will not be distinguishable in the image because the signal collected at any one time is an average over the whole beam section (Figure 12.7). The minimum beam diameter that can be used depends basically on electron lens aberrations and noise factors.

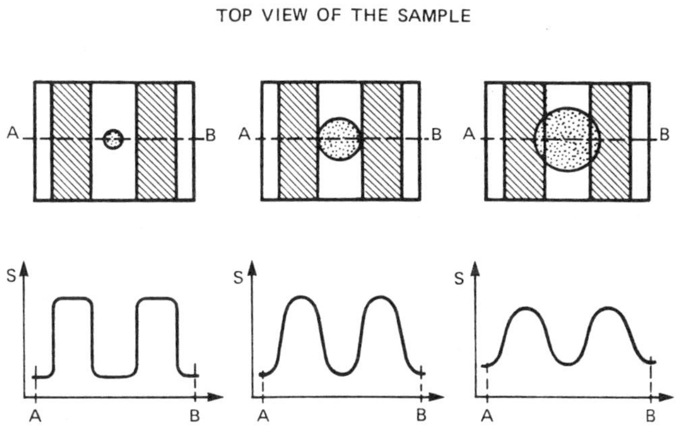

Figure 12.7 Dependence of lateral resolution on beam diameter.

When the beam strikes the sample surface, the penetrating electrons are elastically diffused; this increases the effective beam section and leads to a deterioration in resolution (Figure 12.8). This increase in the effective section depends on the energy of the striking electrons, the type of sample, and the mode of operation.

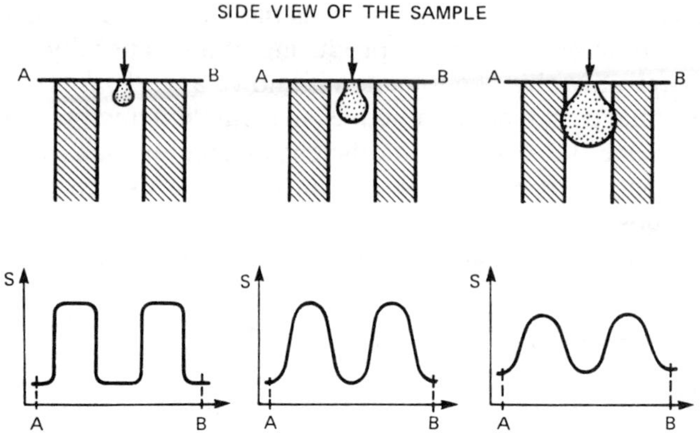

Figure 12.8 Dependence of lateral resolution on emission volume.

To form an image it is essential that any signal detected be at a level sufficiently higher than the level of noise. In SEM, the signal level depends on the acceleration voltage, the beam current, the type of sample, and the operating mode; noise depends mainly on the type of sample, the operating mode, and the detector used. It is often necessary to improve the signal-noise ratio by increasing the current in the beam, which normally implies an increase in its section and hence a loss in resolution. We can outline the way in which these factors restrict lateral resolution for the various operating modes, referring to a mean sample examined in an SEM with a thermionic tungsten gun at an acceleration voltage of 20 kV.

For secondary electron images remember that the secondary electrons originate in two ways: either directly from the incident electrons or indirectly from the backscattered electrons. The first are affected by diffusion in a limited way, whereas the second emerge from a volume with a section considerably greater than the beam section and produce a background that degrades the quality of the image. As a first approximation, however, we can say that resolution is determined only by the secondary electrons generated by the primary incident electrons. Resolution is limited by the signal-noise ratio to values ranging from 10 to 30 nm.

Similar considerations apply for Auger electron images. The resolution, however, is more severely limited by the signal-noise ratio: typical values range from 0.1 to 0.5 μm. For backscattered electron images the resolution is limited by the very pronounced diffusion they undergo inside the sample; resolutions ranging from 40 to 200 nm can be achieved. Conventional amplifiers are used for absorbed electron images. As these normally introduce significant noise, it is necessary to increase the current and hence the beam section; the resolution increases from 0.1 to 1 μm.

In the case of images of electrovoltaic effects, β conductivity, cathodoluminescence, and characteristic X rays, the useful signals are collected from areas considerably larger than the beam. Furthermore, the cathodoluminescence and X-ray signals are weak because of the inefficient processes used to convert the incident electron energy into photons. The resolution for these four types of images hence is limited by the effect of the elastic diffusion and the signal-noise ratio. Typical resolutions fall between 0.3 and 1 μm. Table 12.3 summarizes these comments.

Table 12.3 Typical SEM Lateral Resolutions

Operating mode	*Critical factors*	*Resolution*
Secondary electrons	Signal/noise	30–300 Å
Auger electrons	Signal/noise	0.1–0.5 μm
Backscattered electrons	Elastic diffusion	100–2000 Å
Absorbed electrons	Signal/noise	0.1–1 μm
Electrovoltaic effects	Elastic diffusion	
	Signal/noise	0.3–2 μm
β Conductivity	Elastic diffusion	
	Signal/noise	0.3–2 μm
Cathodoluminescence	Elastic diffusion	
	Signal/noise	0.3–2 μm
X rays	Elastic diffusion	
	Signal/noise	0.3–2 μm

SEM with thermionic gun in W at 20 kV

12.2.4 The Electron Optical System

The function of the electron optical system is to focus on and move a fine beam of electrons over the sample surface. In principle, any electrical or magnetic field having axial symmetry can focus electrons; at present, lenses are mostly the magnetic type. The beam is deflected through two magnetic fields at right angles to one another and to the beam axis.

Apart from the obvious acceleration voltage V, three factors characterize the focused beam: the diameter d, the angle of convergence, and the current I. Different combinations of these three parameters are necessary for the various modes of operation; we will give an idea of how this can be achieved using the electron optical system.

Figure 12.9 diagrams the trajectories of the electrons for a three-lens system; it is assumed that the beam is not deflected by the scanning coils. All three lenses give a reduced image of the surface of the cathode (magnification < 1). The beam diameter at the output of the electron gun is indicated by d_0, the diameter of the two images by d_1 and d_2, and the diameter of the final image by d. If we denote the magnifications of the three lenses by M_1, M_2, M_3 and consider their aberrations to be negligible, the diameter of the beam on the sample surface is given by

$$d = M_1 M_2 M_3 \, d_0 \tag{12.1}$$

The diameter calculated in this way is called a *Gaussian diameter.*

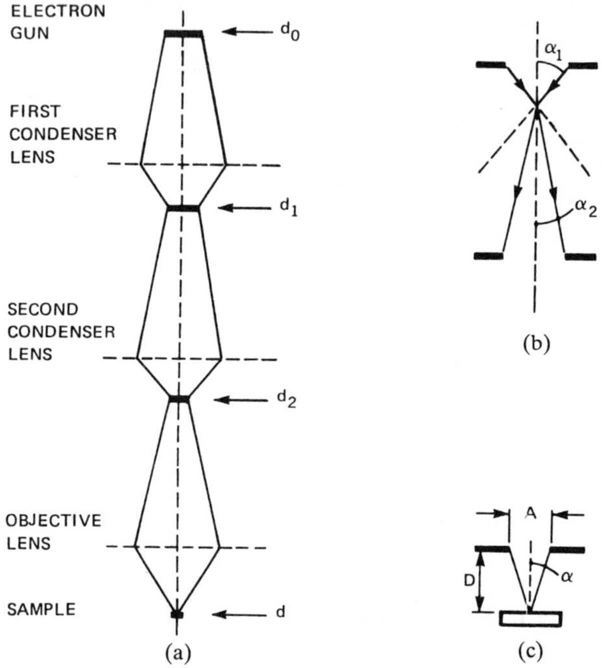

Figure 12.9 (a) 3-lens electron optical system. (b) Reduction of current with reduction of beam diameter. (c) Relation of work distance (D), angle of convergence (α), and diameter of the final aperture (A).

Typical d_0 values range from 25 to 50 μm, whereas each lens is capable of reducing its beam diameter by a factor ranging from 20 to 100.

Figure 12.9b shows that not all the electrons leaving the first condenser lens enter the second condenser lens. If the current in the first lens is increased, the image is formed closer to the lens itself, therefore, α_1 increases whereas α_2 remains practically unchanged, and so the number of lost electrons increases. The same thing happens for the image between the second and third lenses. The overall effect is that a gradual increase of the current in the first two lenses results in a reduction in the beam diameter as well as intensity. The objective lens current is adjusted in such a way as to focus the image on the sample surface.

The final aperture, on the other hand, is designed to control the angle of convergence of the beam on the sample surface (Figure 12.9c). If A is the diameter of the aperture and D the working distance, then (given that α is very small) we have

$$\alpha = \frac{A}{2D} \tag{12.2}$$

The final apertures normally have diameters of 200, 100, or 50 μm. The following relation applies to the beam current I

$$I = \frac{Bd^1\alpha^2}{0.4} \tag{12.3}$$

where d and α have the usual meaning and B is the brightness of the electron optical system involved, defined as the current density per unit of solid angle. Because d and α are fixed, as discussed earlier, I can be calculated if B is known. The value of B depends on the type of electron source and the operating conditions.

In certain applications (stroboscopic observations, measurement of minority carrier lifetime) the electron beam has to be interrupted. This can be achieved in different ways but the most common method is to deflect it with electrostatic plates a long way from the optical axis in the upper section of the column.

12.2.5 Limitations Due to Lens Aberrations

When very small beam sections are required, magnetic lens aberrations have to be considered. These aberrations mean that the focused beam has a larger section than that expected from perfect lens (12.1). Since all three lenses reduce the beam diameter, it is sufficient to consider the final lens. Every point of the intermediate virtual image located between the second and third lenses is expanded on the sample surface in a confusion disc; the different aberrations produce discs with different diameters but that are always dependent on the beam angle of convergence.

It is sufficiently accurate to write that the diameter d of the confusion disc resulting from the overlapping of the various aberrations is

$$d_a^2 = d_s^2 + d_c^2 + d_d^2 = K_s^2\alpha^6 + K_c^2\left(\frac{\delta V}{V}\right)^2\alpha^2 + \frac{1.48}{\alpha^2}\lambda^2 \tag{12.4}$$

where d_s, d_c, and d_d are the diameters of the confusion disc produced by the spherical aberration, chromatic aberration, and diffraction aberration, respectively; K_s is the spherical aberration coefficient, K_c is the chromatic aberration coefficient, δV is the energy interval of the electrons entering the lens, V is their mean energy, and λ is their effective wavelength. This equation shows that there is an optimal value of α (α_{ott}) for which d_a is minimal.

It should also be noted that K_s, K_c, and λ depend on V; hence, the optimal angle of convergence and the minimum diameter of the confusion disc produced by the aberrations will also depend on V. For a commercial SEM working in good conditions, we have

$$\alpha_{ott} = 3 \times 10^{-3} \text{ rad};\ d_a = 5 \text{ nm}$$

This means that, as a result of the aberrations, the minimum final diameter of the beam that can be achieved (regardless of the value of the magnifications M_1, M_2, and M_3 of the three lenses) is about 5 nm. If the three lenses operate to yield a Gaussian diameter d_g, the resultant diameter d_{ott} is

$$d_{ott}^2 = d_g^2 + d_a^2 \tag{12.5}$$

In general, the final lens aberrations can be ignored as long as the values of M_1, M_2, and M_3 produce d_g values > 30 nm.

12.2.6 Limitations Due to Noise

We have seen that, as a result of the final lens aberrations, there is a limit to the minimum beam diameter. We now see that the minimum beam diameter is also limited by noise. There are different sources of noise. First of all, there is the noise associated with the statistical fluctuations in the emission of the electrons by the cathode. Second, there is the noise linked to the interaction between the beam electrons and the sample (for example, the further characteristic statistical fluctuations of the secondary emission process). Finally, noise may be introduced in the detection and amplification of signals; however, the systems currently used are so efficient that this third type of noise can be ignored, at least in secondary and backscattered electron images.

The level of noise that can be tolerated depends on the amount of contrast desired in the final image. It has been found experimentally that the eye cannot distinguish an area of luminosity L from an adjacent area of luminosity $L \pm \Delta L$ unless the signal-to-noise ratio is approximately five times the ratio between L and ΔL. Because $\Delta L/L$ is the relative contrast C, as a criterion for distinguishing the two areas we can write,

$$\frac{\text{signal}}{\text{noise}} \geqslant 5\,\frac{1}{C} \tag{12.6}$$

Let us first consider the incident beam noise; if the number of primary electrons arriving at a certain point of the sample is n, the mean square value of the fluctuations of this number is $n^{1/2}$. At this stage of the image formation process, the criterion expressed in equation (12.6) can be written

$$n \geqslant 25\ (1/C)^2 \tag{12.7}$$

Let us now consider the noise linked to the interaction phenomena; the situation is very complex and normally we adopt the empirical criterion of increasing the factor of equation (12.7) by four times. The distinction criterion hence becomes

$$n > 100\ (1/C)^2 \tag{12.8}$$

and this is to be satisfied for all the points on the image. This is a basic limitation in the image formation process and hence applies to any SEM.

Normally there are 10^6 image spots (1000 scanning lines each with 1000 image pixels) in a square microphotograph with sides of 10 cm; the minimum number of electrons per image required to detect various contrast levels can be calculated. For example, for a contrast of 3 percent to be detected, 10^{11} electrons are required for the whole image; now 10^{11} electrons can be obtained with a beam of 10^{-8} or 10^9 A that completes a scan in 1 or 10 seconds, respectively. It follows that the minimum number of electrons required to detect a given level of contrast in the final image can be obtained with different combinations of I and the overall scanning time.

12.2.7 The Minimum Useful Diameter of the Beam

The minimum beam diameter that can be obtained on the sample surface cannot always be used in practice because the corresponding current is insufficient to distinguish the image from the noise. If we indicate the current required to distinguish a given level of contrast by Ir, we can rewrite (12.3) in the following way:

$$dr = \left(\frac{0.4Ir}{\mathrm{B}}\right)^{1/2} \frac{1}{\alpha} \tag{12.9}$$

Combining equations (12.4) and (12.9) gives the minimum useful diameter:

$$\begin{aligned} d &= d_s^2 + d_c^2 + d_d^2 + d_a^2 \\ &= K_s^2\alpha^2 + K_c^2\left(\frac{\Delta V}{V}\right)^2\alpha^2 + \frac{1.48\lambda^2}{\alpha^2} + \frac{0.4Ir}{\mathrm{B}\alpha^2} \end{aligned} \tag{12.10}$$

12.2.8 The Maximum Useful Magnification

If R represents the minimum distance that the eye can resolve and r the resolution of the electron beam on the sample surface, the maximum useful magnification M is given by

$$M = R/r \tag{12.11}$$

Normally a value of 0.01 cm is assumed for R.

The beam diameter should be chosen on the basis of the magnification it is intended to use. If $d > r$ two consecutive scanning lines on the sample are superimposed and the resolution corresponds to d rather than r. This occurs when the magnification is too high for the beam diameter. *Vice versa,* if $d < r$, there are sample areas not explored; a part of the information is lost and the image can lead to incorrect interpretations. This occurs when the magnification is too low for the beam diameter used.

Putting $d = r$ gives optimal conditions; for example, if a magnification of 10,000 is required, the best resolution that can be obtained is, as a result of (12.11), $r = 10$ nm, and this is also the value of the beam diameter for secondary electron images.

12.2.9 Depth of Focus

The depth of focus F on the surface of an object observed under any microscope is the distance between two planes perpendicular to the optical axis within which the parts still appear to be sufficiently focused; that is,

$$F = r/\mathrm{tg}\alpha \tag{12.12}$$

where r is the resolution and α the angle of convergence. α is small for the SEM and so we have

$$F = r/a \tag{12.13}$$

Comparison with an optical microscope shows that at a resolution of 0.5 μm and for $\alpha = 3.10^{-3}$ rad, FSEM/FOM = 1000. The depth of focus of the SEM makes it very suitable for stereomicroscopy.

12.2.10 Detectors

The electrons leaving the surface of the sample have to be collected for secondary or backscattered electron images. Because these currents are very small (from

10^{-10} to 10^{-12} A), we need a detection-amplification system that gives a linear response, wide passband, high gain, and low noise. The detector most widely used is shown in Figure 12.10.

The sample is usually inclined relative to the beam and faces the detector. The secondary electrons [3] are collected by means of a Faraday cage with a front grid [4] having a voltage of +300 V. The electrons that pass through the grid are accelerated by a voltage of 10 kV and acquire sufficient energy to produce photons when they strike the scintillator [5]. The light is conducted by a light pipe [6] to a photomultiplier [7] located outside the sample chamber. The photomultiplier output current is the signal.

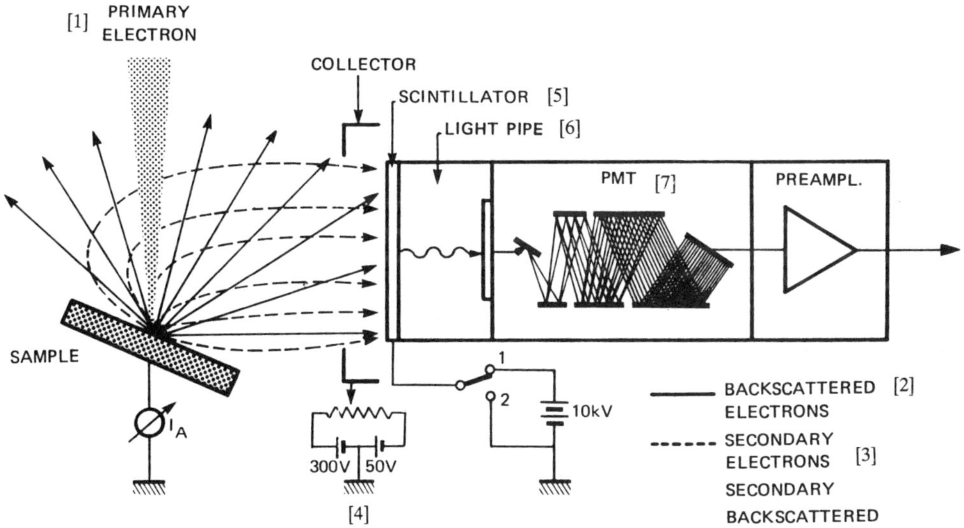

Figure 12.10 Detector for secondary and backscattered electrons.

The collection mechanism is different in the case of backscattered electrons [2]. The backscattered electrons having high energies follow straight trajectories that are practically independent of the grid voltage; so only those backscattered electrons emerging from the sample surface in the direction of the detector will reach the scintillator. Secondary electrons, on the other hand, have low energies: if the grid voltage is −50 V they will all be reflected backwards and none will reach the scintillator; if the grid is at a positive voltage the secondary electrons will have curved trajectories and will reach the scintillator almost regardless of their initial emission direction. In the second case, there is a backscattered electron image; in the first there is a secondary electron image because the large majority of the current detected is due to the secondary electrons.

Some backscattered electron detectors feature semiconductor diodes having a thicker depletion region than the penetration of the electrons (typically 8 to 10 μm). The backscattered electrons create a large number of electron-hole pairs (about 6000 pairs for each electron of energy 20 keV), whereas the contribution of the secondary electrons is insignificant.

The detector used for absorbed electron images is an amplifier with the input resistance connected in series between the sample and ground. Since the values of the absorbed electron current can be very low and change rapidly over a period of time, the amplifier should be a good compromise between a high gain and a large passband.

In the case of images of electrovoltaic effects or β conductivity, the charge collection current is of a few orders of magnitude greater than the beam current; the same type of amplifier can hence be used, but in more favorable conditions.

For cathodoluminescence images a light pipe followed by a photomultiplier or solid state detector is sufficient; the arrangement is similar to that shown in Figure 12.10, which does not include a grid and scintillator. More efficient detectors are required, however, if spectral or time dependent analysis of the radiation emitted is involved. The detectors for X rays and Auger electrons will be described later.

12.3 TYPE OF CONTRAST AND INFORMATION FOR THE VARIOUS MODES OF SEM OPERATION

12.3.1 Secondary Electrons

The contrast in secondary electron images depends on the changes in the number of secondary electrons detected; two effects contribute to this, changes in the number of electrons emitted by the sample and changes in the fraction of electrons collected by the detector.

Let us consider the first effect. The number of secondary electrons emitted *ns* is always proportional to the number *np* of incident primary electrons, and the ratio *ns*/*np* is defined as the coefficient of secondary emission δ. The term δ depends on the energy and angle of incidence of the primary electrons and on the material from which the sample is made. Under normal SEM operating conditions (10 to 50 kV), metals and semiconductors normally have a very similar δ, which is less than 1; for these materials, variations in δ give rise to a weak contrast. The insulators, on the other hand, can have very dissimilar values, which are above 1: however, the contrast that we could expect is often masked by the thin conducting layer deposited on the sample surface to prevent the accumulation of charges. In low voltage operations (0.2 to 10 kV), however, there can be significant contrast due to the diversity of materials. So the coefficient also depends on the angle of incidence (the angle between that normal to the surface at the point of incidence and the direction of incidence) of the primary electrons, and it increases as the angle of incidence increases (Figure 12.11). In fact, the secondary electrons are in general generated closer to the surface and are much more likely to emerge.

As for the second effect, the fraction of secondary electrons collected by the detector depends only on the angle of incidence. In fact, the secondary electrons have an angular distribution approximately proportional to $|\cos\varphi|$, where φ is the angle between the direction considered and that normal to the surface; as the angle of incidence changes, the direction in which there is maximum emission moves relative to the detector, thereby changing the collection efficiency.

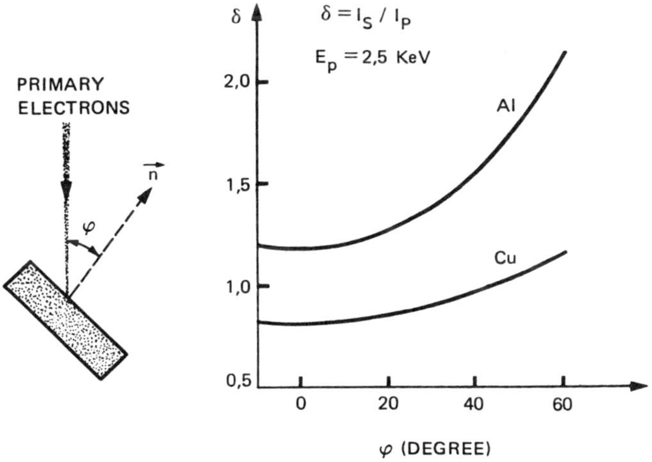

Figure 12.11 Dependence of electron yield on the angle of incidence in the sample.

The combination of these two effects means that the contrast of the secondary electron images depends basically on the local angle of incidence of the beam (and hence on the sample topography), and only secondarily on the sample material. This contrast formation mechanism is illustrated in Figure 12.12; because of the differences in the angle of incidence the number of secondary electrons detected for the various faces is different and the resulting image will appear shaded due to the contrast between the brighter and the less bright face of the mean surface (the detector is assumed to be on the right).

We now examine a few general criteria for interpreting secondary electron images. The image is a representation of how the sample is seen looking from the electron gun along the axis of the electron optical column. There is a strict analogy with the contrast formation mechanism in optical images. The impression of three-dimensionality given by the secondary electron images is due partly to the significant depth of focus of the SEM and partly to this analogy. Information relating to the

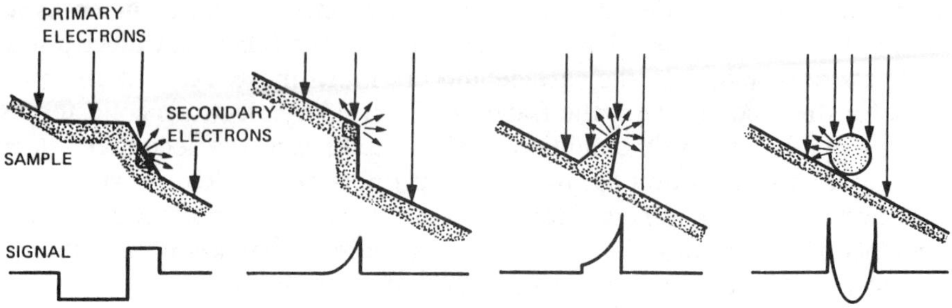

Figure 12.12 Mechanism of topographical contrast formation in secondary electron images.

distances perpendicular to the image plane is not contained in the image itself but is deduced by the observer, who interprets the contrast on the basis of his experience with three-dimensional objects.

As regards the trajectories of the secondary electrons detected, the sample appears as if it were illuminated from all directions and details can also be observed of the areas (depressions, cracks, and so on) from which there would be no straight trajectories toward the detector.

12.3.2 Secondary Electrons with Voltage Contrast

When the sample surface is not equipotential new factors come into play to determine the contrast of the secondary electron images. For a detector of the type described in Figure 12.12, the trajectories of the secondary electrons emitted from a certain area of the sample depend on the difference in potential between the grid and the area considered; so, as the grid voltage is fixed, the variations in potential on the sample surface are translated into variations in the secondary electron trajectories (especially, the lowest energy electrons) and hence into changes in the collection efficiency.

Furthermore, if two of the sample regions are at different potentials, there is a local electrical field between them on the surface that (regardless of the potential difference between the two regions and the grid) affects the trajectories of the lower energy secondary electrons emitted by these regions and hence the collection efficiency.

The overall effect is that in the image the regions at more positive potential appear darker whereas those at more negative potential appear brighter (Figure 12.13). This yields a map of the distribution of the electrical potential superimposed on the sample topography.

The standard detector generally allows potential differences of the order of 200 mV to be detected with a lateral resolution of about 100 nm; higher sensitivities are possible but at the expense of lateral resolution. The contrast obtained with

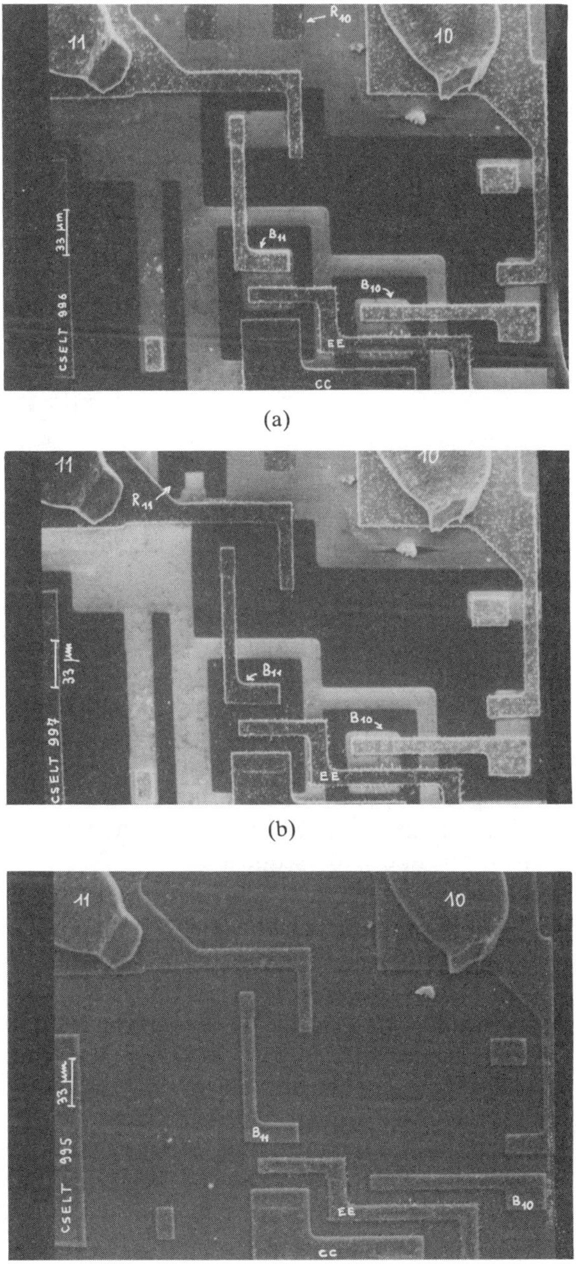

(a)

(b)

(c)

Figure 12.13 Details of ECL integrated circuit: (a) circuit not biased; (b) circuit biased and pin 11 at -5 V; and (c) circuit biased and pin 11 grounded.

the standard detector is not, however, linear, and so the information on the distribution of potential on the surface is only qualitative. Linearization techniques are required if contrast is to be used for measuring voltage, and this involves modifying the secondary electron detection system so that it can analyze the electron energy.

Linking voltage contrast to stroboscopy (interrupting the electron beam periodically) enables a study to be made of electrical potential distributions, which change regularly over a period of time. Two complementary techniques are possible. Sampling at the same phase of the cycle but on different points of the device, a phase of the periodic potential is observed as a static potential. Alternatively, sampling at successive phases on the same point of the device recreates the wave form in the same point. Time resolutions better than 1*ns* are easily obtained.

12.3.3 Backscattered Electrons

The contrast in the backscattered electron images depends on the variations in the number of backscattered electrons detected. Two effects are at work here: changes in the number of backscattered electrons and changes in the fraction of electrons detected by the detector. The number of backscattered electrons is proportional to the number np of the primary electrons, whereas the backscatter coefficient $\eta = nr/np$; η depends on the material of which the sample is made and on the energy and angle of incidence of the primary electrons. Figure 12.14 represents η as a function of the atomic number with angle of incidence θ (normal incidence). Note that η increases significantly as the atomic number increases. If the angle of incidence increases, the distance penetrated by the beam electrons measured along the direction of incidence is the same but the electrons remain on average closer to the surface and are more likely to escape again.

The second effect (changes in the fraction of electrons collected) depends only on the topography of the sample. The backscattered electrons have an angular distribution proportional to $|\cos\gamma|$, where γ is the angle between the direction considered and the direction of incidence; that is, they leave the surface preferably in the direction of incidence. Because the detector (the type indicated in Figure 12.10) is located laterally to the optical axis, only a small portion is backscattered at useful angles. In order to detect this small portion, which is virtually unaffected by the grid voltage, the surface topography should enable straight trajectories toward the detector; the electrons backscattered from faces opposite to the detector, from cavities, and so on do not see the detector. These points can be summarized by saying that the image contrast is due in almost equal measure to the sample material and surface topography.

If, however, a pair of semiconductor detectors is located symmetrical to the beam axis, the two contributions to the contrast can be separated (Figure 12.15).

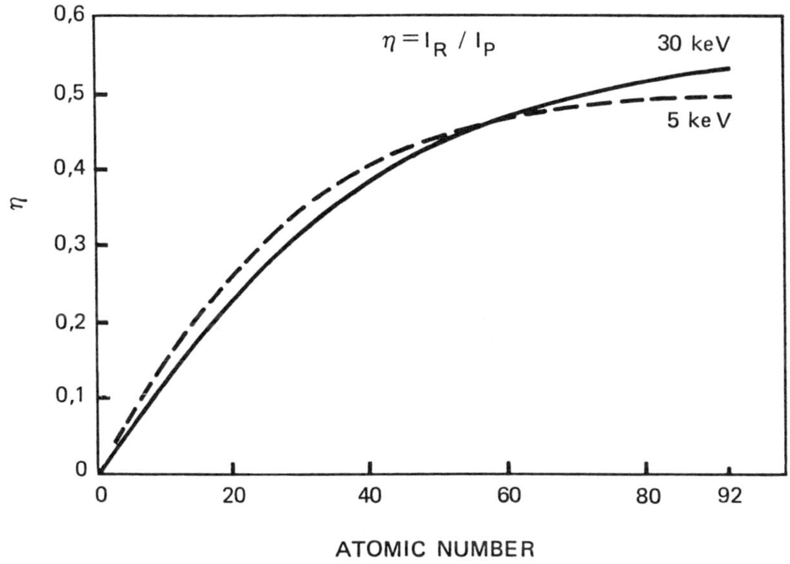

Figure 12.14 Dependence of backscattered electron yield on sample atomic number.

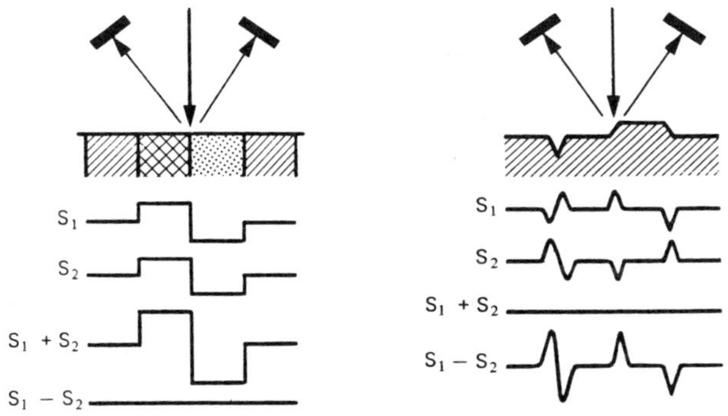

Figure 12.15 Separation of topographical contrast from the effects of composition on backscattered electron image contrast.

The number of backscattered electrons collected at each instant by the two detectors is substantially equal when the surface is flat, regardless of the composition. Hence, the signals of the two detectors vary in the same way according to the changes in the atomic number. When the sample is homogeneous but without a flat surface, the signal of the two detectors varies with opposite phase. If the two signals are added together, the topography contrast is eliminated and the atomic number contrast is reinforced, the opposite occurs if they are subtracted (Figure 12.16). Therefore, there are three types of backscattered electron images: images with atomic number, topography, and mixed contrast.

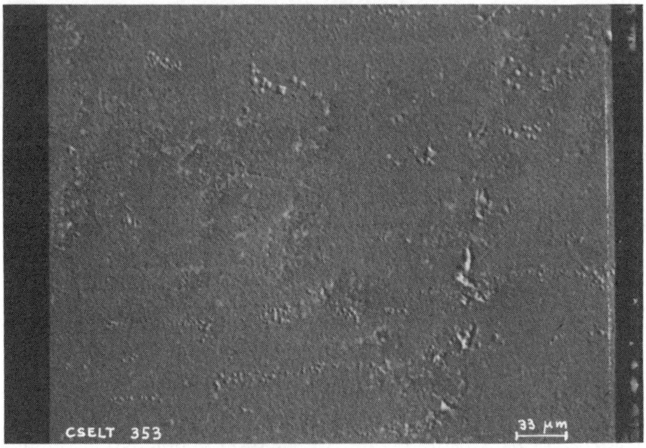

(a)

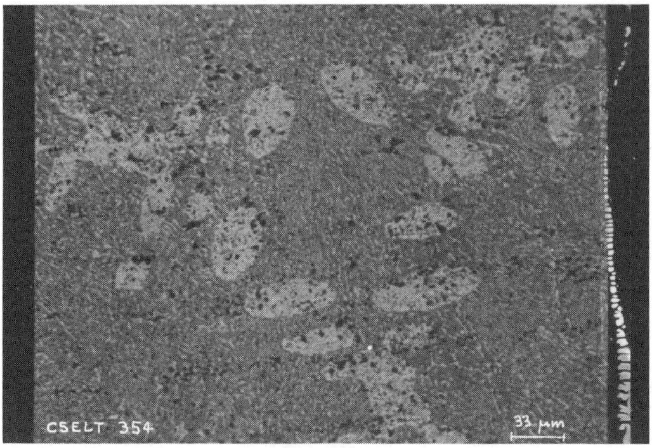

(b)

Figure 12.16 Sn-Pb alloy: (a) topographical contrast; and (b) composition contrast.

The criteria for interpreting images with mixed contrast are the same as those given for the secondary electron images, except that the sample appears as if it were illuminated solely by the position of the detector and only details of the sample areas that see the detector are observed. For images with topography contrast, the criteria given for secondary electron images apply.

In images with atomic number contrast, the areas of the sample with a high atomic number appear brighter than those with a low atomic number; changes can be observed in the average atomic number less than unity. Semiquantitative information thus can be obtained on the space changes of the chemical composition with better lateral resolution in many cases than that from X-ray distribution images.

12.3.4 Absorbed Electrons

If I, i_B, and i_S are the beam current, the backscattered electron current, and the secondary electron current, respectively, normally $I - i_B - i_S = 0$ for a bulk sample so the sample would tend to become charged; this tendency is balanced by the absorbed electron current i_a, which flows in the circuit connecting the sample to ground; hence:

$$i_a = I - (i_B + i_S) \tag{12.14}$$

The absorbed electron signal is proportional to i_a; because I is constant, i_a is complementary to the total current of electrons emitted $(i_B + i_S)$. This does not simply mean that the contrast in absorbed electron images is always simply complementary to contrast in secondary and backscattered electron images. In fact, contrast in the emissive mode depends on the variations in i_B and i_s, and also on changes in the collection efficiency. The contrast is complementary only when the factors causing changes to the collection efficiency in the emissive mode are not predominant.

Because i_B varies mainly due to changes in atomic number and i_s changes as a result of variations to the sample topography, both types of contrast are generally observed at the same time in absorbed current images. This often creates problems of interpretation. If, however, the sample surface is flat, i_s is practically constant, and i_a exhibits a complementary contrast (with the same type of information) to i_B.

12.3.5 Electrovoltaic Effects and β Conductivity

The charge collection processes consist of three mechanisms that operate in sequence: creation of pairs, diffusion of carriers, and the actual collection. Any factor affecting one of these mechanisms in a variable way along the sample can produce contrast. The most important factors of the electrovoltaic barrier effect are nonuniform distribution of traps or surface recombination centers; topographical details, surface films, small precipitates, and microcracks that produce changes in the dissipation of beam energy for variations of η and δ, absorption or scattering at the interface; microcracks that extend into the depletion region or inclusions beneath it and reduce average lifetime due to localized recombinations; dislocations that intersect the depletion regions and cause local variations in trapping or recombination; alterations to the extension and depth of the depletion region from diffusion or abnormal implantation; and finally, traps in the depletion region that delay or prevent collection.

These factors also are significant in the case of Schottky barriers. The contrast obviously depends on the bias applied; especially, if material disuniformities or geometrical factors result in the creation of areas in or near the depletion region with a sufficiently high field to cause current multiplication. In the case of barriers in luminescent materials, the causes of cathodoluminescent contrast can also operate by means of the absorption of radiation in the charge collection region (optical coupling).

A correct interpretation of the contrast in electrovoltaic barrier effect images yields quasi-quantitative information about the position and extension of the barrier, the presence of space charge regions or inversion layers, the presence and the volume of current multiplication, and the presence and electrical activity of crystallographic defects. Moreover, it is possible to carry out localized measurements of the diffusion length and the average life of the minority carriers. It should be said that it is possible to detect the presence of crystallographic defects but it is rarely possible to resolve the individual defects due to the limited spatial resolution (approximately 1 μm); this requires *transmission electron microscopy* (TEM). However, it should also be mentioned that the SEM only detects defects that affect the electrical behavior of the material or device, and this is extremely important because it identifies the significant areas for a subsequent TEM inspection. In complex devices featuring numerous barriers, it is necessary to have a full understanding of the fabrication processes and a clear comprehension of the device's electrical functions.

The most important contrast factor of the bulk electrovoltaic effect is represented by variations in the local concentration of the impurities, which cause a bending of the energy bands (and hence a localized electrical field) and an asymmetric separation of charge carriers of the opposite sign. Information (even quantitative) can be obtained on the position of the regions with nonuniform resistivity and the slope of this nonuniformity, and absolute measurements of resistivity can be carried out; the sensitivities involved are enhanced by greater diffusion lengths of the minority carriers in the material (in Si with Lp = 300 μm, the sensitivity is ~ 10Ω and the spatial resolution ~10 μm).

As regards the β conductivity, because a continuous external bias is applied, there is an electrical field at each point of the sample. This has an impact on the contrast factors affecting a local increase in conductivity induced by the beam and the factors linked to local variations in the electrical field (and, if the current density is constant, to local variations in resistivity). Variations in the local electrical field can be studied in any semiconductor material; in materials where the diffusion lengths are low, the same type of information as the bulk electrovoltaic effect can be obtained with enhanced spatial resolution and sensitivity. Finally, electrically active crystallographic defects can be detected.

12.3.6 Cathodoluminescence

Cathodoluminescent processes consist of four mechanisms operating in sequence: creation of pairs, diffusion of carriers, recombination, and emission of radiation by the sample. Each factor causes local variations in the radiative efficiency, and hence a contrast in the cathodoluminescent images. Some of the commonest factors are

- Presence of a surface film causing variation and an internal reflection of the radiation; in one layer the impurity content is different from the rest of the sample and hence the ratio of radiative to nonradiative recombination is different.
- Local variations in the charge density fixed in the surface layers that cause variations in the ratio of radiative to nonradiative recombination and internal absorption; presence of microcavities or precipitates that cause variations in the density of carriers generated; and scattering of the radiation generated.
- Crystallographic defects that have recombination, trapping, and scattering properties different from the rest of the sample.

The factors listed all concern the presence of some imperfection. Some factors are inherent to the structure of a device: different types and levels of doping in the various parts of the device and so a change in the ratio of radiative to non-radiative recombination; or competition between the charge collection and recombination in the depletion region of a p-n junction. Finally, some contrast factors are linked to the surface topography, which can be recognized by comparison with the images in the emissive mode.

Information (even quantitative) can be obtained from the cathodoluminescent images of the local variations in the radiative efficiency, the presence and activity of crystallographic defects or nonuniformities of the material (segregations, striations), the type and level of doping, and the local temperature of the device. It is also possible to measure the diffusion length of the minority carriers.

12.4 X-RAY MICROANALYSIS

12.4.1 Principles of the Technique

The X-ray microanalyzer with an electron beam is normally indicated in the literature as an *electron micro probe* (EMP) The principle of operation of the EMP is very similar to that of the SEM. A beam of electrons up to less than a micron in diameter is focused in a vacuum column on a point of the sample surface to assess the local chemical composition. These electrons ionize the innermost orbits of the atoms and then generate characteristic X rays; with suitable dispersion, detection, and counting systems, a spectrometer yields the intensity as a function of the wavelength or energy.

The analysis of the spectrum leads to the identification of the elements present (qualitative analysis) in the small volume examined, whereas the comparison of the intensity emitted by an element with that emitted in the same excitation conditions by the same pure element measures concentration (quantitative analysis). Deflecting the beam sequentially on small areas provides the surface distribution of the chemical elements in the sample.

An analysis can be made of any stable element in the vacuum with an atomic number Z greater than 3; the greatest sensitivity is in the range $Z = 12$–30. Lateral resolution normally falls in the range 0.5–3 μm, and the volume analyzed in the range 1–10 μm^3.

The quantitative analysis, using pure elements as a reference, has an accuracy in determining the concentration of 1–3 percent, provided that the concentration measured is greater than 10 percent when Z is greater than 10 and the sample is flat and smooth. The distribution of an element for areas from 20×20 μm and over can be obtained. The detectability limit is normally 50–1000 *parts per million* (ppm).

The basic advantage of the EMP is its capacity to generate X-rays that can be measured in extremely small volumes of material, approaching μ^3, and hence its ability to measure local variations in composition rather than average chemical composition of a whole sample. This is very important in the study of electronic component reliability. Most current microanalyzers combine the functions of the EMP and the SEM.

12.4.2 Electron-Induced Process of X-Ray Emission

In the incident electron energy range (5–50 keV) used in the EMP, three types of interactions have to be considered: elastic interactions with the nuclei, nonelastic interactions with the nuclei, and nonelastic interactions with the electrons.

The elastic interactions with nuclei are by far the most prevalent, because electrons are easily deviated from the nucleus, even from a considerable distance.

Wide angle individual collisions can occur, which give rise to the backscattering of the incident electrons, and multiple collisions can take place, during which the electrons withstand a series of small angle deflections with a gradual deviation from the direction of incidence. They can either escape from the sample surface as backscattered electrons or be diffused inside the sample.

With samples having low atomic numbers, the electrons penetrate more easily and are diffused more deeply. This explains the drop configuration of the volume occupied by incident electrons. On the other hand, samples having high atomic numbers have numerous collisions close to the surface, which makes the volume occupied by the incident electrons take on the shape of a semisphere. In any case, increases in the energy of the incident electrons does not change the shape of this volume but increases the penetration producing an increase in the dimensions.

The dimensions and shape of the volume occupied by the incident electrons determine the dimensions and shape of the volume having greater or equal energy to the excitation energy E_{exc}; characteristic X rays can be generated in the whole of this volume. The lateral resolution of X rays is the diameter of the minimum volume excited (analyzed) and hence will be considerably larger than the diameter of the incident electron beam. For completeness, we should add that the lateral resolution depends also on the secondary fluorescence (Figure 12.17).

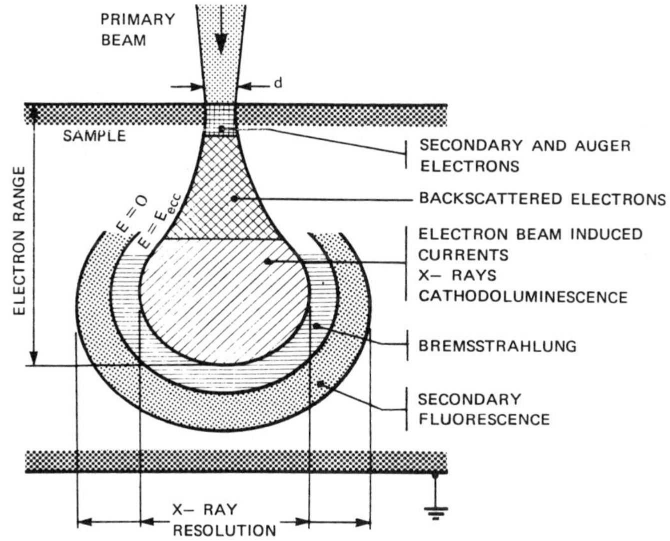

Figure 12.17 Emission volumes of the different types of signal generated by electron bombardment.

Nonelastic interaction with nuclei (deceleration of incident electrons in the coulomb field of the nucleus) causes nondiscrete energy losses and gives rise to continuous X rays, which represent the (unwanted) background of characteristic spectra. The form of this continuous background is a rather complex function of the beam energy and the atomic number of the elements making up the sample.

The elastic interaction with electrons causes discrete losses of energy following the ionization of the atoms, due to the removal of electrons from outer and inner orbits; the characteristic X rays originate with the resettling of the electrons in the innermost orbits. In fact, the stability of the ionized atom is restored by a transfer of electrons from outermost orbits with the emission of radiation of a wavelength inversely proportional to the energy difference between the initial and final states of the transferred electron. Because the energy levels of the innermost layers of an atom are practically independent of the physical state and chemical combination of the element and depend only on its atomic number, then it follows that the wavelength of the X-ray emission lines depends only on the atomic number of the element generating them and hence distinguishes one element from another (Figure 12.18).

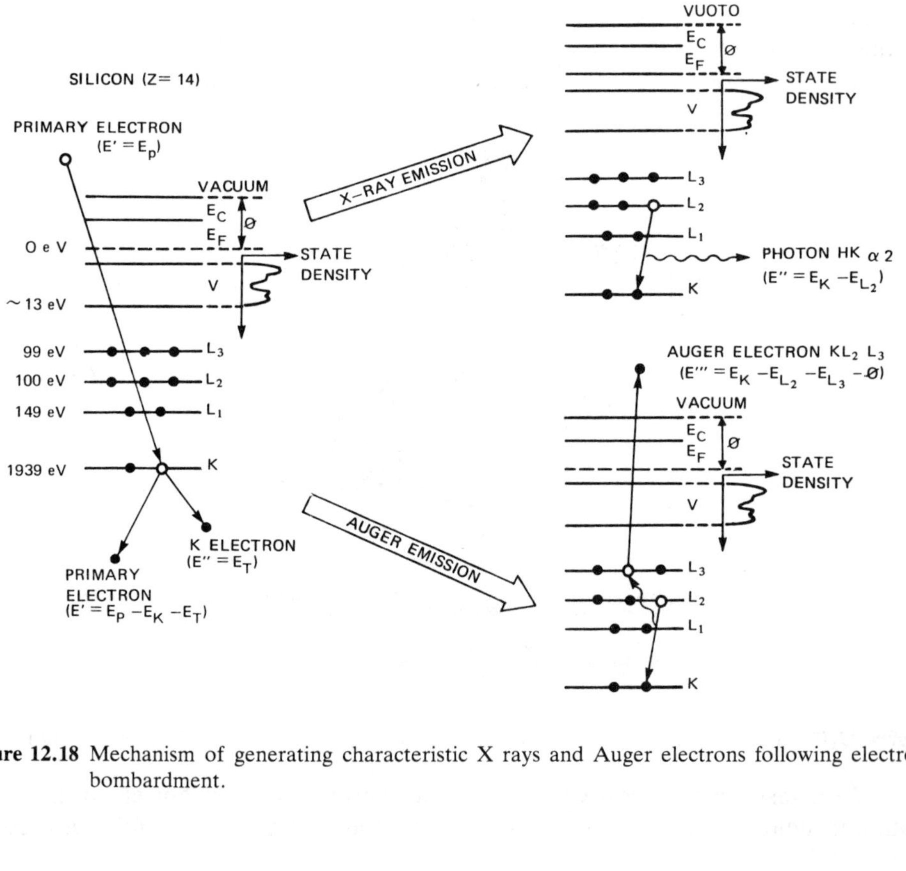

Figure 12.18 Mechanism of generating characteristic X rays and Auger electrons following electron bombardment.

Figure 12.19 shows the energy of the characteristic X rays as a function of the atomic number for various emission lines (remember that the energy E in electronvolts of a photon X is correlated to the wavelength λ in angstroms by the relation $E = 12.398/\lambda$). The terminology used for the emission lines is as follows. The final positioning layer for the transferred electron determines the series (K, L, M) to which the line belongs. So within the series a line is always associated with a transition, with $\Delta n = 1$ (where n is the main quantum number); whereas a line β or γ is associated with transitions with $\Delta n = 2$ or 3, respectively. In other words, each line in a series refers to a transition to the same final layer from a different initial layer. The wavelengths (energies) increase (decrease) moving from series K to series L, to series M, whereas they decrease (increase) moving, in the same series, from lines α, to lines β, to lines γ.

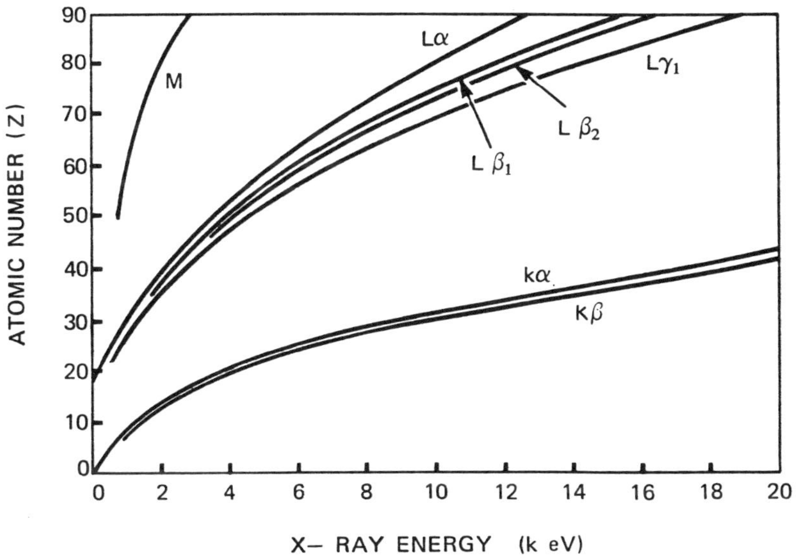

Figure 12.19 Relation between energy of the characteristic X photons and the atomic number of the emitting atom.

The measured intensity of the characteristic lines of a given element is a nonlinear function of the concentration of the element itself; this should be taken into account in the quantitative analysis.

Up to now we have spoken of energy emission by the ionized atom in the form of X photons; alternatively, this energy can be used to reorganize the atom's electron distribution with transitions without emission of radiation: this is the Auger effect (Figure 12.18). Figure 12.20 shows how the probability of this process (*a*) is complementary to the probability of photon emission (ω). The use of the Auger effect for analytical purposes will be discussed.

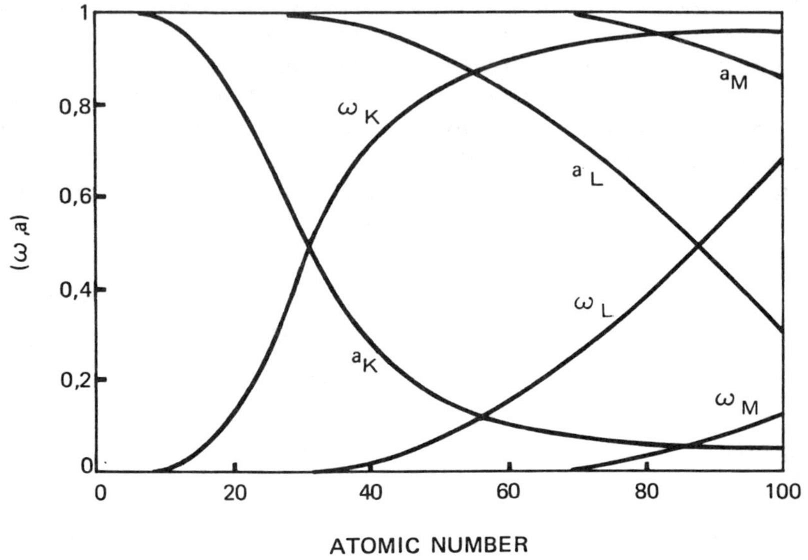

Figure 12.20 Emission yield of *X* photons (ω) and Auger electrons (*a*) as a function of the atomic number.

12.4.3 Wavelength Dispersion Spectrometer

For analysis, it is necessary to separate the various characteristic lines emitted simultaneously from the sample surface under the bombardment of the electron beam and measure their intensity. There are two ways of separating lines from a polychromatic beam: by wavelength dispersion and energy dispersion. We discuss the first method in this section. In wavelength dispersion spectrometer (WDS), the X rays are separated according to wavelength by exploiting the diffraction properties of large individual crystals. When a polychromatic beam impinges on a crystal of interplanar distance d with an angle of incidence θ, only those λ that verify the following relation are subject to Bragg reflection:

$$n\,\lambda = 2d\ \mathrm{sen}\theta \qquad (12.15)$$

where the whole number n is the order of reflection. The most intense reflection is that of the first order. Higher order reflections are not desirable for microanalysis due to the interference that they can cause. The amplitude selection of pulses downstream the detector is used to eliminate this interference.

For the detection it is sufficient to position the detector so that it intercepts the reflected beam forming an angle 2θ with the incident beam. Changing θ from 0° to 90° (and, if necessary, changing the crystal to vary d), the reflected λ changes; so the whole spectrum can be explored sequentially and the intensity of the polychromatic beam can be analyzed.

Among WDS spectrometers, the most widely used detectors are closed gas and gas flow proportional detectors. The X photons penetrate the detector through a special window and interact with the molecules of an inert gas to create a number of primary ion pairs (free electron-positive ion) proportional to their energy.

Since the metallic shell is grounded and the central wire (anode) is at a suitable positive potential (1 to 2 kV), the electrons move toward the anode and the positive ions move toward the external shell. As they move toward the anode, the electrons give rise to a process of multiplication of the number of primary ion pairs (avalanche phenomenon). When the electrons reach the anode wire, they cause a momentary voltage drop, which is then transformed into an electrical impulse having amplitude equal to the energy of the X photons.

In the closed proportional detector, the thickness of the window is such that it is hermetic for the gas contained within; this detector cannot be used for $\lambda > 0.3\ \mu m$ because the window absorbs the X photons. For $\lambda > 0.3\ \mu m$, mylar or expanded polypropylene windows 1–4 μm thick are used; gas leaks are possible with such thin windows, so a continuous flow of gas has to be maintained through the detector.

The pulses produced by the detector are amplified, selected on the basis of their amplitude to exclude the higher order reflections of other lines present in the spectrum emitted by the sample, and counted by a counting chain. The number of pulses counted in a given amplitude range is the measured intensity of the corresponding X line. The most common form of presenting the data consists of tracing the intensity as a function of λ or 2θ on a recorder.

The mechanical structure commonly used is the linear type illustrated in Figure 12.21. The analyzer crystal moves along a straight line passing through the fixed X-ray source (point struck by the electron beam), the center of the Rowland circle rotates about the source, and the detector moves along a noncircular curve, remaining on the Rowland circle.

12.4.4 Energy Dispersion Spectrometer

In the *energy dispersion spectrometer* (EDS), X rays emitted by the sample are distinguished by their energy instead of their wavelength. A block diagram of an EDS is shown in Figure 12.22.

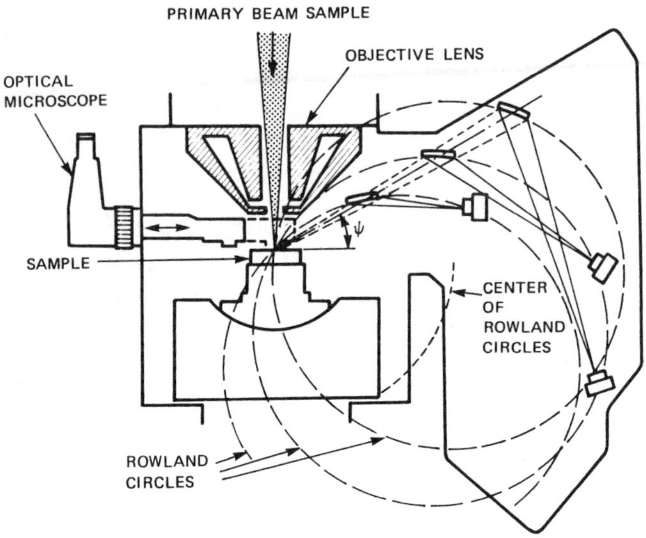

Figure 12.21 Wavelength dispersion spectrometry.

The X rays emitted by the sample strike the detector (Figure 12.22b), consisting of a cylindrical silicon monocrystal, with a diameter falling between 3 and 16 mm and a thickness between 3 and 5 mm. The silicon is doped to form two outer nonintrinsic layers, one *n* type and one *p* type; the inner layer remains intrinsic. The monocrystal is enclosed between two metallic electrodes and a reverse bias voltage of 300 to 900 V is applied to the terminals; under these conditions, in the absence of incident *X* photons, current does not flow in the detector. When an *X* photon penetrates the intrinsic slice, it creates a number of electron-hole pairs proportional to its energy, which are collected at the detector terminals under the action of the applied voltage. The loss of charge carriers during this collection process is negligible, because the intrinsic layer of the Si monocrystal has been treated with atoms of Li so that the imperfections of the crystalline structure (which would otherwise act as traps) are occupied. Moreover, to reduce the background current and allow accurate measurement of the small quantities of charge collected q, the detector works at the temperature of the liquid nitrogen. The collection time is less than a microsecond and hence less than the time interval normally running between the successive instants that two *X* photons are entered.

The next stage, consisting of the FET and its negative feedback circuit, is also kept at the temperature of liquid nitrogen and integrates the charge collected. In other words, it provides, as an output signal, a series of voltage steps in which the height of each step is proportional to q and hence to the energy of the *X*

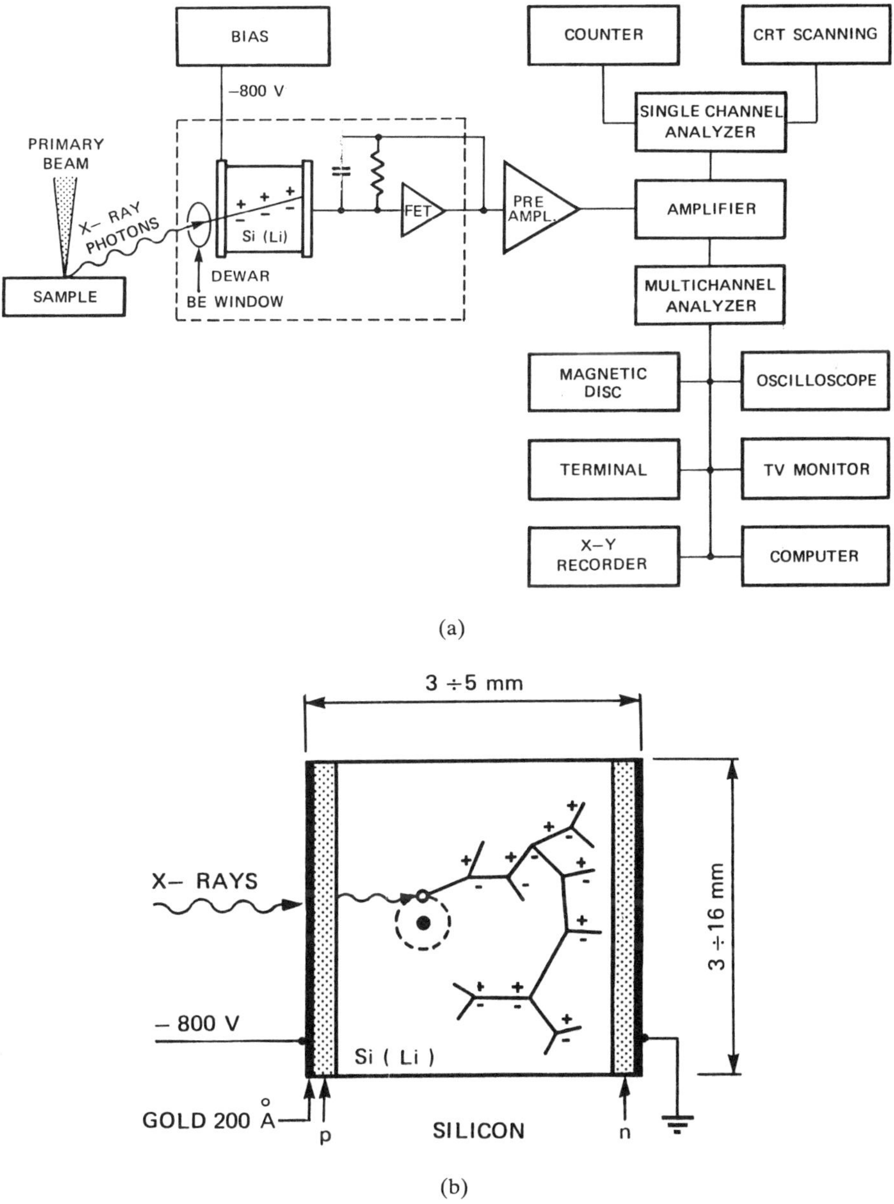

Figure 12.22 Energy dispersion spectrometry: (a) overall block diagram; and (b) detail of detector.

photon that caused it. This signal (amplified and formed in the preamplifier and amplifier stages) is presented finally as a succession of pulses, each of a height proportional to the energy of the corresponding X photon. The pulses are measured and counted by a multichannel analyzer, and the resulting spectrum is usually presented on a TV monitor. A comparison of the performance of the EDS and the WDS is shown in Table 12.4.

Table 12.4 Comparison of Time Required for EDS and WDS Analysis

	EDS (minutes)	*WDS* (hours)
Sensitivity	High ($\geq 10^4 \frac{cps}{nA}$)	Low ($< 10^4 \frac{cps}{nA}$)
Beam current	Low (10^{-9}–10^{-11}A)	High (10^{-7}–10^{-9}A)
Sample contamination and damage	Reduced	Considerable
Focusing problems	Negligible	Considerable
Discrimination between various X-ray sources	Poor	Excellent
Spectral resolution (at 5.9 keV)	Poor (~ 150 eV)	Excellent (< 10 eV)
Spatial resolution	0.5–3 μm	1–3 μm
Peak-background ratio (P-B)	Low (~ 50)	High (~ 500)
Maximum counting speed	10–30 kcps	~ 75 kcps
Analyzable elements	Na → U (Z > 10)	Be → U (Z > 4)
Detection limit	~ 300 ppm	~ 100 ppm
Accuracy of quantitative analysis	Poor at low concentrations	Good

This table shows that the two types of spectrometers are not in competition but complementary. It is hence desirable that an EPA is equipped with both devices to ensure maximum performance. In this ideal situation the EDS is used for a rapid preliminary inspection of all the elements present with $Z > 10$, and then the light elements are detected, the peak interference is resolved, and measurements are carried out for the accurate quantitative analysis with the WDS.

12.4.5 Qualitative Analysis

12.4.5.1 Distribution Images

Although the beam scans the surface of the sample, the "window" relating to the element involved is placed in the EPA; whenever an X photon having sufficient energy to enter the window is detected, a luminous dot is generated at the corresponding spot on the SEM *cathode ray tube* (CRT). The result is an image that corresponds spatially to the other images of the same region of the sample but with the density of the luminous spots in each area linked to the abundance of that given element, as shown in the photograph reproduced as Figure 12.23b.

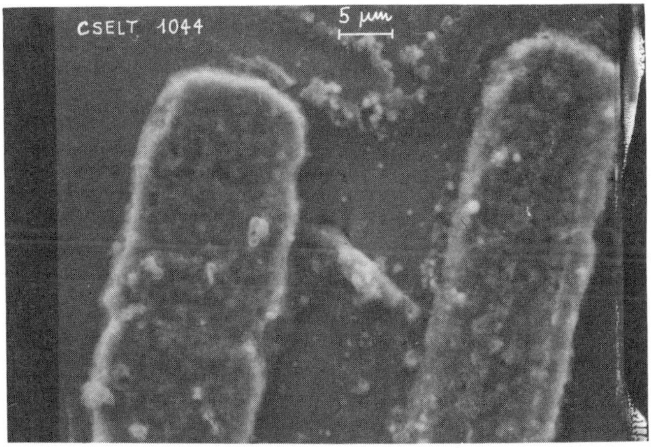

(a)

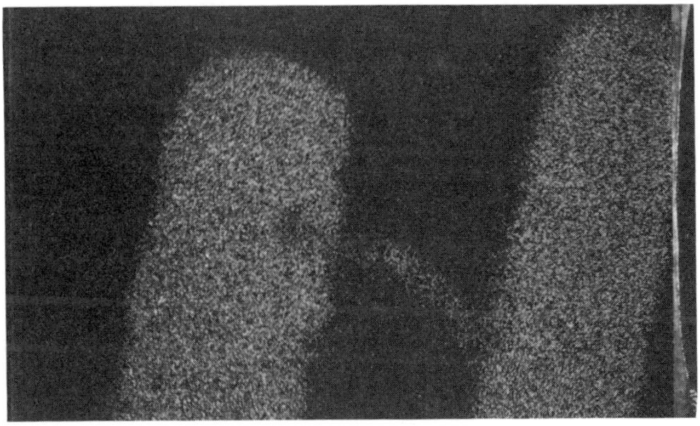

(b)

Figure 12.23 Metalization damage in an integrated circuit subjected to extra voltages: (a) surface morphology; and (b) aluminum distribution map.

The luminous spots, however, are not all due to the *X* photons of the preselected element; some are due to the continuous background photons. So luminous spots may arise in areas not containing the element in question. As a result, the distribution images are not very suitable for detecting elements in low concentrations or in concentrations that change gradually or slightly through the sample. The interpretation of the density of spots in a given area is difficult, just as it is difficult to appreciate and express small fluctuations in a quantitative form.

We conclude by saying that the distribution images provide only qualitative information on the concentration of elements, but the best information about their distribution.

12.4.5.2 Analysis with Scanning over One Line

One limitation of distribution images, involving the difficulty of estimating the relative densities of luminous spots, can be overcome by slowly moving (using from 50 to some hundreds of seconds) the beam along a line on the sample surface while the intensity of the line of a given element is traced directly on the SEM cathode ray tube.

It is also possible here to correlate the variations in the composition at points of the surface, superimposing the *X* scan on the SEM secondary electron image. This enables an accurate assessment of the variations in concentrations from one point to another of the sample.

12.4.5.3 Spot Analysis

Often a full qualitative and quantitative analysis of a given spot on the surface of the sample is required. In this case, the beam is positioned over the spot in question, using the secondary electron image as a reference, and it is kept stationary until the end of the analysis. Meanwhile, the whole spectrum is accumulated along with the EDS, or the WDS is used to progressively explore the various lines present in the spectrum.

The major and minor constituents can be easily identified from the spectrum, and semiquantitative information can also be obtained. In fact, even if the heights of the peaks of the various elements are not a direct, good measure of concentration, a comparison of the peaks of the various elements is often sufficient for semiquantitative purposes.

12.4.6 Quantitative Analysis

We said previously that the intensity of a given characteristic line can lead to the concentration of the element generating the line itself. This could lead us to think that the ratio (usually called the *k ratio*) between the X-ray intensity of the element measured on the sample and that measured on a pure standard, with the same experimental conditions, would give the desired concentration.

This is substantially false because the X-ray intensity emitted is linked in different ways to that generated inside the sample or the standard. The correction factors for changing from k ratios to actual concentrations have been the subject of a large number of theoretical studies and experimental trials.

These corrections, shown in Figure 12.24, should be carried out in iterative mode, because every Z, A, and F term is dependent either on the concentration of the element under study or on the composition of the rest of the matrix (and hence unknown at the beginning). It is therefore necessary to use ad hoc computer programs.

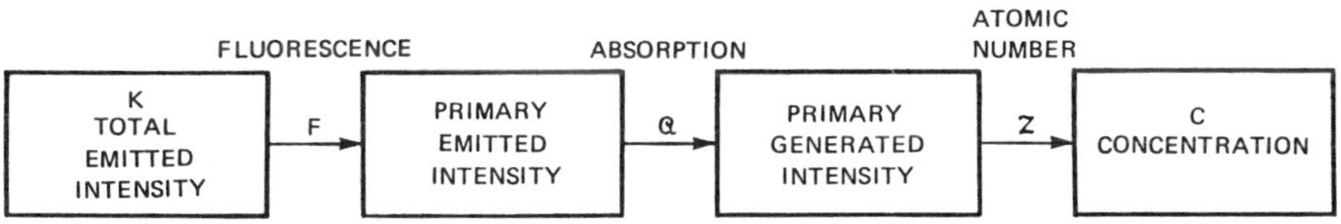

Figure 12.24 Diagram of the Z, A, and F corrections for the quantitative analysis with characteristic X photons.

12.5 AUGER ELECTRON MICROANALYSIS

12.5.1 Electron-Induced Auger Emission Process

The operating principle of *Auger electron spectrometry* (AES) is very similar to the principle of SEM operation. The sample is introduced in an ultrahigh vacuum chamber and struck with a single energy electron beam; electrons are emitted following the interaction of the incident electrons with the surface atoms (the Auger effect). Information on the nature of the atoms present on the surface is obtained by using a spectrometer to measure the energy of the electrons. As well as spot analysis, it is also possible, as in the case of X rays, to obtain images or distribution lines of the elements. Furthermore, linking an ion gun etching to the Auger electron analysis can provide concentration profiles within the sample.

The Auger process for an isolated silicon atom is illustrated in Figure 12.18. The first stage is the ionization of an internal energy level by a primary electron (Figure 12.18 indicates a K level as an example). In the second stage, the atom is deenergized with the transition into the hole of the K layer of an electron coming from a less bonded energy level (L in Figure 12.18), making the energy ($K - L_2$) available for another electron of the same or another level, as well as for the emission of a characteristic X photon. In Figure 12.18, the second electron extracted comes from level L_3; *its kinetic energy is hence approximately* ($K - L_2 - L_3$) and, if this energy is sufficient, the electron is emitted into the vacuum.

The Auger transition used as an example in Figure 12.18 is indicated for convention with KL_2L_3 and clearly the other possible transitions would be indicated with $KL_1\ L_1$, $KL_{2.3}\ L_{2,3}$ and $L_1\ L_{2.3}\ L_{2.3}$

The previous description shows how the emission of an X photon and the emission of an Auger electron are competing events. Figure 12.20 gives the probabilities of decay due to X-ray emission (ω) together with the Auger emission probabilities as a function of atomic number.

We can see how, in the case of the light elements, the probability of X-ray emission is very low and how, with the Auger electrons, it is also possible to analyze the heavy elements.

Fortunately, in practice, only a small portion of the possible transitions have a sufficiently high likelihood of being observed, and so the Auger spectrum in general is not complex and not even difficult to interpret (Figure 12.25). The most probable transitions, and hence the most intense peaks, fall between 10 and 3000 eV.

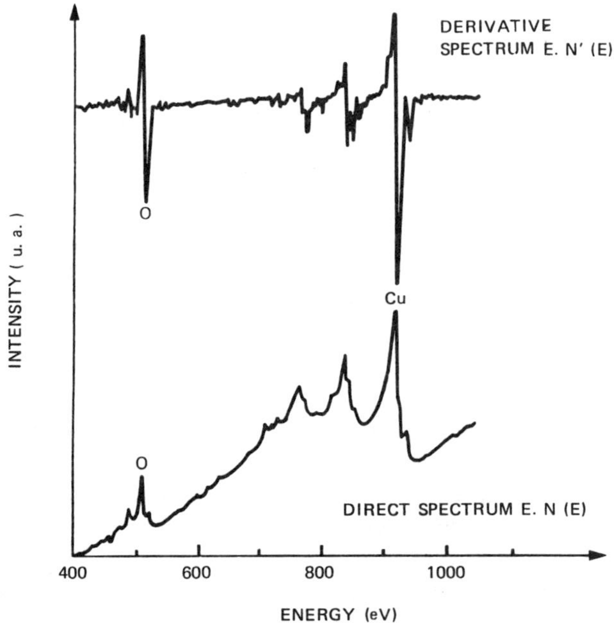

Figure 12.25 Typical Auger spectra.

The previous description refers to the Auger process in an isolated atom, where there are discrete energy levels. In a solid, the discrete levels "widen" to form energy bands, and the external bond levels in particular combine to form the valence band. The Auger processes can still take place, but instead of transitions between levels there are transitions between bands. For many solids, transitions where one or both of the electrons involved in the Auger process come from the valence band are more intense and by convention indicated, for example, with *KLV*, *KVV*. . ., where *V* indicates the valence band. An Auger spectrum is observed where the peak energies are shifted slightly in respect to the isolated atom energies and some peaks have a different form (Figure 12.26).

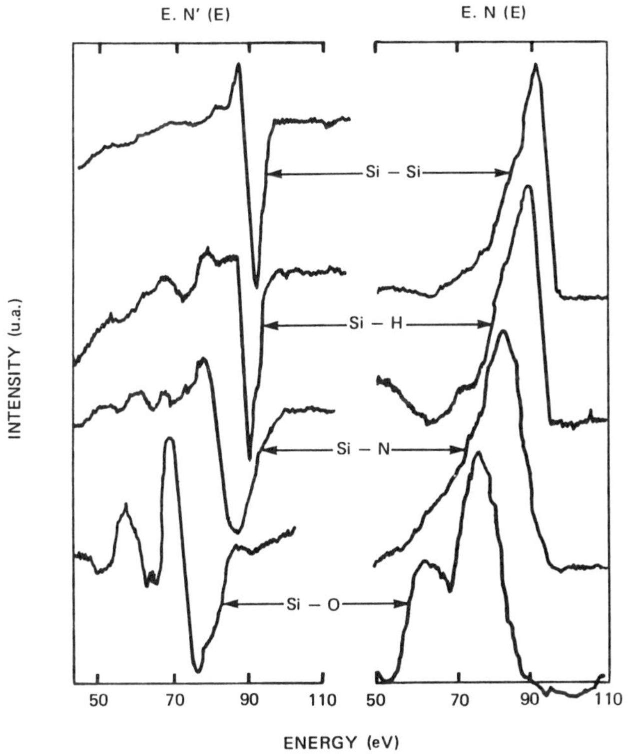

Figure 12.26 Auger spectra from which the type of chemical bond can also be deduced.

It is clear from what has been said that the kinetic energy of the Auger electrons is a function only of the energy levels of the atom or of these levels and the solid's valence band. The Auger spectrum analysis is hence a method of chemical analysis (Figure 12.27). Moreover, apart from a few cases, the analysis of the elements present in the sample can be carried out using transitions at energies below 700 eV. Now, with such low kinetic energies, only those electrons generated within a mean free path from the surface manage to escape without losing energy. For electrons of 20–700 eV, this distance is 0.4–1 nm (see Figure 12.28). Auger spectrometry is hence a method of surface chemical analysis.

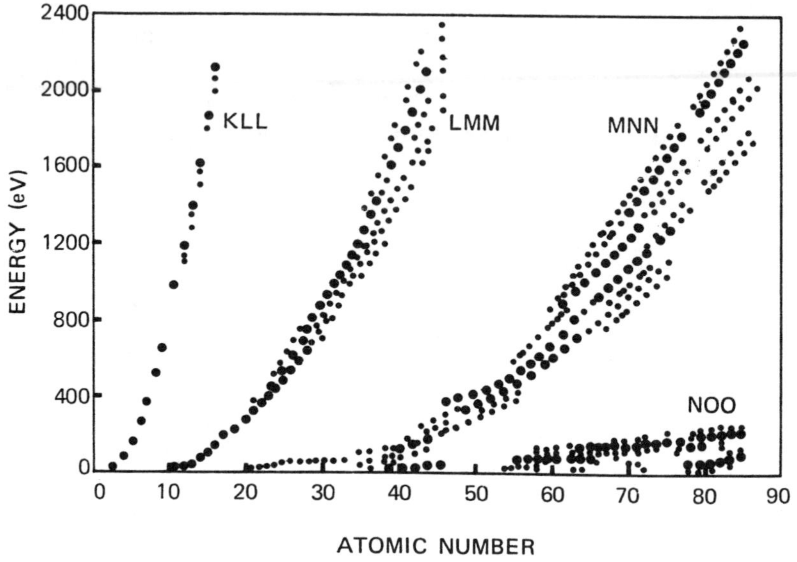

Figure 12.27 Relation between energy of the Auger electrons and the atomic number of the emitting atom.

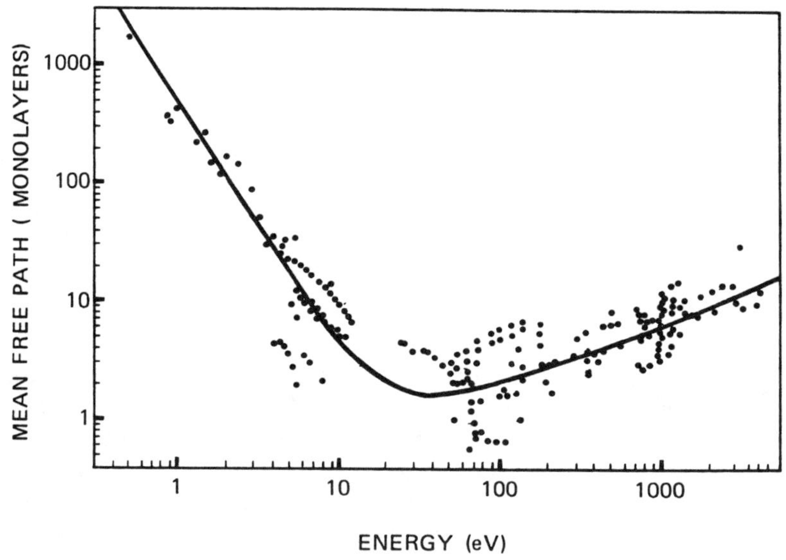

Figure 12.28 Mean free path as a function of the energy of the electrons.

In the intermediate energy region between the secondary electrons and the backscattered electrons without energy loss, the background consists of electrons backscattered after some energy loss; this background is rather uniform, but with small superimposed waves. These waves are due to the Auger electrons and occur at fixed energies (Figure 12.29). The Auger peaks can be extracted from the background by electronically differentiating the signal; after differentiation, the spectrum exhibits evident peaks. By convention, the energy of the minimum of each peak in this $dN/\mathrm{d}E$ presentation, which is normally used, is attributed to the corresponding Auger transition.

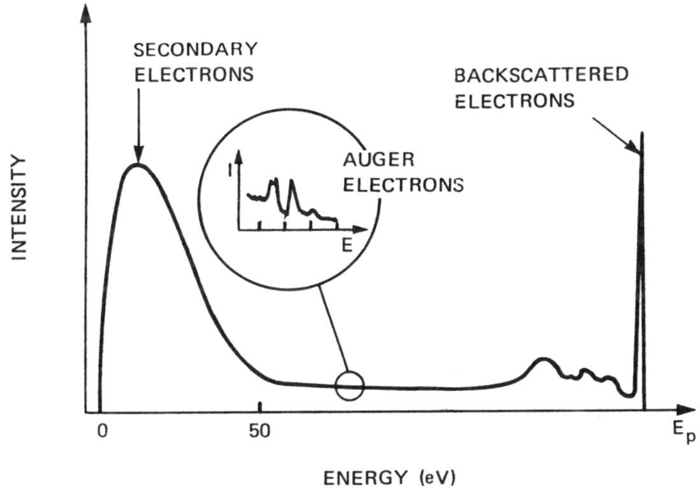

Figure 12.29 Typical energy spectrum of the electrons emitted by a solid following electron bombardment.

12.5.2 AES Instrumentation

The diagram of an Auger spectrometer is shown in Figure 12.30. An electron gun, coaxial with or external to the analyzer, bombards the sample. The electrons emitted enter the *cylindrical mirror analyzer* (CMA) and are deflected by the cylinder brought to a voltage $-V$. As a function of this voltage there is a variation in the energy E of the electrons passing through a small slit and managing to reach the electron multiplier. The interval E of energies that is transmitted by the CMA depends on the resolution in energy $R = \Delta E/E$, which is typically on the order of 0.3 to 0.5 percent. The distribution $N(E)$ is generated by graphing the output signal of the electron multiplier as a function of the voltage $-V$. The spectrum

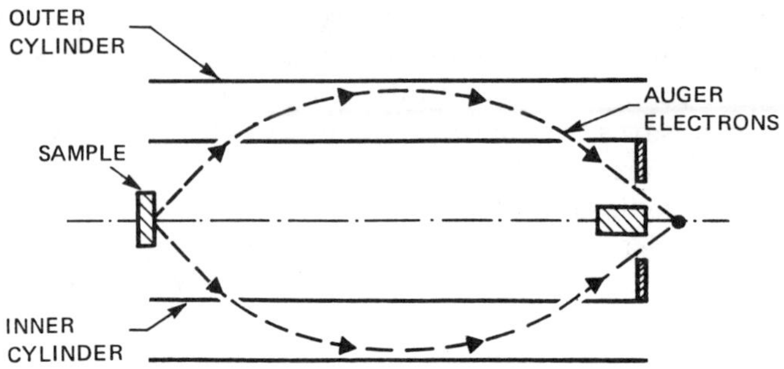

Figure 12.30 Diagram of the cylindrical mirror energy electrostatic analyzer.

$dN(E)/dE$ is obtained by superimposing a small alternating signal ΔV on voltage $-V$ and detecting with a lock-in the output signal in phase with ΔV.

In addition to the electron guns and the electron analyzer, an Auger spectrometer incorporates one or more ion guns for obtaining the concentration profiles, a sample holder for positioning the samples, and a secondary electron detector providing for observation of the topography and centralization of the area to be analyzed.

The high sensitivity of the AES for surface contaminations imposes severe limits on the pressure at which the analysis can be carried out. If sticking factor of 1 is assumed (all the gas particles striking the surface adhere to it) at 10^{-6} torr, a single layer is absorbed in about 1 s. The vacuum installation should therefore allow limit pressures of 10^{-10} to 10^{-11} torr and working pressures of $10^{-}9$ to 10^{-9} torr; these values can be larger when the Auger analysis is carried out at the same time as the ion etch on the surface.

In strict analogy with the X-ray microanalysis, images and distribution lines can be obtained of the elements on the sample surface, by selecting the transmission energy of the CMA equal to the negative excursion of the Auger peak in question and using the lock-in output to modulate the CRT luminosity while the primary electron beam scans the sample surface.

The lateral resolution depends on the beam diameter, signal-noise ratio, and primary electron elastic diffusion (in the case of backscattering). At the moment, lateral resolution is limited by the beam diameter and the signal-noise ratio to typical values of 0.1 to 0.5 μm. It appears difficult, even on the theoretical level, to go below 0.1 μm because the negative effects of the backscatterd electrons (which also generate Auger electrons) cannot be eliminated.

If instead the beam remains on one spot, the Auger spectrum detects the complete chemical composition. This type of analysis is most suitable for identifying differences of composition and contamination on the sample surface. The minimum concentration of an element that can be definitely detected (detectability limit) using Auger electron spectrometry is restricted by the signal-noise ratio. In practice, with a CMA under typical analysis conditions, detectability limits of 100–1000 ppm are achieved for the various elements. Better (lower) detectability limits can be achieved if there is no superimposition of peaks in the spectrum (for which worse resolutions in energy R can be used) and if high primary electron currents and relatively long analysis times can be adopted without altering the sample. However, Auger electron microanalysis (as with X-ray analysis) is not suitable for analyzing traces of impurities dispersed in the volume of a sample.

When, instead, the analysis involves very thin layers of surface or interface contamination, Auger electron microanalysis can be successfully used because, due to its small escape depth values, it can detect about 0.1 percent of a monolayer.

12.5.3 Sputtering-Induced Erosion

One very interesting aspect of Auger spectrometry is its capacity to provide concentration profiles of the various elements, over a surface eroded by sputtering. The concentration profile can be obtained automatically, by electronically measuring the peak-to-peak height of the Auger peaks in question. A multiplexer allows some peaks to be isolated and recorded, with different amplification and modulation, on an $X-Y$ recorder as a function of erosion time. Figure 12.31 illustrates the composition profiles obtained from an Au-Cr-Si multilayer treated for 1 hour at different temperatures.

When an ion impinges on a solid, it can be backscattered by surface atoms or it can penetrate the solid. In the solid, it gradually yields all its energy in a series of primary collisions with the atoms to become an implanted ion, or it is backscattered after yielding some of its energy (Figure 12.32).

The atoms of the solid shifted from their position during a primary collision can be emitted in the vacuum or can cause cascade collisions (secondary or tertiary collisions) with other atoms. When the atoms moved during the cascade collisions are initially close to the surface, they can be emitted in the vacuum. The sputtered atoms can be in a neutral state, ionized, or excited. The average escape depth of the sputtered particles is typically 0.6–1 nm, whereas a relatively small number come from depths of more than 2 nm.

For the purposes of the Auger analysis, the basic parameters of the sputtering process are the sputtering yield and the sputtering rate. In the case of a single atomic solid, the sputtering yield S is the number of atoms emitted in the vacuum per primary incident ion. In the case of a solid composed of two or more types of atoms, it is necessary to distinguish between the total sputtering yield and the differential sputtering yields of the various elements. The total sputtering yield S_T is defined as the total number of atoms emitted in the vacuum per primary incident ion. Differential sputtering yield S_i of the element i is defined as the number of i atoms emitted in the vacuum per primary incident ion.

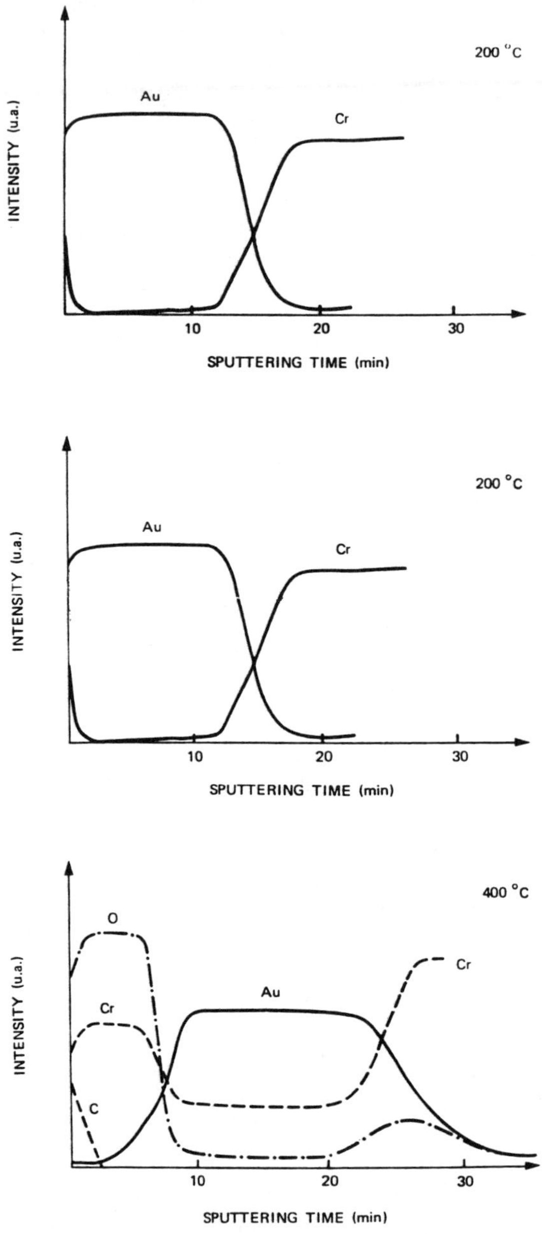

Figure 12.31 Auger profile of composition through the Au-Cr structure after heat treatment.

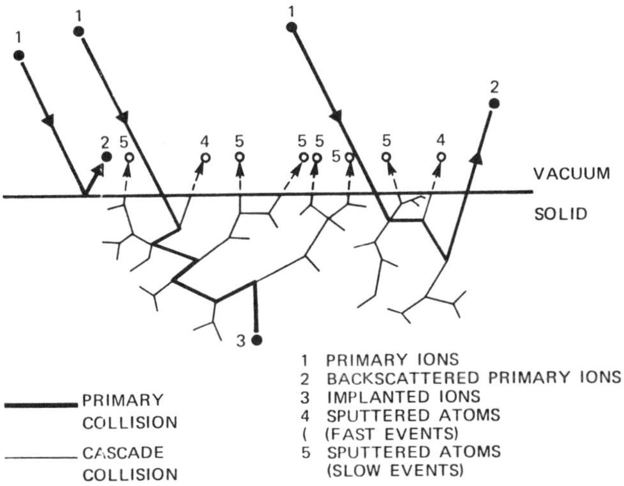

Figure 12.32 Interaction between ions and material.

For single-atom, amorphous, or polycrystalline solids, a theoretical method allows S to be calculated with good accuracy; many experimental data also exist. The parameters determining S are atomic number, electron configuration, and the bond energy of the solid's atoms, surface morphology, and crystal orientation. In general S_i is different from the sputtering yield of the pure element. There are no theoretical expressions of S_i and S_T.

The sputtering rate $\dot{Z}$, or the thickness of the solid layer eroded in the time unit, can be calculated by means of the relation:

$$\dot{Z} = \frac{Z}{t} = 3.6 \times 10 - 4 \frac{J_p Ms}{\rho} \tag{12.16}$$

where $\dot{Z}$ is in m/h, J_p is the current density of primary ions in A/cm^2, M is the atomic mass of the atoms of the bombarded solid in Au, S is the sputtering yield and ρ is the density of the solid bombarded in g/cm^3.

From [16], it can be noted that for a given solid and a given energy of primary ions, the sputtering rate depends only on the current density of primary ions J_p and hence can be controlled by varying the current or the erosion area (for example, by means of scanning). Table 12.5 illustrates the sputtering rates in several typical materials.

The sputtering rate (and hence S) can be determined experimentally by measuring directly, with interferometric methods or with a mechanical stylus, the depth eroded after a very long sputtering time. A second method is to measure

Table 12.5 Sputtering Rate for Ar^+, 1 keV, 100 Å/cm^2 Normal Incidence

Material	Å/min
Ag, Cu	200
Au	160
Al.Mo	40
Fe, Zr	32
Ti, Cr	20
GaAs	260
Si	36
Fe_2O_3	66
SiO_2	40
Al_2O_3	13

the time required to erode a layer of known thickness. It should be noted that these measurements provide only an average value of $\dot{Z}$ (and S) and ignores the possible variations in erosion depth (for example, due to variations in the surface morphology or composition).

Apart from erosion, sputtering in a single-atom solid causes changes in the morphology and surface structure by means of the development of a surface roughness and amorphization, respectively.

In polyatomic structures, it is necessary to consider possible alterations in the chemical composition of the surface layers. Different processes can contribute to this alteration: knock-on, cascade mixing, preferential sputtering, and ion-enhanced diffusion.

Due to the nature of the sample, the composition profiles obtained for sputtering and instrumental factors are generally different than the actual profiles. The degree of likeness depends upon depth resolution. Depth resolution ΔZ is commonly defined as double the standard deviation of the profile through a step composition variation. ΔZ is a measure of the profile enlargement. We offer some general suggestion for obtaining good depth resolution:

- The variation in current density of primary ions should be $< 10^{-3}$ on the diameter of the electron beam.
- It is helpful to use two ion guns in a position symmetrical to the sample or to rotate the sample.
- The effects of knock-on and induced roughness are less if low primary ion energies are used (< 1 keV).
- Most often, the effects of knock-on and cascade mixing are reduced by the use of heavy primary ions.
- The use of reactive types (O, N) as primary ions allows the induced roughness to be reduced.
- The use of an ion beam with almost normal incidence enhances depth resolution (apart from polycrystalline or very smooth samples).

- The most favorable case is that of a flat sample (roughness < 10 nm) that is amorphous or almost amorphous (for example, evaporated films) and homogeneous.
- The influx of the mean free path of Auger electrons is minimal for low energy Auger electrons.
- The influx of the secondary ionization is minimal for samples having an average atomic number that changes only slightly with the depth.
- The influence of ion-enhanced mass diffusion is suppressed by working at low local temperatures (low primary electron density) and with samples that are good conductors of heat and electricity.

In general the depth resolution is 10 percent of the thickness removed, but in ideal circumstances it can reach 3 percent. Figure 12.33 illustrates the depth resolution as a function of the depth at which the interface is found for various materials.

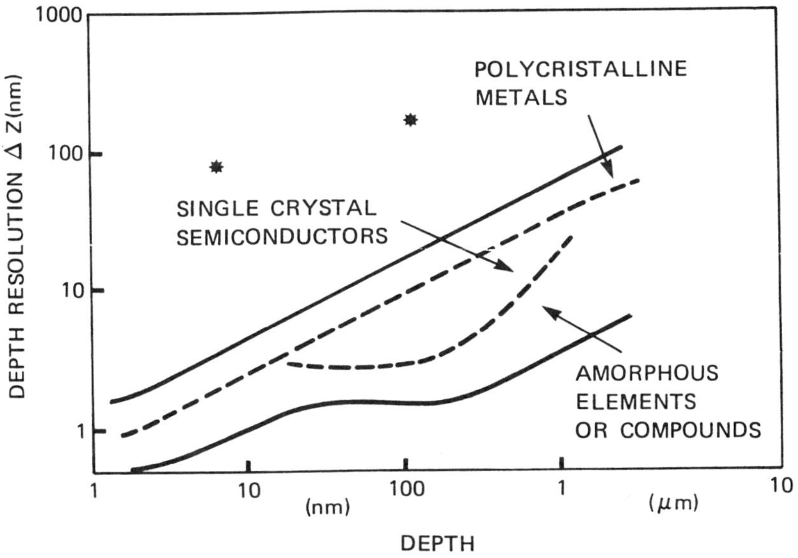

Figure 12.33 Depth resolution as a function of the depth at which the interface is found for different types of material.

12.5.4 Quantitative Analysis

The most widely used method for Auger analysis is that of sensitivity coefficients. If we assume that a relative sensitivity factor F can be attributed to each element, the atomic concentration C of the element can be expressed as

$$C_\alpha = (I_\alpha/F_\alpha) \;/\; (I_i/F_i)$$

where I_i represents the Auger current intensity, F_i the relative sensitivity factor, and i the elements comprising the sample.

The sensitivity factor F_i can be obtained either as a first approximation from published manuals containing the Auger spectra of various pure elements and compounds obtained in the same experimental conditions, or it can be measured from pure standards or compounds. If the matrix of the sample analyzed is similar to that of the samples used to obtain the sensitivity factors the accuracy of this method is 5 to 6 percent; otherwise, it falls to 30 to 40 percent.

Chapter 13
FAILURE ANALYSIS AND TECHNOLOGICAL CHARACTERIZATION

D.E. Riva and G. Benassero

13.1 INTRODUCTION

Reliability cannot be a static discipline. In order to provide the results required of it, and impose a decisive economic and strategic impact, reliability continually must be brought up to date. The major problem stemming from the ongoing technological revolution, and in particular from the push toward ever faster and smaller componentry, is that of adapting investigation techniques and models to shortening lead times, rather than just to the component itself. The classic statistical approach, whereby large numbers of components are tested over lengthy periods, is becoming increasingly impractical, both because of the growing complexity of the devices to be tested, making it difficult or impossible to apply stresses that reflect all potentially critical real-world situations, and because of economic factors, such as the cost of the components and the equipment required to provide suitable signals and bias. Years of experience with reliability-related problems have led manufacturers and users of electronic components to conclude that these problems call for a thorough, in-depth knowledge of the components rather than a "black box" approach, using generalized criteria that make little or no allowance for individual characteristics. For this reason, failure physics has come to occupy a predominant role in the improvement of reliability, because it avoids past errors and raises product quality levels continuously. This approach, which has become common for most of the manufacturers of electronic equipment, originated new qualification procedures for electronic components that are more elastic and effective. It must be so, since qualifying a new component means allowing that component to be used throughout the company. To clarify this concept, a brief description of a typical qualification procedure used in ITALTEL follows:

- Construction analysis of samples representative of the technology under examination with the same packaging and from the same supplier (two functions).
- Environmental tests on devices belonging to the same family with the same packaging and from the same supplier.
- Technological report of the device; that is, construction analysis to verify the technology, mask edition, and internal visual inspection.
- Electrical characterization over the whole range of operating temperatures, also to help write the automated test program.

Furthermore, failure, at any time or stage, requires that the failed components be identified and the failure mode and mechanism investigated. Thus, the "old" quality assessment techniques are still used, but in the new approach analysis of failed components and technological analysis provide the feedback. This has gradually been introduced with a new concept of "incremental" reliability, where rapidly obtained, high-quality data are often more effective than quantitative data that arrive too late or are too difficult to acquire.

This chapter considers construction analysis and failure analysis separately, even if many of the methodologies and analytical techniques described belong to both. This choice was based upon both historical reasons and the differences in their starting points and the objectives.

13.2 FAILURE ANALYSIS

Failure analysis is an *a posteriori* activity, inasmuch as it is carried out on components that have either been proven inoperative or for some reason considered as such. The objective of analysis is to identify the causes of the device's malfunction by promoting one or more degradation mechanisms. The next step is to suggest corrective measures; such measures may include improving the component's manufacturing technology (e.g., by modifying process stages or providing tighter control over certain steps), or by observing different criteria in the particular application for which the device has been selected. The failed device generally comes from the field or from controlled tests such as life tests or step stress tests. In the first case, if the failure is systematic rather than random, repair costs are invariably high, particularly when analysis dictates blanket replacement of all similar components or, should a suitable, equivalent component be impossible to find, that all modules in service be replaced by ones offering a better guarantee against malfunction. Clearly, effective prevention is much to be preferred. In the second case, feedback is more immediate because the purpose of testing is to assess the component's reliability level before it is put into service. In any case, a failed device is a source of valuable information, and the procedure used to obtain this information must ensure that misleading data is not introduced nor useful data destroyed. Such a procedure may be generalized, as shown in the example in Figure 13.1, but in practice all of the operations listed are never carried out at the same time.

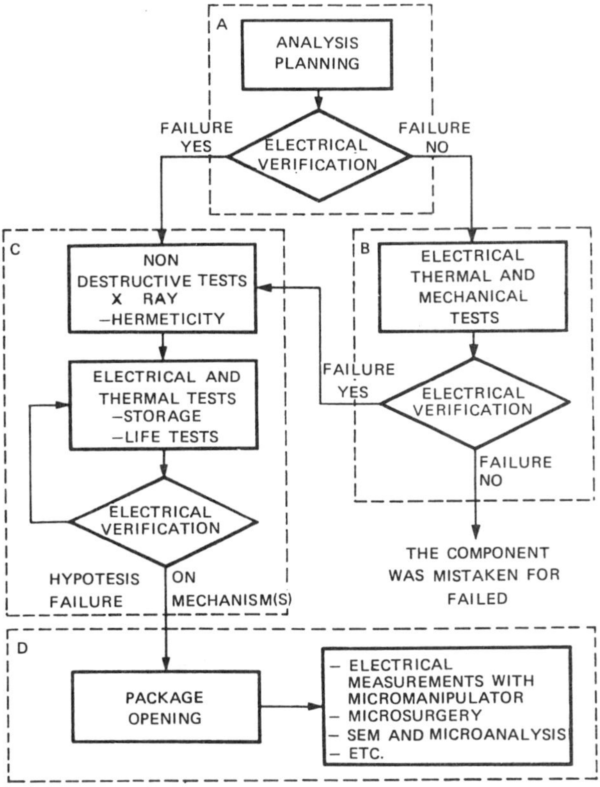

Figure 13.1 Failure analysis flow chart.

Which operations are performed (i.e., the analysis plan as such) depends essentially on component type and the extent of the problem, and thus on the depth of the investigation. Often, factors emerging during analysis cause the analysis plan to be modified. The analysis procedure calls for expertise in a variety of areas, both in applying techniques and in interpreting results. Before starting any activity, it is important that as much information as possible be collected regarding the component under test. Such information includes details of the environment in which failure was detected, as well as information from the manufacturer, even if only a data sheet. Indeed, the manufacurer's data sheet is often the only information available. During analysis, it is of vital importance to document photographically each step before taking further action, because analysis often uses techniques that destory certain details in order to reveal others, and the information provided by the former details must not be lost.

13.2.1 External Analysis

External analysis is the first step and should always be performed with care. Not infrequently, the component exterior (i.e., the package and leads) will present signs that may provide useful information. A low-power (5–50 X) stereoscopic microscope can be used to first check encapsulation integrity, with particular attention to sealing, leads, lead exits, and soldering. In this way, the larger defects, such as cracks, corrosion, or breakage, which are symptoms of poor contact or sealing, can be identified with relative ease and may be correlated with later findings. (For further reference see MIL-STD 883, method No. 2009.)

13.2.2 Failure Check

An operation that might seem trivial but that nevertheless is of fundamental importance, is checking the failure mode, how the malfunction appears from outside. It may be that the failure has been attributed to a component by mistake or that the failure occurs only under certain conditions. If the component is simple, electrical checking may be carried out with the aid of a curve-tracer via direct measurement of the characteristics or by preparing a suitable test circuit to simulate operation. For complex components (LSI, VLSI, and special microwave components), it is necessary to use more sophisticated test equipment and analysis.

Figure 13.2 shows the layout and diagram of a typical TTL circuit. Figure 13.3 contains a few examples of characteristics shown on the curve-tracer for measurements on this circuit. Table 13.1 lists some abnormal characteristics and their probable causes.

It should be borne in mind that a suitable procedure must be followed when applying bias via the curve-tracer. Junctions should be stressed in highly controlled, progressive fashion according to predetermined configurations; namely,

1. All inputs and outputs with respect to ground.
2. All inputs and outputs with respect to supply or supplies.
3. Across inputs shared by the same elements from a logical point of view or in close proximity.
4. Across supply or supplies and ground.
5. If the package is in conductive material and is not connected electrically to an internal reference in the device, across all leads and the package.

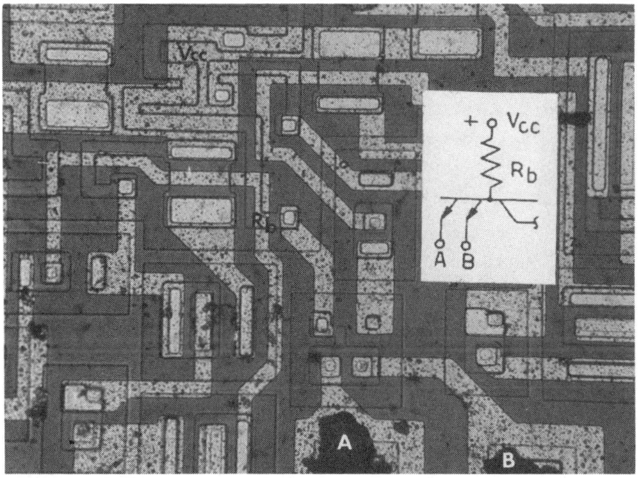

Figure 13.2 TTL logic gate input circuit: layout and circuit diagram.

When a failure cannot be detected under normal measurement conditions, it may sometimes be necessary to apply appropriate stresses to the component. For example, poor contact between an interconnection wire and a pad resulting from a bond detachment in a device provided with a plastic package (i.e., in which the active die, wires, and internal part of the frame are completely immersed in potting resin) can be detected only by heating the component. Bonds resting on the pad lift as a result of the different coefficients of thermal expansion, thus giving rise to failure, which is detected under operating conditions, when the power dissipated by the device causes the temperature of the device to rise.

Other types of failure, such as leakage caused by the presence of mobile contaminant ions, may be identified by subjecting the device to storage at high temperatures with and without applied bias and repeated measurement at the end of this period. The device's different performance may help in forming preliminary hypotheses.

13.2.3 Examination Prior to Opening

Once failure and the failure mode have been identified, the following operations should be carried out: radiographic inspection, particle impact noise detection, testing for leaks, and residual gas analysis.

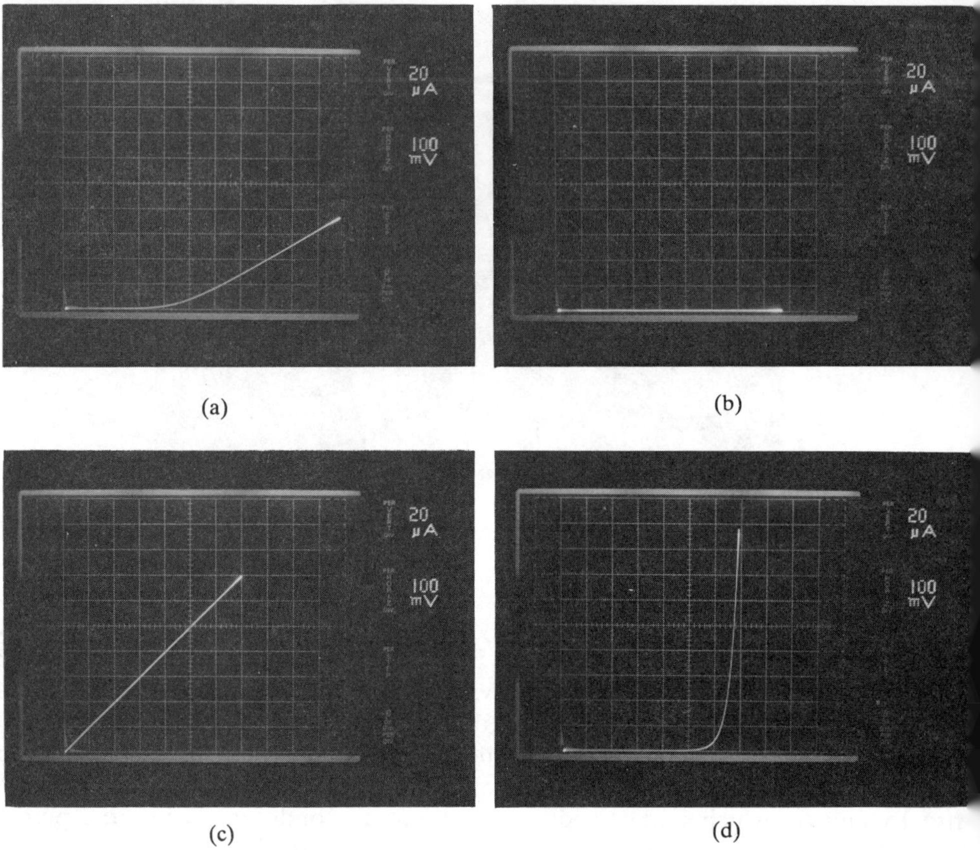

Figure 13.3 Input characteristics of the circuit shown in Figure 13.2, as displayed on a curve-tracer: (a) normal characteristic; (b) open circuit; (c) base-emitter junction short circuited; and (d) Base-resistor short circuited.

Table 13.1 Abnormal Junction Characteristics

Irregularity	*Probable cause*	*Behavior under electrical-thermal stress*[a,b]
Direct characteristic with soft knee	Excessive electrical stress	No change after [a]
Reverse characteristic with high leakage current and low breakdown voltage		
Low level gain degradation and soft direct base-emitter characteristic	Excessive inverse electrical stress across emitter and base	Almost complete restoration after [a] Degradation no longer appears after [b]

Table 13.1 continued

Irregularity	*Probable cause*	*Behavior under electrical-thermal stress*[a,b]
	Surface charges or contamination	Almost complete restoration after [a] Degradation reappears after [b]
	Silicon breakage under solder pad	Further degradation after [a]
Characteristics due to channeling	Surface charges	Almost complete restoration after [a] Degradation does *not* return with [b]
	Contamination by mobile ions	Degradation returns after [b]
Short circuit across emitter and collector, even without damage across emitter and base and across base and collector	Excessive electrical stress with thermal runaway	No change after [a] or [b] punch-through across emitter and collector possibly related to secondary breakdown)
Characteristics with low breakdown voltage with tendency to walkout	Contamination by mobile ions	Almost complete restoration after [a] Degradation reappears after [b]
	Trapped charges	80–90 percent restoration after [a] Degradation normally does *not* return after [b]

[a] Storage at 200°C for 8–24 hr without bias;
[b] Storage at 125°C with bias applied for 168 hr.

13.2.3.1 Radiographic Inspection

Radiography is chiefly used to inspect the internal condition of a device, for example by checking soundness of interconnection wires between the active die and package (see Figure 13.4), die attach uniformity (see Figure 13.5), or the presence of foreign particles of appreciable size inside the cavity of a device. Furthermore, having a picture of the internal structure provides useful information about the device's state prior to opening the package. Test conditions (acceleration voltage, exposure period, and distance from source) are selected according to device material (that of the package in particular) and the sensitivity of the film used. Photographic plates like those used for medical purposes are normally used, but satisfactory results can also be achieved with polaroid plates.

Unfortunately, materials such as silicon and aluminium are practically invisible unless special measures are taken, and steel containers block the interior. Consequently, this type of inspection is not universally applicable. For further reference, see MIL-STD-883 method No. 2012.5.

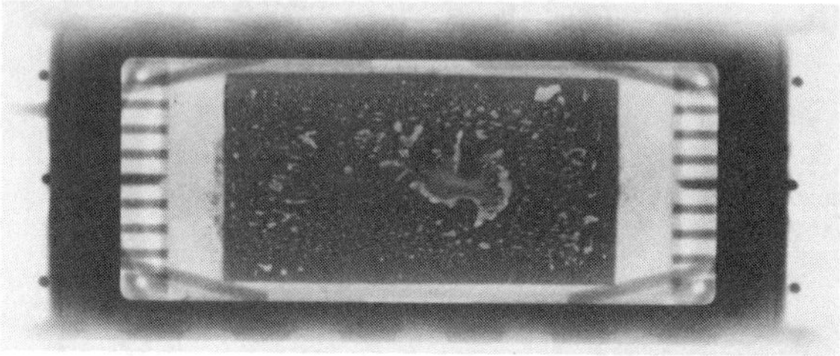

Figure 13.4 Radiographic image of an integrated circuit in a plastic package, showing internal structure.

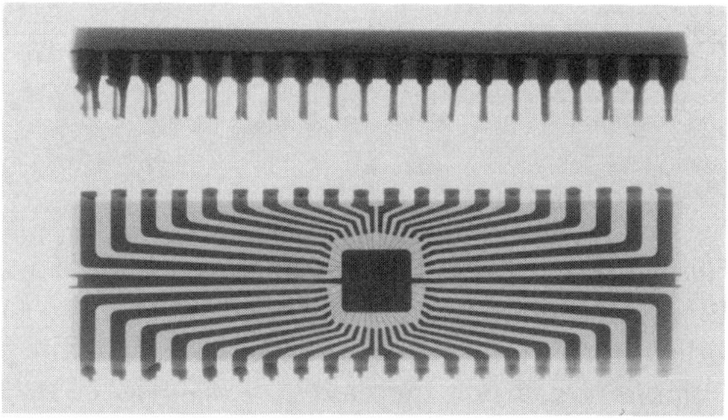

Figure 13.5 Radiographic image of an integrated circuit in a ceramic package, showing die-attach distribution.

13.2.3.2 Particle Impact Noise Detection

Foreign particles inside a cavity device, particularly conductive ones such as pieces of wire or beads of solder can cause malfunctions from short circuits. *Particle impact noise detection* (PIND), is used to reveal the presence of such particles by applying increasing or programmed-frequency vibration to the device; vibration may be accompanied by mechanical shock, which frees any particles that may be blocked inside the device. The PIND equipment schematics are shown in Figure 13.6. The presence of foreign particles is detected by collecting and amplifying the noise caused by their impact against the package walls using an electro-acoustical transducer.

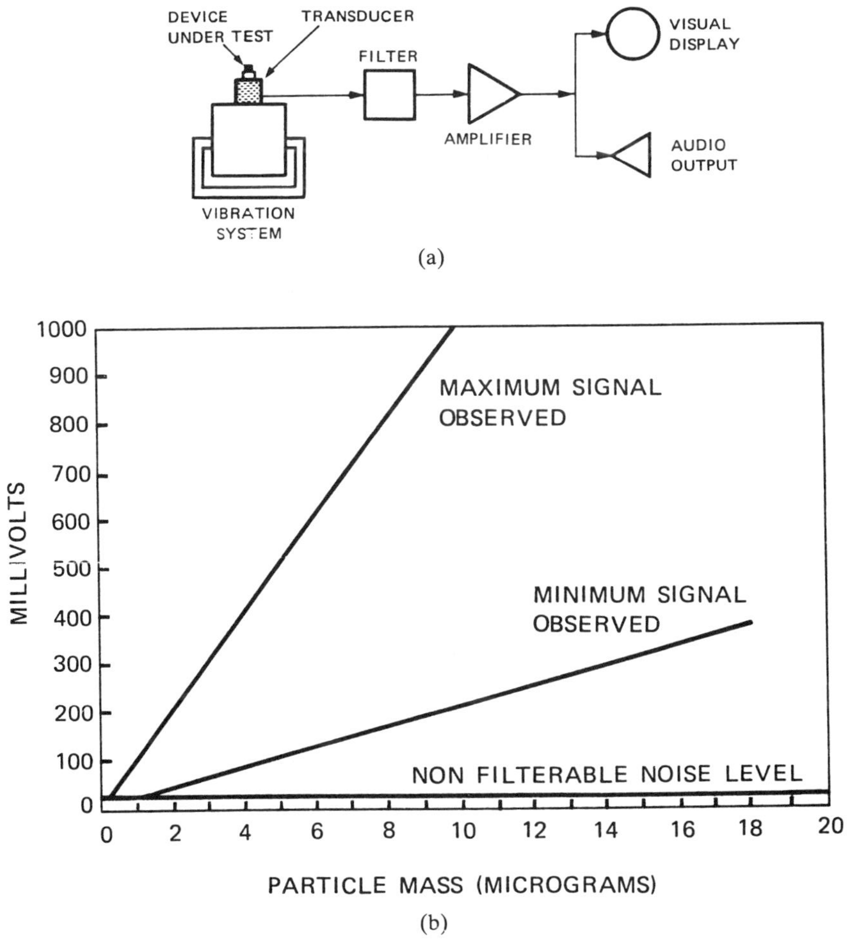

Figure 13.6 PIND (a) equipment schematics and (b) relationship between signal amplitude and particle mass.

It may be useful to repeat electrical measurements and radiographic inspection after PIND. For instance, a short circuit that existed prior to test may be eliminated, since the mechanical stresses applied during PIND removed the cause. For further reference, see MIL-STD-883, method No. 2020.4.

13.2.3.3 Testing for Leaks

Humidity and contaminants on the active die are a potential cause of failure. Sealed cavity packages may lose their impermeability because of poor control over sealing operations or cracks or holes induced by thermomechanical stress. Two basic tests are used to check the seal of devices provided with this type of encapsulation:

- Gross-leak test: In its simplest form, this test consists of immersing the device in a container of liquid, such as Fluorinert FC-43. The container and liquid are then heated to 125°C. If a gross leak is present, a row of bubbles will issue from the device and rise to the surface of the liquid, as shown in Figure 13.7. Alternatively, the device may be pressurized to 0.5 torr and then covered with liquid. Pressure is subsequently raised to 5 atm and maintained for at least 1 hour. The device is then returned to ambient pressure and immersed in the liquid, whereupon the procedure just described is repeated. These methods are applicable to leaks exceeding 10^{-4} and 10^{-5} cm^3/s, respectively. The latter is also the detection limit of another method, which consists in immersing the component in a penetrant dye and pressurizing above 6 atm for at least 3 hours. At the end of this period, the device is thoroughly washed to remove all external traces of dye, which is visible only under ultraviolet light of a suitable wave length. After half an hour in a vacuum below 5 torr, the device is inspected again in the same way. The presence of a leak is denoted by dye on the surface of the device.
- Fine-leak test: This technique uses the equipment shown in Figure 13.8. The test consists of pressurizing in He and measuring leakage with a mass spectrometer. Assuming that the cavity volume does not exceed 0.1 cm^3, He leakage will not exceed 5×10^{-8} cm^3/s after 1 hour at 5 atm. Under different conditions, appropriate formulas must be applied to define acceptable limits. However, since the test cannot detect leaks of more than 10^{-4} cm^3/s, it should always be combined with a gross-leak test. A more precise and sensitive method consists of measuring radioactivity after pressurization in a chamber filled with a gaseous mixture, such as krypton 85 and nitrogen. This test is not used widely because of the high cost of equipment and the risks involved in handling and storing radioactive gases. For further reference, see MIL-STD-883, method No. 1014.5.

Figure 13.7 Gross-leak test: bubbles denote a gross leak, indicating that the device under test is poorly sealed.

13.2.3.4 Residual Gas Analysis

Residual gas analysis is used particularly for failure mechanisms, such as metalization corrosion, which depend on the device's internal environment and moisture content. It can prove especially useful in cases where fine- and gross-leak tests yielded negative results. Residual gas analysis also makes use of a mass spectrometer, but the test is much more critical for a number of reasons.

The device is placed in a vacuum chamber and heated for one or two days in order to eliminate the moisture absorbed by the microcircuit package and the walls of the chamber. The device is opened using a sharp tool made of steel or other hard material and actuated by a special mechanism for remote control from outside the vacuum chamber. Opening the device causes a pressure rise that is picked up by a sensor inside the chamber.

In addition to the difficulties involved in this operation, calibrating the spectrometer to obtain valid and repeatable measurements can be a problem. We should also note that interpreting the test results can be quite complicated, given the variety of possible reactions in the opening environment at the heating temperature. For further reference, see MIL-STD-883, method No. 1018.2

Figure 13.8 Fine-leak test equipment.

13.2.4 Package Opening

As packages used in microelectronics vary widely in shape, dimension and material, opening techniques must allow for all of these factors. Commonly used methods can be broadly classified as either mechanical or chemical. As it is not possible to discuss all of the types of encapsulation, we have chosen to describe the opening methods used for the four major package families.

13.2.4.1 Glass Packages

Glass packages are perhaps the most delicate ones. Only mechanical opening methods can be used, such as lapping parallel to the longitudinal axis after embedding the container in resin. The method is particularly useful for diodes, because it preserves information regarding assembly and internal contacts. The lapping operation must be closely monitored, since material removal must cease as soon as the cavity is reached. After observing the component interior through a microscope, resin may be injected in the cavity to hold internal parts in place before continuing.

13.2.4.2 Metal Packages

Metal packages generally consist of a base soldered to a metal cap. Those used for power transistors (i.e., Jedec Standard TO-3, TO-18, *et cetera*) are typical examples, but metal packages are used to encapsulate components of all types. In some cases, it is possible to force off the cap by inserting a sharp knife between it and the base, particularly if soldered on a preform using Au-Sn or similar alloys. Dimensions and mechanical strength permitting, however, the cap is generally removed on a lathe or by filing down the thickness along the soldered seam.

13.2.4.3 Ceramic Packages

Ceramic packages generally are made from a beryllium or aluminum oxide base (which may consist of several layers) closed with a cover that may be metal or the same material as the base. In the first case, the cover is soldered on a preform; whereas in the second case, a frit seal or resin is used. In both cases, it is advisable to heat the component before opening at a temperature that does not alter the device's internal conditions (for example, by bringing the die-attach to the liquid state); that is, 100–300°C. The package is then opened mechanically by inserting a sharp tool between base and cover or, in the case of flat metal covers, by reducing thickness with a fine-grained file until tweezers can be inserted and the cover pulled delicately back and off.

13.2.4.4 Plastic Packages

Plastic packages are the cheapest and the most widely used—and not only for consumer applications as in the past, since the reliability of these units has increased considerably over the years as a result of the extensive research carried out in various quarters. However, they are still the most critical in terms of opening, because the phenol, epoxy, or silicone resin of which they are made completely covers the frame, the die, and the bonding wires. Only in a few cases is a protective layer of another, cleaner, resin interposed between the die and the plastic encapsulant.

The most frequently applied opening technique uses wet chemical etching agents, such as acids and organic solvents. After placing the component in an oven at a temperature slightly above 100°C for a few hours to eliminate as much as possible of the atmospheric humidity absorbed by the container, the typical procedure consists of ensuring mechanical stability (e.g., by soldering the leads to a suitable support) and reducing the thickness of the plastic in the area where radiographic inspection has shown the die to be located. A small milling cutter, as shown in Figure 13.9, may be used for this purpose. The component or a portion of it is then immersed in nitric acid or fuming sulphuric acid, monitoring the amount of material removed. Finally, the unit is washed in a stabilizing solution of water and sodium bicarbonate. The technique must be performed with caution, because of both the hazards involved in handling acids and the reactions that may take place in the component. In particular, metal parts, and thus the exposed metalizations, tend to corrode in the presence of water or moisture. Consequently, it is not always possible to determine whether the phenomena observed in the device after removing the plastic were already there or introduced during chemical etching.

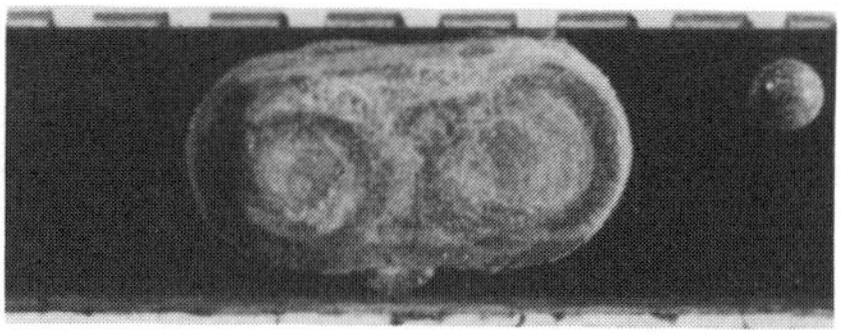

Figure 13.9 Small milling cutter used to reduce thickness of plastic in die area.

Some plastics, including those used for potting hybrid circuits, can be removed by immersion in organic solvents (Unresolve, Desolve, *et cetera*) with temperatures at or near ambience. Opening times, however, can be quite long. Dry chemical etching is unquestionably preferable because of the selectivity it provides in some cases, and because it is far more controllable. It is derived directly from the technique used to strip photoresist during production. The plastic is etched by highly reactive radicals in a gaseous mixture ionized by applying a radio frequency signal. This "plasma" is generated in a cylindrical glass or quartz reactor in which a pressure on the order of several torr is maintained. The equipment schematics are shown in Figure 13.10.

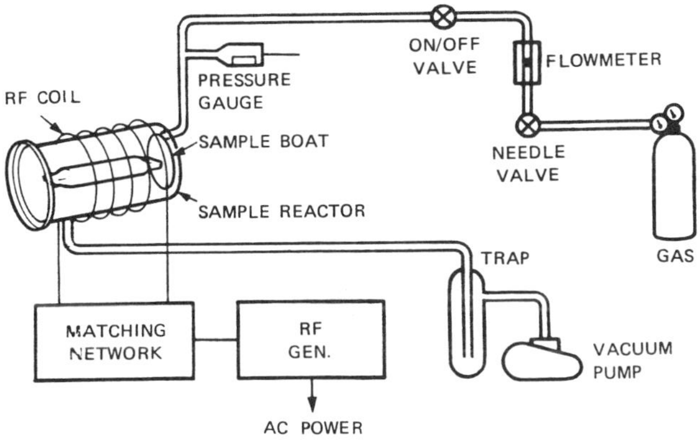

Figure 3.10 Plasma etching system schematics.

Gases used include oxygen (which is highly selective inasmuch as it etches only the plastic without affecting the semiconductor chip) and mixtures of oxygen and fluorine-based gases such as freon (90 percent 0_2, 10 percent CF_4) when the plastic is particularly resistant. In the latter case, since the plasma etches silicon and its compounds more rapidly than plastic, it is advisable to stop the etching process as soon as the die surface is visible and proceed using oxygen alone. The RF power must be kept low because overheating the samples could result in damage or just add artifacts. Figure 13.11 shows two stages in opening an integrated circuit in a plastic package using oxygen plasma etching.

13.2.5 Internal Visual Inspection

After opening, the first operation to be carried out is obviously a thorough inspection of the interior. A stereoscopic microscope can be used to locate macroscopic defects and to check general condition and assembly details. For more detailed examination, a high power (at least 500–1000 X) metallographic microscope is used, together with accessories such as colored filters, polarizer, and a means for illuminating the specimen from the side.

Particularly where the results of electrical measurements have made it possible to identify the general location of the failure, microscopic examination can reveal useful data, from larger defects such as broken or lifted bonds, extended corrosion, fusion of metalizations from high-energy transients, foreign matter or particles trapped in the passivation, down to microscopic defects.

(a)

(b)

Figure 13.11 Opening an integrated circuit in a plastic container by means of oxygen plasma etching: (a) intermediate stage and (b) final result.

Microscopic examination and photographic documentation of the details observed during examination are to be repeated after the subsequent chemical etching and staining steps because new elements for examination come to light during each stage. For further reference, see MIL-STD-883, method No. 2010.7.

13.2.6 Microsurgery

The elements making up an integrated circuit are physically connected to each other. In addition, they are rarely accessible from the exterior. Consequently, two basic problems must be solved before electrical measurements can be made on small circuit parts or elements on the die. First, the part to be examined must be electrically isolated from the rest of the circuit to segregate their measurements. Next, the desired areas must be reached to apply signals and connect measuring instruments. In certain cases, as when physical dimensions of extreme metalizations permit, this is possible using a micromanipulator, as shown in Figure 13.12. This instrument is equipped with mobile heads carrying probes with extremely small-diameter tips (down to 1 μm) that can be positioned on the device while monitoring the operation through microscope eyepieces. To achieve good contact, it is often necessary to remove the top passivation layer, at least in the area concerned. This is done either through chemical etching or an incision with the probe tip.

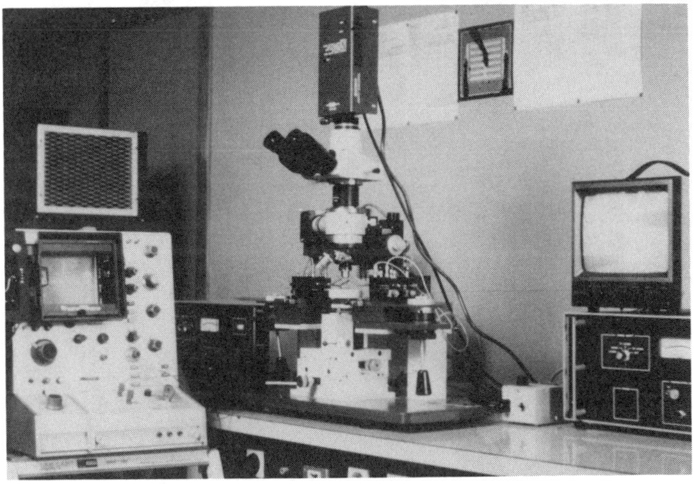

Figure 13.12 Micromanipulator.

An incision may also be made at the point where conductors are to be cut in order to isolate part of the circuit. A few drops of a suitable chemical solution are placed on the exposed metalization to etch and remove it in highly circumscribed areas so that the desired measurements can be carried out on the isolated portion of the circuit. Figure 13.13 shows the results of applying this technique to a transistor of an integrated circuit.

13.2.7 Selective Chemical Etching

To facilitate inspection, the process with which the component was manufactured is reversed. The various layers (surface passivation, interconnection levels, and insulation oxide) are removed one by one through chemical etching. For organic

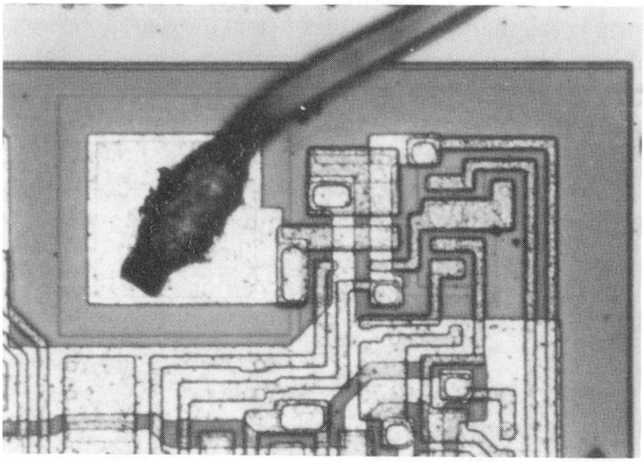

(a)

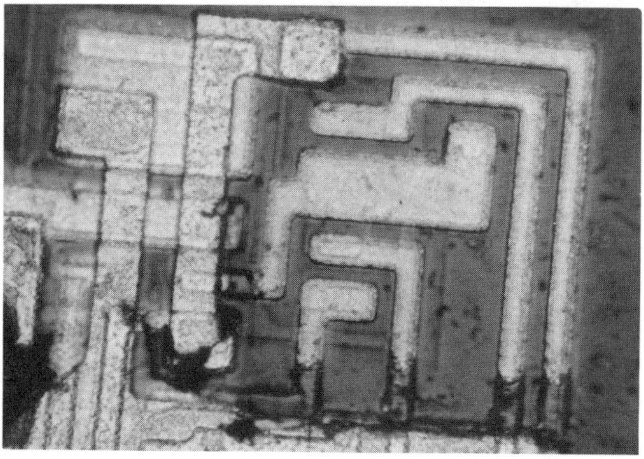

(b)

Figure 13.13 Microsurgery: (a) an integrated circuit transistor (bipolar logic) and (b) isolation by the technique described in text.

materials (such as the polyimide used for protection and insulation) and silicon and its compounds, there are two possibilities: wet etching or dry etching. Table 13.2 shows chemical composition and application for several types of wet etching agents. In certain cases, particularly when highly resistant compounds such as Si_3N_4 are to be removed, it is advisable to use plasma etching, as described earlier, starting from a mixture of oxygen and freon 14 (tetrafluoro-methane of CF_4) according to the following reaction, whereby solid material is converted to gaseous products:

$$SiO_2\,(s) + CF_4(g) \rightarrow SiF_4(g) + CO_2(g)$$

This technique can be controlled better than wet etching, and the relative removal rates of various materials are well known. Several typical values are shown in Table 13.3. This type of etching is not reactive with Al and NiCr but is reactive with gold, which tends to redeposit from the gas phase as fluoride. This method is not applicable for GaAs.

Table 13.2 Examples of Chemical Etching Agents

Material	*Chemical composition*	*Remarks*
SiO_2 passivation	2 parts by weight of ammonium fluoride 1 part by weight of glacial acetic acid	Etching time 2 minutes at ambient temperature
SiO_2 passivation	1 part by weight of ammonium bifluoride 3 parts by weight of glacial acetic acid	Etching time 15 minutes at ambient temperature
Si_3N_4 passivation	50 ml CH_3COOH 50 ml H_2O 25g NH_4F	Etching Si_3N_4 is slow and difficult. Frequent washing and inspection is recommended. The safest method is plasma etching.
Si_3N_4 passivation	30 ml HF 20 ml HNO_3 600 ml H_2O	
Si_3N_4 passivation	4 parts by weight of NH_4F 1 part by weight of HF 2 parts glycerine	
Al metalization	Phosphoric acid at 50°C	
Al metalization	1-normal solution of NaOH tablets in distilled water (40g/l)	
Al metalization	40 ml phosphoric acid 2 ml nitric acid 10 ml H_20	

Table 13.2 continued

Material	*Chemical composition*	*Remarks*
Au metalization	5 g iodine 10 g KI 10 g H_2O 90 g methanol	Etching time approx. 15 minutes
Au metalization	10% potassium cyanide 10% ammonium persulphate 80% H_2O	Heat the solution
Refractory metalization (Ta, Ti, W)	Acetic acid and H_2O with small percentages of HF and HNO_3	
Mo metalization	30% H_2O_2 solution	
Platinum silicide Ni-Cr	Aqua regia 2.25–3.75 molar solution of $FeCl_3$	
Thermally grown SiO_2	1 part HF (49%) 10 parts NH_4F in solution (0.5 kg NH_4F/750 ml H_20)	Removal rate is 7–9 Å/s
Thermally grown SiO_2	NH_4F (50%) CH_3COOH (50)	Removal rate is 700–900 Å/min
Silicon	KOH (20%) at 40°C	Used for complete die removal in order to observe bottom of metalization
Silicon	30% HF (49%) 50% HNO_3 (69.5%) 20% H_2O	Removal rate is 76 μm/min
Copper and copper-base alloys	10 ml H_2O 10 ml NH_4OH 10 ml H_2O_2 (3%)	
Koval-nickel–base alloys	60 ml ethanol 15 ml HCl 5 g $FeCl_3$	
Gold, precious metals	Potassium ferrocyanide (10%)—50 ml Ammonium persulphate (10%)—50 ml	
Molybdenum, tungsten	100 ml H_2O 30 g potassium ferrocyanide 10 g sodium hydroxide	

Table 13.2 continued

Material	*Chemical composition*	*Remarks*
Carbon steel	98 ml ethanol 2 ml HNO_3	
Stainless steel	30 ml glycerine 30 ml HCl 10 ml HNO_3	

Table 13.3 Relative Removal Rates for Various Materials with Fluorine-Based Plasma Etching

Material	*Removal rate*	*Material*	*Removal rate*
Thermally grown SiO_2	1	Monocrystalline Si	15
Vapor phase deposited SiO_2 (with 5–7% P_2O_5)	4	Polycrstalline Si	+100
		Ta, Ta_2O_5, TaN	+100
Si_3N_4	4	Al, NiCr	0

A useful application, shown in Figure 13.14, is that of completely removing the substrate. Here, the back of the semiconductor chip is thinned mechanically and the surface protected with masking material. The semiconductor is then etched away to reveal the underside of the metalization at the contacts, so that contact condition can be inspected. For new devices, this technique is used to check whether metalization-related operations have been carried out correctly; whereas for devices already in service, the technique is used to search for interdiffusion (in which case Al peaks facing the observer would be visible). A similar procedure can be used to thin specimens to prepare them for inspection with a *transmission electron microscope* (TEM).

Metalizations are removed by wet chemical etching, either directly or by undercutting when etching the oxides, particularly if there is more than one level. Certain chemical etching solutions (see Table 13.4) can reveal junctions, in the sense that they etch the doped areas (*n* or *p* according to composition) selectively. This improves visibility of these areas under the optical microscope, where the etched zones can be readily distinguished because of their different color, and under the scanning electron microscope, where the effect of the etching agent is more clearly visible as erosion. The use of these etching agents thus makes it possible to identify the structure of the device's component parts with greater certainty, particularly in the case of integrated circuits.

Figure 13.14 Integrated circuit with entire substrate removed by plasma etching (oxygen + freon) to reveal the rear portion of several contacts.

Table 13.4 Color Etching Agents for Silicon

Composition	*Procedure*	*Applications and remarks*
1 HF 3 HNO_3	15 s under strong light	Used to delineate *p-n* junctions
1 HF 3 HNO_3	10–20 s	Used to delineate junctions
8–12 CH_3COOH	1 min or more	For dislocations
25 HF 18 HNO_3 5 CH_3COOH/0.1 Br_2 10 H_2O 1 g Cu $(NO_3)_2$	2–4 min	Used to reveal dislocations on all planes
100 HF 0.1–0.5 HNO_3	Apply with eyedropper under strong light	*p* = doped material is darker than *n* = doped material
50 ml dilute Cu $(NO_3)_2$ 1–2 drops HF	Apply with eyedropper under strong light	Copper deposits on *n* = doped material (more copper is depositied on material with lower resistivity)
120 cc CH_3COOH 30 cc HNO_3 10 cc HF 5 g $CuSO_4$	Apply with eyedropper under strong light for 10 s	Used to delineate junctions Solution should be used immediately after preparation

Table 13.4 continued

Composition	*Procedure*	*Applications and remarks*
1 HF 1–4 chromic acid (40%)	5–20s	Good for revealing junctions in sections Excellent for revealing dislocations and stacking faults on plane (111).

13.2.8 Microsections

The cross section of a device is always a valuable source of information, both as a means of process control and as a method for highlighting the deep effects of certain failure mechanisms, though the latter application calls for considerable care in performing the operation, as well as a good deal of luck in some cases. The device to be sectioned is secured to a specimen holder by epoxy cement, wax, or other substances suitable for the device's physical makeup. It is advisable to have a number of specimen holders cut to different angles so that sections can be made at 90, 60, and 30° or less, in order to benefit from the thickness "amplification" effect of the specimen's various layers. The specimen holder may be plastic, though metal units are preferred, and its dimensions should be such that the holder can be placed in a SEM observation chamber.

The specimen is then embedded in resin to give it sufficient mechanical strength to withstand the subsequent lapping operation without damage. Epoxy resin is commonly used for this purpose, although a filler may be added to make the resin conductive. The thickness of the structure so obtained is rapidly reduced using coarse grain silicon carbide paper. Once the excess material is removed, progressively finer grades of abrasive paper are used to reach the desired depth. The surface is then finished using alumina powder with particle sizes ranging from a few microns to a few hundredths of a micron in an aqueous solution on glass or a cloth.

Equipment normally used for preparing microsections is shown in Figure 13.15, and the results of sectioning an integrated circuit are shown in Figure 13.16. The section so produced is flat and smooth. The device's structure can be highlighted by subjecting the specimen to oxygen and fluorine-based plasma etching, as described earlier, for a brief period calculated according to the process parameters. This exploits the differing rates of removal of different materials to produce a "step effect" that can be of use in interpreting the section. An example is shown in Figure 13.17. Alternatively or as a further step, the junctions can be revealed using appropriate chemical etching procedures, as described earlier.

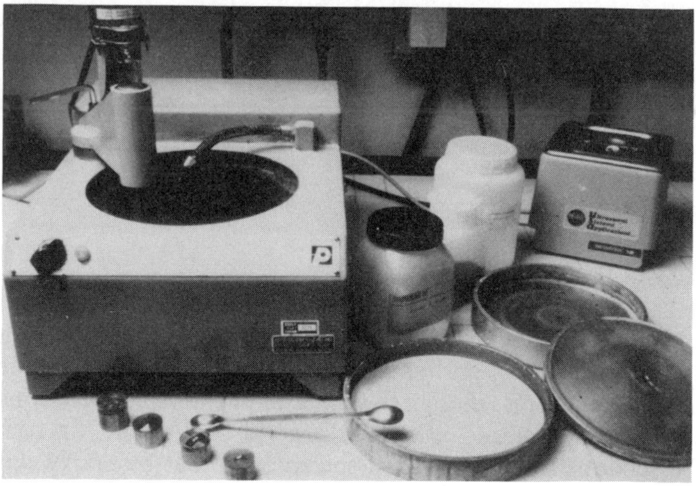

Figure 13.15 Equipment necessary for the preparation of microsections.

13.2.9 Electron Microscopy and Microanalysis

Interaction of beams of accelerated particles and X rays with the material generates signals that can be collected and discriminated to provide precise, extremely useful information concerning the morphology and composition of the bombarded or irradiated specimens. These techniques, which are described in detail in a separate chapter, play a fundamental role in technological investigations of devices. The equipment involved has been refined and improved to the point where these techniques are no longer limited to specialized applications but are now routine.

13.2.10 Infrared Microscopy

The term *infrared microscopy* can be used to define at least three different techniques: thermal emission detection and analysis, infrared emission detection and characterization for optoelectronic devices, and transmission and reflection optical microscopy with an infrared optical source. Temperature distribution on the device's surface is analyzed by using the scanning infrared microscope, which transfers the image of the examined area to a CRT screen and differentiates temperatures by using a gray scale with an extreme white for the hottest areas and black for cool areas, see Figure 13.18. A quantitive analysis can also be performed if a sufficiently precise temperature reference is available. Measurement accuracy depends to a considerable extent on the physical dimensions of the object, and hence

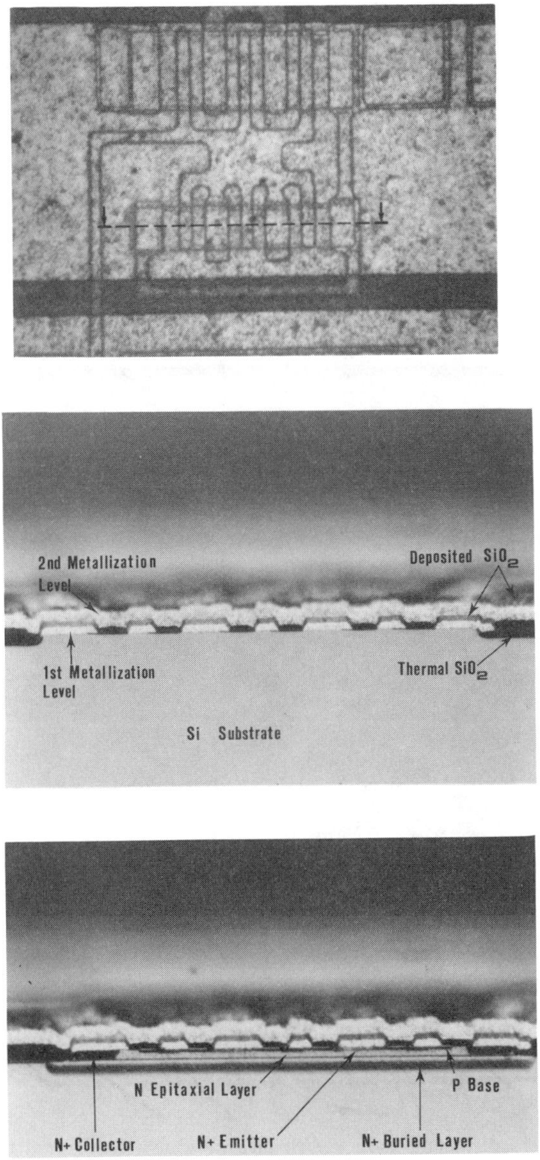

Figure 13.16 Integrated circuit transistor (ECL).

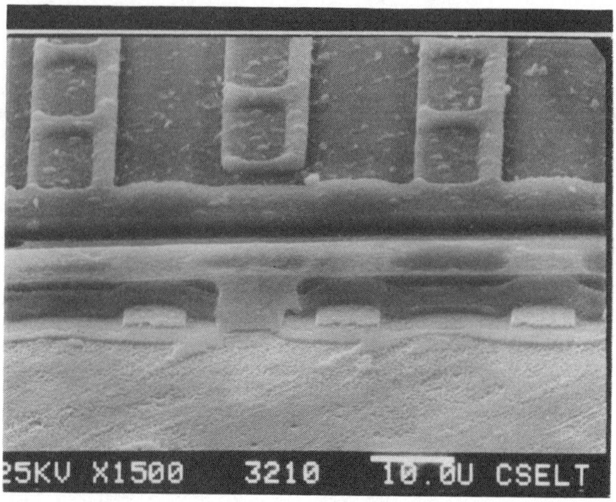

Figure 13.17 Step effect produced by etching a sectioned integrated circuit with oxygen and freon-based plasma.

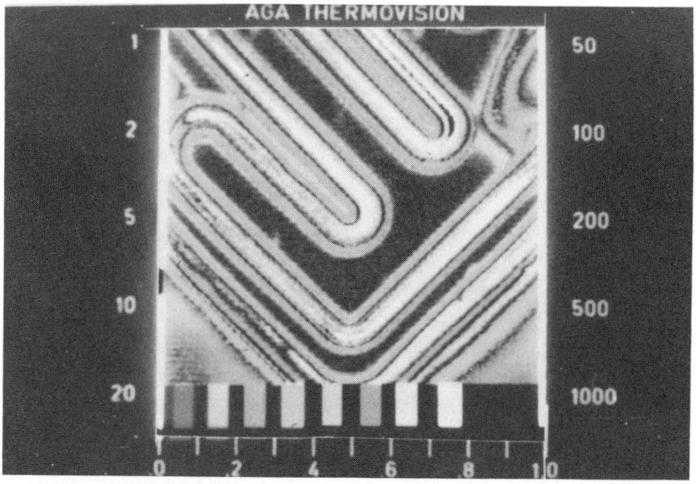

Figure 13.18 Temperature distribution on the surface of a device acquired by a scanning infrared microscope and displayed on a CRT.

on magnification, average temperature, and emissivity. Emission efficiency of infrared radiation varies considerably from material to material. For example, if emissivity of an ideal black body is one, that of crystalline silicon is 0.9 and that of the aluminum metalization used in semiconductor devices is 0.1. To make the specimen surface as uniform as possible in terms of emissivity, it is advisable to cover it with a layer of suitable material, such as carbon. It is also possible to obtain three-dimensional maps of temperature distribution.

Infrared emission can be detected using an optical microscope equipped with suitable optical lenses and a camera operating at wavelength between 700 and 2,000 nm. Defects of emitting devices such as LEDs and lasers can be investigated with this technique. Some examples are reported in the chapter on the reliability of such emitters.

Other information can be obtained by illuminating the specimen with infrared radiation, which makes it possible to examine the internal structure of opaque materials at frequencies falling in the visible spectrum. Typical applications include checking for uniform metalization thickness, small metalization defects, internal crystallographic defects, cracks, inclusions, and homogeneous die attach.

This technique has proven to be effective in evaluating the wire bondings of plastic encapsulated devices and the detection of corrosion along the metal tracks without removing the plastic encapsulation, thus avoiding chemical damage to the active area. Sample preparation is quite simple. The component is lapped from the back after bending back the leads, which can eventually be inserted into a socket for electrical connection to the outside. Lapping is stopped upon reaching the die, and the sample is carefully polished. A prepared sample is shown in Figure 13.19. Figures 13.20, 13.21, and 13.22 show some typical failures detected by this method.

Figure 13.19 Plastic encapsulated IC prepared for rear observation through an IR microscope.

Figure 13.20 Corrosion detected by IR microscope.

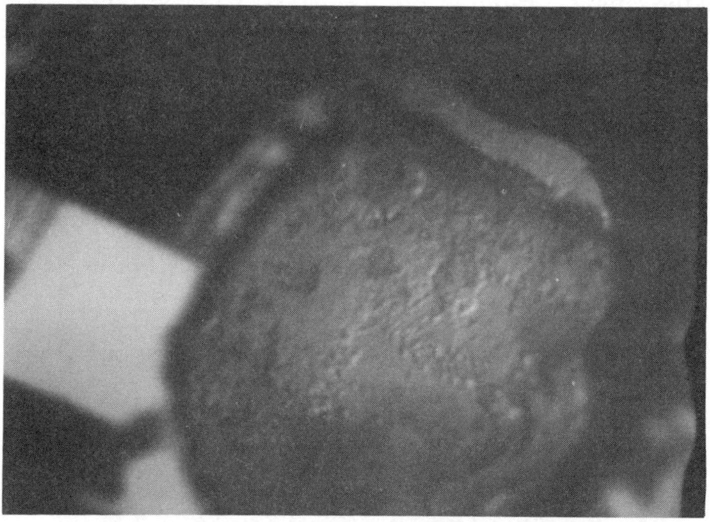

Figure 13.21 Purple plague at a bonding pad of a plastic encapsulated IC.

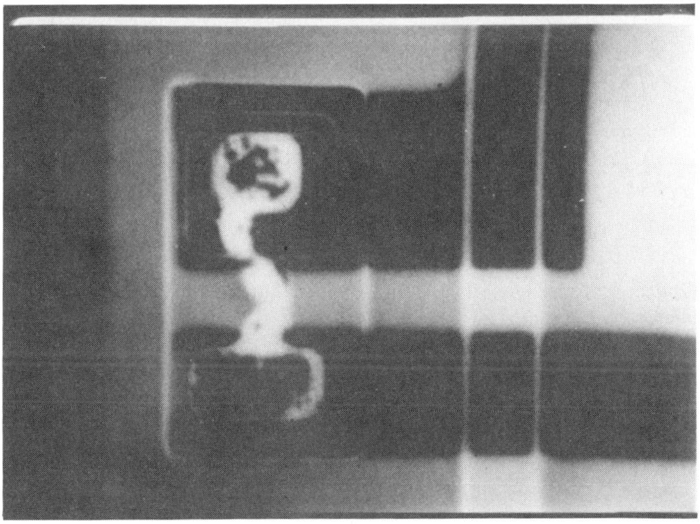

Figure 13.22 Metal filament shorting a diode of an IC due to electrical overstress observed through an IR microscope.

13.2.11 Scanning Optical Microscopy

The *scanning optical microscope* (SOM) constructs an image by causing a spot of light to scan the surface of a specimen. If the light source is a finely focused laser beam, some interesting features such as confocal imaging, differential phase imaging, and *optical beam induced current* (OBIC) imaging become available.

In principle, a point-by-point image of the object under examination can be obtained scanning a point source in a raster-type scan, whereas a large area detector measures the power transmitted by the collector lens (see Figure 13.23a). A confocal scanning microscope uses a point detector that detects light only from the very small region illuminated by the point source and the source and the detector scan in synchrony (see Figure 13.23b).

The term *confocal* means that both lenses are focused on the same point of the object. In practice, scanning the object is much more convenient than scanning the source and the detector together. A practical confocal scanning microscope arrangement is shown in Figure 13.24, where the emitted light of the object is passed through a secondary spatial filter; that is, a small pinhole aperture placed

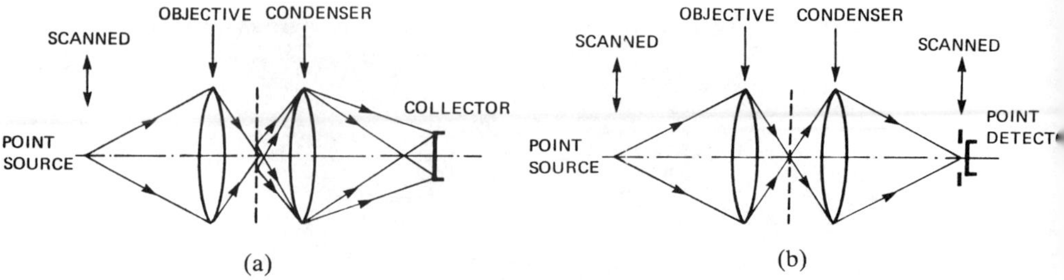

Figure 13.23 Optical arrangements of (a) conventional scanning microscope and (b) confocal scanning microscope.

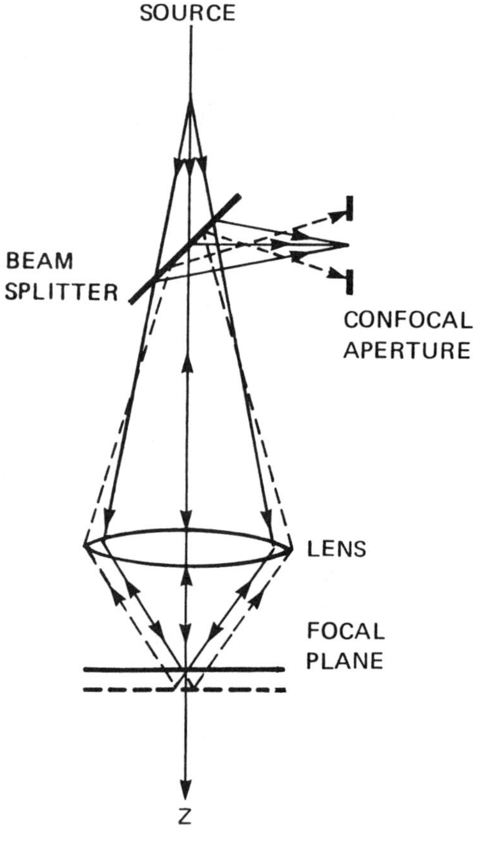

Figure 13.24 The optical arrangement for confocal microscopy. The confocal pinhole blocks light reflected from regions outside the depth of focus.

at a plane conjugate with the object plane. With this arrangement the resolution may be improved by up to a factor of 2 and some new important properties are achieved, such as optical sectioning and surface profiling. This is accomplished by taking advantage of the fact that only in-focus detail is imaged, the remaining of the optical information being rejected.

Another major advantage of the SOM is its *optical beam induced current* (OBIC) mode of operation, which is based on the creation of hole-electron pairs by the finely focused optical source (laser beam). This technique is analogous to electron beam induced current in the scanning electron microscope; but unlike SEM, it is not destructive, does not require high vacuum chambers, and can be applied to passivated samples. OBIC is used for solar cells characterization, measurement of diffusion length and carrier lifetime, imaging of lattice defects, and latch-up diagnostics in CMOS devices.

13.2.12 Scanning Acoustic Microscopy

The *scanning acoustic microscope* (SAM) focuses a radiofrequency electrical signal in a beam of flat acoustic waves at a point in a liquid on the specimen surface. The reflected signal is collected by the same transducer and used to modulate the intensity of an electron beam in a CRT. Scanning the examined area is performed mechanically and synchronized with monitor scanning. The SAM equipment schematics are shown in Figure 13.25.

Although this technique has still to be perfected, its interest lies chiefly in the fact that, by contrast with optical microscopy, information can be obtained regarding the elastic and mechanical properties of the specimen material. In addition, because sound propagates in almost all solids with minimum loss, the technique reveals defects located beneath the surface and provides clearly defined topographic images.

13.2.13 Liquid Crystal Inspection

This technique, which has some similarity to obtaining voltage contrast through a scanning electron microscope, can be used to check electrical signal propagation continuity on the surface of a microcircuit and hot spots due to preexisting failures or local defects. In checking continuity, the technique monitors an electrical field deriving from the bias applied to the microcircuit. In the case of a nematic compound, this field causes the crystals to rotate and thus vary their behavior with respect to a polarized light beam. In searching for hot spots, the technique monitors the effect of the heat that, upon exceeding the crystal transition temperature from the colesteric phase to the isotropic phase, prevents the polarization plane of

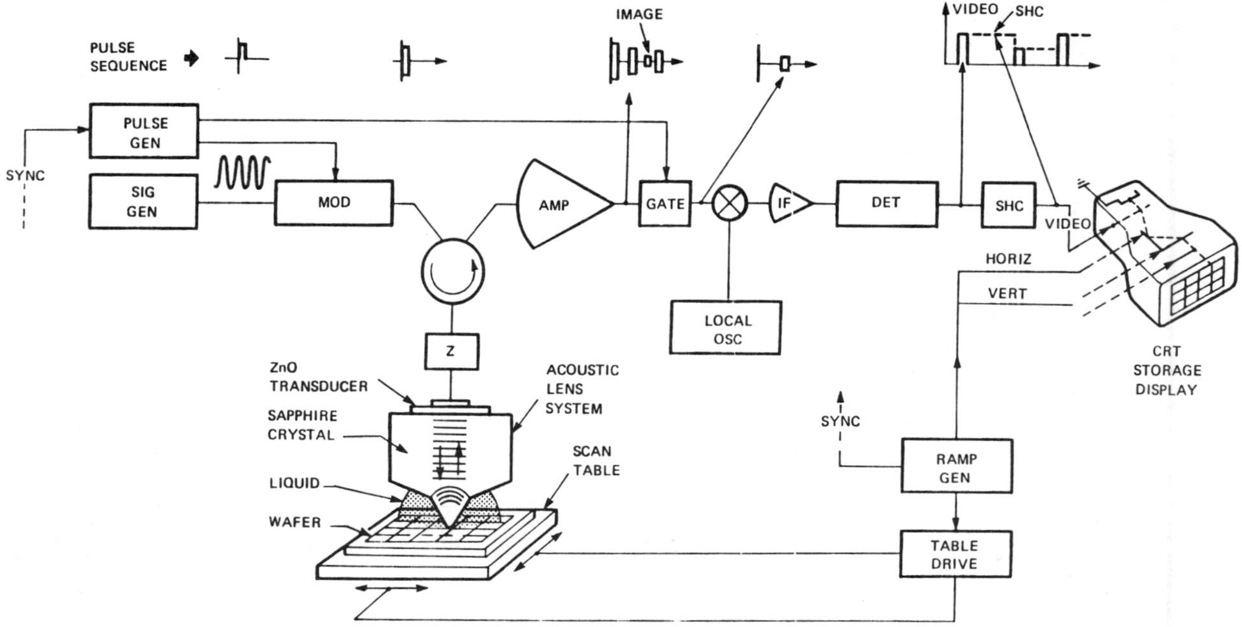

Figure 13.25 Schematics of the scanning acoustic microscope.

transmitted light from rotating in the affected areas. Black spots can then be observed under the microscope.

Specimen preparation is very simple. A drop of liquid crystal is applied to the specimen surface using a syringe to distribute the drop uniformly. A more complex version is to precede this operation with a surface treatment and cover the prepared specimen with a glass or mica slide. The specimen is examined under an optical microscope with polarizer. A polarized illumination system may also be provided. As the method's resolution, in practice, depends on the dimensions of the detail observed, it can be particularly useful for analyzing complex integrated devices.

13.3 TECHNOLOGICAL ANALYSIS

Paradoxically enough, the *a priori* analysis criterion gained currency after the validity of analysis on failed parts had been recognized. The reasons for this are fairly clear. Hypotheses concerning the behavior of a new device can be formulated only after acquiring a thorough knowledge of failure mechanisms and how they are affected by the technology and materials employed. Furthermore, such hypotheses have become especially important now that device complexity often makes it impossible to apply all the stresses required for an exhaustive reliability check.

Technological analysis is thus of major importance in providing qualitative reliability estimates, so much so that it has become obligatory in areas where components must have the highest possible level of reliability; that is, in the aerospace and military fields, applications. The increasing complexity of semiconductor devices has led to longer periods of construction analysis, thus the natural evolution has been toward evaluating the "capability" of the component manufacturer as far as technology and organization, eventually in conjunction with destructive physical analysis.

Analytical techniques are substantially the same as those described earlier, but their purpose is not to identify the reasons for failure but, as thoroughly as possible, to anticipate and forestall failure before the device completes the average life envisaged for it.

13.3.1 Construction Analysis

Construction analysis is used chiefly for preventative assessment, particularly when it is necessary to choose between equivalent components produced by different manufacturers, which can use comparative criterion and technological solutions. The validity of the hypotheses can be verified by subjecting the components to step stress tests oriented toward stressing the most critical points. The flow of operations is shown in Figure 13.26, and it is not unlike that used for failure analysis. However, the accent is placed on certain aspects typical to "reverse engineering," which identifies the parameters involved by reversing the manufacturing process. For complex devices, it can be helpful to extract all or part of the electric circuit, as shown in Figure 13.27.

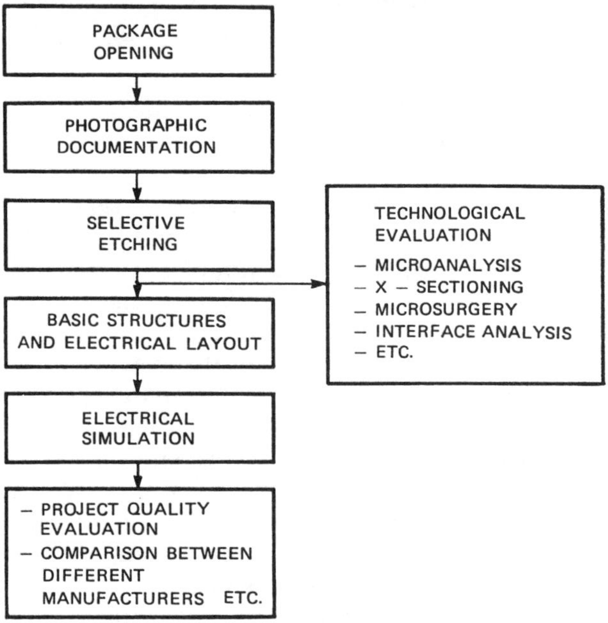

Figure 13.26 Operation flow for construction analysis, or reverse engineering.

In practice, this operation consists of revealing each mask in turn, from the last (highest interconnection level) to the first. The various measurement methods and techniques for etching and highlighting areas with different doping, described earlier, make it possible to distinguish and recognize the various circuit elements and reconstruct the unit. Each step of the device's controlled destruction is documented photographically. At the present time, circuit extraction is chiefly carried out by hand, with the operator sometimes aided by a digitalizer and various graphic utilities. The operator must have a good knowledge of the technology in general and circuit design (including CAD techniques) for the various digital and analog applications.

Efforts are now being made to automate at least some of these operations by adapting the software packages used in design for operation in reverse. The

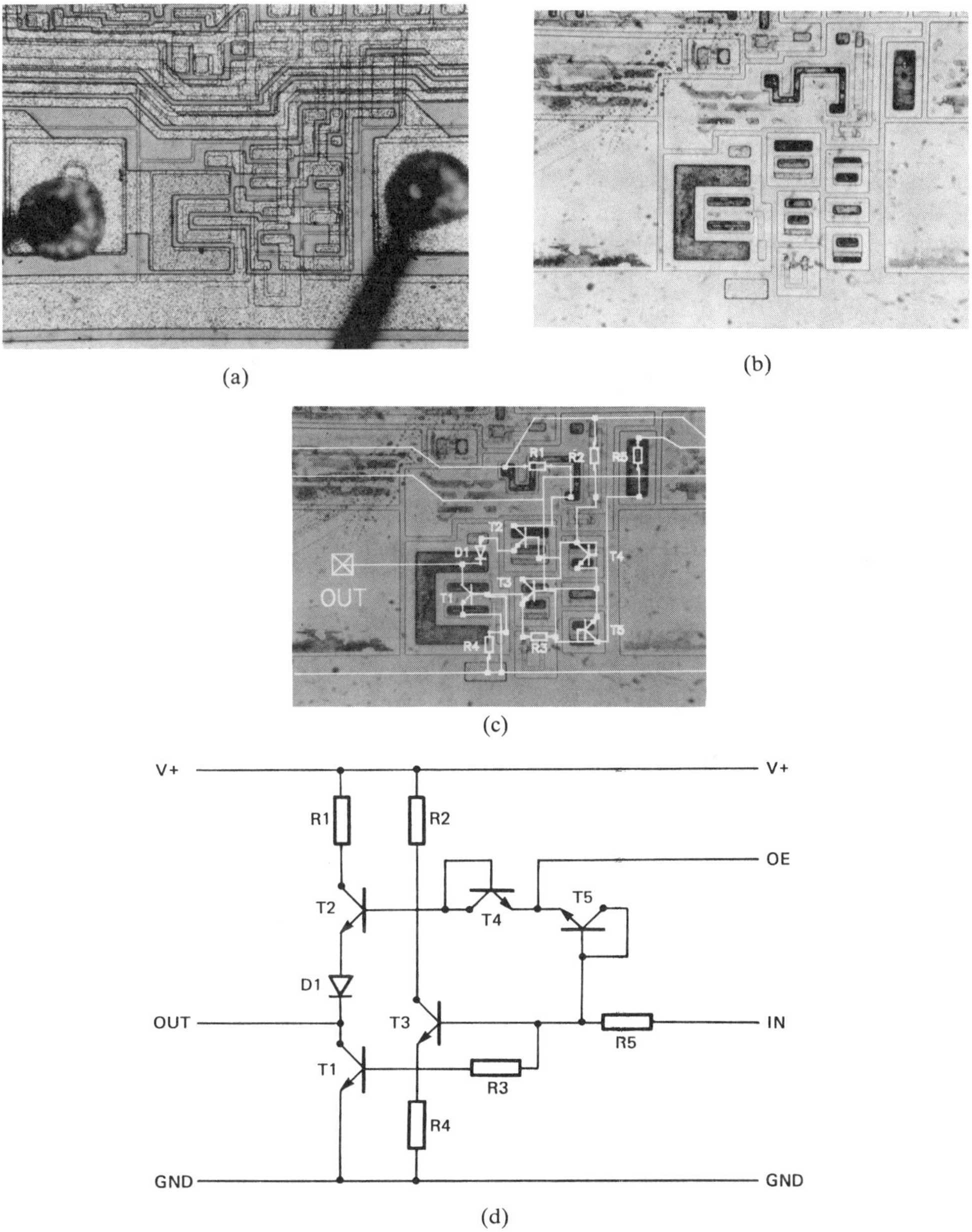

Figure 13.27 The stages of reverse engineering: (a) circuit under test as it appears after opening the component; and (b) after removing passivated and metallized areas and color etching. Electrical circuit (c) superimposed on the corresponding areas of the die and (d) in the functionally ordered version.

obstacles to be overcome are obviously considerable, and they increase as circuit element size is reduced and device complexity is increased.

Once the basic electrical layout has been identified and the technological parameters measured or hypothesized, operation of the various circuit parts can be simulated on the computer, using programs similar to those used in design but modified for reverse engineering. In this way, it is possible to verify the validity of the individual circuit solutions and check signals at inaccessible internal nodes. It may also be possible to simulate malfunctions in order to identify their cause. This technique provides a powerful tool to support design decisions. In addition, by establishing a fruitful dialogue with the component manufacturers, it facilitates product improvement in terms of quality and reliability.

13.3.2 Destructive Physical Analysis

Destructive physical analysis (DPA) is a set of tests used to check incoming lot compliance with acceptance standards. Inspection is carried out on a statistical basis, by sampling a limited number of components from each lot (which is assumed to be uniform), and ensures that components used have the same characteristics as those selected on the basis of these characteristics. From the point of view of reliability, DPA is an extemely important tool. Just as screening procedures eliminate the most obvious defects (infant mortality), so DPA reveals latent defects that cannot be detected during screening but could lead to failure during service. These defects are time-dependent; that is, they cause progressive deterioration of the component's electrical characteristics or structural integrity.

Other information obtained from DPA is not necessarily related to lot reliability, but identifies variations in the manufacturing process (technology, masks, materials, *et cetera*) from the accepted model and the effects of screening and handling on the device's reliability. Naturally, the data acquired from DPA can serve as the starting point for subsequent failure analysis. Above all, DPA provides information that can be used to improve processing methods and testing and screening procedures.

DPA was first used on a regular basis in the aerospace industry, but has also become very popular in commercial electronics, as it often has been demonstrated that costs are amply justified by the benefits to be derived. The costs generally depend on the amount of equipment and time required for analysis, and the time involved depends on the analyst's experience and component's complexity. Table 13.5 gives an approximate idea of time requirements. Equipment and techniques are substantially the same as those described for failure analysis, with the addition of a few specifially designed for DPA, some of which will be briefly described. For further reference, see MIL-STD-883, method No. 5009.1.

Table 13.5 Average Time Required for DPA

Component	*Analysis work hours*
Capacitor	4–8
Connector	3–5
Diode	5–24
Filter	5–10
Inductor	5–12
Integrated circuit	15–30
Resistor	3–5
Transistor	5–24

13.3.2.1 Bond-Pull Test

This test checks the quality of the wire bondings joining interconnection wires between the semiconductor die and package. Data provided by the industry indicate that defective microsoldering is the second most common cause of failure. Testing the mechanical strength of these bonds is thus of considerable importance.

The test consists of placing a hook under the wire halfway between the two bonds and lifting the wire until failure occurs (destructive test) or until a predetermined tension is reached. The force applied is read on a dynamometer scale. The nondestructive test eliminates devices with weak bonds, and the tensile force is selected so as not to damage nondefective devices. For further reference, see MIL-STD-883 method No. 2011.4.

13.3.2.2 Die-Shear Test

This test is used to check that the die-attach operation has been carried out correctly; that is, that semiconductor die adhesion to package is good. For further reference, see MIL-STD-883, method No. 2019.2.

13.4 CONCLUSIONS

In the foregoing pages, we summarized the technological activities related to reliability. For a more detailed treatment of the techniques and practical applications, the reader is advised to consult the bibliography. Spatial limitations have prevented us from offering an exhaustive description of this rapidly growing field, where large numbers of researchers continually contribute new ideas. Developing technologies call for a continuing refinement in techniques and methods of investigation. In addition, they open new areas of research or, paradoxically, restore interest to old problems in the light of new developments.

In such a context, a thorough knowledge of all aspects of the electronic component is of paramount importance in efforts to ensure the reliability of components and the systems of which they will be a part. Furthermore, it is vital that effective answers be made available in a relatively short time.

The facts support this approach. In many cases, the costs involved are not high, if we consider that the major investment is not in sophisticated equipment whose high purchase price can only be justified by constant use. Once the rationale underlying these activities has been grasped, it is possible to manage the problem so as to make best use of available resources. Once the path to be followed is clear, it should be possible to use the equipment available at independent or university research centers specializing in similar investigation to reconstruct the analytical mosaic and obtain clear answers that can be used profitably.

13.5 EXAMPLES OF TECHNOLOGICAL ANALYSIS

The analysis procedure adopted in an actual case is described as an example. The device under test is an asynchronous divider capable of operating in a frequency range between 150 and 1500 MHz and constructed using *emitter coupled logic* (ECL) technology. As the device was selected for use on a telecommunications satellite, it must feature a high degree of reliability. On the whole, the device is fairly simple and not a large scale integration unit. From a technological point of view, however, the device is particularly representative, as was the information which came to light during analysis. The device is shown in Figure 13.28, and its specifications follow.

13.5.1 General Information

Component code:
Component type: Monolithic integrated circuit.
Function: Asynchronous type divide by 4.
Technology: Emitter coupled logic.
Manufacturer:
Remarks: Three unmarked samples were received in a sealed plastic box. Component pins were inserted in nonconductive, colored foam rubber.

External parts
Container: Dark ceramic 14 pin dual-in-line with frit seal. Pins are tinned Kovar (Fe-Ni-Co) alloy.
Marking: None either on top or bottom.

Pin layout: Shown in Figure 13.29, taken from manufacturer's data book of March 1982.
Functional diagram: Two divide by 2 stages in cascade as shown in Figure 13.30.
Note: Inspection after opening showed that, at least for the samples received, V_{EE} is not to be applied to pin no. 5, which is not electrically connected to the die, but to pin no. 7.

Internal parts
Opening method: Mechanical removal of container top.
Description of active part: Rectangular (1×1.6 mm^2) silicon die soldered with Au-Si eutectic to base of gold-plated internal cavity. Interconnections between die and package consist of 1 mil dia. (25.4 μm) Al wire. Ultrasonic bonding on both sides. (See Figures 13.31–13.35 for details.)
Technology: ECL shallow emitter with implanted base and contact P^+ diffusions. Resistors obtained with base diffusions. Single-level metalization with 1.2 μm average thickness and consisting of Al filled with Si and Cu. SiO_2 passivation.

Comments: Surface passivation is cracked at several points, see Figure 13.36. The cracks extend in depth down to metalization. This problem is due to a poor passivating oxide deposition process or to excessive mechanical stiffness of the oxide. Irregularly shaped particles are observed along the metal tracks; EDS analysis shows these to be Cu precipitates (see Figure 13.37). Effective cross section of metalization is reduced near these precipitates, whose presence indicates poor control over the deposition process used for metalizing, where the Cu should be evenly distributed in order to be effective. The analyzed samples also show considerable misalignment of the metalization mask, though this would still seem to be within acceptable tolerance limits.

Figure 13.28 One of the devices, as received.

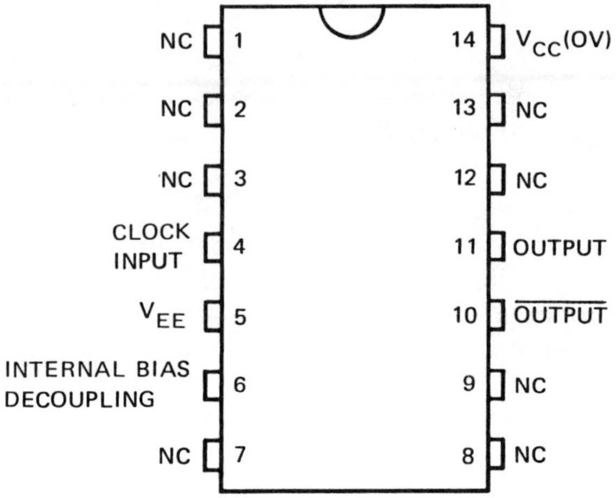

Figure 13.29 Pin layout, as shown in catalogue.

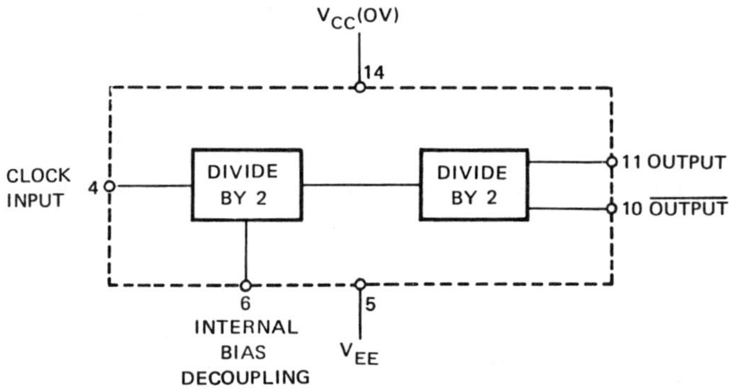

Figure 13.30 Functional diagram of device under test.

13.5.2 Electrical Circuit

The diagrams in Figures 13.38–13.40 show the electrical circuit of the device as revealed during analysis. Numbers used for circuit elements correspond to those in Figure 13.32. Vertical numbers identify resistors, and horizontal numbers identify transistors.

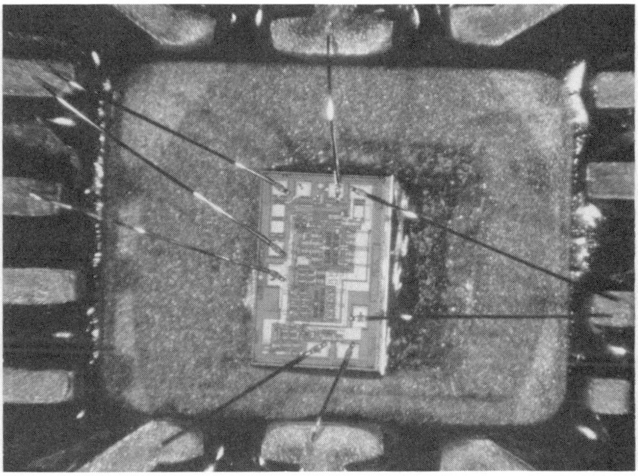

Figure 13.31 Assembly details visible after opening.

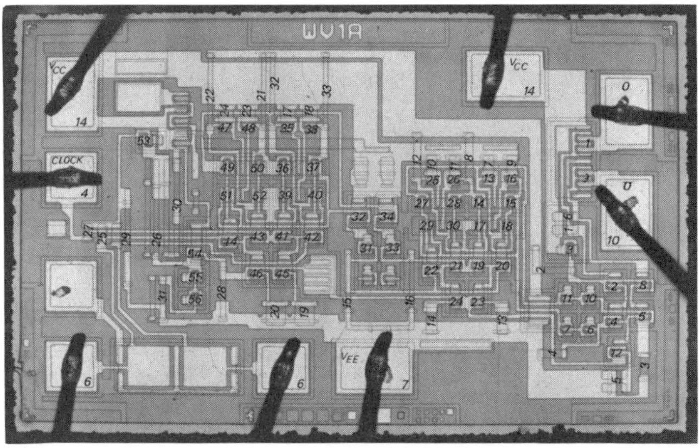

Figure 13.32 Overall view of die.

13.5.3 Conclusions

Of the process defects revealed, the most critical was that involving the metalization. To check the device's resistance to electromigration, which was considered

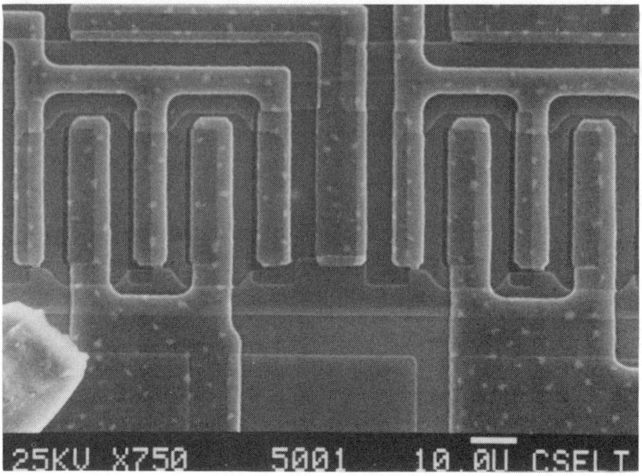

Figure 13.33 Differential output pair.

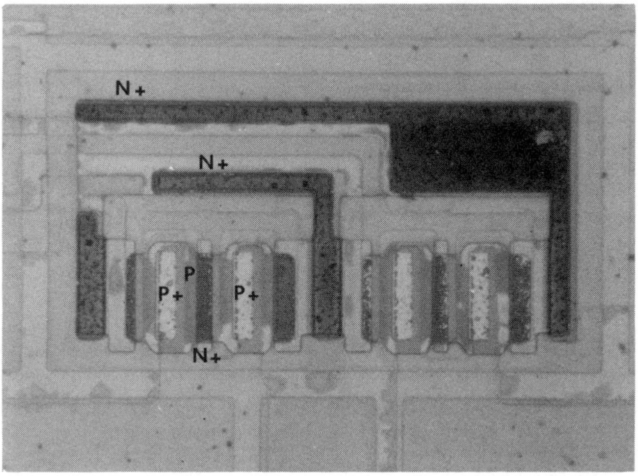

Figure 13.34 The transistors shown in Figure 13.33 after removing metalization and passivation. Doped areas have been highlighted by color etching.

the failure mechanism most likely to occur, tests were carried out at increasing temperature steps with applied bias and controlled frequency operation. After 1500 hours, the devices under test showed no parametric drift, and microscopic

(a)

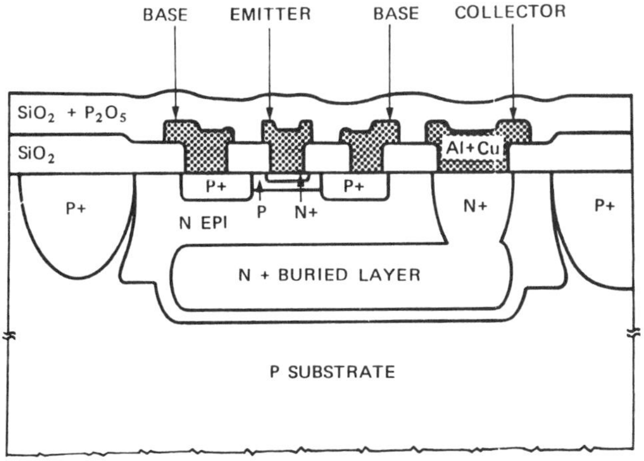

(b)

Figure 13.35 Section through (a) a typical transistor and (b) technological details.

examination after opening revealed no clear signs of degradation, incipient or advanced. Consequently, though stricter process control is advisable, we can conclude that the devices examined could be used for the envisaged application.

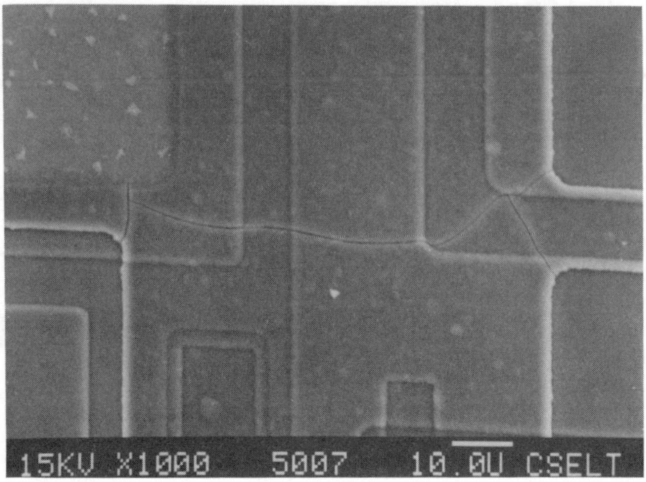

Figure 13.36 Cracks in the surface passivation layer.

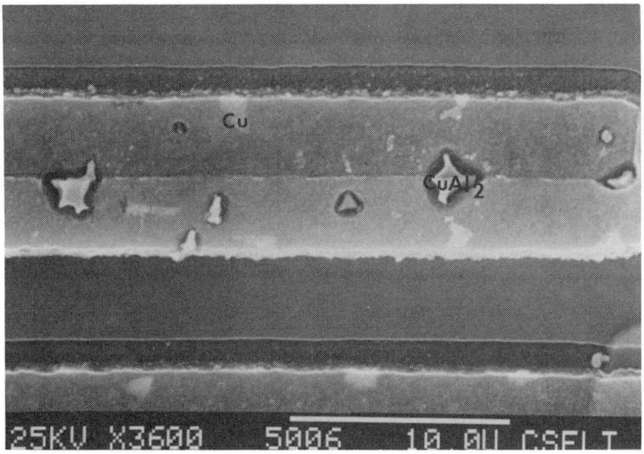

Figure 13.37 Metalization with precipitates. Contact misalignment is clear.

13.6 EXAMPLE OF FAILURE ANALYSIS

As indicated earlier, the main objective of failure analysis is to provide information concerning the failure of a component whenever it may occur and suggest corrective

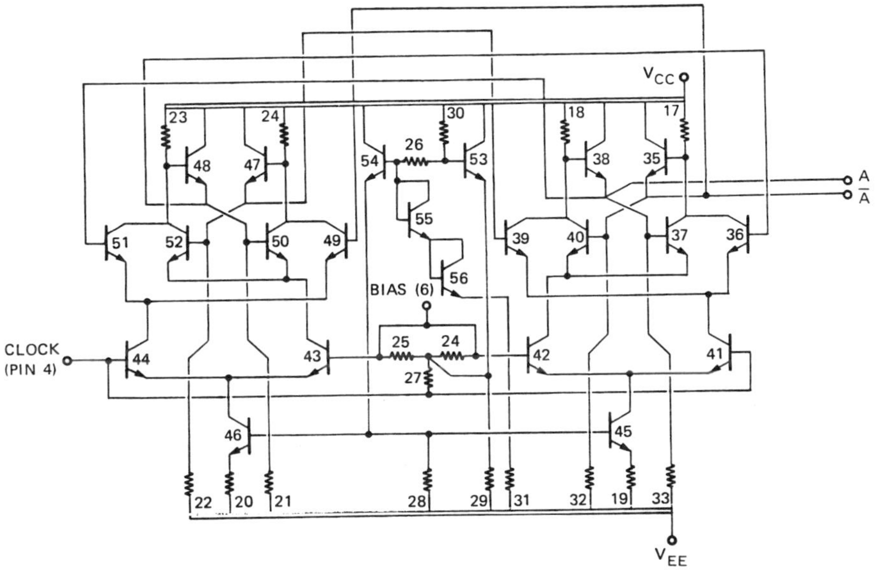

Figure 13.38 First divide by 2 stage.

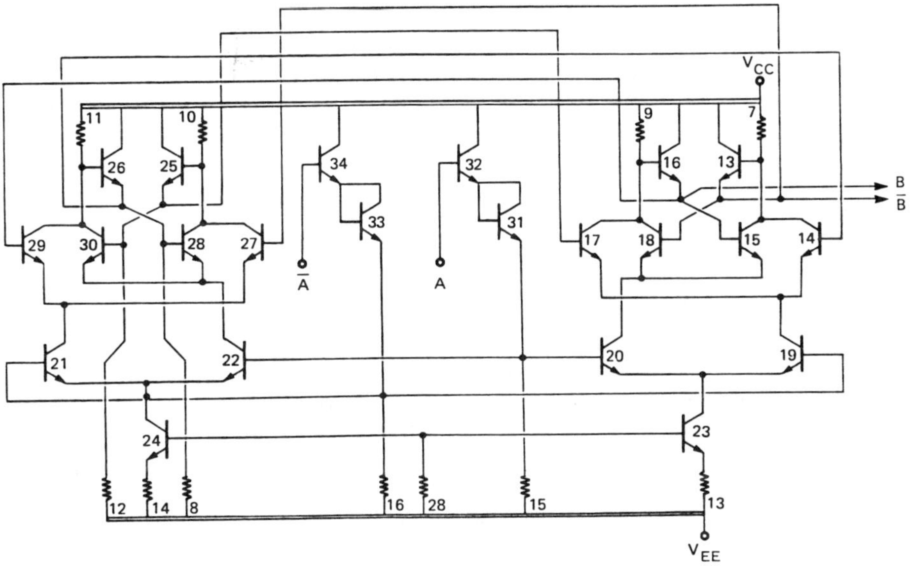

Figure 13.39 Second divide by 2 stage.

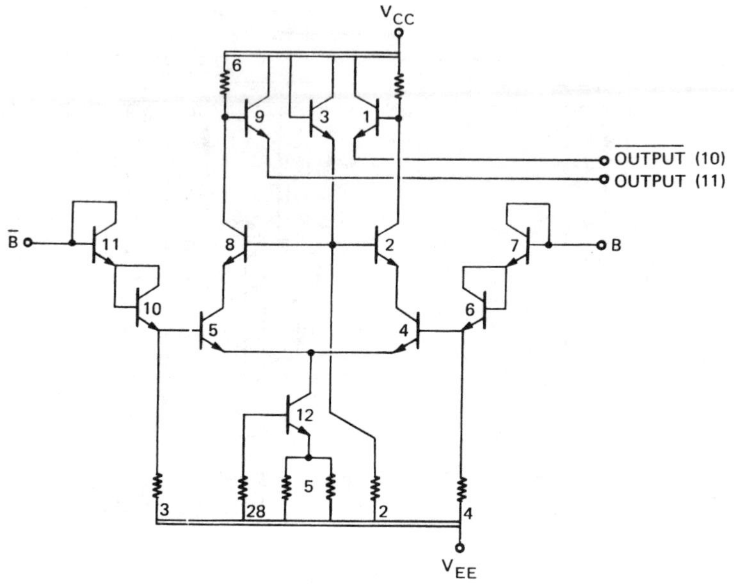

Figure 13.40 Output circuit.

measures. In view of the complexity of the topic and the wide range of components available, we have chosen to provide an example illustrating each step of the procedure, from initial analysis to the final summary report.

What we want particularly to highlight is the method used to conduct the analysis. The basic idea is to collect as much information as possible before employing destructive operations on the component under test. In addition, the example gives an idea of the impact of analysis on the production of a modern manufacturer of TLC equipment.

13.6.1 Steps in Failure Analysis

Failure analysis consists of the following steps: collecting information, formulating failure hypotheses, verifying these hypotheses, and preparing the final report and suggesting corrective measures.

13.6.1.1 Collecting Information (Anamnesis)

This step is of fundamental importance as regards the outcome of analysis and its utility in particular. It consists of finding all information that can be used to guide the course of subsequent analysis and to select corrective measures. A survey is conducted among those who reported failure in the component, although further information is sought from data books and the literature. This data is used to draw up an analysis plan, prepare investigation methods, and provide an idea of the dimensions of the industrial problem. In general, the factors that from an operational viewpoint vary between one analysis and another are the component measuring method and equipment, which should be sufficient for a complete and exhaustive analysis (should this be necessary for technical or economical reasons, as when disputes involving large quantities of components can be foreseen).

13.6.1.2 Formulating Hypotheses

In this step, a flow chart typical of the component family involved is followed (see Figure 13.1) and information is collected systematically step by step. Each failure is evaluated immediately and other failed samples are checked in order to set up a case history. All doubtful aspects must be clarified before passing on to the next step, which may be such as to destroy information (as when opening the component). The hypotheses formulated during this step must then be verified.

13.6.1.3 Verifying Hypotheses

In this step, accelerated simulation or service tests are used to detect irregularities in "new" parts or re-create the failure. This stage is particularly important when failure is due to excessive stress, and it is necessary to determine whether the cause of failure is inherent in the component or the component has been stressed beyond its limits.

13.6.1.4 Final Report and Corrective Actions

The report by the group carrying out failure analysis must clearly indicate the cause of failure, the analyses performed, and the corrective actions to be taken. For this information to be fully effective, it must be forwarded to the user as promptly as possible. If the information is still incomplete, the margin of risk should be indicated. A typical example of this regards components suspected to be inherently defective. In this case, the user must be informed immediately (by telephone, if necessary) so that all such components can be set aside pending further investigation and confirmation.

The report is issued in a suitable form. Details are covered in the appendixes, which may contain a full description of the analysis carried out. The documentation is particularly important as regards experience gained during past analyses. All too often, "epidemics" of similar failures break out as a result of simultaneous process changes by several manufacturers.

Accurate and complete documentation of past cases allows prompt action and enables the company to enact corrective measures whose outcome has already been tested. Documentation must be prepared methodically to facilitate review when new problems come to light. Each item of information or measurement must be collected and catalogued, classified by sample, and retained in files for at least a few years.

13.6.2 Operating Procedure

13.6.2.1 Information Gathering

Analysis should be requested on appropriate forms. These forms should be accompanied by the failed parts and new parts (from same lot as failed parts) or functional parts (from different lots). Specification of the problem in a final report might say, "Rejected parts belong to a well-defined period: parts produced before and after this period are functional. The extent of the phenomenon is considerable, and it cannot be traced to infant mortality. The most likely hypothesis would seem to be a process error on the part of the manufacturer."

Information is obtained from the user, data books, *et cetera.* Reference to the source of information could be: "This type of component is subjected to 100 percent inspection at 70°C, which further confirms that the failure depends on the component's manufacturing process." Or, it could state: "This component is subjected to 100 percent incoming inspection. Failed parts come from equipment burnout and the field, thus the component is presumed to be good originally."

13.6.2.2 Failure Mechanism Hypotheses

External Visual Inspection (stereomicroscope, 40 ×)

The IC was inserted into a socket. It looked clean, marking seemed correct; no mechanical damages due to mishandling or other causes could be detected. The component was not soldered, thus it did not withstand thermo-mechanical stress during mounting on the PC board. Care must be taken during testing in order to prevent oxidation of the external pins which might affect the measurements, or even be the cause of malfunctioning.

Electrical Measurements and Testing

Testing was performed with a SENTRY. Overall malfunctioning with extended damages on the peripheral circuitry accessible from the external pins were noticed.

These results lead to the conclusion that the malfunctioning should not be due to electrical overstress (electrostatic discharge) or mishandling but are process related. The automated equipment used for testing guarantees measurement repeatability.

Hermeticity Tests

The test was performed as per MIL-STD 883, method No. 1014.6, test condition. Al + fluorocarbon bubble test as per MIL-STD 883, method No. 1014.6, test condition C. Measured leak rate was $4 \times 10^{-8} cm^3/s$. The component was hermetic. The component was baked at 125°C for 12 hours. No changes were detected; therefore, ionic dielectric contamination can be excluded as a cause of malfunctioning.

We conclude that malfunctioning is not due to misuse, mishandling, or assembly. It is not attributable to corrosion (the component is hermetic and its operating temperature is 70°C) or ionic contamination either. Defective parts belong to different lots, and this further confirms that failure depends on the component's manufacturing process.

Each subsequent inspection destroys information. Such activity must thus take place only after completing the previous step (for instance, opening the package before leak testing or visual inspection would make it difficult to determine if corrosion were due to poor packaging, the component's internal atmosphere, or to component breakage during assembly).

Package Opening

Package was opened and 40–500 X visual inspection was performed as per MIL-STD-883, method No. 2010. The test yielded no information. One new, defect-free component from the same lot was opened to perfect the method so as not to damage the component under test.

This operation confirmed that failure was due to chip defects and not to assembly or service problems. At this point the customer should be notified not to use the suspect component until the investigation is completed.

Glass Removal

The protective glass was removed by etching it in 10 percent oxygen and 90 percent freon 14 plasma for 20 minutes with 80 W power applied in a 4″ cylindrical reactor. (Time selection was estimated on a disposable sample from the same lot.)

Applied power must necessarily be low so as not to overheat the device under test (not more than 80 to 100 W) and is checked on new, disposable samples.

Internal Visual Inspection

Under the SEM, a number of contacts show Al-Si diffusion with contact sinking and evidence of diffusion on grain boundaries. Also noted were suspected electromigration on tracks with depletion and aluminum grain boundaries uncovered.

Oxygen + freon 14 plasma etching was selective for Si, and it removed the Si diffused in the metalization aluminum as well as the surface vitrification, thus highlighting the phenomenon. The morphological difference was revealed by repeating the etching using a wet agent: 2 parts NH_4F (40%) + 1.5 parts acetic acid.

Metalization Microanalysis

EDS microanalysis was performed at 10 kV acceleration. Metalization consists of pure Al.

The electron beam acceleration potential should not go beyond metalization thickness, otherwise data would be falsified by underlying materials (10 kV means approximately 6 μm penetration in Al). Gold (Au) is present because the sample was metallized (a few dozen Ångströms) to improve visibility under the SEM.

Review of the Situation

Analysis of the last steps carried out (protective glass removal, *et cetera*) shows that the component manufacturer's metal deposition process (evaporation) went out of control. The hypothesis is that the structure or the nature of the metalization is not sufficiently under the manufacturer's control, which implies that all components produced on the same line may share the defect and hence fail in a short time.

13.6.2.3 Hypothesis Verification

Thermal Test

New functional parts from 1983 and 1984 supply lots were baked at 100°C for 168 hours. In the 1983 lots, several parts were rejected after testing. In the 1984 lots, no parts were rejected after testing. This confirms that lots supplied in 1983 are defective.

Hypotheses can be checked more rapidly on new parts because the items to be verified are known *a priori*. In this case, the intention is to check the execution of a technological step. Consequently, the package was opened immediately and the metalization examined after removing the protective glass.

SEM Visual Inspection and Microanalysis

Though originating from the same supplier, the masks of the defect-free component are similar but smaller (area is approximately 20 percent less). Metalization Al is doped with Si to limit Al-Si solid diffusion phenomena.

This confirms the hypothesis that the process is significantly out of control or at least inadequate. In fact, the supplier has revised the aluminum deposition process.

13.6.2.4 Final Report and Corrective Measures

The report showed details of analyses carried out and the conclusions. The report should contain the data essential to understanding the nature of the problem, discussing it with the supplier, and taking corrective actions. If possible, all essential data should be contained on one page, with appendixes for photographs, case histories, X-ray emission spectra, and so on. All data, analyses, and examined samples should be kept, both for discussion with the supplier and to provide case histories for comparison with future cases.

The following corrective actions were suggested:

1. Take components belonging to the lots in which failure was detected out of stores.
2. Return these lots to the supplier.
3. Retrieve and replace defective components on boards rejected during inspection or service.
4. Retrieve and replace defective components on boards in all equipment in the company's possession (in inspection, burnout, or product stores) and installed in service.
5. Check recent samples from the same supplier. Suspend all orders of this product until the component is approved. Step up sampling plans on all lots and failure analysis on rejects, using technological analysis where necessary. Repeat type-approval tests and technological analysis.

In a manufacturing company, this stage is the final purpose of failure analysis. To reach this stage, however, it is necessary to go through the entire analysis procedure. This is true both in errors within the company, and when the failures involve defects caused by the supplier.

BIBLIOGRAPHY

1. J.M. Patterson, "Developing an Approach to Semiconductor Failure Analysis and Curve-Tracer Interpretation," *Proc. of the Int. Reliability Physics Symp.*, 1978.

2. E. Pollino, "Affidabilità dei componenti elettronici e Analisi Guasti nelle Telecomunicazioni," *Elettronica e Telecomunicazioni,* n. 5, 1982.
3. R. Casale, A. Gallesio, C. Guasco, E. Pollino, and D. Riva, "A New Methodological Approach to the Construction Analysis of LSI Integrated Circuits," *Proc. of 'Eurocon '82,* Copenhagen.
4. E. Pollino and D. Riva, "Il ruolo della Fisica dei Guasti per aumentare l'efficacia delle procedure di Controllo Qualità," Proc. of the meeting "La qualità dei componenti e degli apparati elettronci," Caserta, 18 ott. 1984.
5. C. Guasco, E. Pollino, and D. Riva, "Analisi costruttiva: Alla ricerca del punto debole," *Elettronica Oggi,* Dec. 1982.
6. A. Gallesio and E. Pollino, "L'analisi dei dispositivi guasti," *Elettronica Oggi,* Dec. 1982.
7. E. Pollino and D. Riva, "L'approccio tecnologico all'affidabilità dei componenti elettronici," *BIAS 1983,* Milan.
8. R.W. Thomas, "Microcircuit package gas analysis," *Proc. of the Int. Reliability Physics Symp.,* 1976.
9. W. Kinzy Jones, "Plasma Etching as Applied to Failure Analysis," *Proc. of the Int. Reliability Physics Symp.,* 1974.
10. M. Jacques, "Tough Analysis Problems that Have a Solution," *Proc. of ATFA Symp.,* 1978.
11. B.G. Cohen, "Infrared Microscopy," Research Devices Inc. *Technical Bulletin,* n. 1, 1978.
12. P. Angelides, "Precision Cross-Sectional Analysis of LSI and VLSI Devices," *Proc. of the Int. Reliability Physics Symp.,* 1981.
13. C.A. Lidback, "Scanning Infrared Microscopy Techniques for Semiconductor Thermal Analysis," *Proc. of the Int. Reliability Physics Symp.,* 1979.
14. J. Hiatt, "A Method of Detecting Hot Spots on Semiconductors Using Liquid Crystals," *Proc. of the Int. Reliability Physics Symp.,* 1981.
15. G.J. West, "Simple Liquid Crystal Techniques for Fault Site Isolation on Microcircuits," *Proc. of ISTFA,* 1981.
16. D.J. Burns, "Microcircuit Analysis Techniques Using Field Effect Liquid Crystals," *Proc. of the Int. Reliability Physics Symp.,* 1978.
17. C. Guasco, "Tools for Circuit Design Verification," *Proc. of HP Users Conf.,* Brighton, 1986.
18. R.G. Taylor and C.E. Stephens, "Microcircuit Failure Analysis," *British Telecommunications Engineering,* 4, April 1985.
19. B.P. Richards and P.K. Footner, "Failure Analysis in Semiconductor Devices—Rationale, Methodology and Practice," *Microelectronics J.,* 15, 1984.
20. T. Wilson and C.J.R. Sheppard, *Theory and Practice of Scanning Optical Microscopy,* Academic Press, London, 1984.
21. T. Wilson, "Scanning Optical Microscopy and Integrated Circuits," *Microelectronic Engineering,* 7, 1987.
22. A. Dixon, "The Lasersharp Scanning Optical Microscope: an Introduction," Bio-Rad Technical Note No. 301.
23. E. Ziegler and H.P. Feuerbaum, "IC Testing Using Optical Beam Induced Current Generated by a Laser Scan Microscope," *Microelectronic Engineering,* 7, 1987.
24. MIL-STD-883C.

Chapter 14
RELIABILITY PROBLEMS AND STANDARDS
G. Mattana

14.1 INTRODUCTION

Those who face problems related to component reliability (excluding the case where the customer prescribes mandatory criteria) are often compelled to choose among many possible approaches, which are very different from one another: screenings, use of models, historical data, certifications, process control, technological characterization, or accelerated tests. Even this short list, which is certainly not exhaustive, gives an indication of the various methodologies and diverse underlying philosophies. The range of operational problems may be extremely diversified. One problem then is to furnish indications within the sphere of applicability of each technique or, if preferred, to connect the library of methods with the variety of operative problems.

These concepts must be adapted to the specific topic of the connection between reliability and standardization, and therefore it will first be necessary to identify the peculiarity of the electronic components market, then to pick out the main emerging practical problems, and last to survey some possible answers by means of operating tools, among them standardization.

14.2 MAIN CHARACTERISTICS OF THE ELECTRONIC COMPONENTS MARKET

The electronic components market supplies the entire worldwide electronic equipment industry. The annual production of electronic components is worth approximately \$100 billion. As far as international interchange is concerned, this market is very "open."

The producers are far fewer in number than the users. Typically, a semiconductor manufacturer could have several tens of thousands of customers (if most of them were not diverted to distributors), whereas an equipment manufacturer can find at best ten or twenty suppliers for a given device.

Other important characteristics of the component market are the following:

- Mass production with a very high degree of automation.
- Very detailed standards, widely used and having a long tradition especially as concerns general philosophies and methods.
- High proliferation of types: more than 100,000 different components are commercially available.
- Very fast technological evolution and keen competition—even qualitative—in the international marketplace, with a considerable relevance to the learning factor. Due to this fast evolution, problems arise in procuring spare parts designed for a long useful life.
- Only slight wearout, even over a long life of use.
- Generally very low failure rate for established technology, sometimes with a spread of several decades; for new technologies, failure rate often much higher during the start-up phase.
- In numerous significant cases, commercial availability of components that, all other performances being equal, are characterized by different "grades," in term of levels of "expected" reliability.

These characteristics affect the choice of the methods used for quality assurance, as well as the methods for reliability assessment and improvement.

From these characteristics some important implications ensue. The standpoint and temporal vision of the component's manufacturers are quite different than that of the component's users, even more so when the equipment users are "service companies." It is almost impossible to induce manufacurers to make "special" devices—this is against the logic of the market wherein they operate. Innovation not only drives the market but pays off in the long run. However, it is the source of the highest risks in the short run. These risks may pass over the production phase and finally be paid by the customer. Product-related standards may only come *after* a new product has asserted itself; but standard methods, which last very long, are by now indispensable both for evaluating new components and for assuring constancy of conformity and robustness in "current" products.

14.3 PROBLEMS CONCERNING COMPONENT QUALITY AND RELIABILITY

The main problems may be listed as follows:

- How can we ascertain that the product conforms to performance standards, especially those regarding function, product robustness despite stresses, fitness to assembly processes, suitability to storage, transportation, and so on?

- How can we limit the maximum number of defects in production batches that meet performance standards?
- How can we guarantee that, in successive production batches, the product remains within the values specified by standards?
- How can we improve the "design" reliability (the intrinsic "robustness")?
- How can we improve the "production" reliability; that is, eliminate the "freak" items within a "good" population?
- How can we quantitatively verify the product reliability?
- How can we locate the presence of "wear" defects, if any?
- How can we extrapolate or interpolate reliability under conditions different than the test ones?

It is intuitive that, in general, each *different* problem calls for its own *specific* tools. It is also evident that the *use* of such specific tools may be different according to whether it must be applied to consolidated components and technologies, new components and technologies, or components with particular problems. Let us now examine the specific tools whereby these problems are faced.

14.3.1 Initial Compliance to Specifications

How do we verify the compliance of initial performances with specification requirements? Preparation for this test means determining the performances to be verified, the initial limits admitted for such performances, and the tests a random *sample* must withstand without exceeding a maximum number of defects. This operation is called *type homologation.*

The sample consists, generally, of 50–100 pieces. A product that passes the test could be included on the *qualified product list* (QPL) of national or "regional" systems. This operation also ensures the interchangeability of different suppliers within admitted limits.

The *international* regulatory bodies that issue such QPLs are CECC and IECQ (the latter is now gradually starting up). We stress that such operation *assesses conformity to standards;* should conformity to *use* be assessed, the criteria, deciders, and so on would change.

14.3.2 Successive Compliance to Specifications

How do we guarantee that, in successive production lots, the product will remain within the standard values assessed during type homologation? This question is answered by the CECC certification system in the following manner:

- A sample extracted from each production lot must pass some preestablished tests without exceeding a preestablished number of defective pieces.

- Samples periodically extracted from the *process* must pass additional tests.
- The *organization* performing these tests (both initially and periodically) is, in turn, checked.

14.3.3 Limitation of Defects

How do we limit the maximum number of defective pieces present at time zero in individual lots; that is, how do we limit the number of defective pieces in "good" lots? This is the traditional problem of acceptance testing. Today, however, the range of this problem is extended to include "coverage," computer-controlled testing machines, and other topics. In this area, the electronic community sees a controversy between suppliers and users. Suppliers declare an *average outgoing quality level* (AOQL) on the order of magnitude of 100 ppm, and users claim observed AOQL values on the range of 1,000–100,000 ppm [1].

14.3.4 Improvement of Design

How can we improve the "design" reliability of devices? To answer this question we must set up a separation (often difficult and somewhat arbitrary) between *process* quality control problems and *design* improvement problems. Process problems are seen in very large numbers of components. The solution consists of reducing the maximum number of defective units passing individual tests and expanding the quality assurance plan to include more tests. The design problems are solved by submitting a sample to very accelerated tests, to the extent of destruction if necessary, to enhance the product's robustness.

14.3.5 Quantitative Testing

How do we test reliability quantitatively? Is it enough to test components in such numbers that they will generate the wanted reliability value? MIL standards proceeded just in this way during the 1960s. However, the continuous improvement of device reliability made this quantitive approach unfeasible. For example, to assess components of the 10^{-8} level with 5 failures at a 60 percent confidence level, it was necessary to test 630,000 components for 1,000 hours!

The IEC409 II Ed 1981 standard, too, has remained practically unused—and so far unreplaced. This is a *complementary* procedure, although very important, and it cannot substitute for the preceding one. The problem is made difficult by the fact that *the control of weak lots* (even if conforming) is added to the control of *weak pieces in robust lots.*

14.3.6 Improvement of Existing Products

How do we improve reliability once the product has been fabricated? This is a very important topic, particularly during the rising or young phase of the product life cycle, when anomalies of design, process, and control may emerge and the learning speed is the most important factor.

It is well known that initial unreliability, real and potential, has a strong impact on productivity both in the prodution and guarantee period. One solution to this problem is to submit the whole product to appropriate stresses, or purchase (in a dynamic manner correlated with learning and feedback of results) components that have already undergone such screenings.

A cost-benefits problem then arises. On the one hand, technology must be *dynamic;* and on the other hand it is *a function of the failure mechanisms*. A possible solution is to make the product *more selective,* building on the appropriate knowledge. The physics of failure mechanisms, whose purpose is to study *why* failures occur rather than demonstrating *how many* failures have occurred, may guide the adoption of such a choice of selectivity.

14.4 DIFFERENCES AND CONNECTIONS BETWEEN RELIABILITY AND CERTIFICATION

The problems just reviewed may be divided into two groups by their focus: aspects of *reliability* (related to the behavior of objects in time), and aspects of *conformity* (and robustness) at time zero. It is well to identify the connection points as well as the separation between the two groups.

Reliability is considered an aggregate of techniques to improve, build, and predict the future behavior of products. If we are interested in the "punctual" applications (such as improving the robustness of a production lot or seeking why a given failure has occurred), then the connection with certification operations is very weak or nonexistent; but if we want to know the reliability of a mass-produced item, reliability is strictly connected with the control of the entire manufacturing process (and its eventual certification). In fact, it is worth remembering that reliability is a cumulative property that depends on a very large number of factors belonging to design, process, and control as well as to application criteria and stress (during transportation, assembly, and usage). Because the process is continuous for the users, "seeking" reliability without certification is like building on sand.

14.5 PRODUCT STANDARDS FOR ELECTRONIC COMPONENTS

14.5.1 Standard Levels

There are national, regional (e.g., European), and international levels of standards:

- At the individual country level
  CEI (Italy)
  UTE (France)
  VDE-DIN (Germany)
  BS (Great Britain)
  MIL (US, military only)
- At the European level (these standards have largely replaced the standards of individual European countries)
  CECC standards
- At the international level
  IEC standards

IEC standards, very effective and widespread as far as general methods are concerned, do not contain any quality-assurance clauses. New IEC standards including such aspects are now being gradually issued, to be applied to the IEC-Q international certification system.

CECC standards were issued according to IEC standards, but they are complete, with detailed data (dimensions, values, *et cetera*) and quality-assurance clauses. In the future, the CECC sytem could flow into the IEC-Q system.

14.5.2 Types of Standards

The standards for electronic components are product standards, hence they contain product *performance;* methods of *measurement* of such performance; and *criteria for the assessment* of performances and certification, if any. These standards also refer to more general standards, common to many or all products, concerning, for example, units of measure, terminology, or symbology.

A system of standards that contains only performance and measurement cannot guarantee *continual assessment* of performance (this was the situation with IEC standards). At present, all levels are contained in all the main systems. These systems include the CECC (European system for electronic components) started in 1970; the IEC-Q (worldwide system) started in 1983; and the MIL (US military system) that provides the guarantee of periodic assessment almost exclusively within the US.

14.5.3 Structure of a Certification System

The CECC system rests on four main pillars: the system rules; a system of product standards; a system of qualification (on both manufacturer and product); and a supervisory system. The main institutes of the CECC system are illustrated in Figure 14.1.

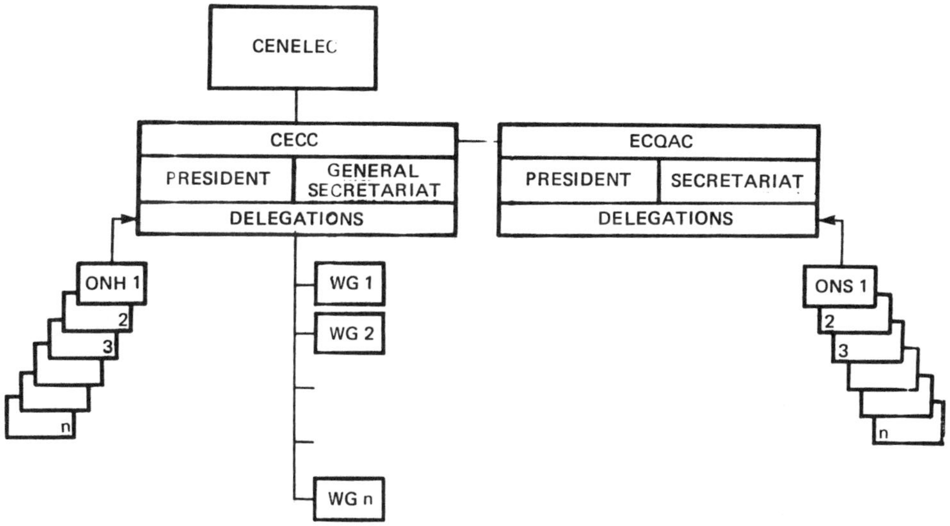

Figure 14.1 Structure of the CECC System.

Note: CENELEC is European Committee for Electrotechnical Standardization; CECC is CENELEC Electronic Components Committee; ECQAC is Electronic Components Quality Assurance Committee; ONH is National Authorized Institution; ONS is National Supervising Inspectorate; WG is Working Group.

14.5.4 Internal Structure of the Standards for Components

In the CECC and IEC systems, the various standards are structured according to the following architecture: sectorial (or general) specifications; sectional or family specifications; and detail specifications. Reference to more general standards concern, for example, mechanical and environmental tests, sampling plans, and so on. The requirements contained in all of these standards are almost always divided into the following groups of tests:

Group A Functional or visual inspection tests (nondestructive, short) to be made on each lot.
Group B Other short tests (maximum one week) to be made on each lot to identify robustness or suitability to particular requirements. These may be destructive.
Group C Tests to be repeated approximately every three months. These guarantee that the product has not changed with respect to the assessments made during type qualification.
Group D Long tests (typically 8–10,000-hour life tests). They guarantee long-term "robustness" and can be used to derive reliability data.
Screening Several stress levels capable of eliminating good but weak pieces, thus improving product reliability (see example on Figure 14.2).

A representation of this architecture is given in Figure 14.3.

14.5.5 Structure of CECC Standards for Integrated Circuits

According to the general diagram in Figure 14.3, CECC standards for integrated circuits are organized as shown in Table 14.1. The *general specification* (No. 90 000) contains, in particular, the "*general aspects,*" the cascade of pertinent standards (IEC and CECC) concerning symbology and terminology, preferred values, markings, designation criteria for purchasing, and so on; *the "quality assessment" procedures,* usually divided into several levels of qualitative assessment (see Tables 14.2 and 14.3); and *the test procedures* for all kinds of tests, including screenings.

The *sectional specifications* contain, among other things, the addition of measurement methods common to all the families of the section considered; and the division of groups A, B, C, and D into subgroups (according to the diagram of Figure 14.3) to conform with the various "assessment levels" (as in Tables 14.2 and 14.3, which are continuously updated though retaining their basic structure).

The *family specifications* and *detail specifications* also contain the *values of the parameters* that must be complied within each test.

14.5.6 The Main IEC Standards for Semiconductors

These are found in several of the IEC's publications. Publication 747 is divided into ten parts concerning specific semiconductor classes (e.g., part 5 concerns optoelectronic devices) or particular aspects common to several classes. Publication

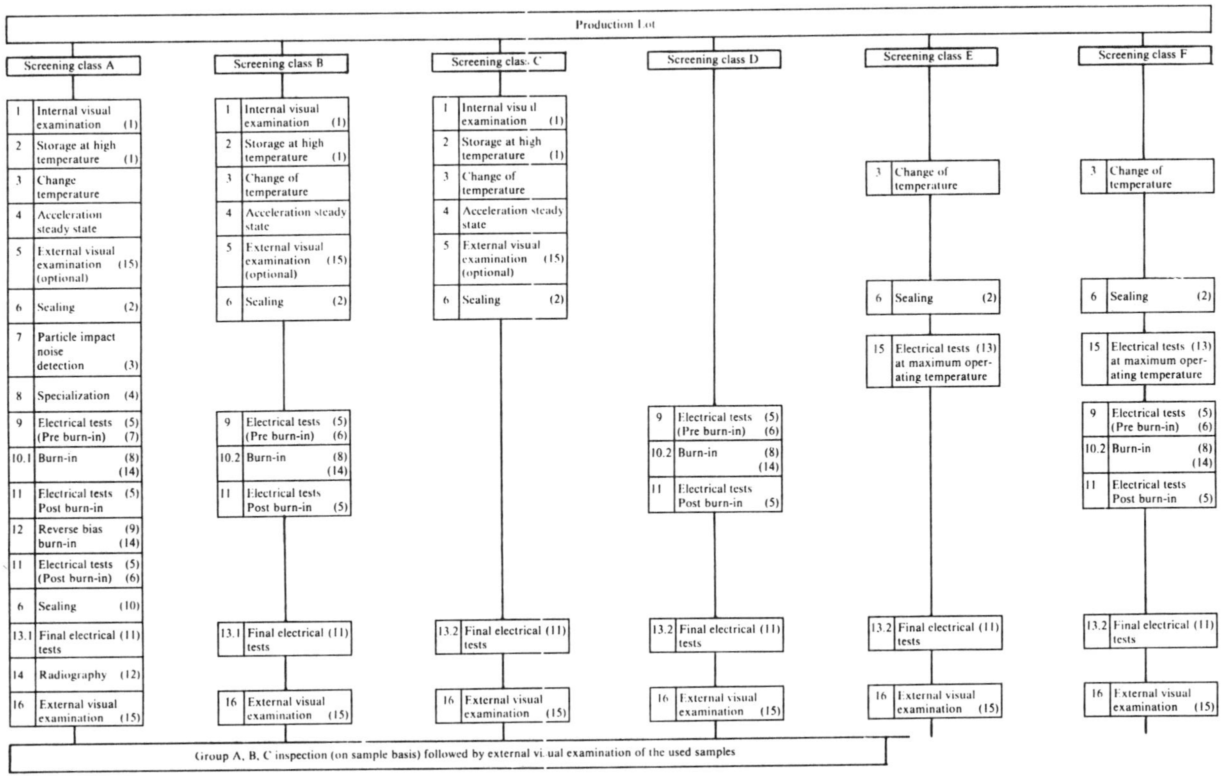

Figure 14.2. Synopsis of the Different "Screening" Classes Specified by Standard CECC 90000, Edition 3 (each class corresponds to a different "basket" of tests, hence to a different degree of severity).

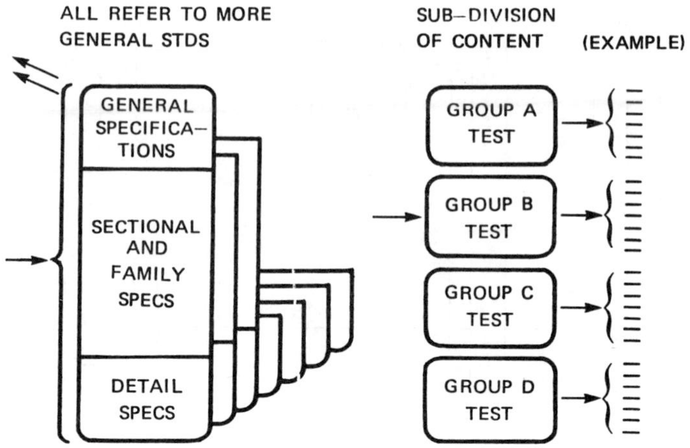

Figure 14.3 Architecture of the standards for components in the CECC and IEC-Q System. Test groups are divided by types of specifications and content.

748-1 (1984) sets general standards for integrated circuits. Publication 249 (1984) focuses on climatic and mechanical tests for semiconductors.

Table 14.1 Architecture of CECC Specs for Integrated Circuits

General Specs	*Sectional Specs*		*Family Specs*	*Detail Specs (as of August 1985)*
CECC 90.000 (3ª Ed. 1985) Monolithic integrated circuit	90101 Digital Monolithic IC	90101	TTL CIRCUITS	—
		90102	TTL SHOTTKY CIRCUITS	63
		90103	TTL, Low-Power Schottky Circuits	309
		90104	CMOS (4000 B e 4000 U B)	223
		90105	ROM MEMORIES	19
		90106	Advanced Low-Power TTL, Schottky	183
		90107	TTL FAST	43
	90200 Analogue Monolithic IC	90201	Integrated voltage regulators	—
		90202	Integrated operational amplifiers	—
		90203	Integrated analogue switching circuits	— —
	90300 Interface Monolithic IC	90301	Integrated line transmitters or receivers	—

Table 14.2 Division into subgroups of Group B Tests (for each test, the possible "assessment levels" and inspection requirements are reported; the Table is derived from standard specification CECC 90103, ed.1)

Tests	*D or ND*	*Test conditions*	*Assessment level*	*Inspection requirements*		
				IL	*AQL*	*Limits*
Sub-Group B1:	ND					
Dimensions 4.3		IEC outline (See IEC 191-2 and see detail spec. for specific type)	R.S.T.V	S-4	1.0	See relevant detail spec.
Sub-Group B2:	D					
Solderability 4.6.10.1		Solder bath No aging required	R.S.T.V	S-3	2.5	4.6.10.1
Sub-Group B3:	ND					
Sealing test (cavity packages)			R.S.T.V	II	1.0	
Fine leak 4.6.9.1		4.6.9.1. Test Qk				4.6.9.1
Gross leak 4.6.9.2		4.6.9.2. Test Qc				4.6.9.2
Sub-Group B4:			R.S.T.V	S-4	2.5	
Change of temperature 4.6.8.1		4.6.8.1 Test Na		n.a	n.a	
		$T = T_{stg\ min\ and\ max}$ (See relevant detail specification)				
Endpoint tests						
(1) for cavity packages: Fine and gross leak test	ND	Test Qk 4.6.9.1 and Qc 4.6.9.2				
(2) for noncavity packages: Damp heat accelerated	D	Test Da 4.6.3.1				
Electrical tests		As for subgroups A2 and A3				as for sub-gr. A2 and A3 of detail spec.
Sub-Group B5:	ND	168 h + 72 h − 10 at $T = T_{amb\ max}$	R.S.T.V	S-4	1.5	
				n.a	n.a	

Table 14.2 continued

Tests	*D or ND*	*Test conditions*	*Assessment level*	*Inspection requirements*		
				IL	*AQL*	*Limits*
Electrical endurance 4.2. of CECC 90 100		See relevant detail specification				
Endpoint tests Electrical tests		As for subgroups A2, A3, A5				As for sub gr. A2, A3, A5 of detail spec.

Table 14.3 Division into subgroups of Group B Tests (for each test, the possible "assessment levels" and inspection requirements are reported; the table is derived from standard specification CECC 90100, Ed.1)

Tests	*D or ND*	*Test conditions*	*Limits to be specified*	*Inspection requirements by assessment levels*					
				R		*T*		*V*	
				IL	*AQL*	*IL*	*AQL*	*IL*	*AQL*
Sub-Group B1: Dimensions	ND	See 4.3	See 5.3.1 of this specification	S-4	1.0	S-4	1.0	S-4	1.0
Sub-Group B2: Solderability See 4.6.10	D	No aging required 4.6.10(1) or 4.6.10(2)		S-3	2.5	S-3	2.5	S-3	2.5
Sub-Group B3: Sealing test See 4.6.9(1) and 4.6.9(2) (cavity packages)	ND			II	1.0	II	1.0	II	1.0
Sub-Group B4: Change of temperature	ND	4.6.8(1) Test Na. Maximum and minimum storage temperature but not exceeding − 65°C to + 175°C		S-4	2.5	S-4	2.5	na	na
Followed by (1) For cavity pack-		Test Q. fine and							

Table 14.3 continued

Tests	D or ND	Test conditions	Limits to be specified	*Inspection requirements by assessment levels* R IL	R AQL	T IL	T AQL	V IL	V AQL
ages: Fine and gross leak test		gross leak test 4.6.9(1) and 4.6.9(2)							
(2) For noncavity packages: Damp heat accelerated	D	Test Da. damp heat, accelerated 4.6.3 (1)							
End point test: Electrical tests as for Sub-Groups A2 and A3		As for Sub-Groups A2 and A3	As for Sub-Group and A3						

14.5.7 The Pros and Cons of a System for Quality Assurance and Certification

The main advantages for such a system are as follows:

- Certainty about the interchangeability of suppliers within the parameters guaranteed in the specification.
- Possibility of substantially reducing or even renouncing acceptance testing.
- Guarantees that components come from production lines on which the envisaged checks are made both lot by lot and periodically.
- Guarantees that the defectiveness of the received lots is no higher than that admitted, in all the envisaged checks.
- Possibility of exploiting all the coherences set up by the system in terms of terminology, measurement techniques and instruments, testing modes, and so on.
- Ability to count on a known quality level, with little or no extra cost, because checks are made where the volumes are larger and statistical inference can be better exploited.
- Possibility of choosing among several levels of assured and certified quality.
- Mandatory preservation of the "fabrication dossier," which reduces the number of disputes and facilitates their solution.

On the other hand, not using such a system may involve the following *disadvantages*:

- Being at the mercy of the market, as only very few big users are in a position to affect the suppliers.
- Being unable to fully exploit the resource of sure, interchangeable, alternative sources.
- Being forced to undertake the very heavy burden of building one's own standards and to manage increasingly more difficult and sometimes dramatic controversies (which may include aspects of metrology, reliability, compatibility with manufacturing processes, and marginal behavior).
- Possible complications in foreign sales.

Generally speaking, not following the standards means renouncing a *resource* that contains a vast know-how and is free, at anybody's disposal.

14.6 MAIN STANDARDS FOR COMPONENT RELIABILITY

14.6.1 IEC Standards for Reliability

The IEC standards on reliability provide a library of methods and guides conceived for all kinds of applications and drawn up in very general terms so as to guarantee the following requisites:

- Maximum generality at the highest structural levels (general methods having very wide validity and applicability).
- Maximum flexibility at the lowest structural levels (wherein all the specific requirements of a particular family of objects can be inserted).
- Internal coherence of the whole system.

These standards provide for different evaluations of reliability: forecasts, extrapolations, verifications on samples, data collection from the field, failures tree, *et cetera* tied to the growth of methods and parameters that, therefore, must be made explicit. IEC standards on reliability (mainly IEC-Publications 271, 300, 409, 605, 706) are applied to components, equipment, and systems.

14.6.2 IEC Standards on Component Reliability

These standards are found in several of the IEC's publications.

Publication 362 is entitled, *Guide for the Collection of Reliability, Availability and Maintainability Data from Field Performance of Electric Items,* which summarizes its contents.

Publication 409 (1981) is *Guide for the Inclusion of Quantitative Reliability Clauses into Specifications of Components (or Parts) for Electronic Equipment.* This standard applies only to the case of *reliability measured by tests,* on finished components, assuming a constant failure rate. Two different approaches are provided:

verification tests (exceeding or not exceeding of a preestablished value, which therefore affects the number of pieces to be tested) and evaluation tests (apart from any objective, which depend on the number of tests). In Standard 409 the various aspects of the procedure are discussed in terms of use of acceleration factors, reference to failure criteria, sample's dimension, and failures admitted for qualification, as well as qualification preservation (different confidence levels, approval preservation or withdrawal, accumulation criteria, and period covered by qualification). A checklist, for the factors to be taken into account whenever a reliability specification is prepared, includes the following:

1. Check of individual lots or series of lots.
2. Conformity or determination tests.
3. Specification of test methods, stress levels, and acceleration factors.
4. Failure criteria.
5. Tests periodicity.
6. Qualification procedure.
7. Determination of the sample's dimension derived from the required failure rate and fixed dimension, or that derived from the tables for AQLs.
8. Criteria for initial qualification.
9. Period covered by qualification.
10. Criteria for preservation, extension, and withdrawal of qualification.

14.6.3 MIL Standards for Component Reliability

During the 1960s, the MIL standards faced the problem of guaranteeing reliability in addition to the other aspects of performance. In fact, the following general standards were issued to which all product standards made reference.

MIL-STD-790, *Reliability Assurance Program for Electronic Parts Specifications,* enacted the concept that only a carefully controlled *process* was able to guarantee quantitative reliability.

MIL-STD-690, *Failure Rate Sampling Plans and Procedures,* provided the plans and criteria for measuring reliability quantitatively through tests. It was based on the assumption of a constant failure rate and provided four procedures: qualification at the initial level of failure rate; extension to lower levels (of failure rate); preservation of the initially reached level; and qualification of the failure rate of individual lots in addition to that of the process.

The MIL-R-38100A-1963 Specification, *Reliability and Quality Assurance Requirements for Established Reliability of Parts,* substantially followed the same criteria, "combining" quality assurance and reliability requirements in the same standard.

14.6.4 The Failure of the Standards on Component Reliability

The just mentioned standards (IEC and MIL) on component reliability have remained substantially unused in the majority of applications, apart from military field. There are many reasons for this, among which are the following:

- The improvement of actual reliability levels drove the "demonstration" costs to astronomical values.
- The technological evolution increased, and the importance of learning made the accumulation assumption very uncertain.
- Model uncertainty made massive recourse to acceleration factors problematic.
- Product standards systematically lagged behind technology.
- Emphasis on the reason *why* failures occur (physics of failure mechanisms) had become more advantageous than showing the "quantum" of mass testing, and allowed a much faster adjustment.

This situation could be highlighted by a simple table, comparing methods suitable for a fast evolution with those suitable for a slow evolution. In fact, the market trend is toward the option of several classes of incremental reliability (screening or other techniques) rather than different failure rate levels.

14.6.5 MIL-STD-883

The *purpose* of this standard, *Test Methods and Procedures for Microelectronics,* is to set up tests not only of an electrical type but also mechanical, environmental, and other types. Strictly speaking, the *applicability* of this standard is limited only to products complying with the specific MIL standards *of that product.* Actually, it is also used outside this sphere but, for example, it cannot be applied to the evaluation of plastic containers, as they are not accepted by MIL standards.

14.6.5.1 Structure of This Standard

The standard is divided into five test blocks. The *first block* (methods 1001, 1002, *et cetera*) describes the mode of testing in *environmental tests* (about 25 different tests). Belonging to this group are the following tests: humidity, high temperature storage, dew point, burnout, characterization for life-reliability, radiation from neutrons, water vapor content in the package, *et cetera.*

The *second block* (methods 2001, 2002, *et cetera*) specifies the testing mode for *mechanical* tests (about 25 different tests). Belonging to this group are not only vibration, shock, and similar tests, but also tests such as visual examination (external and internal) and related assessment criteria, lead solderability, radiographic examination, resistance to solvents, inspection of metalizations by scanning electronic microscope (SEM), and others.

The *third block* (methods 3001, 3002, *et cetera*) specifies the testing mode for *electrical* tests on *digital devices* (about 15 tests). It also includes a classification of sensitivity to electrostatic discharges.

The *fourth block* (methods 5001, 5002, *et cetera*) contains the specific *test procedure* (9 procedures). Belonging to this group are the check of a parameter distribution, procedures for failure analysis in microcircuits, screening procedures, procedures for the qualification and conformity check (with reference to MIL-M-38510 specification), including maximum admitted defectiveness, limit tests, procedures for hybrid and "multichip" microcircuits, and destructive physical analyses. This standard, too, is subject to frequent updating.

14.6.6 Present trends and conclusions

The reality (both technical and market-related) examined in this chapter makes it almost impossible to measure reliability in advance. Technological evolution is very fast—a similar phenomenon can hardly be found in economy history.

Learning plays a determinant role; screenings offer the customer the possibility of purchasing products that are *better* than the standard ones, but responsibility is left to the purchaser. Surely, this is one precaution against failure during the first period of life, but what precautions may be taken against long-lasting problems?

The proliferation and spread of semicustom- and custom-made devices brings back attention to process control (and "capability approval") rather than product control. The considerable improvements in reliability per unit should not be seen as continuously decreasing failure-rate curves; perhaps, they are better described as saw-tooth cures, wherein each point of return corresponds with the introduction of a new technology.

Manufacturers are working to better master the reliability of individual process phases; no doubt, a closer relationship between manufacturers and users will make the cognitive process faster as well as deeper. But, if we limit ourselves to the use of standards, how and how much can they help us? The library of methods grows richer and it plays an important role, but not a determinant one. International certification systems are certainly a prerequisite, but not a guarantee of quantitative reliability for the consumer.

14.7. ARCHITECTURE OF PERFORMANCE AND ASSESSMENT OPERATIONS FOR ELECTRONIC COMPONENTS

Earlier in this chapter, we indicated the main problems connected with components' quality and reliability; we then described the contents of the standards. Comparing these topics two important observations can be made. First, standards cannot adapt themselves to *new* products because of the physiological delay with which they are agreed upon and issued. Second, some operational requirements are entirely covered by standards, among them the procedure for continuous certification on the product qualitative level; other requirements remain completely uncovered, and different strategies must be used for them. It is therefore convenient to establish a comparison between the *architecture of performances* and the *architecture of assessment operations.* Figure 14.4 presents such a comparison, and the following paragraphs illustrate its contents.

14.7.1 Qualification

In general each product is characterized by *n* initial performances. We can imagine them as represented in *n*-dimensional space. These performances may be *electrical,* or focus on *compatibility* with various workings (soldering, bending, washing), and they also may be environmental or focus on mechanical performances, indicating the component's *robustness.* All these performances are, in principle, *described* by a *specification* or *standard.*

Qualification is the operation that assesses conformity to the standards by a sample of the product under examination. Therefore, it must contain *metrology* (the modes of performance testing) and the *criteria for deciding* whether the wanted correspondence is met (e.g., the maximum number of nonconforming items allowed in each test and on all tests). *Metrology* refers to both general criteria and criteria specific to the item under examination. For example, standards IEC-68 and MIL-STD-883 contain references for all the electrotechnical field and for all semiconductors, respectively. Each product (or family) standard contains the mode of testing for each electrical performance.

14.7.2 Reliability of a Product Sample

Initial performance is normally required to be *preserved over time* (i.e., with the product's age) in a known measure (reliability, stability). It is therefore necessary to evaluate, for example, reliability that is a *potential* performance. This particular *operation* may assume different aspects. In theory various approaches can be taken, which are very different from one another:

1. *To test,* possibly using acceleration factors that are known or agreed upon, a number of pieces to demonstrate expected reliability (Standards MIL-STD-690 and the like, and IEC409-1981).
2. *To control "better"* (i.e., with stricter criteria, traceability) the process from which the product comes rather than the product itself. It is logical to expect that such criteria will give rise to a product not only more *reliable* but also

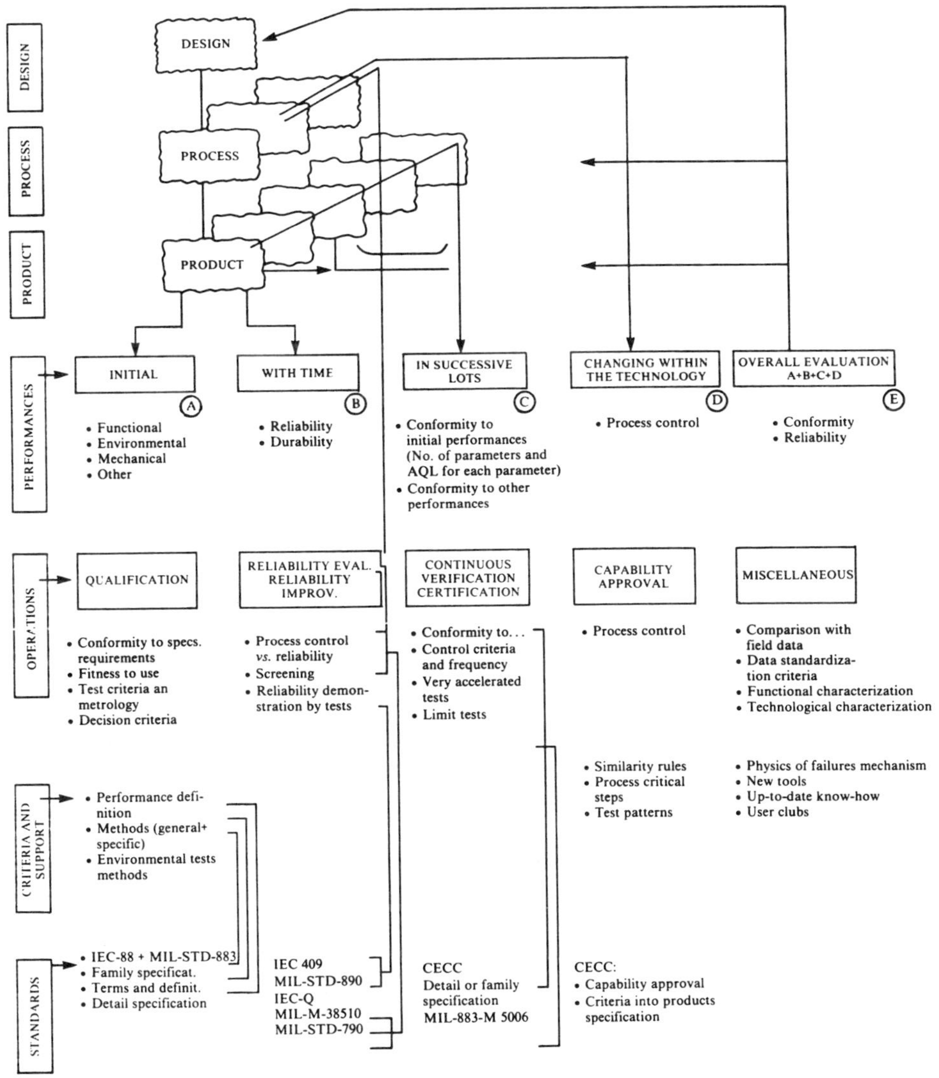

Figure 14.4 The architecture of performances and the architecture of the operations for testing performances.

more *uniform* that, through accumulation, produces more cost effective reliability estimates. (The standards meant for the space industry, and partially CECC standards, follow these criteria.)

3. *To screen the product appropriately* to eliminate the fraction of potential defectives in the process. (In this case, reliability is not measured, but it is improved by a certain measure.)

14.7.3 Continual Process Control and Capability Approval

It is normally required that these performances be preserved in successive production lots. It is specifically required, *with reference to the main performances,* that each lot shall not contain a number of nonconforming pieces larger than the agreed-upon levels (some percent? 100 ppm?) and that the product shall maintain *substantial conformity with the specification also for the other performances.*

The operation whereby this performance is assessed is not only *incoming testing* but also *periodic repetition* of Class B and C tests (i.e., longer tests, sophisticated or destructive) made on a significant sample of the product (see, for example, CECC standards). In some cases it is deemed preferable to control not the variety of product types but the *process from which they originate,* by controlling both the *critical phases* and items that fall within the process limits. A typical case is that of semicustom circuits. In the CECC system, this procedure is called *capability approval.*

14.7.4 Reliability of Nonstop Production

The preceding are only partial schemes; generally speaking, in fact, we ought to know, at the same time, *initial performances* and *behavior with age, both features studied in successive lots,* and *everything for the various items within a technological family.* However, there still is no normalized operation for knowing *all* of these features in the components field. Of course, other operations exist that, properly combined, may set guidelines for action. These are the following (each with its own limitations):

- Process control (more or less extensive) and statistical process control, possibly governed by an independent body; third party assessment or certification.
- Accelerated tests.
- Field data collection, data exchange, and data normalization.
- Laboratory versus field comparisons.
- Technological characterization.
- Electrical characterization.
- Physics of failure mechanisms.
- Dynamic use of procurement.
- Additional tests on purchased products.

But these operations are outside the scope of this chapter.

REFERENCES

1. E. R. Hnatek, "IC Quality—Where Are We?" *IEEE 1987 Int. Test Conf.*

The Authors

Giorgio Baccarani graduated with a degree in electrical engineering in 1967, and in physics in 1969, from the University of Bologna, Italy. In 1969 he joined Bell Laboratories, Murray Hill, NJ, as a member of the technical staff, working in the area of electron-device processing. In 1970 he became a research assistant at the University of Bologna, where he investigated the physical properties of MOS structures and transport effects in semiconductor materials and devices. Since 1973 he has taught a yearly course on quantum electronics, and in 1980 he was appointed full professor of applied electronics at the University of Bologna. In 1982 he was on a one-year assignment at the IBM Thomas J. Watson Research Center, Yorktown Heights, NY, where he investigated the feasibility of a 1/4 μm MOS process from the standpoint of the physical limitations affecting the device performance. Since 1983 he has headed a group involved in numerical analysis of semiconductor devices, acting as partner leader in the context of three succesive EEC-supported projects in the area of CAD for VLSI. His current research interests include numerical-device simulation and integrated circuit design.

Gregorio Benassero was born in Milano, Italy, in 1945. In 1966 he received a degree in electronics from the Istituto A. Beltrami in Milano. He began his activity at ITALTEL as a designer of digital switching networks. In 1973 the reliability service was established, and he turned to the qualification and then failure and construction analysis of electronic devices, being responsible for the failure analysis section since 1980.

Giuliano Boccaletti was born in Reggiolo, Reggio Emilia, Italy, in 1943, and graduated with a degree in electronic engineering from Bologna University in 1969. In the same year he joined SGS (Societá Generale Semiconduttori), now merged into SGS-THOMSON Microelectronics, in Agrate Brianza, Milano, Italy. Since then he has held various positions in product engineering and digital integrated circuit design for both bipolar and CMOS technologies. Since 1978 he has assumed various responsibilities in the central reliability and quality control department, where he is now directing the Q & R programs group and its data bank. Author of various publications on Q & R, he is also in charge of a yearly training seminar on reliability at the SSGRR Stet school in l'Aquila.

Fabio R. Borri was born in Pistoia, Italy, in 1935, and graduated with a degree in electronic engineering from the Bologna University in 1960. After university he attended the F. Mauro managerial courses at the Politecnico di Milano and at the SSGRR Stet school in l'Aquila. Since 1960 he has worked with SGS (Societá Generale Semiconducttori), now merged with SGS-Thomson Microelectronics, one of the world leaders in discrete semiconductors and integrated circuits, in Agrate Brianza, Milano, where he has held various positions including plant quality assurance and reliability manager, production manager, plant manager, and since 1976 director of central reliability and quality control. He has written several papers and has collaborated in the publication of a number of books in the field of quality and reliability of semiconductor devices. He is member of the AICQ (Italian Association for Quality Control) board, and has organized and held several courses on subjects related to quality and reliability.

Marco Cocito was born in Asti, Italy, in 1947. In 1970 he received a degree in applied physics from the University of Torino, and in the same year he joined the technological division of CSELT, there being involved in the development of techniques for the physical-chemical characterization of materials and devices for microelectronics with particular emphasis on the spatial resolution. At present he is head of the analysis and characterization department, and he coordinates activities devoted to the characterization of optoelectronic devices and VLSI customized ICs developed in CSELT. He has authored and co-authored more than fifty papers in Italy and abroad. He is also a lecturer at Scuola Superiore Guglielmo Reiss Romoli and Politecnico di Torino.

Carlo Cognetti received a degree in solid-state physics in 1970. In 1973 he joined SGS-Thomson as technology engineer of the research and development department, where he developed a number of thin-film processes for power device interconnection. Between 1977 and 1980 he was responsible for the technology group for optoelectronics, for the development and preproduction of GaAs light-emitting diodes and lasers (fiber optic telecommunication). Between 1980 and 1986 he was leader of power IC technology and packages design group for the SGS-Thomson monolithic microsystem division. At present, he is responsible for corporate package development, where new signal and power packages are designed and preproduced. He holds 3 patents and has authored a number of technical papers related to semiconductor processing and packaging.

Giovanni D'Espinosa was born in Palermo in 1935. He graduated with a degree in electronic engineering in 1960 from Palermo University. Author of several technical papers and a teacher at Ancona University, his professional experience includes two years as adjoint professor at Palermo University, testing manager of semiconductor division for Raytheon in Palermo, SCR and TRIAC device testing manager at Dino Olivetti, and new products, materials and processes qualification manager at SGS. He is now acting as manager of EWS-testing-assembly for SGS-Thomson VLSI division.

Fausto Fantini graduated with a degree in electrical engineering in 1971 from the University of Bologna, Italy. In 1973 he joined the quality and reliability (Q & R) department of Telettra S.P.A. in Vimercate (Milano), where he worked on the reliability of semiconductor devices and established the laboratory of failure analysis. From 1979 to 1987 he was responsible for the Q & R of the plant in Bologna. In 1987 he was appointed associate professor of electronics at the Scuola Superiore di Studi Universitari e di Perfezionamento S. Anna in Pisa, and since then he has taught a course in the technology of solid-state devices. He has also lectured at the University of Padova in 1987-88, and is responsible for the course in microelectronics at the University of Parma. His research interests include various aspects of semiconductor device physics and reliability, in particular corrosion, electromigration, and metal/GaAs interactions. He has authored or coauthored two books and over seventy research articles and international conference papers and organized two summer sessions on failure physics. He is a member of IEEE, ISHM, AICQ (Italian Quality Association), and SIF (Italian Physical Society).

Manuela Finetti graduated in physics from the University of Bologna, Italy, in 1974. That year she joined the LAMEL Institute of the Italian National Research Council, where she been working on the electrical properties of ion-implanted laser-annealed silicon, and in the development of high-efficiency solar cells for concentrators. From 1981 to 1983 she was a research fellow at the California Institute of Technology, Pasadena, where she was involved in the study of diffusion barriers for application in low resistance metalization schemes on silicon. In 1985 she spent six months at the Semiconductor Laboratory of the Technical Research Center of Finland, where her interests included the characterization of test structures for contact resistivity measurements and the analysis of electromigration in thin films. She then went back at LAMEL to coordinate the activity on contact and interconnect characterization and reliability, partially funded by the EEC under the ESPRIT program, and by the Italian National Research Council, under the Progetto Finalizzato Materiali e Dispositivi per l'Elettronica a Stato Solido.

Giorgio Fioravanti was born in Turino in 1932. He received a degree in electrotechnical engineering from the Polytecnico di Turino. For the Italian Westinghouse Company he conducted research and development activities related to ac-dc conversion facilities and their regulation and production circuits. At the Societa Generale Semiconduttori (SGS) he workd on quality and reliability. At present he is responsible for the reliability of packages and assembly processes in the monolithic microsystem division of SGS.

Vittorio Ghergia was born in Milano in 1939. He received his doctorate in electronic engineering from the Politechnico di Torino in 1965. After joining CSELT in 1966, he specialized in thin-film, thick-film, and semiconductor integrated technologies. Since 1975 he has been responsible for the material and component section, and then in 1981 for the technological division of CSELT dealing with environmental testing, component reliability, chemical, and physical analysis, and he carries out material research on optoelectronic components in 1.3–1.5 μm wavelength region. From 1978 to 1986 he also taught electronic technologies, as associate professor at the Politechnico di Torino, and in 1988–1989 at the University of Parma. He is author or co-author of some twenty papers and the book *Optical Fibre Communication,* published by McGraw-Hill in 1981. He has lectured on optoelectronic components and materials at the Scuola Guglielmo Reiss Romoli STET group school.

Emilio Ghio was born in Setri Levante, Italy, in 1954, and graduated with a degree in physics from University of Genoa, Italy in 1978. After post-graduate work at the University of Genoa, he joined SGS-Thomson in 1979. From 1979 to 1983 he worked on the development of technology for bipolar integrated circuits. From 1983 to 1986 he was with the corporate quality and reliability group, mainly involved in reliability study of new packaging materials and processes. Since 1986 he has been with the central research and development, CMOS process development group. His work has been concerned with the development of new processes for gate array and ASIC applications.

Lonzi Leonardo was born in S. Giovanni in Marignano, Italy, in 1942, and graduated with a degree in electronic engineering from the University of Bologna in 1973. In 1974 he joined SGS-Thomson, carrying out activities in testing and design of bipolar integrated circuits, and working in the reliability field since 1980. In 1981 he was appointed by the monolithic microsystems division responsible for product reliability and new processes qualification. He researched and studied several subjects in the semiconductor reliability field, the most interesting of them being electromigration in aluminum film, intermetallic formation in Gold-Aluminum joints, provisional failure models, and corrosion mechanisms in plastic packages.

Angelo Messina was born in Milano in 1941. After graduating with a degree in chemistry he started his activity in 1960 at Centro Sperimentale Cellulosa in Milano in research programs on macromolecular chemistry. In 1968 he was given responsibility for the chemical laboratory of quality service in ITALTEL, where he has been working on surface chemical and electrochemical treatments and PC board fabrication technology development. In 1976 he received a degree in physics from the University of Milano. At present he is involved in process qualification concerning physical-chemical issues in quality management.

Paolo Montangero was born in Torino in 1957. In 1985 he graduated with a degree in physical chemistry from the University of Torino and joined the reliability department, where he started investigating the reliability of optoelectronic devices for telecommunication, being deeply involved in failure analysis techniques and electro-optic characterization. He is a lecturer on optoelectronic device reliability at SSGRR (L'Aquila) and at Politecnico di Torino.

Michele Muschitiello was born in Bitonto, Bari, Italy in 1961 and received his degree in industrial electronics in 1980. In 1984 he joined the microelectronic area of Tecnopolis—CSATA, where he is currently responsible for electrical characterization and failure analysis of electronic design. He worked on ESD effects on integrated circuits, latch-up in CMOS, failure analysis techniques.

Agnese Piccirillo received the degree in physics from the University of Torino, Italy, in 1969. She joined CSELT (Centro Studi e Laboratori Telecomunicazioni) in 1969. For three years she worked in the mathematics and traffic division on the definition of mathematical models for the telephone network. She then moved to the reliability department of the technological division. She was involved in silicon device degradation mechanism studies and in definition of failure rate models for mean-time-to-failure (MTTF) evaluation. From 1978 she worked in and had charge of optoelectronic device reliability activities. Since 1986 she has been responsible of III-V compound technology. She is the author of several papers in both silicon and optoelectronic device reliability and the technological field.

Emiliano Pollino was born in Torino in 1947. He received a degree in physics from the University of Torino in 1970. In the same year he joined CSELT (components and materials section) to set up an activity on electromechanic and electronic component reliability. Since 1980 he has been in charge of the reliability section of the technological division of CSELT, dealing mainly with reliability testing, failure analysis, and technological characterization of devices used in telecommunication. Author or co-author of some twenty technical or review papers, Emiliano Pollino is editor and co-author of the Italian book on semiconductor reliability published by SSGRR (L'Aquila, Italy) in 1986. He has lectured on component quality and reliability at SSGRR and at the Politechnico di Torino, where in 1989 he teaches "Electronic Technologies" as associate professor.

Danilo E. Riva was born in Torino, Italy, in 1954. He holds a degree in applied physics from the University of Torino. In 1978 he joined the reliability department of the technological division of CSELT, where he started investigating the reliability of advanced electronic devices, mostly high-reliability for applications, being deeply involved in reverse engineering and construction analysis activities. In 1980 he was qualified as an inspector for space components by the European Space Agency. He is a lecturer on semiconductor device technology and reliability at the Politecnico di Torino and Scuola Superiore Guglielmo Reiss Romoli.

Gianni Clerico Titinet was born in Varallo Sesia (Vercelli), Italy in 1960, received his degree in electronic engineering in 1985 from the Politecnico di Torino with a thesis on reliability of GaAs power MESFETs developed in CSELT laboratories. He joined then CSELT reliability department where he is involved in reliability, characterization, and modeling of active microwave devices. He holds a patent on a test fixture for active microwave devices and is co-author of a number of technical papers. He lectures on failure physics at SSGRR (L'Aquila) and at Politecnico di Torino.

Enrico Zanoni was born in Legnago, Verona, Italy, in 1956 and graduated with a degree in physics from the University of Modena, Italy, in 1982. After two years of study at the physics department of the University of Modena, he joined the University of Bari in 1985, and the University of Padua in 1988. He is currently a full-time researcher in the electronic engineering department at the University of Padua. He worked in the field of reliability physics of electronic devices, and is co-author of more than fifty papers on this subject. His interest include: study of ESD and latch-up phenomena, hot-electron effects, reliability of GaAs MESFETs, and failure analysis techniques. He is member of the IEEE, the Italian Physical Society (SIF), the Scientific Instrument Society (SIS), and the Italian Electron Microscopy Society (SIME).

Index